ATM NETWORKS

Author's Profile

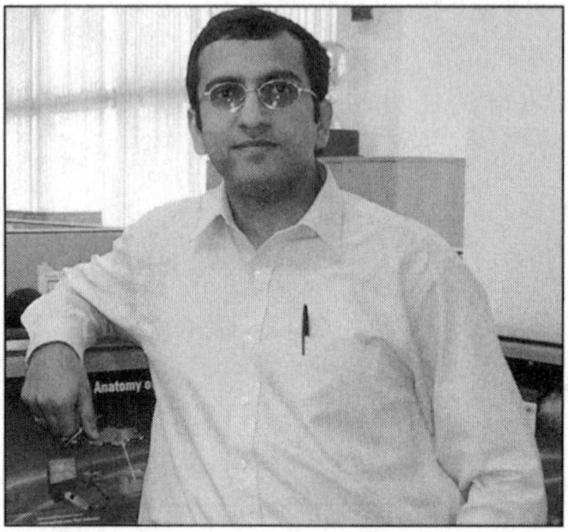

Sumit Kasera is Senior Technical Leader at Flextronics Software Systems (formerly Hughes Software Systems), India. He has a B. Tech degree in Computer Science and Engineering from IIT, Kharagpur, India, and an M.S degree in Software Systems from Birla Institute of Technology and Science, Pilani, India. His current area of interest is software development for GSM, GPRS, and EDGE access network. Sumit is also experienced in software development for networking protocols like ATM, TCP/IP and 3G UMTS. He has been an active participant in various technical forums like ITU-T, ATM Forum and 3GPP, where he has presented papers and conducted seminars; and also participated in the review of 3GPP specifications.

Sumit is the author of books—*ATM Networks: Concepts and Protocols, Communication Networks: Principles and Practice, 3G Networks: Architecture, Protocols and Procedures.*

ATM NETWORKS

Concepts and Protocols

Sumit Kasera

Senior Technical Leader
Flextronics Software Systems
Guragon, India

McGraw Hill

New York Chicago San Francisco Lisbon London Madrid
Mexico City Milan New Delhi San Juan Seoul
Singapore Sydney Toronto

The McGraw-Hill Companies

Cataloging-in-Publication Data is on file with the Library of Congress.

1 2 3 4 5 6 7 8 9 0 DOC/DOC 0 1 3 2 1 0 9 8 7 6

ISBN-13: 978-0-07-147732-1
ISBN-10: 0-07-147732-2

This book was first published in India in 2006 by Tata McGraw-Hill.

The sponsoring editor for this book was Stephen S. Chapman and the production supervisor was Richard C. Ruzycka. The art director for the cover was Brian Boucher.

Printed and bound by RR Donnelley.

This book was printed on acid-free paper.

To

My mother Sneha Lata Kasera
for making me what I am

Sumit Kasera

Preface

Raison d'être

It goes without saying that we now live in a networked world: a world where networking technologies play important roles in our lives. Be it the railway booking counter, or automated banking, be it the ubiquitous Internet or the global wireless telephony systems, a life without networks is inconceivable.

Among the plethora of networking technologies that have emerged in the last two decades, ATM is considered a breakthrough one. The popularity and pervasiveness of ATM lies in the fact that it seamlessly integrates local area network and wide area network—a concept also referred to as single network for desktop-to-the-core. Moreover, ATM provides a single platform for voice, video and data, thereby leveraging the process of network convergence. Apart from this, ATM also provides quality of service. All these factors are considered unique selling points of ATM technology.

When I first became interested in ATM technology and wanted to know more on the subject, I found to my surprise that there were hardly any books that provided a conceptual treatment of ATM. This, despite the fact that ATM technology attracted widespread interest. Most of the books were a shadow of ATM standards, derived from ITU-T and ATM Forum publications. Moreover, they had too many underlying assumptions to provide any insight into the topic. The concepts were shrouded in abbreviations and jargons. It resulted in people being more concerned with whether ATM was a 'telecommunication network' or a 'data communication network', than with understanding what either of them meant. I also observed that authors of technical books wrote with the assumption that technical stuff ought to be complicated because it is technical. I believe however, that everything is more or less simple unless made otherwise.

Given this, I felt a need to write a book on the subject that was simple, and easy to understand; a book that solved the problems I had faced when I was studying; a book that provided a comprehensive overview of ATM, and a book that offered fewer assumptions. This provided the motivation to write.

The Book

This book tries to cover all the important topics related to ATM. To this end, the book is organized into five parts (as shown in Figure P.1).

Part 1 introduces some of the important topics/concepts to the readers. To start with, Chapter 1 looks at the technological and marketing drivers that impact a given technology. This is useful in the sense that it provides a perspective on why ATM is what it is. Chapter 2 looks at the important concept of transfer mode and explains the benefits of asynchronous mode. Chapter 3 provides an overview of ATM and summarizes contents of the whole book.

Part 2 covers the three layers of ATM protocol reference model. This includes the physical layer covered in Chapter 4, the ATM layer in Chapter 5 and the ATM Adaptation Layer (AAL) in Chapter 6.

Part 3 Details the core concepts of ATM, which include parameterization of traffic and services in ATM networks (Chapter 7), elements of traffic management (Chapter 8), switch design and architecture (Chapter 9), addressing (Chapter 10), point-to-point and point-to-multipoint signalling (Chapter 11), routing based on PNNI protocol (Chapter 12), AAL2 signalling (Chapter 13), ATM Network management (Chapter 14) and ATM Security (Chapter 15).

Part 4 covers the interworking aspects of ATM. Chapter 16 describes the ATM network architecture and important interfaces defined for interworking. ATM in LAN environment and LAN Emulation (LANE) are explained in Chapter 17. Classical IP over ATM provides mechanisms to transfer IP datagrams over ATM networks. This standard also defines the means for address resolution. This technique is explained in Chapter 18. Another technique for IP over ATM is Multiprotocol over ATM (MPoA). The MPOA technique is also explained in Chapter 18.

Part 5 covers the application aspects of ATM. This includes ATM in MPLS Networks (Chapter 19), Voice over ATM (Chapter 20), ATM in DSL Networks (Chapter 21) and ATM in Third Generation (3G) Networks (Chapter 22).

Website

For greater interaction with readers after the publication of the book, the authors have created a website http://atmbook.tripod.com/. This website offers the following:

- Preface
- Table of Contents
- Errata
- Feedback and Review Comments
- References
- Other Related Material

Readers are encouraged to visit the website and use the available material.

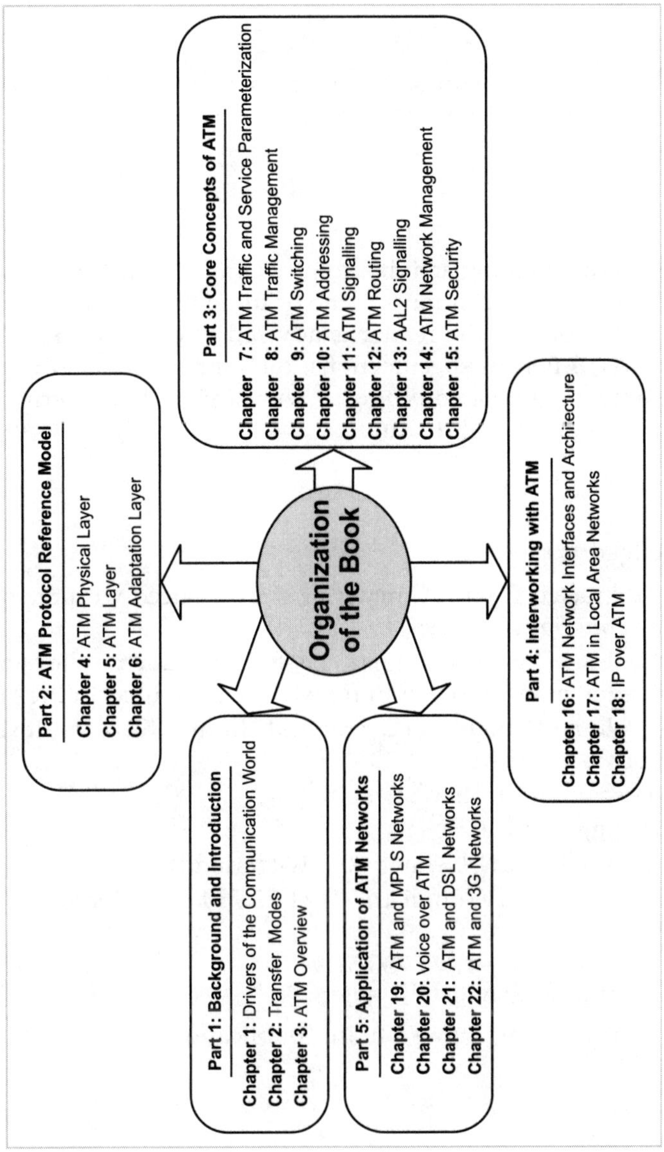

FIGURE P.1

Organization of this Book

Suggestions

Your comments, feedback and constructive criticism are valuable to me. So, please free to drop an email at s.kasera@lycos.com. I would be glad to incorporate your comments in the subsequent edition of the book.

Notes to Readers

Since standardization is an ongoing activity, some of the standards quoted in this book have been superseded by newer specifications. Some of these have been explicitly mentioned in the reference section. For others, the reader is advised to cross-check with the appropriate governing body. For ATM Forum specifications, the reader could browse www.atmforum.com. For ITUT recommendations, could go to www.itu.ch. For Internet RFCs, the reader could browse http://www.ietf.org/rfc.html.

Copyrights

SUMIT KASERA

Acknowledgements

I would first like to thank my organization—Flextronics Software Systems, (formerly Hughes Software Systems), Gurgaon, India—for providing me the right environment for learning and growth. It goes without saying that Hughes Software Systems is one of the best communication software companies in the country. The tremendous support I got from the company in terms of resources, time and motivation is sincerely appreciated. I also thank various persons with whom I am/was associated in the company. In particular, I thank Dr Ravi Mittal, Mr. Arun Punj, Mr. Vinod Sood and Mr. Gautam Brahma.

I would like to thank the technical reviewers, Rajiv Gupta, Nishit Narang, and K. R. K. Mohan, who painstakingly reviewed the whole manuscript and provided valuable comments.

I thank Nishit Narang again for contributing three chapters of this book, viz. 'ATM and MPLS Networks', 'ATM and DSL Networks' and 'ATM Network Architecture and Interfaces'.

I thank Gurpreet Singh, Yogesh Garg, Paras Shah and L. Sreenivasan for providing constructive suggestions.

Then, I would like to thank the entire team at Tata McGraw-Hill for helping in producing and marketing this book.

I also thank my alma mater Indian Institute of Technology (Kharagpur, India), and all its professors for providing me the necessary technical foundation to write a book.

In addition, I would like to acknowledge the role of my family in making this book a reality. Among all the family members, my mother Sneha Lata Kasera's contribution stands out and hence the dedication. I would also like to thank my father J. P. Kasera, wife Manisha, baby Raj (who is unaware of the happenings in this world!), sister Smita, brother-in-law Gaurav and brother Rajiv for their continued support.

SUMIT KASERA

Contents

PART 4

Interworking with ATM 321

Abbreviations

1G	First Generation
2G	Second Generation
3G	Third Generation
3GPP	Third Generation Partnership Project
AAL	ATM Adaptation Layer
AAL1	ATM Adaptation Layer 1
AAL2	ATM Adaptation Layer 2
AAL3/4	ATM Adaptation Layer 3/4
AAL5	ATM Adaptation Layer 5
ABR	Available Bit Rate/Area Border Router
ACAC	Actual Connection Admission Control
ACR	Allowed Cell Rate
ADR	Average Data Rate
ADSL	Asymmetric DSL
ADT	Assured Data Transfer
ADTF	ACR Decrease Time Factor
ADTR	ACR Decrease Time Rate
AES	Advanced Encryption Standard
AESA	ATM End System Address
AFI	Authority and Format Identifier
AINI	ATM Inter-Network Interface
AIS	Alarm Indication Signal
ALCAP	Access Link Control Application Part
AMI	Alternate Mark Inversion
AN	Access Network
ANS	ATM Name System
ANSI	American National Standards Institute
API	Application Programming Interface
APS	ATM Protection Switching

ARP	Address Resolution Protocol
AS	Autonomous System
ATM	Asynchronous Transfer Mode
ATMARP	ATM Address Resolution Protocol
ATU-C	ADSL Terminal Unit for the Central Office
ATU-R	ADSL Terminal Unit for the Remote site
AuC	Authentication Centre
AvCR	Available Cell Rate
AW	Administrative Weight
BAsize	Buffer Allocation size
BD	Burst Duration
BECN	Backward Explicit Congestion Notification
BER	Bit Error Rate/Bit Error Ratio
BGP	Border Gateway Protocol
BGT	Broadcast and Group Translator
B-ICI	BISDN Inter-Carrier Interface
BIP	Bit Interleaved Parity
B-ISDN	Broadband Integrated Services Digital Network
BLER	Block Error Rate/Block Error Ratio
BLER	Block Error Result
B-NT1	Network Termination 1 for B-ISDN
B-NT2	Network Termination 2 for B-ISDN
BOM	Beginning of Message
BPS	Bits Per Second
BR	Backward Reporting
BRI	Basic Rate Interface
BSC	Base Station Controller
BSI	British Standards Institute
BSS	Base Station Sub-system
BT	Burst Tolerance
B-TA	Terminal Adapter for B-ISDN
B-TE1	Terminal Equipment 1 for B-ISDN
B-TE2	Terminal Equipment 2 for B-ISDN
BTS	Base Transceiver Station
BUS	Broadcast and Unknown Server
CAC	Connection Admission Control
CAM	Content Addressable Memory
CAS	Channel Associated Signalling
CBR	Constant Bit Rate
CC	Country Code
CCITT	Consultative Committee on International Telegraph and Telephony
CCR	Current Cell Rate

CCS	Common Channel Signalling
CDF	Cut-off Decrease Factor
CDM	Code Division Multiplexing
CDMA	Code Division Multiple Access
CDV	Cell Delay Variation
CDVT	Cell Delay Variation Tolerance
CER	Cell Error Ratio
CES	Circuit Emulation Service
CID	Channel Identifier
CLEC	Competitive Local Exchange Carriers
CLIP	Calling Line Identification Presentation
CLIR	Calling Line Identification Restriction
CLP	Cell Loss Priority
CLR	Cell Loss Ratio
CMIP	Common Management Information Protocol
CMR	Cell Mis-insertion Ratio
CN	Core Network
CNM	Customer Network Management
COLP	Connected Line Identification Presentation
COLR	Connected Line Identification Restriction
COM	Continuation of message
CPCS	Common Part Convergence Sub-layer
CPCS-UU	CPCS User-to-User Indication
CPE	Customer Premises Equipment
CPI	Common Part Indicator
CPN	Customer Premises Network
CPS	Common Part Sub-layer
CRC	Cyclic Redundancy Check
CRM	Cell Rate Margin
CRM	RM-Cells before Cut-off
CRV	Call Reference Value
CS	Circuit Switched/Convergence Sub-layer
CSI	Convergence Sub-layer Indication
CSMA	Carrier Sense Multiple Access
CSMA/CD	Carrier Sense Multiple Access with Collision Detection
CSU	Channel Service Unit
CTD	Cell Transfer Delay
DBCES	Dynamic Bandwidth Circuit Emulation Service
DBU	Dynamic Bandwidth Utilization
DCC	Data Country Code
DCE	Data Circuit-Terminating Equipment
DES	Data Encryption Standard
DFA	DXI Frame Address

Diffserv	Differentiated Services
DIR	Direction
DLCI	Data Link Connection Identifier
DLL	Data Link Layer
DLPI	Data Link Provider Interface
DNS	Domain Name System
DSAID	Destination Signalling Association Identifier
DSL	Digital Subscriber Line
DSLAM	DSL Access Multiplexer
DSP	Domain Specific Part
DSS	Dynamic Structure Sizing
DSU	Data Service Unit
DTE	Digital Terminal Equipment
DTL	Designated Transit List
DTMF	Dual Tone Multiple Frequency
DWDM	Dense Wavelength Division Multiplexing
DXI	Data Exchange Interface
ECF	Establish Confirm
EDGE	Enhanced Data Rates for Global Evolution
EFCI	Explicit Forward Congestion Indication
EIA	Electronic Industries Association
EIR	Equipment Identity Register
ELAN	Emulated LAN
EM	Element Manager/Error Monitoring
EOM	End of Message
EPD	Early Packet Discard
ER	Explicit Rate
ERQ	Establish Request
ESF	Extended Superframe
ERV	End-point Reference Value
ESI	End System Identifier
ETSI	European Telecommunications Standards Institute
FA	Frame Alignment
FAST	Frame-based ATM over SONET/SDH Transport
FATE	Frame-based ATM Transport over Ethernet
FAX	Facsimile
FCAPS	Fault, Configuration, Accounting, Performance and Security management.
FCS	Frame Check Sequence
FDDI	Fiber Distributed Data Interface
FDM	Frequency Division Multiplex
FDMA	Frequency Division Multiple Access
FEBE	Far End Block Error

FEC	Forward Error Correction/Forwarding Equivalence Class
FECN	Forward Explicit Congestion Notification
FERF	Far End Reporting Failure
FIFO	First-in-First-out
FM	Fault Management
FPM	Forward Performance Monitoring
FRS	Frame Relay Service
FRTT	Fixed Round Trip Time
FSM	Finite State Machine
FTP	File Transfer Protocol
FTTC	Fiber-To-The-Curb
FUNI	Frame-based UNI
GC	General Purpose Communication Channel
GCAC	Generic Connection Admission Control
GCRA	Generic Cell Rate Algorithm
GFC	Generic Flow Control
GFR	Guaranteed Frame Rate
GGSN	Gateway GPRS Support Node
GMSC	Gateway MSC
GPRS	General Packet Radio Service
GSM	Global System for Mobile Communications
GTP	GPRS Tunnelling Protocol
HDLC	High-level Data Link Control
HDSL	High-Bit-Rate DSL
HEC	Header Error Control/Header Error Check
HLR	Home Location Register
HO-DSP	High-Order Domain Specific Part
HOL	Head-of-Line
HSSI	High Speed Serial Interface
HTTP	Hyper Text Transfer Protocol
HW	Hardware
IAD	Integrated Access Device
ICD	International Code Designator
ICIP	Inter-Carrier Interface Protocol
ICIP_CLS	ICIP Connectionless Service
ICR	Initial Cell Rate
ID	Identifier
IDI	Initial Domain Identifier
IDP	Initial Domain Part
IE	Information Element
IEEE	Institute of Electrical and Electronic Engineers

IETF	Internet Engineering Task Force
IGRP	Interior Gateway Routing Protocol
IISP	Interim Inter-Switch Protocol
ILEC	Incumbent Local Exchange Carrier
ILMI	Integrated Layer Management Interface
IM	Input Module
IME	Interface Management Entity
IMS	IP Multimedia Sub-system
INE	Internetworking Network Elements
IOTA	Identifiers for Organizations for Telecommunications Addressing
IP	Internet Protocol
IPoA	IP over ATM
IPv4	Internet Protocol Version 4
IPv6	Internet Protocol Version 6
IPX	Internetwork Packet Exchange
ISDN	Integrated Services Digital Network
ISH	Interworking Specific Header
ISO	International Organization for Standardization
ITU-T	International Telecommunication Union-Telecommunication Standardization
IVR	Interactive Voice Recognition
IWF	Interworking Function
LAG	Local Address Group
LAN	Local Area Network
LANE	LAN Emulation
LAP-B	Link Access Procedure-Balanced
LAP-D	Link Access Procedure for D-channel
LCN	Logical Channel Number
LCT	Last Conformance Time
LD-CELP	Low Delay Code Excited Linear Prediction
LDP	Label Distribution Protocol
LE-ARP	LAN Emulation Address Resolution Protocol
LEC	LAN Emulation Client
LECID	LEC Identifier
LECS	LAN Emulation Configuration Server
LES	LAN Emulation Server
LGN	Logical Group Node
LI	Length Indicator/Indication
LIJ	Leaf Initiated Join
LIS	Logical IP Sub-network
LLC	Logical Link Control
LMI	Local Management Interface
LNNI	LANE Network Network Interface
LOS	Loss of Signal

LSCES	Low Spread Circuit Emulation Service
LSP	Label Switched Path
LSR	Label Switching Routers
LUNI	LAN Emulation User Network Interface
M3UA	MTP3. User Adaptation Layer
MA	Maintenance and Adaptation
MAC	Message Authentication Code/Medium Access Control
MADR	Maximum Allowable Data Rate
MAN	Metropolitan Area Network
maxCR	Maximum Cell Rate
maxCTD	Maximum Cell Transfer Delay
MBS	Maximum Burst Size
MCR	Minimum Cell Rate
MD5	Message Digest 5
ME	Mobile Equipment
MFS	Maximum Frame Size
MIB	Management Information Base
MID	Multiplexing Identifier
MIPS	Millions of Instructions Per Second
MOS	Mean Opinion Score
MPC	MPOA Client
MPLS	Multi-Protocol Label Switching
MPOA	Multi-Protocol Over ATM
MPS	MPOA Server
MS	Mobile Station
MSC	Mobile Switching Centre
MSS	Maximum Segment Size
MTP	Message Transfer Part
MTP3b	Message Transfer Part 3 for broadband
MTU	Maximum Transmission Unit
NADH	North American Digital Hierarchy
NBAP	Node B Application Part
NBMA	Non-Broadcast, Multi-Access
NDC	National Destination Code
NDIS	Network Driver Interface Specification
NE	Network Element/Network Entity
NHC	NHRP Client
NHRP	Next Hop Resolution Protocol
NHS	Next Hop Server
NI	No Increase
NID	Network Interface Device
N-ISDN	Narrowband ISDN

NM	Network Management
NMS	Network Management System
NNI	Network-Node Interface/Network-Network Interface
NPC	Network Parameter Control
NR	Network Operator
nrt-VBR	Non-real-time VBR
NSAP	Network Service Access Point
NT	Network Termination/Non-Transparent
N-UNI	Network side of UNI
OAM	Operations and Maintenance/Operation, Administration and Management
OC-x	Optical Carrier level 'x' (e.g. OC-3)
ODI	Open Data-link Interface
OOF	Out of Frame
OS	Operations Systems
OSAID	Originating Signalling Association Identifier
OSF	Offset Start Field
OSI	Open Systems Interconnection
OSI-RM	OSI Reference Model
OSPF	Open Shortest Path First
OUI	Organizationally Unique Identifier
OAM&P	Operations, Administration, Maintenance and Provisioning
P2P-CDV	Peak-to-Peak Cell Delay Variation
PAD	Padding
PARIS	Packetized Automated Routing Integrated System
PBX	Private Branch Exchanges
PC	Personal Computer
PCM	Pulse Code Modulation
PCR	Peak Cell Rate
PDA	Personal Digital Assistant
PDH	Plesiochronous Digital Hierarchy
PDU	Protocol Data Unit
PG	Peer Group
PGID	Peer Group Identifier
PGL	Peer Group Leader
PHY	Physical Layer
PID	Protocol Identifier
PIM	Parallel Iterative Matching
PLCP	Physical Layer Convergence Protocol
PLL	Phase Locked Loop
PLMN	Public Land Mobile Network
PLP	Packet Layer Protocol
PMD	Physical Medium-Dependent

PNNI	Private Network Node Interface
POH	Path Overhead
POI	Path Overhead Identifier
POTS	Plain Old Telephone System
PPD	Partial Packet Discard
PPP	Point-to-Point Protocol
PRI	Primary Rate Interface
PS	Packet Switched
PSTN	Public Switched Telephone Network
PT	Payload Type
PTI	Payload Type Indicator/Identifier
PTSE	PNNI Topology State Element
PTSP	PNNI Topology State Packet
PVC	Permanent Virtual Circuit/Connection
QoS	Quality of Service
R&D	Research and Development
RAI	Remote Alarm Indication
RAM	Random Access Memory
RAN	Radio Access Network
RANAP	Radio Access Network Application Part
RCC	Routing Control Channel
RDF	Rate Decrease Factor
RDI	Remote Defect Indication
RED	Random Early Discard
RFC	Request for Comments
RID	Rate Increment Factor
RIF	Rate Increase Factor
RIP	Routing Information Protocol
RM	Resource Management
RMCSN	Reported Monitoring Cell Sequence Number
RNC	Radio Network Controller
RNS	Radio Network Subsystem
RSA	Rivest, Shamir and Adleman
RSC	Regional Switching Center
RSN	Receive Sequence Numbers
RSVP	Resource Reservation Protocol
RTP	Real Time Protocol
rt-VBR	Real-time VBR
SA	Security Association/Security Agent
SAAL	Signalling ATM Adaptation Layer
SAID	Signalling Association Identifier

SAP	Service Access Point
SAPI	Service Access Point Identifier
SAR	Segmentation and Reassembly
SCCP	Signalling Connection Control Past
SCR	Sustainable Cell Rate
SCTP	Stream Control Transmission Protocol
SDH	Synchronous Digital Hierarchy
SDSL	Symmetric DSL
SDT	Structured Data Transfer
SDU	Service Data Unit
SEAL	Simple and Efficient Adaptation Layer
SECBC	Severely Errored Cell Block Count
SECBR	Severely Errored Cell Block Ratio
SEL	Selector
SF	Superframe
SGSN	Serving GPRS Support Node
SHA1	Secure Hash Algorithm 1
SLA	Service Level Agreement
SMDS	Switched Multi-megabit Data Service
SME	Security Message Exchange
SN	Subscriber Number/Sequence Number
SNAP	Sub-Network Attachment Point
SNMP	Simple Network Management Protocol
SNP	Sequence Number Protection
SOHO	Small Office/Home Office
SONET	Synchronous Optical Network
SPE	Synchronous Payload Envelope
SRTS	Synchronous Residual Time Stamp
SS	Supplementary Service
SS7	Signalling System No. 7
SSCF	Service-Specific Co-ordination Function
SSCF-NNI	Service-Specific Co-ordination Function for NNI
SSCF-UNI	Service-Specific Co-ordination Function for UNI
SSCOP	Service-Specific Connection-Oriented Part
SSCS	Service-Specific Convergence Sublayer
SSM	Single segment message
SSN	Send Sequence Numbers
ST	Segment Type
STC	Signalling Transport Converter
STDM	Statistical Time Division Multiplexing
STM	Synchronous Transfer Mode
STS	Synchronous Transport Signal
SVC	Switched Virtual Circuit/Connection

SW	Software
TAT	Theoretical Arrival Time
TAXI	Transparent Asynchronous Transceiver Interface
TBE	Transient Buffer Exposure
TC	Transmission Convergence
TCP	Transmission Control Protocol
TCR	Tagged Cell Rate
TDM	Time Division Multiplexing
TDMA	Time Division Multiple Access
TE-2	Terminal Equipment 2
TIA	Telecommunication Industries Association
TLV	Type-Length-Value
TMN	Telecommunication Management Network
TR	Trial Trace
TS	Technical Specification
UBR	Unspecified Bit Rate
UDP	User Datagram Protocol
UE	User Equipment
UMTS	Universal Mobile Telecommunications System
UNI	User-Network Interface
UP	User Plane
UPC	Usage Parameter Control
USIM	Universal Subscriber Identity Module
UTOPIA	Universal Test and Physical Interface for ATM
UTP	Unshielded Twisted Pair
UTRAN	Universal Terrestrial Radio Access Network
UU	User-to-User
UUI	User-to-User Indication
U-UNI	User Side of UNI
UUS	User-to-User Signalling
VBR	Variable Bit Rate
VC	Virtual Circuit/Virtual Channel
VCC	Virtual Channel Connection
VCI	Virtual Circuit Identifier/ Virtual Channel Identifier
VCIP	VCI Present
VCL	Virtual Channel Link
VDSL	Very-High-Rate DSL
VF	Variance Factor
VLR	Visitor Location Register

VLSI	Very Large Scale Integration
VoATM	Voice over ATM
VoDSL	Voice over DSL
VoIP	Voice over IP
VP	Virtual Path
VPC	Virtual Path Connection
VPI	Virtual Path Identifier
VPL	Virtual Path Link
VPN	Virtual Private Network
VTOA	Voice and Telephony Over ATM
WAN	Wide Area Network
WCDMA	Wideband Code Division Multiple Access
WDM	Wavelength Division Multiplexing
WWW	World Wide Web

ATM NETWORKS

PART 1

Background and Introduction

Networking requirements are primarily classified into two broad categories, viz. one for voice communication and the other for data communication. Voice communication requires a fixed bandwidth and low delay, and is tolerant to occasional information loss. On the other hand, data communication is bursty in nature. Further, though data is tolerant to higher transit delays, it is susceptible to information loss. Consequently, prior to the deployment of ATM, the networking fraternity was divided into two groups—the *telecommunication world* and the *datacommunication world*. The telecommunication world was backed by the international standards bodies like the International Telecommunications Union-Telecommunication (ITU-T) and catered to services required for voice conversation. The services included basic telephony and facsimile (fax). The datacommunication world was backed by bodies like the Internet Engineering Task Force (IETF), the Institute of Electrical and Electronic Engineers (IEEE), and International Standards Organization (ISO), and catered to services required for data transfer. The services included data transfer in LAN and WAN environment.

However, during the last decade or so, the distinction between a telecommunication network and a datacommunication network has become very hazy. Two classical examples in this regard are packet-based voice and integrated services network. The following paragraphs elaborate upon this.

Till recently, voice was carried by only using circuit-switched networks. Packet-based networks were considered unsuitable for voice because of their unpredictable and high-delay characteristics. However, with the advent of Internet telephony, this notion has changed. Specialized software packages (like Voice-over-IP or VoIP), compression and silence suppression schemes, and echo cancellation devices have considerably improved the voice quality over the Internet. Moreover, since Internet charges are based on connectivity, and not on distance, as against basic telephony, where the charge is distance-related, Internet telephony offers a cheap alternative for long-distance voice conversation. Although the quality of voice-over Internet is somewhat inferior, the price differential makes Internet telephony a very attractive proposition indeed.

The popularity of Internet telephony indicates the aggressiveness of datacommunication companies to capture the voice market, a market hitherto monopolized by telecommunication companies. In order to counter the aggressiveness of datacommunication companies, telecommunication vendors are also trying to expand their realm from voice-only services to integrated services. The development of Integrated Services Digital Network (ISDN) by the telecommunication body ITU-T is an indicator of this move. The primary motive behind the development of ISDN was to provide a digital pipe for integrated services (voice, data and video). However, as the ISDN attempt has not been very successful, the next attempt of ITU-T to provide a universal platform is through Broadband ISDN (B-ISDN). Like ISDN, B-ISDN aims to provide integrated services, albeit at higher data rates.

The aforementioned arguments put forward only a very small part of the various developments in the communication world. A number of technological advances like better processing power, high-speed optical links, pervasive use of mobile telephones and the Internet have drastically altered the communication landscape. This part of the book tries to capture these trends and builds a case for having ATM networks. It also provides an overview of ATM networks.

Chapter 1 discusses at the *technological* and *marketing drivers* that impact a given technology. This is useful in the sense that it provides a perspective as on ATM is what it is.

Chapter 2 describes at the important concept of *transfer mode* and explains the benefits of asynchronous mode. In this chapter, various transfer modes like circuit-switching, packet routing and virtual circuit-switching are explained. A comparison is also made between different transfer modes.

Chapter 3 provides an *overview of ATM*. It also summarizes the contents of the whole book. In particular, the chapter discusses the motivation for having ATM, its definition, the precursor technologies to ATM (including X.25, frame relay and ISDN), the basic principles of ATM, the ATM protocol reference model, the core concepts, the interworking aspects and the application areas of ATM networks. Some other topics like the relation between ATM and B-ISDN, ATM standardization aspects and the ATM market outlook are also elaborated in this chapter.

Chapter 1

Drivers of the Communication World

1.1 Introduction

We often hear people say, "Technology is changing very fast". This is indeed true. A product that takes three years to develop may become obsolete even before it reaches the market. In the case of a technology which takes a couple of years to develop, one may find that the services for which it was originally intended, no longer exist. Programming languages and tools which existed until recently, have now become outdated. A Personal Computer (PC) or a microprocessor that was considered state-of-the-art a few years ago may be completely out of use today.

In essence, the communication industry is in a state of flux and changing rapidly. Such change in technology cannot be attributed to a single factor. There are many factors or *drivers* that alter the face of the communication world. This chapter attempts to identify drivers that have significantly affected the world of communications. Such drivers are categorized as *technological drivers* and *market drivers*. In brief, technological drivers impart direction to emerging technologies, while market drivers determine the popularity of such technologies in the market. The technological and market drivers are explained in the following sections. The chapter concludes by describing how these drivers affect the growth and deployment of ATM networks.

1.2 Technological Drivers

Technological drivers are factors that impart direction to emerging technologies. In order to understand this, consider the example of fiber cables. Prior to the emergence of fiber, copper was the primary means of data transmission. The low-speed and poor-quality copper cables severely restricted the rate of data transfer. Thus, the maximum speed of a networking technology was inevitably linked to the maximum speed of the underlying transmission medium (i.e. copper). The evolution of fiber completely revolutionized the situation. Fiber provides virtually infinite bandwidth with low error rates, thus breaking the transmission medium bottlenecks which older technologies had to face. In essence, the emergence of fiber is an important technological driver that alters the transmission medium limitations.

Over the last two decades, four key technological developments have significantly altered the face of communication and provided shape to communication technologies. These developments are:

1. Improvements in the processing speed,
2. Improvements in the transmission media,
3. Growth of the Internet, and
4. Growth of mobile networks

Each of these is explained in the subsequent sub-sections (see Figure 1.1).

FIGURE 1.1

Technological Drivers

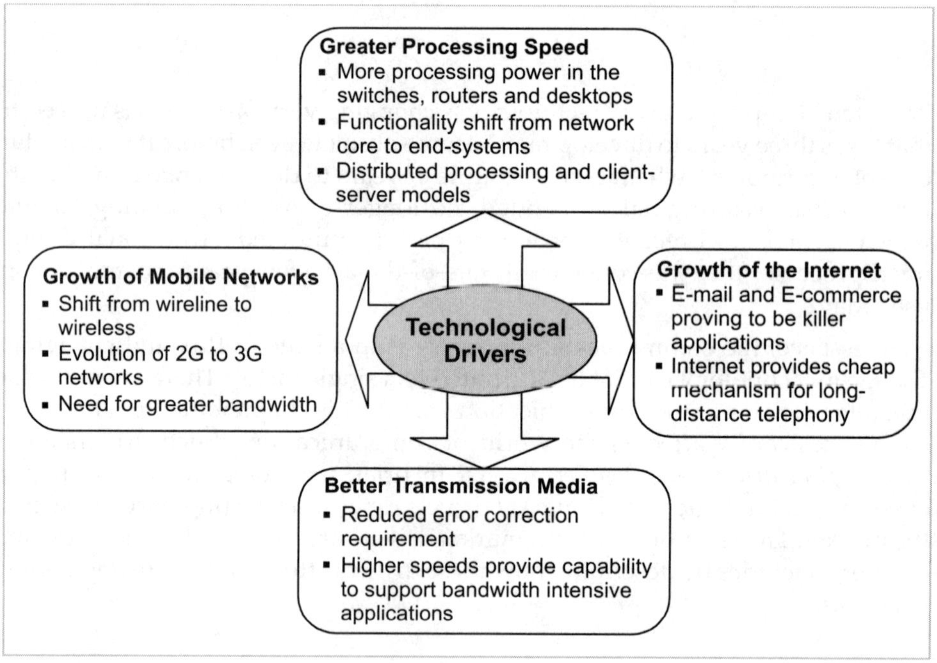

1.2.1 Improvements in Processing Speed

In any network, the maximum speed of data transfer is determined by the slowest entity. If the intermediate nodes (i.e. switches/routers) are slow, then no matter how fast the transmission link is, the data transfer speed gets limited by the processing speed of the switches/routers.

Thus, the growth of high-speed networks is inevitably linked with the improvements made in the processing speeds of microprocessors.

During the last two-and-a-half decades, microprocessor speed has increased by leaps and bounds. About three decades back, Gordon Moore had observed that each new chip contained roughly twice the speed of its predecessor, and that each chip was released within 18–24 months of the release of the previous chip. That observation of Gordon Moore, now better known as Moore's law, is still applicable for the present generation microprocessors. Processing speed is still doubling every 18 months, thereby resulting in an exponential growth. Figure 1.2 highlights the number of instructions executed (MIPS) and the number of transistors used in different Intel microprocessors. As shown in the figure, the number of transistors in a microprocessor has increased logarithmically. In 1971, the 4004 processor had merely 2,250 transistors. This figure has increased to a staggering forty-two million transistors found in Pentium 4 processors. This has been facilitated by the improvements made in the field of Very Large Scale Integration (VLSI). Note that the Y-axis is marked in the logarithmic scale. Hence, a linear curve implies an exponential relationship between the time elapsed and improvements made in the processing power.

FIGURE 1.2

Growth of Intel Microprocessors

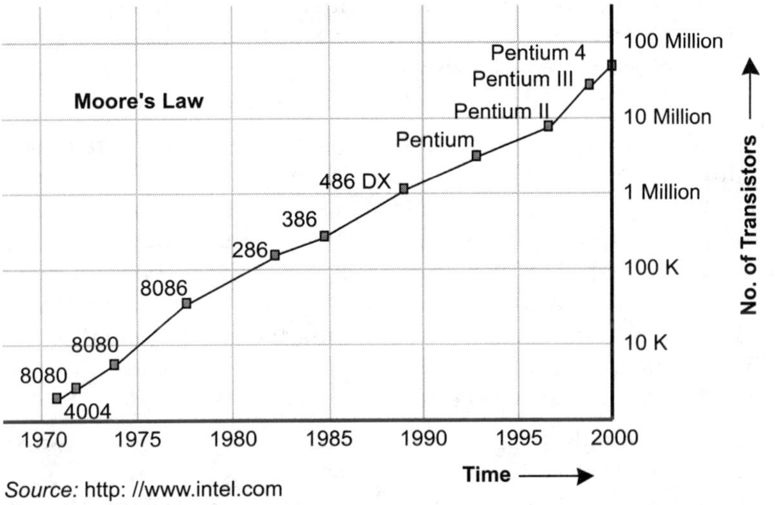

Source: http: //www.intel.com

The exponential growth in microprocessors has meant that processing power is not restricted to high-end servers. On the contrary, microprocessors are cheap, powerful and readily available. This availability of processing power has three key influences in the nature of communication networks. First, the switches and routers have become powerful and it is now easier to switch data at much higher speeds. The high-speed processors have made it possible to support high-speed OC-3 (155 Mbps), OC-12 (622 Mbps) and even OC-48 (2.4 Gbps) links.

Second, there has been a gradual shift from a centralized networking environment to a distributed environment. In a distributed work environment, the network intelligence is distributed across different entities, which communicate with each other at a peer-to-peer level (i.e. without any hierarchical relationship). The client-server model is an outcome of this distributed work environment. Using the client-server model, a large number of clients can share the critical resources and applications residing at the servers. Since all clients are not active at the same time, such sharing brings about significant benefits.

The third development is the bearing that the growth of microprocessors has on the underlying assumptions of communication networks. Earlier, when end-systems were supposed to be dumb devices, the network was wholly responsible for the reliable transfer of user information. However, as the end-systems have become more powerful, this assumption no longer holds true. Thus, most of the processing that was earlier under the purview of the network (like flow control and sequence numbering) has shifted from network core towards end-systems. This has made present-day networks leaner and faster.

1.2.2 Improvements in the Transmission Media

The widespread use of fiber optic cables is one of the developments that have revolutionized the networking fraternity. Optical fiber has now emerged as the transmission medium of choice. In North America alone, there are around 50 million miles of fiber. And thousands of new fiber cables are laid every day. The popularity of fiber stems from the fact that it offers several benefits over traditional copper cables.

First, fiber can support bandwidths which are much higher than bandwidths supported by any other media. The maximum bandwidths of fiber is in the range of gigabits (1 gigabit = 1000 megabits). By using advanced multiplexing techniques like Dense Wavelength Division Multiplexing (DWDM), the bandwidth supported can be extended up to few terabits (1 terabit = 1000 gigabits). Considering the fact that the total bandwidth requirement for data in 1997 was to the tune of 250 Gbps, a single fiber link is capable of satisfying a lot of demand. In fact, switching systems capable of exploiting the full potential of fiber are yet to be developed. Thus, it can be safely presumed that at least fiber-optic cables will not cause speed bottlenecks in the days to come.

Besides providing higher bandwidth, fiber is immune to electro-magnetic interference from external entities. In other words, signals from power sources, electrical storms or stray electro-magnetic sources have no effect on the data that is being carried. This is because fiber-optic bandwidths and electro-magnetic bandwidths are non-overlapping in the frequency domain.

Another major advantage of using fiber is that its Bit Error Rate (BER) is very low. The low BER value of fiber can be attributed to the immunity of fiber against external interference. Typical values of BER fiber are to the order of 10^{-8} to 10^{-12}. At a speed of 10 Mbps and a BER of 10^{-10}, this turns out to be roughly, one bit error per 15 minutes, which is quite reasonable. The BER value of a transmission medium is important because it directly affects the design of a networking protocol. If the BER value is low, as is the case with fiber, limited error-detection and error-correction functionality needs to be incorporated in the packet headers. This reduces the header length considerably as also the processing time involved in detecting and correcting bit errors. Moreover, for very low BER values, it can be assumed that there will be very little packet loss. Hence, sequence numbering and retransmission procedures can be simplified. In contrast, if the BER is high, the header has to carry adequate error-detection and error-correction bits, thereby resulting in unnecessary bandwidth and processing overheads.

Fourth, not only are the error rates in fiber low, but the errors are mostly limited to a single-bit error. This aspect of fiber is also very important. Detecting and correcting single-bit errors is much easier than, say, correcting multiple-bit errors. Thus, the error-correcting codes for fiber-based technologies are of relatively smaller length. Moreover, the error-detection and error-correction algorithms used in fiber-based technologies are much simpler as compared to technologies which use copper as their transmission media.

Fifth, the attenuation in fiber is very low. Attenuation is the signal loss, measured in decibels, of a signal transmitted over a distance. For analog signals, attenuation means a loss of amplitude, while for digital signals, it means degradation of the signal. Due to attenuation, a signal becomes so weak that it cannot be detected correctly at the receiving end. In order to circumvent this problem, repeaters are required that can regenerate signals by either amplification (for analog signals) or regeneration (for digital signals). Since the signal attenuation of fiber is very low (~ 0.2 dB/km), fiber cables can be laid for long distances without the need for any repeaters. The reduced requirement for repeaters translates into reduced equipment and maintenance costs.

Finally, since fiber is a guided media (i.e. unlike air, which is unguided), any attempt to tap the transmitted signals can only be done at the cost of a signal loss. This signal loss can easily be detected at the receiving end. Thus, tapping optical signals entails taking a great risk. This aspect of fiber makes it an attractive proposition for defence and other banking organizations where security is a matter of prime concern.

To conclude, fiber cables offer so many benefits that it will gradually replace copper and will provide a high-quality and a high-bandwidth transmission medium. The high bandwidth will provide the means to support bandwidth-intensive applications like digital video and multimedia conferencing. The popularity of such applications will further increase the demand for bandwidth, thereby creating greater need for high-speed fiber links.

1.2.3 Growth of the Internet

During the last two decades, the Internet has seen an unprecedented growth. In fact, the exponential growth of the Internet is termed as 'one of the biggest things to happen this century in the history of communication networks'. In 1981, there were only 200-odd Internet hosts. By 1992, this number rose to over one million. By early 2003, the number was a stupendous 170 million hosts. Figure 1.3 depicts the exponential rise in the number of Internet hosts during the last decade of the twentieth century.

The growth of the Internet is important for a number of reasons. First, the Internet has brought with it new kinds of applications—the e-mail for example, that has antiquated the conventional means of communication. E-mail has now become the preferred mode of

FIGURE 1.3

Approximate Growth of Number of Internet Hosts

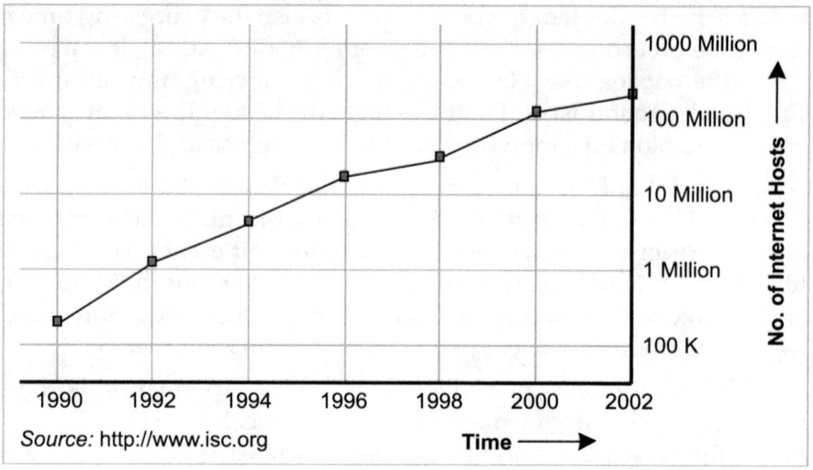

communication, especially in developed countries. The ubiquity of the e-mail warrants that a person has to be connected to the Internet, be it at home, at office or on the move. Apart from e-mail, E-commerce or transaction over the Internet is another application that has revolutionized the way business transactions are done.

Second, the Internet has made remote access to corporate networks possible. With the help of a modem and a dial-up connection, an employee can access the file systems of his/ her organization with considerable ease. Thus, the Internet has indirectly facilitated telecommuting.

Third, a service like Internet telephony has drastically reduced the cost of voice communication over long distances. Although the voice quality over the Internet is relatively poor, better software solutions have improved the quality of voice. Since Internet telephony is cheap, it has become a serious contender to challenge the monopoly of telecommunication networks.

Finally, the Internet has leveraged the deployment of high-speed networks which can provide Quality of Service (QoS). This is because new Internet applications like streaming audio/video are bandwidth-extensive and require some minimum service guarantees.

Overall, the growth of the Internet and the growing popularity of applications with ever-increasing bandwidth requirements demand higher Internet speeds. This, in turn, warrants that the underlying transport network be able to support such high speeds.

1.2.4 Growth of the Mobile Networks

Like the Internet, another area that has witnessed tremendous growth during the last decade is that of mobile cellular networks. In 2003, the mobile subscriber base reached 1.3 billion users, surpassing the seemingly invincible fixed-line user base of 1.2 billion. It is expected that by 2007, the number of mobile users will be a staggering two billion; by that time, one in every three persons will have a mobile phone.

As the growth of moblie networks is reaching saturation levels, and as service providers and users migrate towards Third Generation (3G) networks, there is continuous pressure on the underlying infrastructure to be able to cope with the increased subscriber base. This pressure is further increased by the fact that new multimedia applications are being developed for the latest mobile phones and the air interface in 3G networks is capable of supporting much higher bandwidths (in the order of few hundreds of Kbps). Overall, this development is also fostering the need for high-speed switching and transport networks.

1.3 Market Drivers

Being a superior technology may not necessarily imply that the technology is also popular. Beyond the technological realm lies the unpredictable psyche of the people. And to become successful, a technology needs to be not only efficient, but also needs to conquer this unpredictable psyche of the masses. This section elaborates upon some of the factors that help in making a technology both commercially viable as well as an attractive option. These factors are also termed as *market drivers*, because they help in leveraging the popularity of a technology in the market. A market driver includes the following factors:

1. Need for flexible services,
2. Cost,
3. Inertia of the installed base,
4. Support of the industry,
5. Open standards, and
6. Changing management structure.

Each of these is explained in the subsequent sub-sections (see Figure 1.4).

1.3.1 Need for Flexible Services

In all spheres of life, people today demand flexible solutions that can match their needs—that too, without having to pay for their disparate service requirements. With respect to the communication world, flexibility implies:

1. The ability to use network resources whenever wanted, and
2. To be charged only for the resources used.

As an example, electricity distribution network is highly flexible. People can use their electrical gadgets any time during the day. They are also allowed to switch on all their electrical appliances, if they so want. However, they are charged only for the units they consume, not for the duration for which they have their electric connection.

The need to provide similar capabilities in communication networks is a problem which has been long felt. This need has led to the development of flexible networks that can provide similar service capabilities. The Internet is a good example of this. An Internet subscriber can establish a connection whenever he or she wishes, and is charged only for the duration for

FIGURE 1.4

Market Drivers

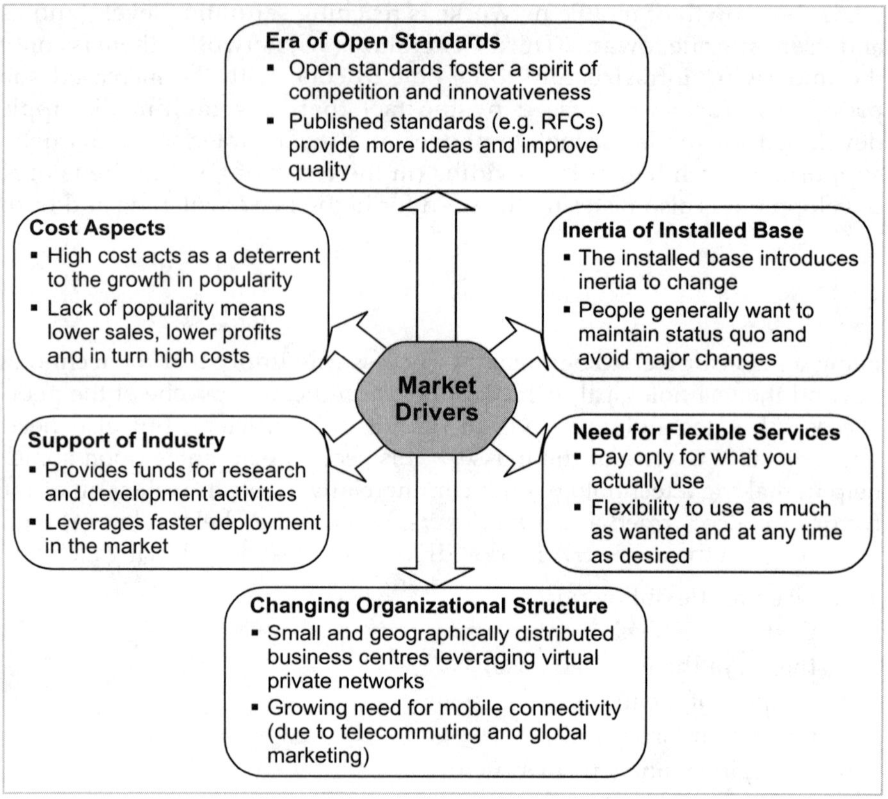

which the connection is active. Thus, both aspects of flexibility are fulfilled by the Internet architecture.

Providing flexibility is generally a trade-off between sub-optimal resource utilization and inferior service. Providing greater flexibility mandates that the network keep sufficient spare bandwidth to allow subscribers to use it anytime. As an example, telephone networks in the US have an in-built capacity to handle the severest of loads. Thus, most of the time, the network resources are underutilized. However, the positive side is that subscribers can rest assured of getting a telephone call established anytime. In some underdeveloped countries, this is not the case. There, the number of subscribers is generally much more than the number of active calls that the networks can handle simultaneously. During peak hours, a lot of connection requests are rejected. Thus, it is evident that by improving resource utilization, some flexibility is sacrificed, which leads to inferior service.

In order to recognize the importance of flexibility, consider the example of leased lines. Leased lines provide dedicated bandwidth at fixed costs, irrespective of the utilization of the link. Thus, leased lines are very inflexible. In contrast, frame relay and ATM networks offer the capability to dynamically establish and release connections. This provides greater flexibility, as bandwidth does not get locked, as was in the case of leased lines.

To summarize, providing flexibility is an important concern from the market's point of view. Designers must bear this in mind while developing new technologies and new protocols.

1.3.2 Cost Aspects

A very important aspect in the success of any product is the pricing and cost aspect. The market forces of demand and supply determine the equilibrium price. In the initial stages in the life-cycle of a product when demand is low, the economies of scale are hard to come by. Thus, it becomes very difficult to lower the prices. The higher prices, in turn, deter the users and keep potential buyers away. This is the vicious cycle of cost and demand, as depicted in Figure 1.5. A good example in this regard is the use of mobile phones in India. When the mobile service was introduced for the first time, they were priced exorbitantly at Rs 16 per minute. This made the service an elite one, away from the reach of the masses. It was after many years of struggle, when the prices were lowered to around Rs 2 per minute, that the mobile service sector showed an exponential growth.

FIGURE 1.5

Vicious Circle of High Cost and Low Demand

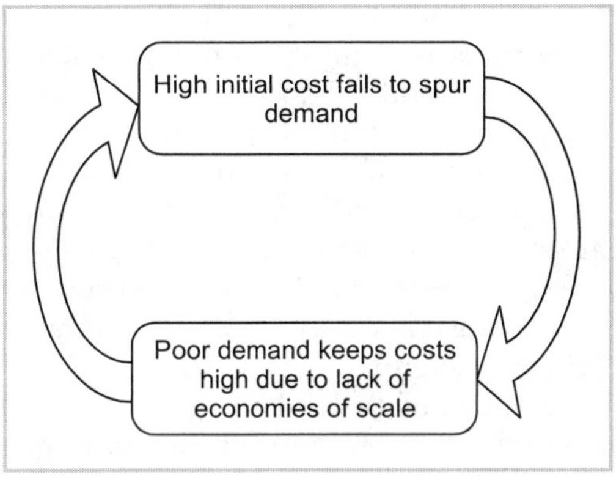

A related issue is the relative comparison of two products on the cost vis-a-vis the features they offer. If the price differential does not justify the feature difference, then people will continue to use what they are using, unless significant gains are clearly visible. An example here could be the migration of users from Second Generation (2G) mobile service to Third Generation (3G) services. Unless clear benefits are seen, people seem content with the 2G services.

1.3.3 Inertia of the Installed Base

One of the biggest hurdles in the acceptance of a new technology is the financial and emotional cost involved in the process of migration. The financial cost includes components like the cost of building new infrastructure, loss of old infrastructure and the cost involved in training the staff. Very rarely do we find the introduction of a new technology resulting in the complete overhaul of the network. The exorbitant fixed costs involved in the process make the process economically unfeasible. This mandates that every new solution has provisions for reusing the existing infrastructure. Financial overheads are one of the very important reasons why technologies like ATM are yet to live up to their potential.

Apart from the financial aspects, there is always an emotional touch associated with every process of change. Since people generally like status quo, introduction of a new technology is always met with fierce resentment. The only solution for preventing this resentment lies in educating the people and creating awareness. However, this process may not always succeed.

In essence, the inertia of installed base of existing networks prevents widespread migration to newer technologies. For example, it is not easy to replace the existing LAN networks (Ethernet and token ring) with ATM networks, unless there are distinct advantages associated with the process. Thus, migration problem is one factor that drives older technologies and prevents introduction of newer ones.

1.3.4 Support of the Industry

Research and Development (R&D) centres are the breeding grounds for innovative ideas, techniques and new technologies. However, any form of research needs funds for its sustenance and the industry is the only source of funds. Even research workdone in top academic institutions is partially or fully sponsored by the industry. Thus, for the growth of any communication technology, it is important that there be sufficient financial and technical backing by the industry. From the point of view of industrialists, they invest in only those areas where they are confident of recovering their costs and making profits. Once certain areas are identified, the industry provides funds to the research community, while the latter provides technical inputs to the former. The industry then markets the technology, which brings in revenue. A part of this revenue is pumped into research, which further improves the capabilities of the technology. This symbiotic relationship between the industry and the research community results in a multiplier effect, which is beneficial for both sides.

As an example of this symbiotic relationship, consider the case of ATM technology. More than a decade back, many people postulated that ATM was going to be 'the hottest technology of the future'. At that time, it was also conjectured that the first set of ATM switches would be ready anytime in the late 1990s. But the extensive research work carried in the field of ATM (especially in fields related to switching and traffic management) resulted in an ATM switch being ready much before the predicted date. This was made possible mainly because of a very strong industrial backing that ATM got during its infancy. This, however, is only one side of the story. The boost ATM received resulted in mass awareness and this made ATM even more popular. This resulted in new markets, new demands, and subsequently, new products and services. The above is an example of *quid pro quo* where, backed by the industry, the research community works towards the development of a new technology. In return, the industry is benefited by getting greater revenues by selling new products and services.

1.3.5 Era of Open Standards

Today's networked world is moving towards an era of *open* standards. Open standard implies that standards/recommendations are fully published and available to all and communication takes place using internationally accepted protocols (i.e. multi-vendor solutions are interoperable).

The standardization process and timely availability of standards are very important because of several reasons. First of all, in the pre-standardization era, a company was tied to the components of a single vendor. Standardization, however, has completely altered the scenario. Now, a single network may have components sourced from 5 to 50 different vendors. Interoperability between multi-vendor components is guaranteed by means of standards. Various interoperability meets are also held to demonstrate the stability and efficacy of the standards.

Secondly, evolving proprietary standards is usually a risky affair. This is because if a standard from a governing body supersedes the proprietary standard, then the latter becomes useless. Thus, companies always try to ensure that their products are interoperable with products of any other company. If well-defined standards are used, such problems are precluded.

Most importantly, open standards result in innovation, competitiveness and greater efficiency. In this regard, the Request for Comments (RFC) in which Internet standards is documented, is a case in point. The RFCs are available in a text format freely on the Internet for anyone to read and analyse. People make their comments, raise doubts and even suggest improvements. These comments/suggestions are incorporated in the RFCs and new versions released, thereby making the older version obsolete. RFCs are easy to read and implement, and it is not an exaggeration to say that the simplicity and effectiveness of RFCs is one of the major reasons why the Internet has blossomed over the last decade.

Although standardization has its merits, it also has a few drawbacks. The biggest drawback is that it delays the introduction of new products and services. Since companies are wary of evolving proprietary standards, they usually adopt a wait-and-watch policy. In case the publication of a standard is exceedingly delayed, the launch of a new product also gets delayed. In the process, competing technologies gain an edge over time. All this is detrimental to the growth of a technology.

Secondly, frequent revision of standards leads to instability. Instability of protocols requires that products based on those protocols also be frequently modified, which can be quite irritating for the engineers. Moreover, frequent revisions create a fear of technological obsolescence in both the developers and the users. This leads to a disinclination both to develop solutions as well as to use them.

1.3.6 Changing Organizational Structure and Practices

In the old days, the organizations used to be of mammoth size. The networking requirement was limited to connecting the computers, mainframes and servers within the same organization. Local Area Networks (LANs) fulfilled this requirement. However, things have now changed and the organizations are now far more geographically dispersed than they used to be. This geographically dispersed organizational set up results in a need to interconnect various business centers. Typically, leased lines are used to connect two LANs. In an organization that has multiple LANs, leased lines result in a meshed topology. The drawback of this scheme is that the addition of another LAN requires many new leased lines to ensure connectivity of new LAN to existing LANs. In contrast, the current trend is to get a single connection from a public carrier and use the carrier's public network to connect to several places. This concept, where a network extends across different LANs and uses a public network, is referred to as Virtual Private Network (VPN).

Apart from leveraging the market for VPNs, the new organizational structures have altered the movement of network traffic. Earlier, the traffic was predominantly local, resulting mainly from the mainframe–terminal communication. However, with Wide Area Network (WAN) providing external connectivity, the percentage of local network traffic has dropped.

Another important development is that the work staff has become more mobile. Moreover, people prefer to work from home (telecommuting). In order to support mobility, there are requirements on two fronts. First, there is a demand for notebooks and laptops, and second, there is a need to provide network connectivity to the mobile users. Providing mobile connectivity requires developing new protocols for mobile communication. The concern is to provide not only connectivity, but also 'secure' connectivity. This favours the growth of remote access protocols, dial-up connection protocols (through modems), authentication protocols and a host of other protocols. Along with new protocols come service providers providing new services.

In essence, the new management order demands a new networking environment, new products and new services, which leverage hitherto unexplored markets.

Conclusion

The chapter focussed on the technological and market drivers of the communication world. Each driver has some influence on the way ATM has been designed and has evolved. The rapid growth of microprocessors enables ATM to build high-speed switches capable of forwarding millions of cells per second. Thus there are standards in ATM for carrying cells at OC-48 (2.4 Gbps) rates.

The assumption that fiber is the underlying transmission link, reduces the need for error-detection and flow control in ATM. This has made the link layer functionality in ATM quite simple.

The growth of the Internet provides the necessary application base for the use of ATM technology. In fact, one of the important uses of ATM lies in the Internet backbone, wherein ATM is used to provide high-speed data pipes.

The growth of mobile networks has also provided a boost to ATM networks as ATM has been chosen as the transport technology for both the Access Network as well as the connectivity between Access and Core Network.

Among the market drivers, providing a flexible and scalable network is one of the prime goals of ATM. Thus, ATM is viewed as a technology that can be used from the desktop to the core.

The relative high costs and the inertia of the installed base constitute a deterrent to the deployment of ATM. This is specially reflected In the LAN environment wherein Ethernet and Gigabit Ethernet remain the popular choices, ahead of ATM. However, in the WAN environment, ATM has found its takers.

As discussed, the industry has backed ATM standardization and deployment contributing to its moderate success.

ATM retains the philosophy of open standards and has a large and rich set of standards from bodies like ITU-T and ATM Forum. While the ATM Forum standards are available in the public domain from its website, this is not true for ITU-T standards, which have to be purchased. Still, ATM Forum continues to play an active role in the growth of ATM technology.

The changing organization structure is favourable for ATM as it is very well capable of providing seamless networking and VPN services.

Review Questions

Q 1. *What are the current technological trends? How do these trends impact the growth and deployment of ATM? Do these trends themselves get impacted by developments in ATM technology? If yes, explain with examples how this is possible.*

Q 2. *Comment upon the market drivers and their relation to the deployment of ATM networks.*

Q 3. *What is the role of standardization in the success of a networking technology? Discuss this with special reference to the standardization bodies for ATM and the role they play.*

Further Reading

There is no specific reference for this chapter. The materials collected from various sources are mentioned at appropriate places in the chapter (e.g. Intel and ISC websites).

Chapter 2

Transfer Modes

2.1 Introduction

The fundamental function of a network is to transfer user information from one entity to other. In order to understand how the transfer of information takes place, it is important to understand the notion of *transfer mode*, which is the central theme of this chapter. Transfer mode refers to the techniques used to transmit, switch and multiplex information. In other words, transfer mode is a means of packaging, sending and receiving information on the network. In current literature, interpretation of the term 'transfer mode' is subject to wide variations. According to some authors, Asynchronous Transfer Mode (ATM) and Synchronous Transfer Mode (STM) are the only two possible transfer modes. For others, circuit-switching and packet-switching are the two basic transfer modes. Still others consider telegraphy and telephony also to be variants of transfer mode.

According to this book, there are various ways of classifying transfer mode. If one looks at the way information is packaged, the two basic techniques are *circuit switching* as embodied by telecommunication networks and *packet switching* as embodied by frame relay networks and the Internet. In circuit-switching, a dedicated circuit (or a channel) is established from source to destination and the information is sent as a bit stream through the fixed bandwidth channel. This technique is used primarily to carry voice. In packet switching, user information is carried in variable-sized or fixed-sized packets. These packets are also called frames, cells or datagrams. Each packet has a header and a payload. The header carries address information and is used to make switching decisions and to forward the packet to the next hop. The payload carries the actual user information. This technique is used predominantly to carry data.

If one looks at the way information is transferred and resources reserved, there can be three different techniques, viz. *circuit switching*, *virtual circuit switching*, and *routing*. In circuit switching, the information is carried over a dedicated connection and resources are reserved for the duration of the connection. In virtual circuit switching, virtual circuits are established and the packets are forwarded by using the virtual circuit numbers. A variant of this technique is *cell switching* in which fixed-sized small cells are used instead of variable-sized frames. In routing, each packet header carries a well-defined destination address. This destination address field is used at each intermediate hop to forward the packet. This form of forwarding is also referred to as *datagram forwarding*.

The following sections discribe the circuit switching, routing and virtual circuit switching techniques, each of which can be viewed as a distinct transfer mode. Cell switching is discussed under the section on virtual circuit switching.

2.2 Circuit Switching

Circuit switching is used primarily in telecommunication networks to carry voice. In circuit switching, a *dedicated connection* is established between two end-systems. The connection may be a physical connection using copper cables, or it may be a wireless link using radio waves or satellite links. Irrespective of the nature of the connection, a dedicated non-sharable channel is reserved between the source and destination end-systems for the entire duration of the connection. The dedicated connection is established using a process called *call-setup*. Call-setup implies finding a path from the source to the destination and establishing a connection between them. Data transfer takes place only after call-setup. After the data transfer is over, the call is cleared, and all resources reserved for the call are freed. In essence, information transfer in a circuit switched network is a three-step process call-setup, data transfer and call clearing.

A typical switch has 'm' incoming links and 'n' outgoing links. The number of incoming links may or may not be equal to the number of outgoing links. However, if they are equal, the switch is said to be a *symmetric*. Each incoming link receives a frame that contains 'N' slots. The frame is switched by using a switching table or a translation table. Each time slot on each frame is switched to a new time slot on an outgoing link, as per the translation table. No header information is necessary to perform switching. The position of the slot implicitly identifies the source and destination end-systems. This information is obtained when the connection is established. It is at the time of connection establishment that the entries in the translation table are filled. The entries are removed when the connection is cleared.

The circuit switching technique primarily relies on *Time Division Multiplexing* (TDM) to carry user information from one node to another. Time Division Multiplexing is a technique in which the individual input channels are multiplexed onto a single physical channel. The multiplexed channel is divided into *frames* and each frame is further sub-divided into *slots*. Each input channel gets a fixed time slot in each frame. This ensures fairness because input channels equally share the time slots. However, this method is wasteful because an allocated slot is wasted if the input channel has nothing to send for that slot.

Figure 2.1 illustrates the TDM technique in which four input channels are multiplexed onto a single output channel. In each frame, a slot is reserved for each of the input channels *A*, *B*, *C* and *D*. Irrespective of whether the input channel has valid data or not, the output channel reserves the slot for the input channels. Since the allocation of time slots is static, this wastage is unavoidable. Additional intelligence needs to be incorporated into the multiplexers to avoid this wastage. *Statistical Time Division Multiplexing* (STDM) is one of the ways for reducing this wastage.

FIGURE 2.1

Time Division Multiplexing

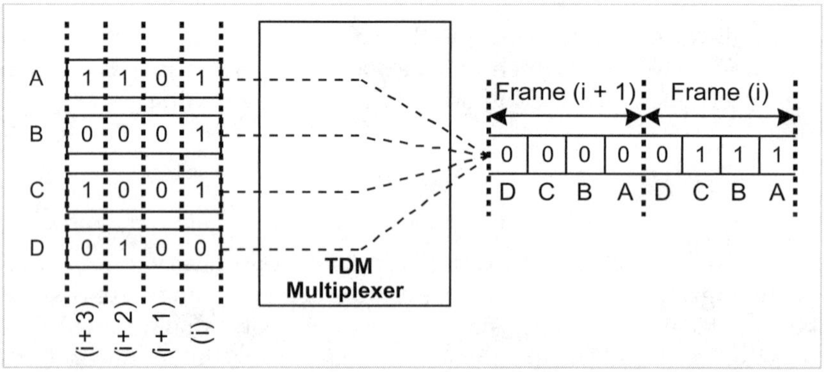

In the STDM technique, no slot is reserved for an incoming channel. Rather, a slot is assigned as and when user data is received. In order to allow the proper sharing of bandwidth, resource reservation takes places through the use of statistical methods. These methods entail calculation of the probability distribution functions of all the incoming connections and prediction of the requirements of each of them. However, *a priori* calculation of these distribution functions is not easy and hence, only a rough estimate is made. The request for a new connection is accepted only if there are ample resources throughout the path of the connection. In case the network cannot support new connections, the connection request is rejected. This ensures that each multiplexer is able to support all the incoming connections. The STDM is also referred to as 'asynchronous TDM'.

2.3 ROUTING

In contrast to circuit switching, routing is a *hop-by-hop datagram forwarding* technique in which the destination address field of the packet header is used to make forwarding decisions. The packets are typically large in size (~1000 bytes) and are called *datagrams*. In this technique, there is no concept of connection and each datagram is routed to the destination independent of other datagrams in a connectionless manner. *Datagram forwarding, datagram switching* and *layer 3 forwarding* are some of the other terms used for routing. Among all the terms, *routing* is the most commonly used one. Hence, the nodes forwarding the datagrams are referred to as routers.

As an example of routing, consider Fig. 2.2, which depicts how datagrams are routed from host A1 towards host A3. A1 forwards the packets destined for A3 to its nearest router (i.e. router R1). When a packet from A1 reaches R1, R1 uses the destination address field (A3 in this case) to forward it further. For routing purposes, every router maintains a routing table with each entry having an association between a destination address and the link to which it is to be forwarded (i.e. outgoing interface on which the received packet is to be forwarded). This entry is used to make routing decisions. In case no entry is found, datagrams are forwarded to a default route. Since R1 has an entry for destination address A3, it forwards the datagram on the link R1–R2. When R2 receives a datagram, it uses its own routing table to forward the datagram further. In this way, each intermediate router forwards the datagrams hop-by-hop, using the destination address field until the datagram reaches the destination host.

FIGURE 2.2

Example Depicting the Mechanism of Routing

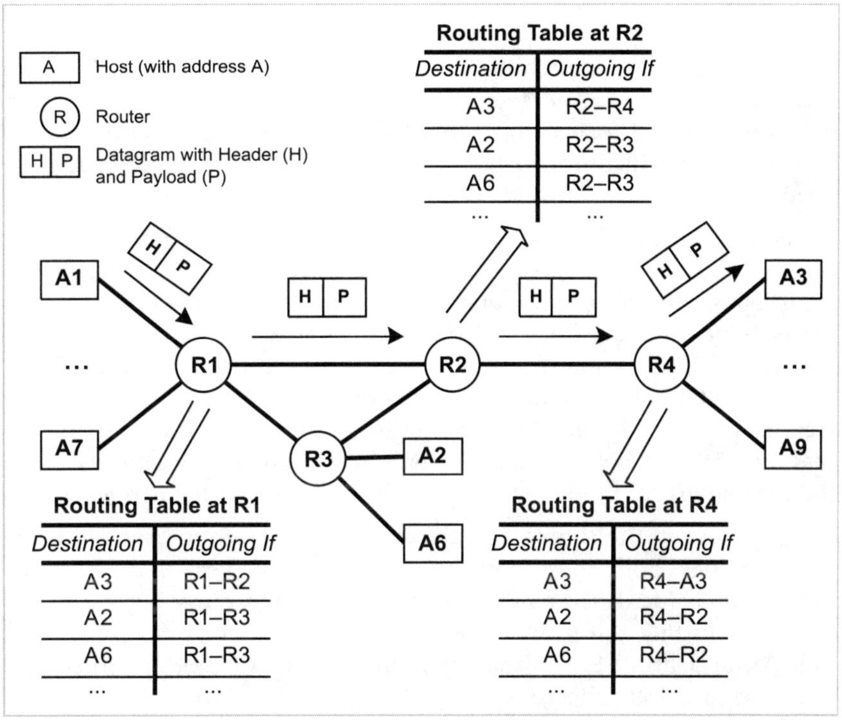

Datagram routing lays the foundation of Internet architecture and its popularity seems to be rising unabated with the exponential growth of the Internet. However, this growing popularity causes a lot of problems related to the routing table look-up. As the number of hosts increases, there is almost a linear growth in the number of entries that a router must maintain. As the number of entries increases, so does the processing time associated with finding the outgoing link for a datagram, and forwarding it. This has led to new forwarding paradigms like *Multi-Protocol Label Switching* (*MPLS*), in which the concepts of virtual circuit switching and datagram routing are merged, resulting in a faster and scalable routing architecture. Virtual circuit switching is detailed in the next section.

2.4 Virtual Circuit Switching

Two techniques have been discussed so far. In circuit switching, there is a dedicated connection which is used to transfer information. In routing, there is no concept of connection nor is there any resource reservation; each packet carries the complete destination and a router routes the packet in a best-effort fashion. Virtual circuit switching lies between these two extremities. In this technique, information is transferred by using *virtual circuits*. Virtual circuits are logical channels (or circuits) between two end-systems that are identified by small identifiers or labels. For the sake of discussion, let us call this label *Virtual Circuit Identifier (VCI)*.

In virtual circuit switching, each packet that is transmitted by the sender carries a VCI. When this packet reaches the next hop, the VCI field of the packet header is used to forward it further. Before forwarding, this field is overwritten with a new label. This process of overwriting an incoming label with a new label is called *virtual circuit switching* or *label swapping*.

As in circuit switching, the connection in virtual circuit switching goes through three phases *call-setup*, *data transfer* and *call clearing*. The virtual circuit is established in the call-setup phase. In order to understand how a call is setup, consider a simple example in which three connections are established A-E (VC1), D-C (VC2), and A-E (VC3) (see Fig. 2.3). For the first connection, A sends a message to B, requesting it to establish a connection with E. B returns a VCI value (34) and forwards the call to E. A, upon receiving the VCI value 34, fills it in its VC table. Meanwhile, the message forwarded by B reaches E. E accepts the request and returns a VCI (47) to B. E also enters the returned VCI in its VC table. B, upon receiving the message from E, maps the VCI 34 with VCI 47. This is how VC1 is established.

As shown in the figure, the VC table has two important fields an *incoming VCI* and an *outgoing VCI*. Both these collectively identify a part of the virtual connection. The total connection is identified by the set of incoming and outgoing VCIs present in all intermediate nodes, including the source and the destination. For example, VC1 has two distinct parts, link AB and link BE. Link AB is identified by VCI 34 and link BE by VCI 47. Therefore, {(AB, 34) (BE, 47)} collectively identifies the virtual circuit VC1.

After VC1 is established, the virtual circuit tables at A, B and E contain only the first row entries. After the virtual connections VC2 and VC3 are made, the complete virtual circuit tables for nodes A, B and E contain entries as shown in Fig. 2.3. Note that the dashed column entries in the VC table indicate that the node is either an ingress node or an egress node.

This is as far as establishing virtual circuits is concerned. If the data transfer between A and E for the first connection is now considered, then each packet sent from A for this virtual circuit will have a VCI 34 written in its header field. When this packet reaches node B, the

FIGURE 2.3

Example Depicting Establishment of Virtual Circuits

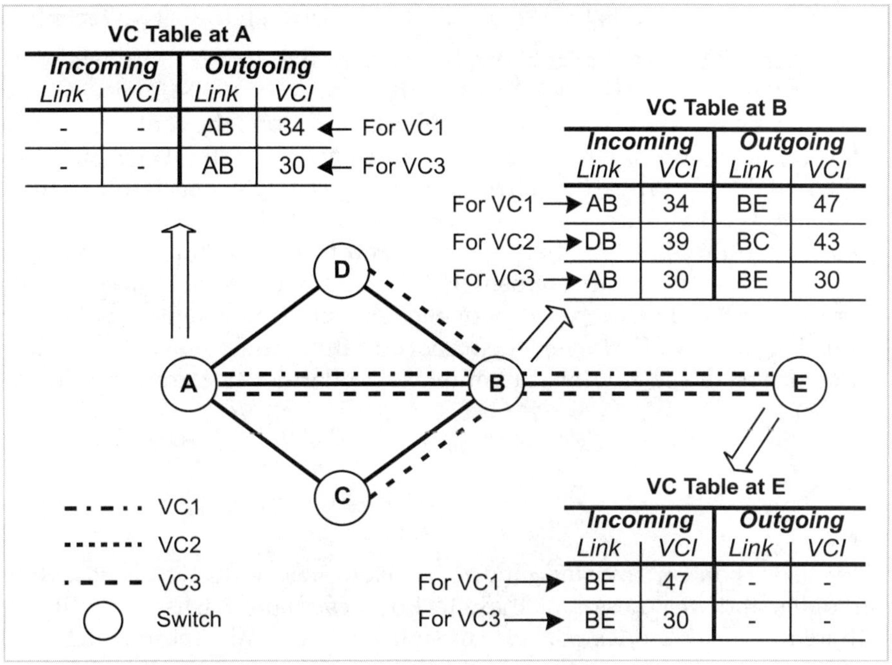

virtual circuit table at node B is searched for the tuple (AB, 34). Here, AB refers to the link on which the packet is received, and 34 refers to the VCI value contained in the received packet. After a row entry is found, the VCI value in the packet is swapped with the value written in the outgoing VCI column of that row (i.e. VCI 34 is replaced with VCI 47). The packet is then forwarded on the outgoing link as per the entry in the virtual circuit table (i.e. BE). This is how label swapping and packet forwarding take place.

One important thing to note about VCIs is that they are *locally unique* and *not globally unique*. For example, consider VC1 in Fig. 2.3. The value of VCI in link AB is not the same as VCI in link BE, i.e. the same value of VCI is not maintained end-to-end. However, this does not imply that a VCI value used in one portion of the link cannot be used in other part. For example, for VC3, the VCI value for both the AB and BE links is 30. This is acceptable as long as two virtual circuits on the same incoming link have different VCIs. The Sharing of VCIs across different links is allowed.

The discussions on virtual circuit switching and routing beem to indicate that little separates datagram routing from virtual circuit switching. The few differences centre around the following aspects:

1. The global uniqueness of complete addresses in datagram routing versus local uniqueness of small VCIs in virtual circuit switching, and
2. Packet routing over a connectionless network versus label swapping over a virtual circuit.

Section 2.5 highlights the other differences between these two techniques.

The virtual circuit concepts are not new to the data communication world. In fact, this concept has been in use in X.25 networks since the 1960s. The virtual circuits in X.25 are identified by a 12-bit Logical Channel Number. Among the contemporary technologies, frame relay employs virtual circuit concepts. In frame relay, the Data Link Connection Identifier (DLCI) field identifies the virtual circuits. The latest technology to use virtual circuit switching is ATM. Virtual circuit switching in ATM is also referred to as *cell switching*. The most important difference between cells and frames is that ATM cells are small and have a fixed size, whereas frames are usually large and have variable sizes. Otherwise, conceptually, there is very little difference between virtual circuit switching in frame relay and cell switching in ATM. However, the important thing to note here is that this single conceptual difference results in a number of other dissimilarities (e.g. transmission delay and switching speeds).

2.5 Comparison of Transfer Modes

The discussion on transfer modes is incomplete without an assessment of the relative strengths and weaknesses of these techniques. Table 2.1 lists the differences between four different transfer modes, viz. circuit switching, virtual circuit switching, cell switching and routing. To classify one of them as the best option is unjustified as different scenarios warrant different solutions.

Since all switching techniques, except datagram routing, are connection-oriented, most of the advantages and disadvantages of connection-oriented services are applicable to these switching techniques. The advantages of connection-oriented services include *low propagation delay, sequenced data transfer* and *reliable service*. On the downside are factors like latency associated with call-setup and *inefficient resource utilization* due to dedicated allocation. However, despite being connection-oriented, circuit switching, virtual circuit switching and cell switching do not have identical characteristics.

As mentioned earlier, circuit switching is mainly used to carry voice. Dedicated channels ensure that there is always enough bandwidth in the network to carry voice samples. A connection request for voice call can be refused, but once it is accepted, the voice transfer is of good quality. Although allocating dedicated channels provides better services and is thus preferable, it comes at the cost of sub-optimal resource utilization. This is because even though the resource for a call remains allocated for the entire duration, it is actually used when there is useful data to transmit. For example, consider the case of a telephonic conversation in which a person normally talks for only about 40 per cent of the time. The remaining time, there is nothing to transmit, and hence the reserved channel carries no useful data. This is gross wastage of bandwidth. For data communication, a circuit-switched network is more wasteful because data transfer is essentially bursty in nature. There are sharp bursts followed by prolonged periods of inactivity. If resource is reserved for maximum burst, then for a major period (which is nearly 90-95), the link will be under-utilized or unutilized.

Table 2.1 Comparison of Transfer Modes

Circuit Switching	*Virtual Circuit Switching*	*Cell Switching*	*Routing*
Connection-oriented; a dedicated connection is established before data transfer	Connection-oriented; a virtual circuit is established, before data transfer	Connection-oriented; a virtual circuit is established before data transfer	Connectionless; no connection is established
Sequential delivery No header overheads	Sequential delivery Label overheads	Sequential delivery Small cells cause higher overheads	Out of order possible Overheads of carrying complete address information
Fixed resource allocation	Statistical resource allocation	Statistical resource allocation	No resource reservation
Stream oriented	Packets or frames oriented; variable size packets or frames	Cell-oriented; small fixed size cells	Packet (datagrams) oriented; variable size packets
Highest reliability	Reliable	Reliable	Best-effort service
Low delay and jitter	Not very good delay/jitter behaviour	Controlled delay/jitter	Unpredictable delay/jitter
Mainly for voice traffic	Mainly for data traffic	For all kinds of traffic	Mainly for data traffic
Hardware-based switching	Hardware or software based switching	High-speed hardware based switching	Hardware or software-based switching
Very fast, but inefficient resource utilization	Relatively slow, but better resource utilization	Fast and optimal resource utilization	Slowest of all, but robust and flexible
Ex: Plain old telephone systems (POTS)	Ex: X.25 and Frame relay networks	Ex: ATM networks	Ex: TCP/IP networks (i.e. Internet)

By virtue of being connectionless, datagram routing is ideal for carrying *bursty data traffic*. No resource is reserved and packets are forwarded as and when they arrive. This precludes the possibility of wasted bandwidth. However, this also means that during heavy traffic conditions, the quality deteriorates rapidly. There is no guarantee that adequate bandwidth will be available and that the packet will not be dropped. Also, because of the lack of bounds on the delay involved in the transfer of datagrams, this method is not very popular for voice.

Virtual circuit switching is a *compromise* between dedicated resource allocation and connectionless switching. By using *statistical multiplexing*, the resources are shared in such a manner that neither is the capacity under-used, nor does the quality of a connection suffer because of lack of resources. However, this form of switching assumes that the resource requirements of a connection can be predicted *a priori*, which is not always true. Moreover, unlike a router, a virtual circuit switch does not maintain information on all possible destinations. Informations or taining to only established virtual circuits is maintained at each switch. This reduces the size of the VC table and makes it look up faster.

Cell switching is the latest form of virtual circuit switching. In this technique, small fixed-sized cells are used instead of variable-sized packets. The small cell size ensures that the delay *is predictable* due to which the voice quality can be maintained. For carrying data, multiple cells are carried, instead of a large packet.

Conclusion

This chapter provided an overview of the basic transfer modes. The comparison between different transfer modes can be reduced to the difference between connection-oriented and connectionless services. The trade-off here is between reduced resource utilization at better service guarantees, versus higher resource utilization at the cost of possible loss in service quality. It is said that ATM offers the best of both worlds, i.e. connection-oriented statistical multiplexing resulting in better resource utilization along with small cell size resulting in ability to carry voice and real-time traffic by providing controlled delay and jitter response.

Review Questions

Q 1. *What are transfer modes? Why are they important in the context of networking techniques? What is the basis of classifying these transfer modes?*

Q 2. *What is circuit switching? What are its advantages and disadvantages?*

Q 3. *What are the essential differences between the two?*

Q 4. *What is the difference between virtual circuit switching and cell switching?*

Q 5. *Compare the four transfer modes, viz. circuit switching, virtual circuit switching, cell switching and routing on the following aspects, with proper reasoning and justification:*

- *Nature of connection*
- *Resource Reservation*

- *Resource utilization*
- *Reliability*
- *Ability to carry voice*
- *Header overheads*

Further Reading

For details of the topics covered in this chapter, a plethora of options are available. [Gen A. Tanenbaum] and [Gen W. Stallings] remain two very popular texts. Apart from this, few other books [Gen S. Keshav], [ATM M. Prycker], [ATM W. Goralski] and [Gen S. Kasera] also provide good coverage on the topic.

Chapter 3

Overview of ATM

3.1 Introduction

This chapter provides an overview of ATM. It also summarizes the gamut of topics covered in this book. To start with, the motivation for having ATM is discussed in the next section. Thereafter, the definition of ATM is provided, which is seen to encompass three essential components, namely transfer mode, cell-based transfer and asynchronous transfer. These three terms are also explained in this chapter. This is followed by a study of the events that led to the genesis of ATM. Then, the precursor technologies to ATM, including X.25, frame relay and ISDN are explained. After this, the basic principles of ATM are explained. The ATM protocol reference model, comprising Physical Layer, ATM Layer and ATM Adaptation Layer (AAL), is then elaborated. Thereafter, the various chapters in Part 3 (Core Concepts), Part 4 (Interworking Aspects) and Part 5 (Application of ATM Networks) are summarized. A few other topics like the relation between ATM and B-ISDN, ATM standardization aspects and the ATM market outlook are also discussed.

3.2 Motivation for ATM

Today, one of the biggest problems related to networking is that there are just too many networks, with each one being tailored for a specific service requirement. For transferring voice in public environment, there are the telecommunication networks. In corporate environment, voice is transferred using Private Branch Exchanges (PBX). For transferring data in LAN, there are Ethernet, Fast Ethernet, Gigabit Ethernet and token ring networks. In a WAN environment, there are the X.25 networks, frame relay networks and the ubiquitous

TCP/IP-based Internet. For video distribution services, there are the cable networks. For wireless voice and low-rate data applications, there are cellular networks.

It is desirable that all these networks be replaced by a single platform, capable of handling the disparate requirements. A single platform leads to better network management and reduced operational costs. Besides this, a single platform imparts greater flexibility to the network infrastructure. This is not the case for service-specific networks, which are very inflexible. For example, consider the traditional telecommunication network where, in order to provide voice communication, the bandwidth is partitioned into 64 Kbps channels. Even if advances in compression techniques permit voice communication at lower rates, the inflexible infrastructure does not provide this flexibility.

Instead, if a single integrated platform is used, a platform that is independent of the needs of the applications, the above problem is ameliorated. ATM is developed to fulfill this need for providing a single integrated platform for a very diverse class of network services.

In the context of a 'single integrated platform', the following aspects need to be considered two:

1. A single platform that provides end-to-end connectivity, and
2. A universal platform that integrates different network services.

The following sub-sections elaborate upon these two aspects.

3.2.1 Seamless Networking

Seamless networking refers to the use of a single platform for end-to-end communication. In seamless networking, the geographical distance between communicating entities is hidden from the end-user. Networking is generally not seamless in the sense that the geographical distance is reflected in the time it takes to fetch data from a remote site. For example, in a LAN environment, there are Ethernet and token ring LANs, which operate at 10–1000 Mbps. In contrast, WAN links are characterized by low speeds (ranging in the order of few Kbps to few Mbps) and high costs. The response time of fetching data in the two environments is markedly different.

As long as the traffic movement is limited to LAN, such speed differences are not significant. But over the years, the behaviour of traffic movement has altered and there is a shift of traffic from the Intranet to the Internet. What used to be an occasional foray outside one's own network domain is now a necessity. All this makes the LAN–WAN difference crucial.

ATM is developed to alleviate this problem by providing a common medium of transmission in both LAN and WAN environments. In order to understand why this is possible, one has to look at the limitations of the current LAN and WAN technologies. LAN technologies like Ethernet and token ring are broadcast-based technologies. In order to ensure proper data transfer, these technologies use Medium Access Control (MAC) protocols, which puts an upper limit on the maximum length of cables. This maximum limit varies with the type of physical medium used and is in the order of a hundred metres. This fact renders these technologies useless for a WAN environment, wherein larger distances are involved. Similarly, most of the WAN technologies (like X.25 and frame relay) are not suitable in a LAN environment. This is because both X.25 and frame relay provide data transfer rates below 2 Mbps, which is too low for a LAN environment.

ATM does away with both these shortcomings because it is a point-to-point based technology and is capable of providing gigabit data rates. In essence, the need to have a common LAN/WAN transmission medium is a major motivation behind the development of ATM technology.

3.2.2 Universal Platform

The growth of ATM is also leveraged by the requirement to have a universal platform that can support different types of applications. Primarily, the networking requirements are classified into two broad categories, viz. voice communication and data communication. Voice communication requires a fixed bandwidth and low delay, and is tolerant to occasional information loss. On the other hand, data communication is bursty in nature, and though data is tolerant to higher transit delays, it is susceptible to information loss. This basic difference in the requirements for voice and data communication makes the design of a universal platform difficult. Thus, prior to the deployment of ATM, the networking fraternity was divided into two groups—the telecommunication world and the datacommunication world. The telecommunication world is backed by the international standards bodies like the International Telecommunications Union-Telecommunication Standardization (ITU-T) and caters to services required for voice conversation. The services include basic telephony and facsimile (fax). The datacommunication world is backed by bodies like the Internet Engineering Task Force (IETF), the Institute of Electrical and Electronic Engineers (IEEE) and International Standards Organization (ISO), and caters to services required for data transfer. The services include data transfer in LAN and WAN environments.

However, during the last decade or so, the distinction between a telecommunication network and datacommunication network has become very hazy. Two classical examples in this regard are packet-based voice and integrated services network. The following paragraphs elaborate upon the two networks.

Till recently, voice was carried only by using circuit-switched networks. Packet-based networks were considered unsuitable for voice because of their unpredictable and high-delay characteristics. However, with the advent of Internet telephony, this notion has changed. Specialized software packages (like Voice-over-IP or VoIP), compression and silence suppression schemes, and echo cancellation devices have considerably improved the voice quality over the Internet. Moreover, since Internet charges are based on connectivity, and not on distance, as against basic telephony where the charge is distance-related, Internet telephony offers a cheap alternative for long-distance voice conversation. Although the quality of voice-over Internet is somewhat inferior, the price differential makes Internet telephony a very attractive proposition indeed.

The popularity of Internet telephony indicates the aggressiveness of datacommunication companies to capture the voice market, a market hitherto monopolized by telecommunication companies. In order counter the aggressiveness of datacommunication companies, telecommunication vendors are also trying to expand their realm from voice-only services to integrated services. The development of Integrated Services Digital Network (ISDN) by the telecommunication body, ITU-T is an indicator of this move. The primary motive behind the development of ISDN was to provide a digital pipe for integrated services (voice, data and video). However, as the ISDN attempt has not been very successful, the next attempt of ITU-T to provide a universal platform is through B-ISDN. Like ISDN, B-ISDN aims to provide integrated services, albeit at higher data rates.

The above example indicates that the two worlds will ultimately converge into one, a phenomenon termed as '*convergence*'. ATM is developed to expedite the process of convergence and to provide a synergy between the telecommunication and datacommunication worlds.

3.3 Definition of ATM

ATM stands for 'Asynchronous Transfer Mode'. The ITU-T recommendation [ITU-T I.113] defines ATM as: "A transfer mode in which the information is organized into cells; it is asynchronous in the sense that the recurrence of cells containing information is not periodic." This definition encompasses three basic terms, viz. transfer mode, cell-based transfer and asynchronous transfer. The following sub-sections elaborate upon each of these terms.

3.3.1 Transfer Mode

As discussed in Chapter 2, ATM fits in between the extremes of circuit switching and packet switching because it uses a small-sized frame (53 bytes). By using a small-sized frame (precisely, a cell!), ATM retains the speed of circuit switching while still offering the flexibility of packet switching. This is why ATM is also referred to as fast packet switching technology.

3.3.2 Cell-based Transfer

Information in ATM is 'organized into cells', which means that the lowest unit of information in ATM is a cell. A cell is a fixed size frame of 53 bytes, with 5 bytes of header and 48 bytes of payload. The header carries information required to switch cells, while the payload contains the actual information to be exchanged.

Figure 3.1 illustrates the concept of cell-based transfer. Information from various sources is multiplexed and segmented, resulting in a stream of ATM cells. Each cell is transmitted and received independent of other cells. A cell is identified by the labels carried in the header. Here, *label* refers to the Virtual Channel Identifier (VCI) and Virtual Path Identifier (VPI) fields. These fields identify the virtual circuit to which a cell belongs. In essence, ATM is a cell-based technology, which employs virtual-circuit concepts to forward information streams.

FIGURE 3.1

Cell-based Transfer

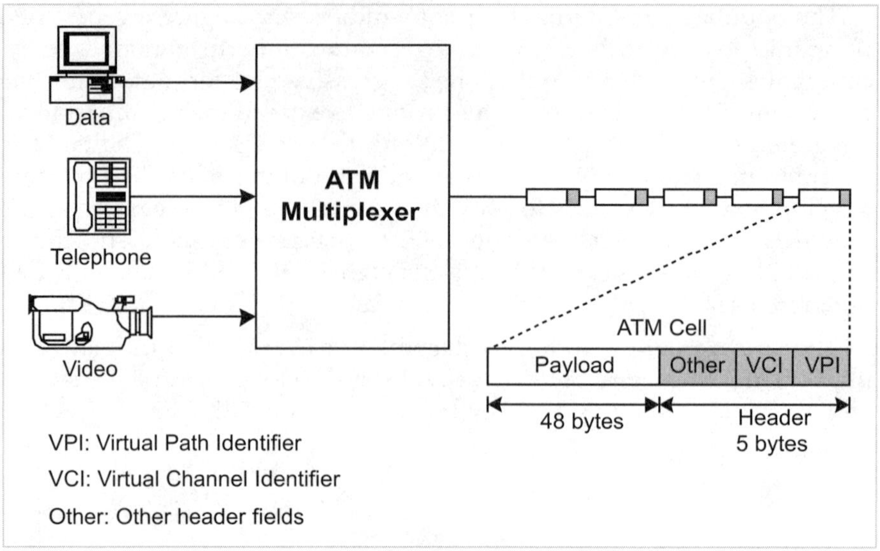

3.3.3 Asynchronous Transfer

ATM is an asynchronous transfer mode. There is considerable confusion regarding the term 'asynchronous' in ATM. While the [ITU-T I.113] definition provided earlier ("asynchronous in the sense that the recurrence of cells containing information is not periodic") clarifies the meaning of the term 'asynchronous', the connotation of 'asynchronous' is still not very clear.

Usually, the terms *'synchronous'* and *'asynchronous'* refer to the way data is transmitted. In synchronous mode, the transmitter and receiver clocks are synchronized and frames are sent/received periodically. Time division multiplexing, where each time slot is reserved for a particular voice channel, and wherein frames recur at 125 micro second interval, is a good example of synchronous transfer. In asynchronous mode, timing information is derived from the data itself, and that transmitter is not compelled to send data periodically (though it is allowed to do so). The RS-232 protocol, wherein start/stop bits are used to indicate the start/end of transmission, is an example of asynchronous data transfer.

With this knowledge, where does the term 'asynchronous' fit into the definition of ATM? There is more confusion if SONET is the underlying transmission medium for ATM because SONET operates synchronously. The answer lies in realizing that the term 'asynchronous' is used for the ATM layer and not for the physical layer; that is, the multiplexing of cells on to the physical medium is asynchronous, not the transmission of cells. Unlike TDM, in ATM, no slot is reserved for a logical channel and cells are transmitted as and when they arrive. This makes the transfer of cells for a particular channel non-periodic, which is why ATM is an asynchronous transfer mode.

Figure 3.2 depicts the asynchronous version of the TDM scheme presented in the previous chapter. Recall that in TDM, each information unit is identified implicitly by the slot position

FIGURE 3.2

Decoupled Multiplexing of ATM Cells

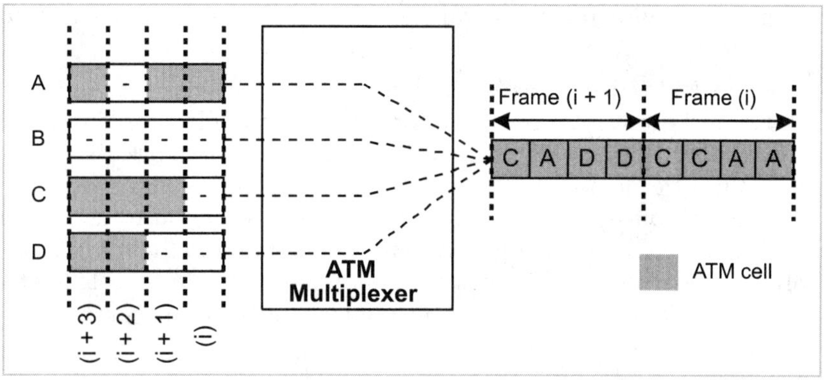

in which it is carried. For example, data from Source A is carried in slot 1, data from Source B is carried in slot 2 and so on. Thus, whenever any source has nothing to send, the slot goes empty, thereby wasting bandwidth and reducing efficiency.

However, this is not the case in ATM, because there is no coupling between the slot position and the source number. This decoupling results in more efficient utilization of link bandwidth because the link does have to carry empty slots. Besides, ATM also does not couple the application data rate with the underlying transport mechanism. In order to understand this, note that to carry voice, TDM partitions bandwidth into 64 Kbps channels. This is an example wherein the application data rate (64 Kbps) is tied to the underlying transport mechanism (64 Kbps channels). The effect of this coupling is that even if technological advancements permit voice transfer at lower data rates, the widespread deployment of transport infrastructure hinders the use of this advancement. ATM does not have this drawback as information is packaged into independent cells, and depending upon the application requirements, the number of cells transferred can be altered. This makes ATM much more flexible.

However, the flip side of the above strategy is that ATM requires each cell to explicitly carry information about its virtual channel. This information is carried as VPI/VCI fields in the ATM cell header, accounting for an overhead of 10.4 per cent (5/48).

When both the strategies are compared, it is observed that the overhead of cell header is more than compensated by better link utilization and gain in flexibility. This is why the switching of cells in ATM takes place asynchronously.

3.4 GENESIS OF ATM

In today's world, technological advancements are observed in almost all the fields. This is particularly visible in the field of telecommunications where there is ever-growing demand for higher bandwidth and rich applications. In order to deliver new services such as video-conferencing and video on demand, as well as to provide more bandwidth for the increasing volume of traditional data, a new technology was developed that provided a common format for services with different bandwidth requirements. This technology is Asynchronous Transfer Mode (ATM).

Before ATM, there was inadequate support for linking geographically distinct networks. As demand grew, growing attention was given to fiber cable without adequate standardization backing. In parallel, there was a need for a network that could carry voice, data and video. In order to address this need, ITU-T and other standards groups started work in the 1980s to establish a series of recommendations for the transmission, switching, signalling and control techniques required to implement an intelligent fiber-based network that could solve current limitations and would allow networks to be able to efficiently carry services of the future. This network was termed Broadband Integrated Services Digital Network (B-ISDN).

B-ISDN is an attempt by telecommunication companies to provide a single network for all types of communication (including voice, video and data). The term broadband is formally defined as "a service or a system that requires transmission channels capable of supporting rates greater than the primary access rate" (the primary rate depending upon the hierarchy used is equal to 1.544 Mbps or 2.048 Mbps). Informally, broadband technologies refer to an array of high-speed technologies including Synchronous Optical Network/ Synchronous Digital Hierarchy (SONET/SDH), Digital Subscriber Line (DSL), ATM and Gigabit Ethernet.

The effort to provide a unified network for different applications started with the development of ISDN (now called N-ISDN). However, the inherent limitations of N-ISDN networks forced ITU-T to look for alternatives that could satisfy the increased bandwidth demand without sacrificing efficiency. ATM technology was developed to fulfil the requirements for B-ISDN services. In fact, ATM derived its name because it was an alternative to Synchronous Transfer Mode (STM). STM networks like Plain Old Telephone Systems (POTS) and Narrowband-ISDN (N-ISDN) networks are based on TDM frame structures. STM, and ATM were the two prospective candidates when the transfer mode of B-ISDN was being decided. Since N-ISDN used STM, it was thought appropriate to use STM for B-ISDN also. However, owing to the inefficient bandwidth utilization and inflexibility of the STM, its alternative, i.e. ATM, was chosen as the transfer mode for B-ISDN.

In 1988, ATM was chosen by ITU-T (then CCITT) as the preferred transfer mode for broadband-integrated services digital network (B-ISDN). By 1990, decisions had been made to base B-ISDN on Synchronous Optical Network/Synchronous Digital Hierarchy (SONET/ SDH) and ATM.

In order to attain quick standardization and the backing of the industry, the ATM Forum was established in October 1991, which issued its first specifications eight months later. The ATM Forum was formed to accelerate the user of ATM products and services through a rapid convergence of interoperability specifications. In addition, the Forum promotes industry co-operation and market awareness.

3.5 Precursor Technologies

ATM is often described as a breakthrough technology because of its ability to provide a single integrated platform with comprehensive QoS guarantees. Nonetheless, ATM owes its origin to three technologies, namely X.25, frame relay and ISDN.

X.25 (to some extent) and frame relay (to a greater extent) have influenced the design of ATM technology. The basic similarity between X.25, frame relay and ATM is that they are all virtual-circuit-based technologies and use small labels to forward packets, frames and cells, respectively. The important difference between them is that while X.25 and frame relay use large variable-sized packets, ATM uses a small, fixed-sized cell. This single difference leads to many secondary differences like higher switching speeds, higher bandwidth and greater capability to support real-time traffic in ATM.

While virtual-circuit concepts were derived from X.25 and frame relay, the intent or basic motivation of providing a truly integrated platform came from ISDN. In fact, the rather poor response of ISDN triggered the development of a high-speed alternative, which turned out to be ATM.

3.5.1 X.25

In the early 1970s, end-system communication with public networks was done the using proprietary protocols. The publication of [ITU-T X.25] was an attempt of the ITU-T to standardize the whole procedure. X.25 provides a standard interface and standard frame format for data communication between an end-system and a public network. Formally, X.25 is an interface specification between a Data Terminal Equipment (DTE) and Data Circuit-Terminating Equipment (DCE). As against popular belief, X.25 is not packet-switching technology; it is just an interface specification. What the world views as an X.25 public data network, is a collection of various standards (including X.25) like X.21, X.28, X.28, X.29, X.3,

3.5.1.1 X.25 Protocol Stack

The X.25 protocol stack strictly follows the OSI-RM layering model. Figure 3.3 shows the different layers of X.25, and the relationships of these layers with the OSI model. At the physical layer, a number of standard interfaces are defined. Prominent among these are the [ITU-T X.21] and [ITU-T X.21 bis] standards. X.21 defines an interface between DTE and DCE for synchronous operations. X.21 bis is designed for DTE for interfacing with synchronous V-series modems.

At the link layer, X.25 uses a protocol called Link Access Procedure-Balanced (LAP-B). The term 'balanced' here refers to equal control between communicating entities during data transfer. In unbalanced transfer, there is a master and a slave, wherein the master controls the way the data is exchanged between the two. The LAP-B protocol is an adaptation of a very common ISO data link layer standard, the High-level Data Link Control (HDLC).

FIGURE 3.3

Protocol Layering in X.25

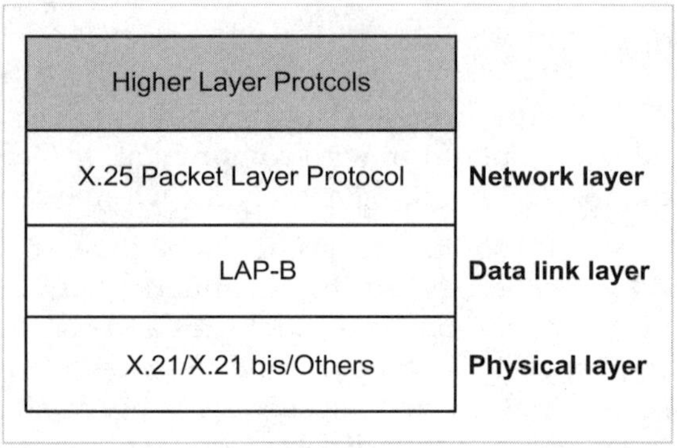

The X.25 Packet Layer Protocol (PLP) resides at the network layer. This protocol manages virtual connections between the DTE and DCE. An X.25 virtual connection is identified by using a 12-bit field Logical Channel Number (LCN) present in a PLP packet's header. The 12-bit LCN field allows a total of 4095 virtual circuits.

3.5.1.2 X.25 Features

X.25 was designed for slow and unreliable links. Thus, X.25 has a lot of error and flow control functions incorporated in the protocol. The packet's payload is protected by using a Frame Check Sequence (FCS). The FCS is used for error detection in the received frames.

In order to prevent packet loss due to unreliable delivery, each X.25 packet is duly acknowledged by the receiver. For this purpose, sequence numbers are used. There are two types of sequence numbers, Send Sequence Numbers (SSN) and Receive Sequence Numbers (RSN). SSN indicates the sequence number of the packet sent, while RSN indicates the sequence number of the next expected frame. Collectively, these sequence numbers provide reliable delivery.

In essence, the link-by-link flow control, sequence numbering and error checking considerably slow down packet forwarding. This is why X.25 is applicable for low-speed links (56 Kbps to 2 Mbps). However, the elaborate procedure is also advantageous when the link is unreliable and a high level of integrity is warranted.

3.5.2 Frame Relay

With improvements in the speed and quality of links, observers felt that a lot of functions of X.25 could be removed. This would have significantly increased the rate of packet forwarding. Frame relay was designed to serve this purpose. In fact, frame relay is said to be the X.25 of the 1990s, because it is a very simple protocol and provides much higher speed as compared to X.25.

3.5.2.1 Frame Relay Protocol Stack

Since frame relay is a simple protocol, its protocol stack is also simpler as compared to X.25. Figure 3.4 shows the layering in frame relay. As shown in the figure, there are two distinct planes—the user plane in which data is transmitted, and the control plane in which connection establishment messages are exchanged.

FIGURE 3.4

Protocol Layering in Frame Relay

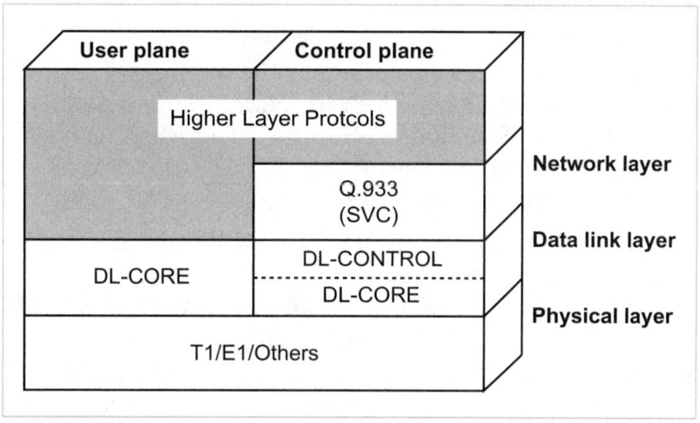

If the user plane of the frame relay is compared with that of X.25, one would observe that the link layer and network layer are merged to form the DL-CORE layer. DL-CORE combines the functionality of LAP-B and PLP protocols of X.25 and provides a subset of these functionalities. The DL-CORE protocol is detailed in [ITU-T Q.922]. In essence, the whole of the frame relay's protocol suite in the user plane is limited to the DL-CORE layer. Almost any higher layer can reside over it. Also, DL-CORE can reside over a variety of physical layer standards (e.g. T1/E1).

The control plane is used to establish and release dynamic virtual connection, also referred to as Switched Virtual Connection (SVC). In the control plane, [ITU-T Q.933] resides at the network layer. Q.933 is a signalling protocol based on the ISDN signalling protocol [ITU-T Q.931]. At the data link layer, besides DL-CORE, another sub-layer DL-CONTROL is used. DL-CONTROL provides reliable data delivery in the lines of LAP-B. The physical layers for both control plane and user plane are same.

3.5.2.2 Frame Relay Features

Frame relay gets its name because it simply relays frame. A frame received on one link is simply relayed on the appropriate outgoing link. This is done by using small labels, with each of them called the Data Link Connection Identifier (DLCI).

Frame relay is quite different from X.25. In X.25, link-by-link flow control is done. This is absent in frame relay, at least in the data plane. The error detection using FCS is retained in frame relay. At the network layer, an X.25 PLP packet is encapsulated in a LAP-B frame, which requires encapsulation and decapsulation. Since frame relay merges the network layer and the link layer, this encapsulation/decapsulation is unnecessary. All this makes frame relay much faster than X.25. Additionally, frame relay provides congestion notification features using two bits in the frame header—the Forward Explicit Congestion Notification (FECN) and the Backward Explicit Congestion Notification (BECN). It also provides the discard eligibility feature, which facilitates the marking of low-priority frames. These marked frames are discarded during periods of congestion.

Frame relay shares certain features with ATM. Both technologies use virtual-circuit concepts. The latter relays frame while the former relays cells (this is why ATM is also called a cell relay technology). Both technologies assume intelligent end-systems and thus retain bare minimum functions within the network. This is why both technologies do away with time-consuming flow control functions. Lastly, congestion notification and discard eligibility features are common to both technologies.

These similarities notwithstanding, there are some key differences too. Frame relay is designed for links with speed up to 2 Mbps (though higher link speeds can be supported). ATM is a scalable technology and offers gigabit speeds. Moreover, frame relay is a technology designed primarily to support data applications. In contrast, ATM is designed to support various types of applications including voice, video and data. Of late, frame relay is also being used to carry voice. For this, some extensions are made to the existing protocol so that a large frame can be segmented into smaller frames to meet delay and jitter requirements for voice. This modification minimizes the differences between frame relay and ATM. Besides technological differences, ATM standards are many times more comprehensive and elaborate than frame relay standards. Table 3.1 summarizes the similarities and differences between ATM and frame relay.

3.5.3 Integrated Services Digital Network (ISDN)

ISDN evolved due to some serious shortcomings in the Plain Old Telephone Systems (POTS). First, POTS are voice-only networks and it was desirable to have a single integrated platform. Second, and more importantly, the last mile (i.e. the local loop) in POTS is analog, which leads to poor quality transmission. ISDN was developed to remove these shortcomings. In essence, ISDN is an end-to-end digital network, which supports voice, video and data in an integrated fashion. One of the biggest advantages of a digital network is that computer data can be directly exchanged using it. Otherwise, an intermediate modulation/demodulation step, using a modem, is required.

3.5.3.1 ISDN Protocol Stack

ISDN technology is the brainchild of ITU-T. The complete I-series of the ITU-T recommendation is devoted to integrated networks (ISDN and B-ISDN). Figure 3.5 shows the protocol stack of ISDN. Although only two planes are shown in the figure, the ISDN protocol stack is a three-dimensional model, consisting of a management plane also (which is explained later).

TABLE 3.1

Similarities and Differences between Frame Relay and ATM

Similarities	*Differences*
1. Virtual-circuit based technology	1. Frame relay has a variable-sized frame, whereas ATM has a fixed-size cell.
2. Allow establishment and release of SVCs/PVCs	2. Frame relay supports moderate date rates (~2 Mbps) while ATM supports very high data rates (~ Gbps).
3. Assume reliable links, smart end-systems	3. Frame relay is designed primarily for data, while ATM is designed for data as well as voice, video and multi-media applications.
4. No flow control	4. Reference model in frame relay is two-dimensional, whereas it is three-dimensional in ATM.
5. Congestion notification and discard eligibility feature (Note 1)	5. As compared to frame relay, standardization in ATM is more elaborate.
	6. Frame relay is a virtual leased line technology for WAN interconnectivity. ATM has a much wider scope (LANs and WANs).
	7. Frame relay uses single label DLCI, whereas ATM uses two labels VPI/VCI.

Note 1: A minor difference is that frame relay supports both forward and backward congestion notification, while ATM supports forward congestion notification only.

FIGURE 3.5

Protocol Layering in ISDN

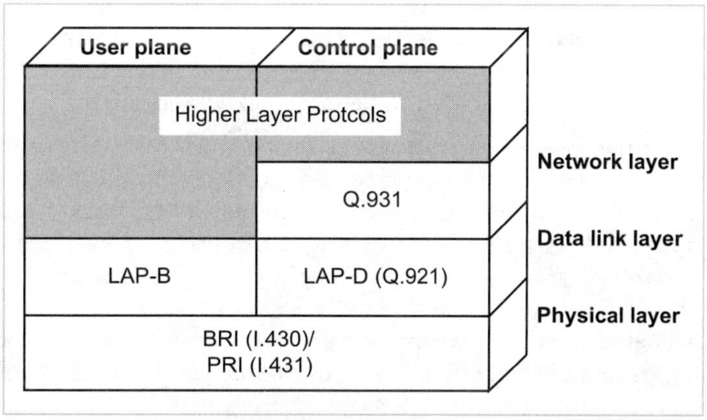

At the physical layer, two interfaces are defined—the Basic Rate Interface (BRI) and the Primary Rate Interface (PRI). The BRI interface, defined in [ITU-T I.430], is a 144 Kbps channel, consisting of two bearer channels (2B) and a signalling channel (D). The bearer channels are used for information transfer at 64 Kbps, while the D channel is used for signalling (16/64 Kbps) and a part for user data (9.6 Kbps). The PRI structure is defined in [ITU I.431] and has two variants, one for the US and the other for Europe. In the US, PRI interface is a 1534 Kbps channel, consisting of 23 bearer channels and one signalling channel. In Europe, PRI interface is a 1984 Kbps channel, consisting of 30 bearer channels and one signalling channel.

At the link layer, LAP-B and LAP-D (link access procedure for D channel) are used. In the control plane, [ITU-T Q.931] resides at the network layer and is used for signalling. There is no restriction on higher-layer protocols.

3.5.3.2 ISDN Services

As compared to POTS, ISDN offers a lot of attractive services. For example, consider the example of Internet access. Through an analog modem, the maximum speed of an Internet connection is around 34.4 Kbps. Using a BRI instead, a 128 Kbps channel can be obtained, implying a quadruple improvement in speed. Other services of ISDN include voice, fax, video-conferencing and LAN interconnection.

3.6 Basic Principles of ATM

This section elaborates upon the basic principles of ATM. In particular, the following aspects of ATM are discussed: fixed and small-sized packet, connection-oriented approach, virtual channels and virtual paths, negligible error and flow control, and quality of service.

3.6.1 Fixed and Small-Sized Packet

In the late 1980s, when the ATM cell format was being discussed, two important issues were the subjects of a fierce debate. The first issue was to decide whether the ATM cell would be of fixed size or of variable size. The second issue was related to finding the right cell size, provided it was decided to keep the size fixed.

At the time the decision was made, and even now, almost all packet-based technologies (including X.25, frame relay, and Ethernet) employ a variable-sized frame/packet. The actual length is indicated in one of the fields of the packet header. A variable-sized packet, as compared to a fixed-sized packet, offers two key benefits. First, since the packet size is variable, there is no bandwidth wastage and only the requisite amount of data is transmitted. This is not true for fixed-sized packets in which irrespective of the size of the higher layer packet, the packet transmitted always remains constant. So, when the number of bytes to be sent is less than the (fixed) size of the packet, the packet contains padding bytes, which carry no useful information.

Second, if the packet size is fixed and if the higher layer PDU does not fit into one packet, an intermediate segmentation and reassembly layer is required (see Figure 3.6). For variable-sized packets, in contrast, this intermediate segmentation and reassembly layer is not required. This statement is explained with the help of Figure 3.6. As shown in the figure, for variable-sized frame-based technologies like Ethernet, an IP datagram fits directly into an Ethernet frame. This condition is true, provided the datagram is less than the maximum payload that the Ethernet can carry. Since an Ethernet frame can carry up to 1500 bytes of payload, segmentation and reassembly are generally not required. However, assuming a small cell size (say 53 bytes), ATM requires an intermediate layer that segments IP datagrams at the transmitter and reassembles them at the receiver. If a very large cell size is chosen to prevent segmentation and reassembly, a majority of the cells will carry partially filled cells. This leads to inefficient bandwidth usage, which is not acceptable. To conclude, for a variable-sized packet, there is no bandwidth wastage and no need for intermediate segmentation and reassembly layer. For fixed-sized packet, if the packet size is too small, a segmentation and reassembly layer is required, and if the size is too large, there is a lot of bandwidth wastage.

FIGURE 3.6

Variable-sized versus Fixed-sized Packet

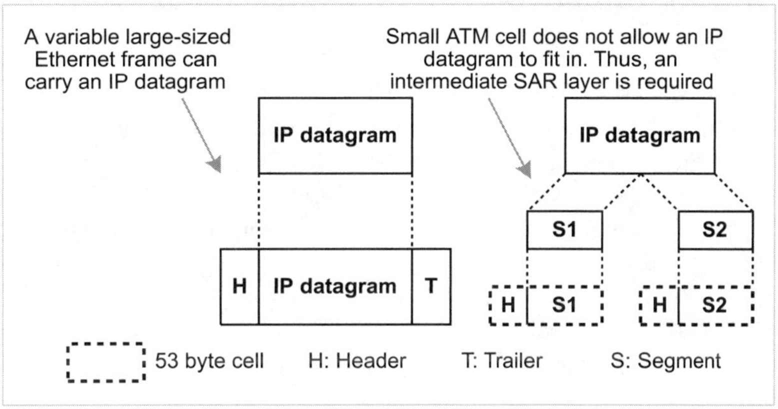

Despite the drawbacks, a fixed-sized cell was preferred over variable-sized packet. Since ATM was designed for both voice and data communication, it was mandatory to keep transit delay within reasonable bounds. In this regard, a fixed-sized cell, as compared to a variable-sized packet, offered many advantages like predictable and lesser switching delay and simpler memory management in switch buffers. Keeping variable size packets meant that each cell carried the length field. This introduced extra processing in intermediate switches because the length field had to be processed and appropriate size buffers had to allocated. Since guaranteeing bounded delay was a greater concern than minimizing overheads, a fixed-sized cell was chosen.

Once it was decided that the cell would be of a fixed size, choosing the right size became critical. For a large cell which was only partially filled, there was bandwidth wastage. A small cell size meant less propagation and switching delay. Moreover, a smaller size cell required less time to get filled. For example, to fill a 53 bytes cell using a source of 64 Kbps, 0.75 ms is required. If the cell size was increased, the packetization delay also increased proportionately.

In fact, the whole world was divided into two camps. The Europeans were in favour of 32 bytes because they thought that having smaller payloads would ensure transit delays below a threshold, thereby obviating the need for echo cancellers. The US and Japan were in favour of 64 bytes because this would have resulted in higher transmission efficiency. Ultimately, a compromise was made and cell size was fixed at 53 bytes, with 48 bytes of payload and 5 bytes of header.

In essence, the decision of variable size or fixed-size cell, and choosing the cell size, was a battle between the telecommunication world and the datacommunication world. The former group was concerned about delay while the latter was concerned about transmission efficiency. In the end, the telecommunication world seems to have triumphed. This was expected because ATM was primarily designed by the telecommunication body, ITU-T.

3.6.2 Connection-oriented Approach

ATM is a connection-oriented technology. The connection-oriented approach of ATM is characterized by the following factors:

- ATM uses virtual connections to switch cells from one node to another. The virtual circuits are identified by VPI and VCI. The VPI/VCI fields are carried in the ATM cell headers and are called connection identifiers.
- Cells belonging to a virtual connection follow the same path. Thus, cell sequence is maintained implicitly.
- The bandwidth allocated to a virtual connection is assigned at the time of the connection setup and is based on the requirements of the source and the available capacity.
- Each virtual connection is provided with a Quality of Service (QoS). The QoS concept is detailed in Chapter 7.
- The connections are bi-directional in nature. The bandwidth allocated in forward and backward directions may or may not be same (symmetric/asymmetric connection). The bandwidth in the backward direction can even be zero (thus rendering the connection uni-directional).
- The sames VPI/VCI value is used across a link in both the directions.

The virtual connections in ATM are established either *statically* or *dynamically*. A static connection is one in which the end-points of the connection are defined in advance and the bandwidth is predetermined. Such static connections are established through subscription and are referred to as Permanent Virtual Connection (PVC).

Dynamic connections, on the other hand, offer much more flexibility. They are established when there is data to be transferred and released when the data transfer is over. Such dynamic connections are referred to as Switched Virtual Connection (SVC). The establishment and release of SVCs takes place by using signalling procedures, which is explained in Chapter 11.

Between the two, PVCs are preferred when there is relatively stable data flow between two end-points. SVCs are useful when the communication is for a very short duration only. Public carriers providing services over ATM networks generally offer PVC connections only. Of late, carriers have started offering SVC connections also.

3.6.3 Virtual Channels and Virtual Paths

ATM uses VPI/VCI fields of a virtual connection to forward cells. At each intermediate node, VPI/VCI values are swapped and new values filled. This process is also referred to as cell relaying, cell switching or cell forwarding. Figure 3.7 shows VPI/VCI swapping performed by an ATM switch. The VPI/VCI pair of an incoming cell is used as an index to search the VPI/VCI translation table. In the VPI/VCI translation table, a separate entry is maintained for each established virtual circuit. For example, the table in Figure 3.7 maintains entries for two virtual circuits. The values in the table get filled when the connection is established. For a PVC connection, each switch between source and destination is configured using some management tool. For an SVC connection, this connection takes place automatically via an exchange of signalling messages.

VPI/VCI Translation Table

Incoming		*Outgoing*		
VPI	*VCI*	*VPI*	*VCI*	*Port*
12	5	23	7	3
15	7	26	9	2

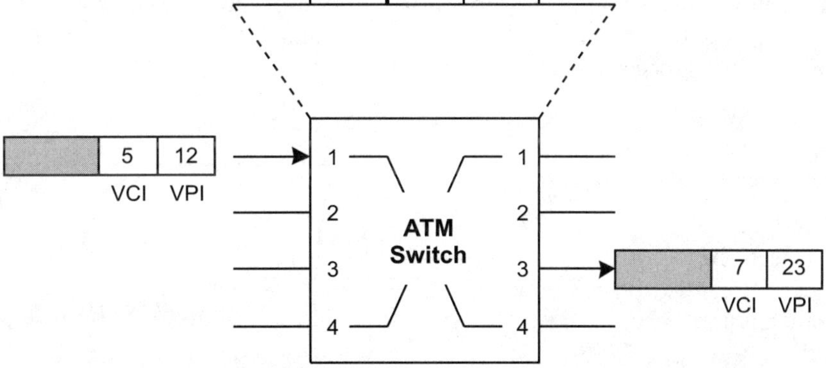

FIGURE 3.7

VPI/VCI Translation in an ATM Switch (Cell Switching)

Through this table, the new VPI/VCI values are extracted and filled in the incoming cell's header. The entry in the table also provides the port number on which the cell is to be forwarded. For example, in Figure 3.7, a cell with VPI = 12 and VCI = 5 arrives on port number 1. This VPI/VCI value is used to obtain the new VPI/VCI values, VPI = 23 and VCI = 7, and also the port number (3 in this case) on which the cell is to be forwarded.

The VPI/VCI swapping in ATM is not as simple as explained. Actually, the virtual circuits or connections in ATM are classified into two categories, namely Virtual Channel Connection (VCC) and Virtual Path Connection (VPC). This establishes a two-level hierarchy in the forwarding of cells. The following paragraphs elaborate upon the concept of VCC and VPC as defined in [ITU-T I.113].

- **Virtual Channel (VC) and Virtual Channel Connection (VCC):** VC is a logical concept used to describe the transport of ATM cells based on a unique identifier value called the VCI. A VCC is a much more formal term, defined as: "A concatenation of virtual channel links, which extends between two end-points, and which is switched upon the combined VPI/VCI value." Specifically, a VCC extends between two points where the adaptation layer is accessed. A virtual channel link refers to a logical connection—either between two intermediate nodes or between an intermediate node and an end-point.
- **Virtual Path (VP) and Virtual Path Connection (VPC):** VP is a logical concept used to describe the transport of ATM cells based on a unique identifier value called the VPI. A VPC is a much more formal term, defined as: "A concatenation of virtual path links, which extends between the point where the VCI values are assigned and the point where those values are translated or removed." A VPC is switched by using the VPI values only.

Figure 3.8 shows the relationship between VC, VP and transmission path. A transmission path contains one or more virtual paths, while a virtual path contains one or more virtual channels. Switching is done either at the virtual path level or at the virtual channel level. In the first case, only the VPI field is used for switching and the process is referred to as VP switching. A switch capable of VP switching is called a VP switch or a VP cross-connect. VP switching is useful when a number of connections have to be established between two nodes. The aggregation of multiple virtual channels into a virtual path connection, makes the forwarding process easier, as only the VPI field is looked at and replaced.

FIGURE 3.8

Relationship Between the VC, the VP, and Transmission Path

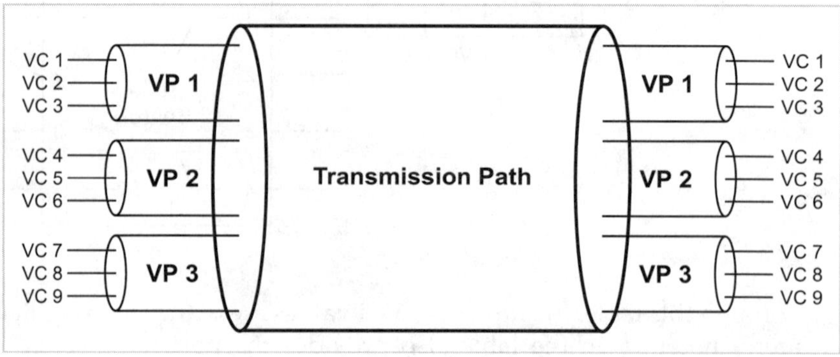

In the second case (i.e. switching at the virtual channel level), both VPI/VCI fields are used for switching and the process is referred to as VC-switching. A switch capable of VC-switching is called a *VC switch* or a *VC cross-connect*.

Figure 3.9 illustrates the concept of VP and VC switching. In this figure, the lower part represents a VP switch where only the VPI is changed. This is the reason why the VPC with VPI = 7 has only its VPI value changed to 8. The VCI values of this VPC (i.e. VCI = 31 and VCI = 32) are not changed by the VP cross-connect. The upper half of the figure represents a VC cross-connect. A VC cross-connect changes both the VPI and VCI values of a VCC. For example, the VCC with VCI = 32 and VPI = 4 is changed to VCI = 35 and VPI =1. Also note that VCI values can be reused in different virtual paths. For example, in the given figure, even though VCI =31 is used twice, there is no conflict because the VPI values are different.

FIGURE 3.9

VP and VC Switching [ITU-T I.311]

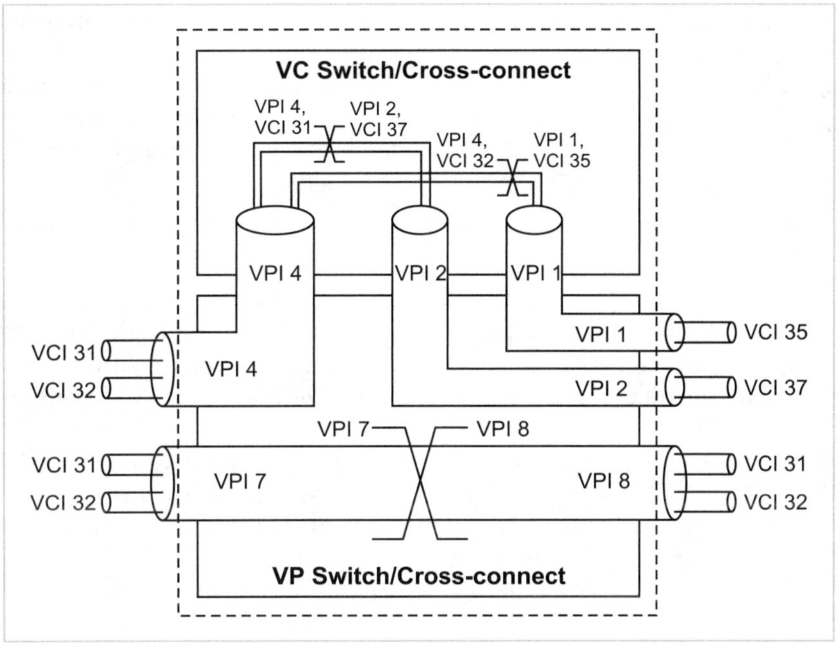

It may not seem very obvious as to why there are two identifiers (VPI and VCI) in ATM at the very first place. After all, both X.25 and frame relay, which are virtual-circuit based technologies, use only a single label. However, there are many advantages of having a two-level hierarchy of identifiers. The first advantage is that a two-level hierarchy allows virtual paths to be distinguished on the basis of QoS. Thus, the resource may be allocated on the basis of virtual paths, and depending upon the nature of the connection request, a virtual channel from a particular virtual path can be granted. This reduces the workload on the switch and makes the granting/rejecting of connections easier.

In order to understand the second advantage, consider the scenario in which two private networks, A and B, connect via a public ATM network. In this case, by having a single virtual path connection between networks A and B, the two networks can establish multiple virtual channels. This gives them the flexibility to add, modify and delete virtual channels without bringing this to the notice of the service provider. The service provider also does not bother about the number of channels established because the charging will be based upon the parameters of the virtual path connection.

Apart from the advantages mentioned above, VP switching (which is based upon VPI values only) is faster as the VCI value is not considered. This facilitates the aggregation of similar flows (i.e. virtual connections) over a single virtual path.

3.6.4 Negligible Error and Flow Control

In ATM, significant processing has been pushed out of the network domain and very little error control/flow control is done within the network. Error correction is limited to the one-byte Header Error Control (HEC) field, which is a part of the five byte cell header. The HEC protects the cell header only. There is no error correction for the payload. Moreover, even if cells are lost or discarded due to transmission errors, no retransmission takes place. It is left for the higher layers to retransmit lost packets.

The decision to keep negligible error control/flow control is based on the following statements:

- Throughput is inversely related to the processing done within the network. In order to achieve high throughput, it is necessary to have minimal error control/flow control.
- Most of the real-time applications (like voice conversation and video-conferencing) do not permit any kind of flow control. This is because a delayed real-time packet is totally useless.
- It is assumed that the majority of underlying transmission mediums for ATM will be fiber. Since error rates in fiber are low and since they seldom occur in bursts, it is sufficient that only the cell-header be protected by one-byte CRC and that the cell payload be left unattended.
- Today the end-systems have much more computing power than what they had a decade back.

All the aforementioned points imply that a high-speed technology supporting real-time services must have minimal error and flow control.

3.6.5 Quality of Service (QoS)

One of the unique capabilities of ATM technology is its ability to provide Quality of Service (QoS). The QoS of an ATM connection refers to limiting the cell loss, the transit delay, and the delay variation induced by the ATM network in transiting cells of that connection. In simple terms, to guarantee QoS is to ensure that a connection gets what it requires in terms of network resources. For example, guaranteeing QoS for voice conversation (of top quality) means ensuring that the transit delay is not more than 50 ms and that the data rate is 64 Kbps for Pulse Coded Modulation (PCM) voice.

It is not an exaggeration to say that the ability to guarantee QoS is one of the biggest strengths of ATM technology. In fact, ATM gets an edge over its competitors (like frame relay) because of its ability to provide QoS.

3.7 ATM Protocol Reference Model

In the simplest of forms, the ATM reference model looks somewhat like that shown in Figure 3.10. The lowest layer, like that in any reference model, is the *physical layer*. The physical layer performs functions typical to the OSI physical layer. The second layer is the *ATM layer*, which performs important functions like cell multiplexing/demultiplexing and cell relaying. There is no true parallel of this layer in the OSI model. The same is true for the *adaptation layer* also. The adaptation layer performs a wide variety of functions necessary for the transport of information streams belonging to different applications. For example, for voice applications, the adaptation layer performs functions like source clock recovery, compression and silence suppression. For data applications, this layer performs segmentation and reassembly functions. For video applications, coding and decoding of video signals is performed. Not all these functions are actually performed and the specifics vary with the type and implementation of the adaptation layer used.

Figure 3.10

Simplistic form of ATM Reference Model

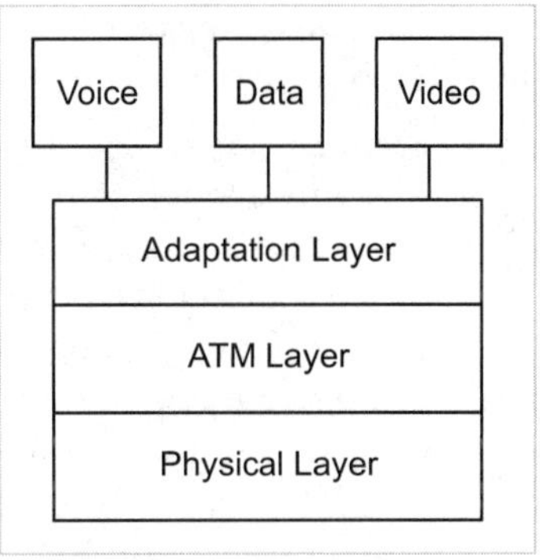

The actual ATM reference model is defined in ITU-T recommendation [ITU-T I.321]. Figure 3.11 depicts the ATM reference model. This model is a three-dimensional model, comprising the three planes, viz. the *user plane*, the *control plane*, and the *management plane*. The user plane provides for the transfer of user information. The control plane performs connection control functions, which includes the establishment, monitoring and release of virtual connections. The management plane is responsible for maintenance and management-related functions. The functions of this plane are categorized into layer management functions and plane management functions. Layer management handles

FIGURE 3.11

ATM Reference Model [ITU-T I.321]

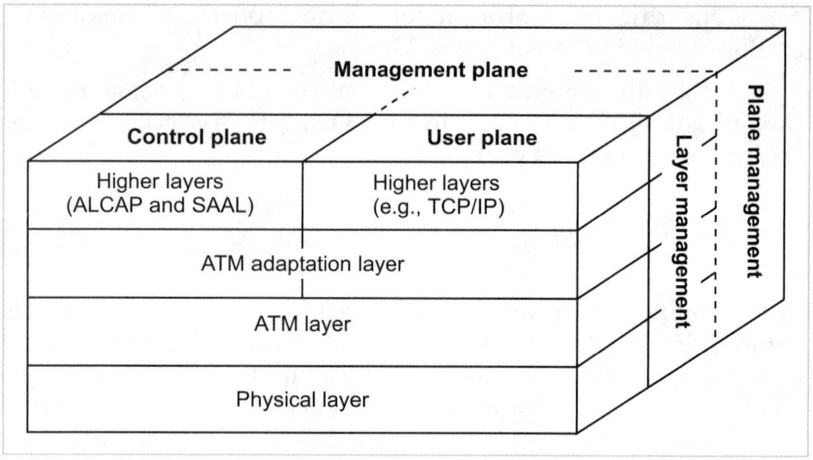

Operation and Maintenance (OAM) flows, specific to the layer concerned. This layer also handles the resources and parameters residing in its protocol entities. Plane management provides co-ordination between all the planes. Hence, plane management does not contain a layered structure.

Each plane in the ATM reference model is further divided into a number of layers. However, only the lower three layers are specified by the ITU-T. The higher layers are left undefined because they will probably comprise the plethora of protocols existing today (e.g. IP/IPX). The lower three layers of the ATM reference model were briefly discussed earlier. A more elaborate explanation is as follows (see Table 3.2):

- **Physical layer:** The functions of this layer are categorized into two parts, viz. Physical Medium Dependent (PMD) functions and Transmission Convergence (TC) functions. PMD functions are specific to the underlying physical medium. Some of the PMD functions include bit transfer, bit alignment, electrical-optical transformation and line coding. Transmission convergence functions include cell rate decoupling, cell delineation, cell header processing, scrambling/descrambling and frame generation and recovery. Chapter 4 describes the ATM physical layer functions and options in detail.
- **ATM layer:** The core functions of the ATM layer include cell multiplexing/de-multiplexing, cell VPI/VCI translation, and cell header generation/extraction. Besides this, other functions include generic flow control, Usage Parameter Control/Network Parameter Control (UPC/NPC), congestion notification, and connection assignment and removal. Chapter 5 describes the ATM layer functionality in detail.
- **ATM Adaptation Layer (AAL):** Depending upon the type of application, AAL performs a wide variety of functions. Some of these functions are already explained above. The AAL layer is divided into the Segmentation and Reassembly (SAR) sub-layer and the Convergence Sub-layer (CS). Chapter 6 describes the functions of different sub-layers of AAL and also provides a detailed description of different types of AAL including AAL1, AAL2, AAL3/4 and AAL5.

TABLE 3.2

Functional Overview of Core Layers [ITU-T I.321]

A	Convergence Sub-layer	Service-specific Convergence Sub-layer (SSCS)
A		Common Part Convergence Sub-layer (CPCS)
L	Segmentation and Reassembly Sub-layer	Segmentation and Reassembly (SAR)
	ATM Layer	Cell Multiplexing/Demultiplexing Cell Construction Cell VPI/VCI Translation Interpretation of Pre-defined Header Values Support for Multiple QoS Classes Usage Parameter Control (UPC) Explicit Forward Congestion Indication Generic Flow Control
P H Y S I	Transmission Convergence Sub-layer	Cell Rate Decoupling Cell Delineation Cell Header Processing Transmission Frame Adaptation Scrambling/Descrambling
C A L	Physical Medium Dependent Sub-layer	Encoding Timing and Synchronization Transmission

3.7.1 Quality of Service (QoS)

For deployment of a technology based on a particular reference model, it is necessary to know where the technology fits in the existing setup. Very rarely does one find a completely new network architecture along with a new set of applications being proposed at once. The preferred approach is to define a technology that performs only a subset of the total functions. This makes it mandatory for the new technology (in current context, the ATM) to interwork with existing technologies. This interworking requires that the position of the new technology, relative to OSI-RM, be absolutely clear. For example, if ATM is to support IP, it must be clear whether ATM operates at the network layer, the data link layer or the physical layer. Only when this is clear can one say how a synergy between ATM and IP is achievable.

However, the peculiar characteristics of the ATM reference model make it very difficult to fit into the OSI model. The difficulty arises due to the fact that IP acts as an *overlay network* over ATM network. Such an overlay model was not considered in the OSI model. Still, attempts are made to achieve some sort of mapping between the two models.

The physical layer is the easiest to map as it performs similar functions in both the models. However, the physical layer in the ATM reference model has a much wider scope. In fact, the Transmission Convergence (TC) sub-layer performs some of the functions of a typical data link layer (e.g. cell delineation).

The ATM layer, according to some, lies in the lower part of the data link layer. This view stems from the assumption that the ATM layer is the connection-oriented counterpart of shared-medium protocols like Ethernet and Token ring. According to others, the ATM layer is comparable to the network layer because of the cell header format and elaborate network layer protocols.

The AAL layer is the most difficult one to map. Again, some consider it to be a part of the data link layer. This opinion is applicable for the control plane wherein AAL acts as the reliable layer for higher layers. For the user plane, it is more appropriate to place AAL in the lower half of the transport layer because AAL works end-to-end.

The following sub-sections give examples of where the ATM stack can fit in different models.

3.7.2 ATM Stack as MAC Layer

Due to the peculiar nature of the ATM reference model, ATM can fit into the existing protocol stacks in a wide variety of ways. In one of the early adoptions, the ATM can be used as a Medium Access Control (MAC) interface like Ethernet and Token ring. This configuration, called LAN Emulation, is used in emulating broadcast-based LANs by using an ATM network (see Chapter 17 for details). LAN emulation provides a means for traditional LAN technologies like Ethernet and Token ring to co-exist with ATM LANs in a heterogeneous LAN environment. Using LAN emulation, two non-ATM end-systems can communicate not only with each other, but also with end-systems directly attached to an ATM network. This configuration is depicted in Figure 3.12.

FIGURE 3.12

ATM Stack Till MAC Layer

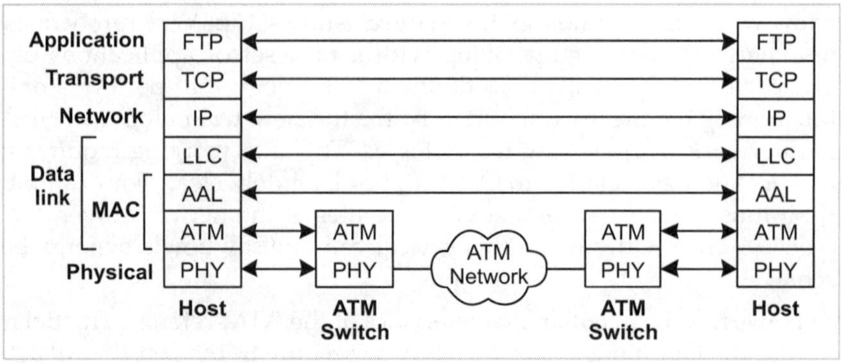

3.7.3 ATM Stack as Data Link Layer

While the MAC layer option is good to start with, it embeds the ATM layer quite deep in the protocol hierarchy and is not useful in exposing real capabilities to the actual user. Another approach that is amongst the most common positions of an ATM stack is the one depicted in Figure 3.13. In this configuration, ATM provides data link layer functionality. The transport layer and network layer in this case are TCP and IP, respectively. This position of ATM is similar to the way X.25 and frame relay protocol stacks interface with TCP/IP. Since, like them, ATM is also a virtual-circuit based transport technology, this position is abundantly used. The popularity of this configuration is also due to the widespread use of TCP/IP protocol suite and due to the need to provide high-speed link layer services to the network layer.

FIGURE 3.13

ATM Stack Till Data Link Layer

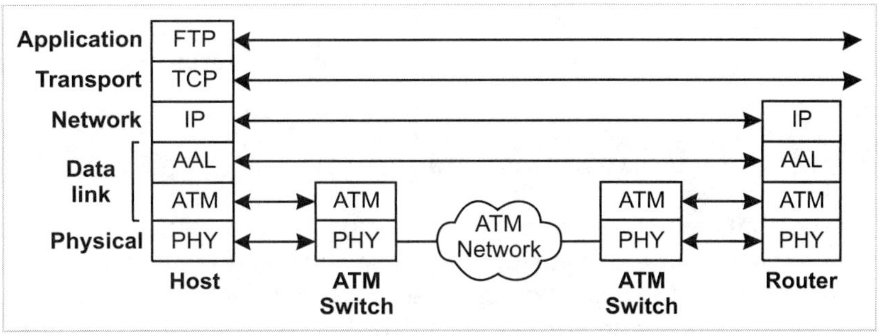

3.7.4 Other Options for ATM Stack

There are two other alternatives. First, ATM can sit directly below a transport layer (like TCP). However, the synergy between TCP and IP is so well established that no one can risk altering this combination. A second alternative is to use the ATM stack till the transport layer. This is the most ideal case, because in this case, applications can directly utilize the strengths of ATM. However, it will take some time before this happens.

3.8 Core Aspects of ATM

This chapter describes a number of core aspects of ATM. Each of the aspects briefly described below are subsequently described in Part 3 of this book.

3.8.1 ATM Traffic and Service Parameterization

ATM networks provide Quality of Service (QoS). But what are the parameters that define the QoS? How does one define the service quality in ATM? Are all users of ATM networks provided with similar quality? Or are there means to differentiate between different users? The answer to these questions lies in two sets of parameters defined for ATM. The first set of parameters, called traffic parameters (or traffic descriptors), is from the user's perspective and provides quantitative measurements of the traffic characteristic. Traffic description includes parameters like Peak Cell Rate (PCR), Sustainable Cell Rate (SCR), Maximum Burst Size (MBS), Burst Tolerance (BT) and Minimum Cell Rate (MCR).

The other set of parameters, called service parameters (or service descriptors), is from the network perspective and includes parameters that describe the quality with which the network is carrying the user traffic. Service descriptions include Cell Loss Ratio (CLR), Cell Transfer Delay (CTD), Cell Delay Variation (CDV) and Cell Error Ratio (CER), among others.

On the basis of the above parameters, a number of service categories are defined in ATM. These include Constant Bit Rate (CBR), Real-time Variable Bit Rate (rt-VBR), Non-real-time Variable Bit Rate (nrt-VBR), Unspecified Bit Rate (UBR), Available Bit Rate (ABR) and Guaranteed Frame Rate (GFR).

ATM traffic descriptors, service descriptors and service categories are described in detail in Chapter 7.

3.8.2 ATM Traffic Management

Traffic management is the act of managing network traffic, providing service guarantees to user connections and ensuring the optimal utilization of network resources. In ATM, six key elements of traffic management are described that include Traffic Contract Management (including connection admission control), Traffic Shaping (using leaky bucket techniques), Traffic Policing (including usage parameter control and network parameter control), Priority Control (using a bit in ATM cell header), Flow Control (including ABR flow control) and Congestion Control (including cell discard techniques and congestion notification).

ATM Traffic Management is described in detail in Chapter 8.

3.8.3 ATM Switching

Switching is the process of transfer of packets from input ports to output ports. In ATM, switching is a complicated process because of the the high speeds at which ATM switches operate (few Mbps to few hundred Gbps), large number of inputs (~1000) that ATM switches must support, variety of data rates that ATM switches support and the traffic management features that ATM switches support.

An ATM switch buffers and routes cells from input ports to output ports. The function of an ATM switch is divided according to the three planes—the user plane, the control plane and the management plane. In the user plane, the primary concern of an ATM switch is to transfer user cells from incoming links to outgoing links. The ATM switch, however, remains transparent to the user information carried in the 48-byte payloads of every cell. The only part of the cell inspected and processed is the 5-byte cell header. In the control plane, the switch is responsible for the establishment, monitoring, and release of switched virtual circuits. All these control plane functions are performed by exchanging signalling messages. In the management plane, an ATM switch performs fault management and performance management functions.

The ATM switches can adhere to any of the three classes of switching architectures, viz. shared-memory architecture, shared-medium architecture and space-division architecture.

ATM switching functions and ATM switch architectures are described in detail in Chapter 9.

3.8.4 ATM Addressing

Addressing refers to the process of assigning unique numerical identifiers to entities for the purpose of locating or identifying these entities. In ATM, the addresses are also referred to as ATM End System Address (AESA). An AESA identifies the location of one or more ATM interfaces. AESA is based on the generic Network Service Access Point (NSAP) format. As per this format, an NSAP address is divided into two parts—the Initial Domain Part (IDP) and the Domain Specific Part (DSP).

AESA format and other aspects of ATM addressing are described in detail in Chapter 10.

3.8.5 ATM Signalling

Signalling is used to dynamically establish, monitor and release connections. ATM standards provide elaborate signalling support for the dynamic establishment and release of Switched Virtual Circuits (SVC). Signalling assumes a greater significance in ATM owing to the extensive traffic contracting and Connection Admission Control (CAC) mechanisms involved during the connection establishment phase. The information necessary for traffic contracting is exchanged between the user and the network through these signalling messages. The signalling information includes QoS parameters (like cell delay variation and cell loss ratio) and traffic descriptors (like PCR, SCR and MBS). Further, the VPI and VCI values that identify a connection in ATM are indicated by the network to the user through signalling messages. ATM signalling standards exist for both point-to-point as well as point-to-multipoint signalling models.

ATM signalling for both point-to-point as well as point-to-multipoint connections is described in detail in Chapter 11.

3.8.6 ATM Routing

Routing is the act of forwarding packets from source to destination using intermediate network elements called 'routers'. In the context of ATM, routing is seen in context with signalling. Signalling is used to establish switched virtual connections. The purpose of routing in ATM is to route the signalling messages to the destination. Thus, routing of signalling messages for the first time is comparable to routing IP datagrams. The only difference is that routing in ATM is restricted to signalling messages (i.e. restricted to call-setup period), while in IP, each datagram is routed. However, the core issues governing the design of routing protocol remain more or less the same for both IP and ATM. To provide the means to route signalling messages, the ATM Forum has come up with Private Network-to-Network Interface (PNNI) protocol.

ATM Routing using PNNI is described in detail in Chapter 12.

3.8.7 AAL2 Signalling

Apart from ATM signalling, another important signalling protocol in ATM is the AAL2 signalling protocol. Note that ATM signalling provides the means to establish/release ATM Virtual Channel Connection (VCC) or Virtual Path Connection (VPC). In the context of AAL2, the AAL2 signalling protocol encompasses the control plane functions to establish, clear and maintain AAL2 connections. An AAL2 connection can be viewed as the logical concatenation of one or more AAL2 links between two AAL2 service end-points. An AAL2 link is a communication path between two AAL2 nodes that is uniquely identified by a Channel Identifier (CID) that allows multiplexing of many AAL2 channels within an ATM virtual connection. In order to establish and release these AAL2 channels, a dynamic protocol is necessary. The requirement is fulfilled by AAL2 signalling. A similar requirement does not arise in other AALs because AAL1 and AAL5 do not have multiplexing support. The AAL3/4 has a multiplexing identifier but that can be ignored, as AAL3/4 is not very popular.

AAL2 signalling is described in detail in Chapter 13.

3.8.8 ATM Network Management

Network management is the act of monitoring and modifying network operations for the purpose of detecting and solving problems, improving performance, and collecting reports and statistics. In ATM, the focus is fault management and performance management through the use of special cells, termed as Operations and Management (OAM) cells. Apart from OAM procedures based on OAM cells, there is the Integrated Layer Management Interface defined by the ATM Forum, which specifies how the Simple Network Management and ATM Interface MIB are used to provide any ATM device with status and configuration information.

These network management protocols for ATM are described in detail in Chapter 14.

3.8.9 ATM Security

As organizations grow and unscrupulous elements try to disrupt the proceedings, there is a clear need to protect networks from various kinds of attacks. In the context of ATM, security is a relatively new concept. Due to the use of virtual circuit switching, typical threats for a networking technology were not found to be important in the context of ATM networks. However, in recent times, there have been increasing security violations and there are growing concerns for secured corporate environment. Given this, the ATM Forum has worked towards incorporating security features in ATM networks as well. The security features include authentication, confidentiality, and data integrity.

The security protocols for ATM are described in Chapter 15.

3.9 ATM Network Architecture and Interfaces

The ATM network interfaces are of different types (see Figure 3.14). Prominent among these is the User-Network Interface (UNI). Two variations of UNI exist, namely *private UNI* and *public UNI*. The private UNI is the interface between an ATM end-system and an ATM switch, both of which lie completely within a private ATM network. A public UNI provides the means for a private ATM network to interface with a public network. Public UNI can also be an interface between an ATM end-system and a switch belonging to a public ATM network. The prominent differences between public and private UNI pertain to the nature of physical interfaces supported, administrative control operations (like traffic policing) and management complexity. The UNI signalling procedures are discussed in Chapter 11.

Figure 3.14

ATM Interfaces

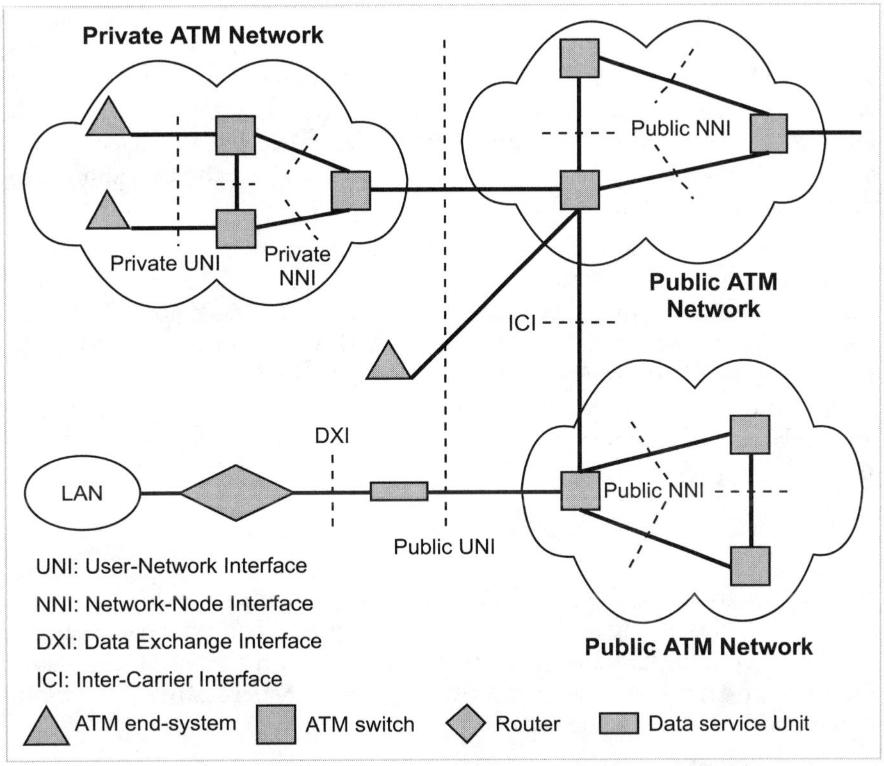

As UNI provides an interface between an ATM end-system and an ATM switch, Network-Node Interface (NNI, also known as Network-Network Interface) provides an interface between two ATM switches. NNI can again be private NNI (i.e. the switches lie within a private ATM network) or public NNI (i.e. the switches lie within a public ATM network). The differences between public and private NNI are similar in flavour to those between public and private UNI. The private NNI interface is also referred to as PNNI interface. The PNNI signalling and routing procedures are discussed in Chapter 12.

There is another specification called the ATM Inter-Network Interface (AINI) that is based on PNNI. The goal of AINI 1.1 protocol is to facilitate the interworking of two networks running PNNI internally in disjoint PNNI routing domains. Thus, AINI can be viewed as inter-domain routing protocol akin to Border Gateway Protocol (BGP) used in the Internet. AINI is very briefly explained in Chapter 16.

Besides UNI, PNNI and AINI the BISDN Inter-Carrier Interface (B-ICI) is the interface between two public ATM networks. Standardization of B-ICI helps in achieving inter-operability between two public ATM networks belonging to different carriers. BICI and its multi-service aspects are explained in Chapter 16.

While the UNI, NNI and B-ICI define a cell-based interface between the network entities involved, ATM 'cell tax' issues lead to the specification of frame-based ATM interfaces for providing data services. Amongst the initial specifications in this regard was the Data Exchange Interface (DXI). DXI defines a standardized mechanism to interface between frame-based devices (e.g. IP-based routers) and ATM network elements. The DXI specifications were later evolved to a more elaborate specification, called Frame-based UNI (FUNI). FUNI can be considered as an enhancement on the DXI, and in fact, the two share a lot of commonalities. Both DXI and FUNI are elaborated in Chapter 16.

Besides the DXI and the FUNI specifications, two other specifications were aimed at standardizing frame-based interface between network elements supporting conventional transmission techniques like Ethernet and SONET. These specifications were called Frame-based ATM over SONET/SDH Transport (FAST) and Frame-based ATM Transport over Ethernet (FATE). As the names suggest, FATE defines a mechanism to carry frame-based ATM traffic over Ethernet, while FAST defines the same for SONET/SDH. FAST and FATE are explained in Chapter 16.

3.10 Interworking with ATM

As regards interworking, an important fact that must be kept in mind is that migration towards ATM is a long and steady process. Thus, during the early phases of the deployment of ATM technology, ATM must interoperate with the existing link-layer protocols (e.g. the Ethernet) and network-layer protocols (e.g. IP). For successful interoperability, it is mandatory to have well-defined standards that govern such interoperability.

In an attempt to provide well-defined standards, ATM Forum and IETF have approved a number of interworking standards like LAN Emulation (LANE), Classical IP over ATM and Multi-Protocol over ATM (MPOA).

LAN emulation provides a means for traditional LAN technologies like Ethernet and Token ring to co-exist with ATM LANs in a heterogeneous LAN environment. Using LAN emulation, two non-ATM end-systems can communicate not only with each other, but also with end-systems directly attached to an ATM network. ATM in LAN environment and LAN Emulation (LANE) are explained in Chapter 17.

Classical IP over ATM provides the mechanisms to transfer IP datagrams over ATM networks. This standard also defines the means for address resolution. This technique is explained in Chapter 18.

Another technique for IP over ATM is Multi-Protocol over ATM (MPoA). MPoA uses LAN emulation to techniques and the Next Hop Resolution Protocol (NHRP) to provide the means to establish a direct VC connection between end-systems belonging to different sub-nets. The MPOA and NHRP techniques are explained in Chapter 18.

3.11 Applications of ATM Networks

A discussion of ATM is incomplete without a mention of the applications of ATM technology. Part 5 of this chapter highlights the application areas of ATM. Owing to its unique features, ATM has a potential for widespread use, four key application areas are considered in this book. These include application of ATM in Multi-Protocol Label Switching (MPLS) networks, Voice over ATM solutions, application of ATM in Digital Subscriber Line (DSL) and application of ATM protocols in Third Generation (3G) networks.

3.11.1 ATM and MPLS Networks

A recent industry development for the interworking between IP and ATM is Multi-Protocol Label Switching (MPLS). MPLS is a new industry development standardized by the IETF. MPLS is being standardized as a technology that can best be described as a synergistic approach between IP and ATM. MPLS merges the flexibility of the IP routing protocols with the speed that ATM switches provide to introduce fast packet switching in frame-based IP networks.

IETF started the MPLS standardization process with the intention of introducing label switching within IP networks. While the initial focus was on IP networks alone, the MPLS framework is flexible enough to work with multiple protocols. Due to its close resemblance to the ATM protocol, MPLS has many possible applications when used in conjunction with ATM. Amongst the two main areas of research involving ATM and MPLS is the support of IP over ATM using MPLS, and the concept of ATM-MPLS network interworking.

An overview of MPLS along with the role of ATM in MPLS networks is provided in detail in Chapter 19.

3.11.2 Voice Over ATM

ATM provides convergence in the sense that both voice and data can be carried together. For voice, the ATM Forum has defined a number of techniques for Voice and Telephony over ATM (VTOA). The initial standards by ATM Forum were based on AAL1 while the newer ones are based on AAL2. AAL1 provides Circuit Emulation Service (CES) including its low-speed variant Low Speed Circuit Emulation Service (LSCES) and more efficient Dynamic Bandwidth Circuit Emulation Service (DBCES). Apart from circuit emulation, ATM also provides trunking solutions based on AAL1 and AAL2.

The different solutions for VTOA are described in Chapter 20.

3.11.3 ATM and DSL Networks

For decades, residential users have been using the dial-up modem to access the Internet via the Public Switched Telephone Network (PSTN). However, with the theoretical upper bound of 56 Kbps on the bandwidth available using dial-up modems, other broadband access technologies are being explored. Not all solutions proposed for broadband access are easily deployable. Most solutions for broadband access involve huge infrastructure costs in the installation of new network elements and the networking of these elements by cables. This means that the solutions are not cost-effective, and are beyond the reach of the common users.

Amongst the proposed solutions for broadband access, Digital Subscriber Line (DSL) is being seen as a technology that will overcome the above limitations of cost to offer a cost-effective broadband access solution. DSL proposes to use the existing infrastructure of the Plain Old Telephone Systems (POTS) to provide a high-rate access solution. DSL is designed to provide typical data rates of up to 2 Mbps over the existing copper cables installed for the POTS network.

An overview of DSL along with the role of ATM in DSL networks is provided in detail in Chapter 21.

3.11.4 ATM and 3G Networks

In recent times, there has been a gradual migration of voice users from fixed wireline networks to wireless mobile cellular networks. The wireless networks too have evolved from Second Generation (2G) networks like Global System for Mobile communications (GSM) and General Packet Radio Service (GPRS) to newer Third Generation (3G) networks like Universal Mobile Telecommunications System (UMTS). As the networks evolve, there is associated requirement for higher bandwidth and better service capabilities. Towards this end, the ability of ATM to provide speed transfer with service guarantees emerges as a suitable choice for transfer of voice and data in wireless networks.

An overview of 3G along with the role of ATM in 3G networks is provided in detail in Chapter 22.

3.12 B-ISDN and ATM

ATM is the preferred transfer mode for B-ISDN services. B-ISDN standards (published by ITU-T) define the scope and placement of ATM in the current networking environment. The standards prescribe various functional entities that can exist in B-ISDN network architecture and also provide the interfaces with which two B-ISDN entities interface with each other. The standards also list possible B-ISDN services that must be supported now and in future. These aspects, which capture the relationship between B-ISDN and ATM, are elaborated upon in the following sub-sections.

3.12.1 B-ISDN Reference Configurations

In order to introduce a new technology, one of the primary requirements is to define the type of entities or devices that will exist in the network and the functional behaviour of these entities. An associated requirement is to define the interfaces between these entities. This requirement is fulfilled by ITU-T recommendation [ITU-T I.413]. Basically, [ITU-T I.413] is derived from recommendation [ITU-T I.411]. [ITU-T I.411] defines ISDN reference configuration. In fact, [ITU-T I.413] merely adds a subscript B (indicating 'broadband' capability) to the existing ISDN reference points and makes it applicable for B-ISDN.

[ITU-T I.413] defines the functions and interfaces between various B-ISDN entities. This is done by using the notion of reference points and functional groups.

A reference point (indicated by black dots in Figure 3.15) refers to a particular type of interface and includes the description of bit rates and frame formats. Four reference points are defined—namely R, S_B, T_B and U_B. As shown in Figure 3.15, each reference point is an interface between two functional blocks. For example, S_B is the interface between B-TE1 and B-NT2. Similarly, other reference points can be described. Note that subscript *B* is missing from reference point *R*, because it may or may not have broadband capabilities. Out of these four reference points, only SB and TB are subject to standardization.

FIGURE 3.15

B-ISDN Reference Configurations

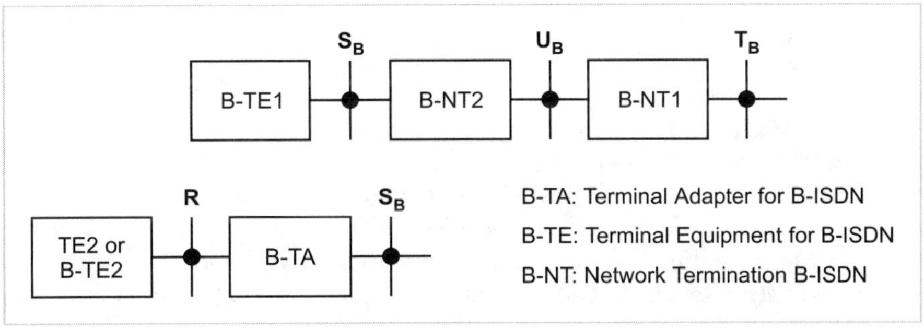

A *functional group* (indicated by rectangle boxes in Figure 3.15) refers to a set of functions performed by a B-ISDN entity. Six functional groups are defined, namely B-NT1, B-NT2, B-TE1, TE2, B-TE2 and B-TA. These functional groups are explained below.

- **B-NT1 (Network Termination 1 for B-ISDN):** B-NT1 performs functions equivalent to the physical layer (like line transmission termination and transmission interface handling).

- **B-NT2 (Network Termination 2 for B-ISDN):** B-NT2 performs higher layer functions (like multiplexing, buffering, signalling and switching) and some physical layer functions (like cell delineation).
- **B-TE1 and B-TE2 (Terminal equipment for B-ISDN):** These refer to a user equipment with B-ISDN capability. The difference between BTE-1 and BTE-2 is that while BTE-1 uses a standard B-ISDN interface, BTE-2 requires a B-TA (Terminal Adapter) because it has a non-standard B-ISDN interface.
- **TE-2 (Terminal equipment 2):** TE-2 refers to a user equipment without B-ISDN capability (e.g. ISDN terminal).
- **B-TA (Terminal adapter for B-ISDN):** B-TA refers to the equipment that connects a non B-ISDN equipment to a B-ISDN network.

Figure 3.16 shows two sample reference configurations. Figure 3.16(a) shows a PC (without an ATM interface) that uses an intermediate bridge (with an ATM interface) to get connected to an ATM network. This corresponds to TE-2—R—B-TA—S_B configuration (the PC here is TE2 and the bridge B-TA). Figure 3.16(b) shows a PC with an ATM interface card. This configuration corresponds to BTE-1-SB configuration (the PC here is B-TE1).

FIGURE 3.16

Sample Reference Configurations

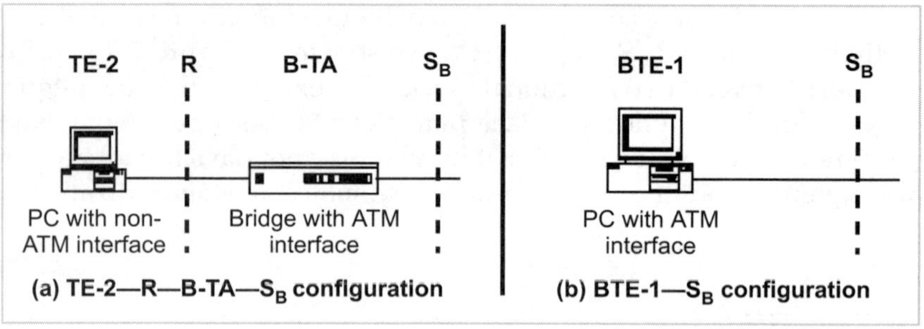

3.12.2 B-ISDN Service Aspects

ATM is designed to support a wide spectrum of broadband services. In order to define the scope and nature of broadband services, ITU-T has published a recommendation [ITU-T I.211]. I.211 classifies broadband services under two broad categories—interactive services and distributive services. These service categories are explained below (also see Figure 3.17):

- **Interactive services:** Interactive services usually involve a bi-directional information flow. The connections are generally point-to-point, though in some cases, they may be point-to-multi-point. Interactive services are further classified into the following subcategories:
  - **(a) Conversational services:** These services provide the means for communication in real-time. The information flow can be bi-directional symmetric, bi-directional asymmetric, and even unidirectional in some cases. The store-and-forward technique should not be applied to these services because of real-time communication.

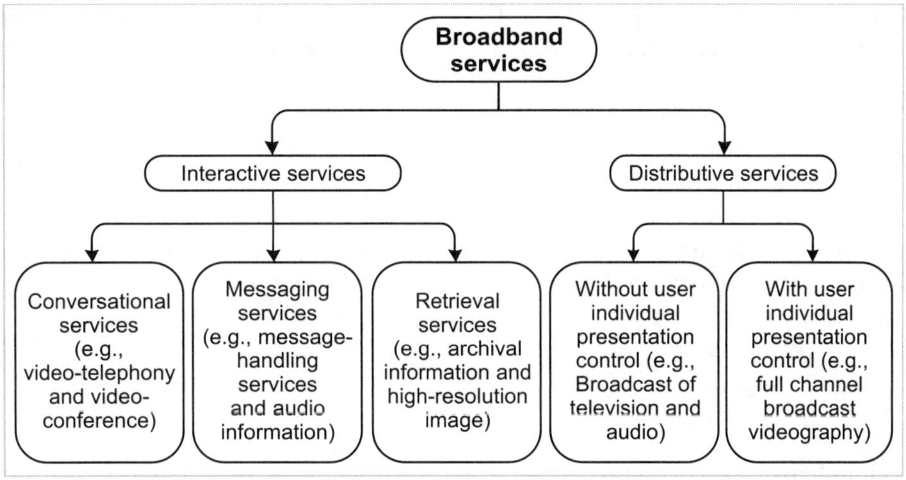

FIGURE 3.17

Classification of Broadband Services

Examples of conversational services include video telephony, video-conferencing and high-speed data transmission.

(b) **Messaging services:** These services offer communication between individual users via storage units with store-and-forward, mailbox and/or message handling functions. Examples of messaging services include message-handling services and mail services for moving pictures (films), high-resolution images and audio information.

(c) **Retrieval services:** These services are used for retrieving stored information maintained for public use. The information is retrieved only at the command of a user; moreover, the user can control the time at which the information flow starts. Examples of this category include retrieval services for films, high-resolution image, audio information and archival information.

- **Distribution services:** Distribution services involve the distribution of information from a central source to a large number of users (broadcast). Distribution services are further classified into the following sub-categories:

(a) **Services without user individual presentation control:** These include those broadcast services wherein the user has no control over the content and the start/end of information flow. Examples of these services include broadcast for television and audio programmes.

(b) **Services with user individual presentation control:** These services also provide the means for information distribution from a central source to a large number of users. However, unlike the previous category, these services allow users to control the start and order of presentation. Full channel broadcast videography is one example of such services.

Now that the B-ISDN service categories have been explained, some comments are in order. I.211 only identifies the current and future broadband services; i.e. it only provides a list of possible services that are envisioned. The standardization and complete specification of these services has not yet been done. Complete specification is possible only after a thorough examination of the needs of users by market research. I.211 also does not mention as to how these services will be implemented. For example, if one examines some of these services, e.g. multi-site video-conferencing, he or she will realize that these are currently not supported by ATM technology. However, I.211 does provide some general guidance to support these services.

In essence, [ITU-T I.211] is a guide for the development of future broadband applications and services. It identifies all possible services (constant rate and variable rate, connectionless and connection-oriented) that can be theoretically supported. But what will be finally supported is a matter of conjecture and speculation.

3.13 ATM Standardization

A lot of effort has been put in by various quarters to standardize ATM procedures. Standardization helps in the faster deployment of technology, fosters competition and promotes interoperability. Recognizing this need, various networking companies have joined hands with standards bodies like ITU-T to expedite the process of ATM standardization. In fact, ATM Forum, one of the approving bodies, is a consortium of over 600 ATM vendors. The following sections summarize the key organizations working on ATM standardization.

3.13.1 ITU- Standards

The International Telecommunications Union—Telecommunication standardization sector, better known as ITU-T, is a United Nations-sponsored organization. ITU-T is responsible for studying technical, operating and tariff questions and issuing recommendations with a view to standardizing telecommunications on a worldwide basis.

Alongside with the ATM Forum, ITU-T has approved the bulk of the ATM standards. The standardization process began in late 1980s and the early 1990s with the publication of [ITU-T I.113]. By the end of 1991, ITU-T standards (total 13 in number) covered most aspects of the ATM network operations. At the time of writing, there are more than 50 ITU-T standards related to ATM. The emphasis of most of the ATM-related ITU-T standards is upon network operations. The underlying idea behind most standards is to provide a network infrastructure that can then be used for specific applications. In contrast, the ATM Forum specifications (as discussed in the next sub-section) take into account the end-users' viewpoint.

3.13.2 ATM Forum Standards

ATM Forum is a consortium of ATM vendors, formed with the objective of accelerating the use of ATM products and services through a rapid convergence of interoperability specifications. The ATM Forum also promotes industry co-operation and awareness. ATM Forum's origin dates back to October 1991, when it was formed by four companies: Northern Telecom (Nortel), Sprint, CISCO, and Network Equipment Tech. Currently, it has over 600 members.

The activities of the ATM Forum are distributed across the following committees:

- **Technical committee:** This committee works with other worldwide standards bodies for selecting appropriate standards, resolving differences among standards and recommending new standards when existing ones become obsolete or inappropriate.
- **Marketing committee:** The ATM Forum has three marketing committees, one each for Asia-Pacific, Europe and North America. These committees provide marketing and educational services for leveraging the growth of ATM technology.
- **User committee:** This committee acts as a bridge between the technical committee and the end-users, thereby ensuring that the technical specifications meet the actual end-user needs.

3.13.3 Internet Engineering Task Force (IETE) Standards

The IETF is a large and open international community of network designers, operators, vendors and researchers, concerned with the evolution of Internet architecture and the smooth operation of the Internet. It is the main body responsible for the development of new Internet standard specifications. Its mission includes identifying problems of the Internet, proposing solutions for those problems and recommending new protocols and architecture for the Internet.

The actual technical work of the IETF is done in its working groups. The working groups are divided, on the basis of functional areas to which they belong. In total, there are eight functional areas—viz. application areas, general area, Internet area, operations and management area, routing area, security area, transport area and the user services area.

The working groups of IETF have standardized a number of ATM-related RFCs. These mostly relate to the interaction of IP with ATM (i.e. how IP is carried over the ATM network), SNMP-based network management, multicasting and resource reservation protocols.

3.14 ATM Market Study

As the Business Communication Review (1999) [BCR R. Cochran] said, "First everyone loved ATM, then everyone hated it. Today, practicality has set in; people buy it." This sentence succinctly describes the progress of ATM technology. Today, as the figures prove, the growth potential of ATM is really strong. The revenue for ATM falls into two broad categories, viz. equipments and services.

ATM equipment, which contributes to the bulk of ATM revenues, includes switches and access equipments. In 2000, the total market for ATM equipment was worth $5.7 billion. The ATM equipment market encompasses the following segments:

- **Carrier switch:** The carriers use ATM mainly for building core network backbones. Indeed, the ability to provide stringent QoS guarantees and to carry information in an integrated fashion, makes ATM quite popular with the carriers.

- **ISP switch:** Here again, ATM is used for providing backbone transport.
- **Enterprise WAN switch:** Enterprise WAN switches are used by corporate organizations to build their private ATM networks. For this segment, ATM's competitive edge lies in providing support to voice, data, video and multimedia applications.
- **WAN access equipments:** These equipments are used for aggregating low-speed ATM and non-ATM traffic for access onto either a public or a private ATM network. Examples of WAN access equipments include CPE-based equipments and central office equipments.
- **LAN equipments:** In the LAN environment, ATM's QoS and other attractions get mired in protocol overheads. Nonetheless, this segment includes ATM hubs, switches, uplink modules, etc.

The growth of the ATM services' market is sluggish as compared to the equipment market but was still worth $5 billion in 2004. The services market is expected to pick up in the near future and reach nearly $8 billion by the end of 2007. The ATM service offerings are differentiated according to the following factors:

- Port speeds offered (e.g. T1, T3, OC1, OC3, etc.),
- Service class offered (CBR, rt-VBR, nrt-VBR, UBR and ABR), and
- SVC capability (majority of the carriers offer PVC connections only).

Conclusion

The early days of ATM were marked by lot of media hype and baseless speculations. In those days, ATM was labelled as the panacea for all networking problems, which unfortunately ATM is not. The most striking criticism against ATM is that it is a jack of all trades and master of none. In other words, ATM equals neither the efficiency of datacommunication networks nor the quality of telecommunication networks. This criticism is justified to some extent. Since ATM cells are small and fixed-sized, the cell header overhead is nearly 10 per cent. Experts refer this 10 per cent overhead in the ATM technology as the cell-tax. The burden of cell-tax is much more critical in a WAN environment where the bandwidth is costly. This cell-tax (or percentage overhead) fares poorly against the less than 5 per cent overheads of other datacommunication technologies like frame relay and X.25.

When compared to telecommunication networks, no packet or cell-based technology (including ATM) can match the voice quality of circuit-based transfer. Although the cell size in ATM is deliberately kept small, there is inherent delay and jitter in the cell-switching process because of buffering. This makes the voice quality inferior to pure circuit-based transfer.

Another important problem of ATM is the lack of broadcast facility. In ATM LANs, this drawback is severely felt while carrying out MAC address resolutions. Although point-to-multipoint connections can be established in ATM, these connections have significant overheads. In contrast, address resolution in Ethernet or Token ring-based LANs is straightforward.

The final point raised against ATM is its complex structure. This complexity is best reflected in ATM signalling protocols like [ITU-T Q.2931]. Complex structures have a detrimental effect on the growth of a technology because it makes implementation and modification difficult. In contrast, simple protocols are easy to implement and modify. In this regard, the popularity of the TCP/IP protocol suite immediately comes to one's mind. Their success can be attributed to the simplicity of their standards (RFCs). These criticisms notwithstanding, rationality has crept in and people have started realizing the actual strengths and capabilities of ATM and trying their best to exploit it.

REVIEW QUESTIONS

Q 1. *Why is it said that ATM is a desktop-to-the-core networking solution? What are the challenges in making this statement true in reality?*
Q 2. *Why is ATM asynchronous?*
Q 3. *How does ATM provide convergence of various data types? What is the role of the adaptation layer in this?*
Q 4. *What are the important differences between frame relay and ATM?*
Q 5. *How is the ATM protocol reference model different from the OSI reference model? What is the relation between the two models? If OSI model is used, what are layers in which ATM resides?*
Q 6. *In the context of IP over ATM, what is an overlay network?*
Q 7. *How is information transferred in ATM?*
Q 8. *Why is flow control absent in ATM? What problems would have been encountered had flow control been introduced?*
Q 9. *Why are two levels of identifiers used in ATM cell relaying?*
Q 10. *What are application areas where ATM networks can be used?*

FURTHER READING

A plethora of options are available for further study on ATM networks. [ATM W. Goralski], [ATM D. McDysan], [ATM M. Prycker], [ATM U. Black] and [ATM R. Handel] are some of the popular texts available on ATM. [ATM IBM Redbook] is also a very good reference text available for free download from the Internet. Apart from this, [ATM A. Alles] provides a very good summary of ATM. http://www.cse.ohio-state.edu/~jain/refs/atm_book.htm also provides a list of references for ATM.

PART 2

ATM Protocol Reference Model

The ATM protocol reference model is a three-dimensional model as discussed in Chapter 3. The reference model has three planes, viz. a *user plane*, a *control plane* and a *management plane*, with each corresponding to a set of functionality required of the protocol stack.

The user plane (or U-plane) is concerned with the transfer of user information. At the transmitting side, this plane is responsible for packing user information into cells and transmitting the cells by using the underlying physical medium. At the receiving side, this plane performs the reverse operation and derives the higher layer information in exactly the same format as it was received from the user at the transmitting side.

The control plane (or C-plane) is responsible for establishing and releasing connections between a given source and destination. When a new connection is established, the control plane establishes a mapping at the intermediate switches between incoming VPI/VCI and outgoing VPI/VCI. The mapping thus derived is used to switch cells. When the same connection is released, the control plane removes the mapping stored within the intermediate nodes.

The management plane (or M-plane) is responsible for managing the individual layers in the protocol stack and providing co-ordination between the layers. The management plane is divided into layer management and plane management. Layer management is responsible for managing each of the layers, including its administration, maintenance and configuration. The plane management, which cuts through all the layers, is responsible for co-ordination amongst different planes.

Irrespective of the planes, the three lower layers, viz. the *ATM Adaptation Layer* (AAL), the *ATM layer* and the *physical layer*, form the core of the ATM protocol stack and exist in all the three planes.

The lowest layer is the physical layer, which is concerned with carrying cells over the physical media across the network. Physical layered is covered in Chapter 4.

The next layer is the ATM layer, covered in Chapter 5. It is this layer where the core functionality of ATM resides. It is the highest layer that deals with cells. It receives 48-byte packets from AAL, attaches a 5-byte header and sends it to the lower layer. Apart from the generation and multiplexing of cells, a host of other functionalities are implemented at this layer.

The ATM Adaptation Layer (AAL) lies above the ATM layer and is responsible for handling different types of data and mapping the requirements of the applications to the services provided by the lower layer. In order to support a variety of applications, four AALs are defined. These are AAL1, AAL2, AAL3/4 and AAL5, with each serving the requirements of a separate class of applications. The AAL layer is covered in Chapter 6.

Chapter 4

ATM Physical Layer

4.1 Introduction

The ATM physical layer, similar to an OSI physical layer, is responsible for carrying bit streams across the network. The ATM physical layer, however, has a much wider scope in the sense that it performs some of the functions that are usually performed by a data link layer (e.g. cell delineation).

The functions of physical layer are categorized into two parts, namely Transmission Convergence (TC) functions and Physical Medium Dependent (PMD) functions. The TC functions include header error control, cell rate decoupling, cell delineation, scrambling/de-scrambling and frame adaptation. The PMD functions are specific to the underlying physical medium. Some of the PMD functions include timing and synchronization, encoding and bit transmission. The functions of the two sub-layers are detailed in the following sub-sections.

4.2 Transmission Convergence (TC) Sub-layer

The TC sub-layer performs the functions that are required independent of the underlying physical medium. The functions of this sub-layer are (also see [ITU-T I.432]):

- Header Error Control
- Cell Rate Decoupling
- Cell Delineation
- Cell Scrambling
- Transmission Frame Adaptation

Each of the functions is explained below.

4.2.1 Header Error Control (HEC)

The TC sub-layer provides an error control mechanism for the cell header. This mechanism, referred to as Header Error Control (HEC), provides single-bit error correction and multiple-bit error detection capabilities. In the five-byte cell header, the fifth byte (called the HEC field) is used for error control. The HEC field contains the checksum calculated over the first four bytes of the cell header using a generator polynomial $G(x) = x^8 + x^2 + x + 1$.

In brief, the operations involved in HEC are as follows:

- The sender computes the checksum over the four bytes of the header and places it in the fifth byte.
- The receiver recomputes the checksum (called the *syndrome*). Any difference between syndrome and the received checksum is an indication of an error.

HEC procedures specify two modes of operations in which the receiver operates, viz. the *correction mode* and the *detection mode*. The error control procedures taken on the received cell depend upon the mode which the receiver is currently in (see Figure 4.1). In the correction mode, the receiver provides for single-bit error correction. In the detection mode, only error detection is done and all cells with errored headers are discarded.

FIGURE 4.1

HEC (Receiver Modes of Operation)

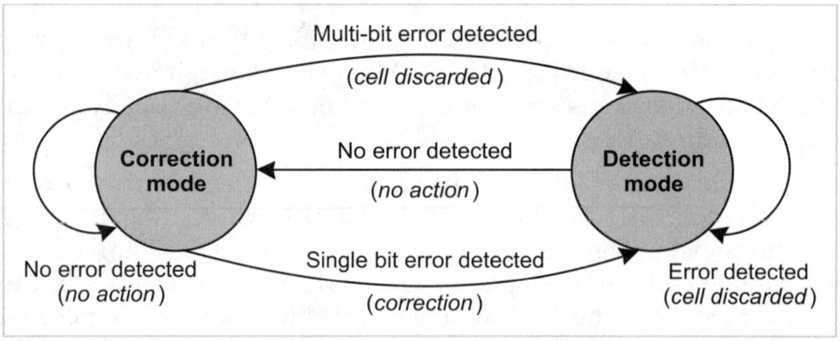

Initially, the receiver is in the correction mode. For each incoming cell, the receiver calculates the syndrome for the cell header. If the syndrome matches with the received checksum, which indicates that no error has occurred, the receiver remains in the correction mode. On the other hand, if there is a mismatch, two possibilities exist. If the difference between the syndrome and the received checksum indicates a single-bit error, the error is corrected and the receiver changes its mode to the detection mode. However, if the difference indicates multiple-bit errors, the receiver discards the cell and changes its mode to the detection mode.

Once the receiver enters the detection mode, it does not correct any single-bit errors. Thus, any cell with a checksum mismatch is discarded, irrespective of whether the single-bit error has occurred or multiple-bit errors have. When the receiver in detection mode receives a cell with no error (i.e. the syndrome is zero), the receiver changes its mode to the correction mode.

There is a specific reason for not correcting consecutive single-bit errors. Since single-bit errors are rare in the current fiber networks, the existence of consecutive single-bit errors is an indication of a burst error or malfunctioning of some other component in the network. In cases like this, it is better to drop the cells because of a possibility of incorrect correction.

4.2.2 Cell Rate Decoupling

If cells are transferred over a synchronous physical medium (like SONET), a fixed number of cells have to be periodically transmitted. In other words, irrespective of whether there are user cells to transmit or not, in order to maintain synchronization between the sender and receiver, a fixed number of cells have to be transmitted per second. If, at any time, there are no user cells to be transmitted, the Transmission Convergence sub-layer inserts dummy cells (or idle cells) to fill the time slots allotted for user cells. Inserting idle cells allows the rate of outgoing cells to match the link rate. This process is referred to as cell rate decoupling. In order to enable a receiver to distinguish between user cells and idle cells, the idle cells contain a specific bit pattern, which is invalid for user cells.

The first four bytes of an idle cell are 0x00 0x00 0x00 0x01, while the fifth byte is a valid HEC. The remaining 48 bytes contain the recurring bit pattern 01101010. Note that the first bit of the fourth byte is set to 1. A value of 1 of this bit (with other fields in the first four bytes set to 0) indicates that the cell is an idle cell and must be removed by the physical layer at the receiving end. In the next chapter, while discussing the cell header format, a counterpart of idle cells at the ATM layer is discussed. These counterpart cells, called the unassigned cells, have identical format excepting for the first bit of the fourth byte, which is set to 0. Thus, the first bit of the fourth byte distinguishes an idle cell at the physical layer from an unassigned cell at the ATM layer.

4.2.3 Cell Delineation

In order to identify the start and end of a cell, the Transmission Convergence sub-layer delineates the boundaries of the cells from a stream of incoming bits. Once the beginning and the end of one cell are identified, the position of subsequent cells is known, because the end of one cell identifies the beginning of the next.

The usual method to identify the start and end of frames is to have special start and end flags. In order to avoid user data from emulating start and end flags, bit stuffing is done. For understaning this, consider a start and end flag of 01111110. Now, it is possible that the payload contains a similar bit pattern. In order to prevent ambiguity at the receiver end, for each consecutive run of five 1's, a bit '0' is added (i.e. stuffed) by the sender. This extra bit is removed by the receiver and thus the interpretation remains unambiguous. This method is quite common and is used in most of the data link protocols. However, the method has its inherent disadvantages—in that it amounts to increased overheads. Hence, a rather unique method is adopted in ATM, whereby the HEC field is used for cell delineation. In essence, this method relies on the fact that the fifth byte of a cell is a function (i.e. checksum) of the previous four bytes. Putting it differently, if a given byte proves to be the checksum of the previous four bytes, it can be said with some probability that the cell boundaries have been identified. A detailed explanation of the method is given below.

In order to delineate cell boundaries, a starting point is chosen in the incoming bit stream. This starting point is assumed to represent the beginning of a cell. The syndrome is then calculated for the next five bytes by using the generator polynomial ($x^8 + x + x + 1$). If the calculated syndrome is not equal to the value stored in the fifth byte, the starting point is shifted one bit at a time, until the syndrome becomes equal to the fifth byte from the starting point. When the calculated syndrome matches the fifth byte, the possible beginning of a cell has been identified. Note that the starting point may not be a beginning of the cell if the checksum of the previous four bytes is emulated by the bytes contained in the payload.

In order to overcome the above problem, cell scrambling is done at the source. Scrambling randomizes the payload data, thereby reducing the possibility of user data emulating the syndrome. This improves the robustness of the HEC cell delineation mechanism. Although scrambling reduces the possibility of false delineation, it does not completely eliminate the chances of such a pattern occurring in the user data. Such a pattern may occur at a receiver due to a bit-error during transmission.

In order to confirm that the starting point is actually the beginning of a cell, the receiver verifies the header of subsequent cells. If for a few subsequent cells, the syndrome matches the supposed checksum field, the receiver is quite sure about the position of cell boundaries and stops the process of searching for cell boundaries. The receiver does not search for the cell boundary till it receives cells with a checksum mismatch. However, when a checksum mismatch occurs, there are two distinct possibilities—either there is an error in the cell header or the boundary of the cells currently identified is no longer valid (due to a possible mismatch in the clock speed).

In order to handle the first possibility, the receiver waits for a few more cells with checksum mismatch before concluding that the boundaries of the cells are actually lost. The receiver then starts the same process again, i.e. shifting the bits one bit at a time and calculating the syndrome.

The cell delineation procedure defines three possible states, viz. HUNT state, SYNCH state and PRESYNCH state, that a receiver can be in. Figure 4.2 shows these three possible states and the state transitions. When the receiver starts, it is in HUNT state. In this state, the receiver is searching for cell boundaries (bit-by-bit). When it finds a correct HEC with syndrome matching the fifth byte, it moves into the PRESYNCH state. In this state, the receiver verifies DELTA number of cells. If the HECs of all the DELTA cells are found correct, the receiver moves to the SYNCH state, a state wherein cell boundaries are known. However, if any cell among the DELTA cells has an incorrect HEC, the receiver changes its state back to the HUNT state.

In the SYNCH state, if the receiver receives less than ALPHA number of cells with incorrect HEC, it still remains in the SYNCH state. This ensures that a rare bit error does not restart the cell delineation procedure. However, after ALPHA consecutive cells with incorrect HEC are received, the receiver moves to the HUNT state and restarts delineation procedures.

FIGURE 4.2

Cell Delineation State Diagram

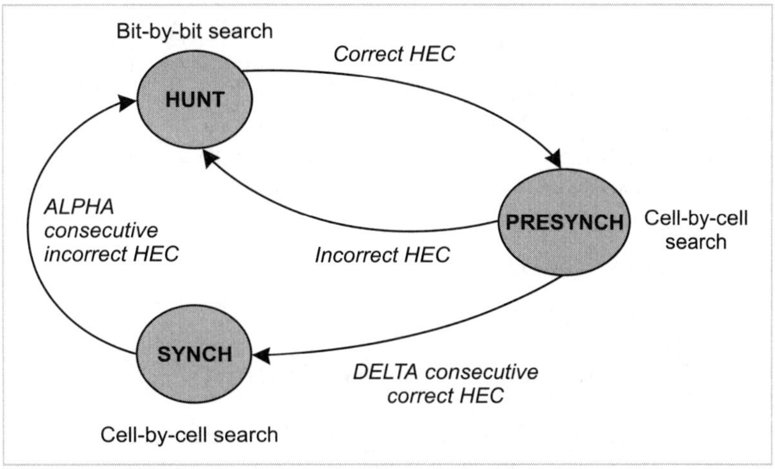

The values of ALPHA and DELTA are chosen so as to make the cell delineation process as robust and secure as possible. The robustness against false misalignments due to bit-error depends upon ALPHA, while the robustness against false delineation during re-synchronization depends upon the value of DELTA. For SDH-based physical medium, the recommended value of ALPHA is 7 and DELTA is 6, while for cell-based physical medium, the recommended value of ALPHA is 7 and DELTA is 8.

4.2.4 Cell Scrambling

Cell scrambling is a mechanism whereby the undesired patterns in data are changed at the source and the changes are reverted back at the destination. In essence, cell scrambling is done to improve the security and robustness of the HEC delineation algorithm, as described earlier. The choice of the cell scrambling algorithm depends upon the choice of the physical transport used. For details of cell scrambling, the reader is referred to [ITU-T I.432].

4.2.5 Transmission Frame Adaptation

The cells may be carried on the physical layer either within a frame or just transported over the physical link as a bit sequence. When a framed transport is used, a structure is defined at the physical layer. The cells are mapped into this structure and carried across the network. This mapping of cells into a framed physical layer structure is called transmission frame adaptation. At the destination, the cells are extracted from the structure and delivered to the upper layer.

Further, the receiver has to keep track of the frame generated by the source. Initially the destination has to recover the frame structure and identify its beginning and end, on the basis of certain bit-patterns present in the frame. In the following sections, a number of framed formats are considered, and the mapping of ATM cells onto the frame is discussed.

4.3 Physical Medium Dependent (PMD) Sub-layer

The Physical Medium Dependent (PMD) sub-layer functions depend upon the underlying physical media. PMD functions relate directly to the physical representation of the bits on the physical medium and the interpretation of the bits at the receiver. The two important functions of this sub-layer are:

- Timing and Synchronization
- Encoding and Transmission

The two functions are discussed below.

4.3.1 Timing and Synchronization

In order to enable the receiver to correctly receive and interpret the bit-patterns sent by the sender, a timing relationship must exist between the sender and the receiver. For example, consider a bit that remains on the wire for some duration of time during which the receiver must read it. In such a scenario, the relationship determines at what duration and at what rate the bits should be sampled from the physical link. This relationship between the sender and the receiver may be either explicit or implicit. With an explicit timing relationship, both the sender and the receiver maintain a clock of common frequency. Thus, the time duration for which one bit remains on the wire is known to both the sender and the receiver and is determined by the clock frequency. Implicit timing relationship is obtained by a special encoding scheme like the Alternate Mark Inversion (AMI). Maintaining this timing and synchronization falls within the purview of the PMD sub-layer.

4.3.2 Encoding and Transmission

The bits may be put on the physical link in a number of ways. On a copper wire, the representation of 1's and 0's may be in several forms. In order to uniquely represent 0's and 1's, either two different voltage levels (binary encoding) or two different frequencies, may be used. In fiber-optic cables, the representation of the 0's and 1's is through the presence or absence of photons. It is the PMD sub-layer that defines the encoding of 0's and 1's and the transmission of these bits.

4.4 Physical Layer Standards for ATM

A number of physical layer standards are defined by the ATM Forum and ITU-T, which is listed in Table 4.1. Among the standards listed, the standard [ITU-T I.432] defines the TC sub-layer functions that have already been covered in this chapter. Most of the other standards specify the means to map ATM cells onto the underlying physical layer structure. The various standards are discussed in the following sections. As a general approach, for each physical layer standard explained, the corresponding physical layer structure is first explained. For

TABLE 4.1

ATM Physical Layer Standards

Standard	*Description*
[ITU-T I.432]	Defines TC sub-layer functions (like cell delineation and cell scrambling). Also, specifies means to map ATM cells on a 155.52 Mbps and 622.08 Mbps interface.
[ATMF DS1]	Specifies means to map ATM cells on DS1 (T1) at interface rate of 1.544 Mbps.
[ATMF DS3]	Specifies means to map ATM cells on DS3 (T3) at interface rate of 44.736 Mbps.
[ATMF E1]	Specifies means to map ATM cells on DS1 (T1) at interface rate of 2.048 Mbps.
[ATMF E3]	Specifies means to map ATM cells on DS3 (T3) at interface rate of 34.268 Mbps.
[ATMF FRA E1/T1]	Specifies means to map ATM cells on fractional E1/T1 at interface rate of n*64 Kbps.
[ATMF UNI 3.1]	Specifies means to map ATM cells on SONET STS-3c at interface rate of 155.52 Mbps.
[ATMF PHY 2.4G]	Specifies means to map ATM cells on SONET STS-48c at interface rate of 2.4 Gbps.
[ATMF UTOPIA Lx]	Defines a common, standard interface between ATM and PHYSICAL layers of ATM subsystems. The 'x' here refers to four levels, 1, 2, 3 and 4.
[ITU-T G.804]	Defines means to map ATM cells on PDH interfaces like DS1, DS3, E1 and E3.

example, before explaining the STS-3c related ATM standard, an overview of SONET/SDH standard is first provided.

In general, there are two main classes of such standards, ones that use Plesiochronous Digital Hierarchy (PDH) and others that use Synchronous Digital Hierarchy (SDH). PDH includes TDM-based hierarchy, which is followed in North America, Japan and Europe. SDH refers to digital hierarchies for optical fibers. In the PDH-based system, there are two ways to carry ATM cells. The first method is to directly map ATM cells in the payload part, without any regard to the frame format. This technique is followed in DS1 standard, among others. The second method is to define another framed structure, which is then carried in the payload part. An example in this regard is the DS3 standard, details of which are provided in Section 4.6. In the SDH-based system, ATM cells are carried in the payload part of a well-defined framed structure. Again, details are provided in Section 4.10.

Here, a clarification regarding the terms 'plesiochronous', 'synchronous' and 'asynchronous' is in order. A synchronous network, as the name suggests, is one in which the digital transitions occur at exactly the same rate. However, there can be a phase difference between different transitions of two signals. In order to achieve synchronization between different network elements, all the clocks are derived from a single clock, which is also called the primary reference clock.

In a plesiochronous network, digital transitions of two signals occur at 'nearly' the same rate. The term 'plesiochronous' is derived from a Greek word 'plesio' meaning 'nearly', hence the above definition. For example, two network elements on the same network can derive their clocks from two different reference clocks. The two reference clocks may be extremely accurate, but are separated by a small difference. The difference lies within a given threshold, and is called the 'plesiochronous difference'.

An asynchronous network is one in which the digital transitions do not occur at the same nominal rate. There can be few variants in asynchronous networks. For example, the timing information can be derived from different clocks whose difference exceeds the plesiochronous difference. Alternatively, start and stop bits can be used to provide timing information to the receiving end.

4.5 DS1 ATM Physical Layer Interface

The DS1 interface (also popularly referred to as a T1) operates at 1.544 Mbps over UTP-3 cables. Figure 4.3 shows the format of a DS1 frame. As shown in the figure, a DS1 frame is 193 bits long, which is transmitted every 125 microseconds giving a data rate of 1.544 Mbps (193/ 0.125 ms). The first bit, called the F-bit, is used for frame alignment, performance monitoring and other purposes. The remaining 192 bits carry the DS1 payload. The payload is divided into 24 voice channels, wherein each channel occupies one byte. The one byte corresponds to a voice sample of a particular user. The DS1 frames are themselves organized in a hierarchical structure. Twelve DS1 frames form a Superframe (SF) while 24 frames (or two superframes) form an Extended Superframe (ESF).

Figure 4.3

DS1 Frame Format

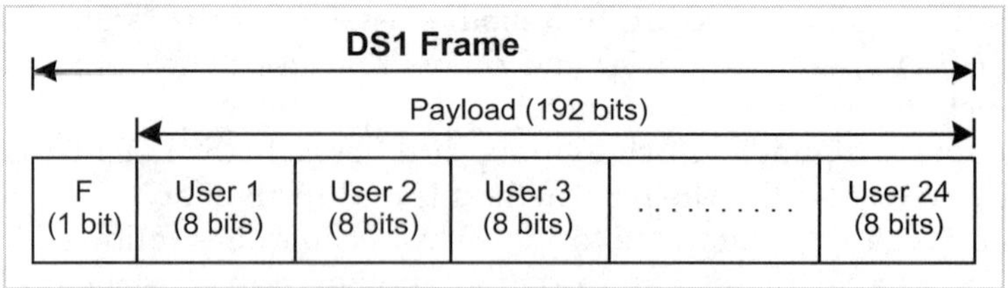

The above frame format is an example wherein DS1 carries channelized information, i.e. where information is divided into channels. However, ATM cells are not carried in this fashion. Rather, cells are directly mapped into the payload portion, without regard to the DS1 channels.

The procedure for mapping ATM cells in a DS1 frame is detailed in [ITU-T G.804] and [ATMF DS1]. Figure 4.4 shows how ATM cells are mapped in a DS1 frame. Since the number of bytes in a DS1 frame is 24, while the size of an ATM cell is 53 bytes, one cell spans over

FIGURE 4.4

Mapping ATM Cell in DS1 Frames

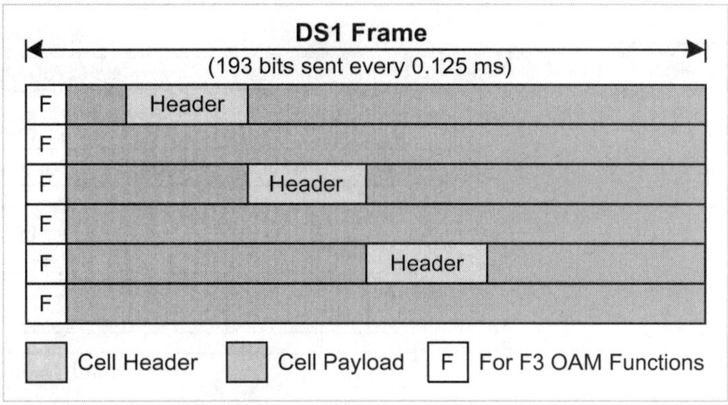

multiple DS1 frames. ATM cells are byte-aligned in the DS1 frame and the effective cell transfer rate is 1.544 * (192/193), i.e. 1.536 Mbps. The overhead per DS1 frame is just one bit. This bit provides various F3 level OAM functions (e.g. loss of frame alignment).

As per [ATMF DS1], TC functions like cell rate decoupling and cell delineation are performed as described in [ITU-T I.432]. However, payload scrambling is not allowed. The use of the single-error detection functionality, using an HEC mechanism, is optional.

4.6 DS3 ATM Physical Layer Interface

DS3 systems have emerged as a popular transport mechanism to carry information between the customer premises and public networks. In essence, DS3 (also known as T3) and DS1 (which was described earlier) form a part of the North American Digital Hierarchy (NADH). NADH is a TDM hierarchy used in the US and Japan for bundling voice channels. Table 4.2 shows the NADH hierarchy. As shown, the DS0 is at the lowest level in the hierarchy, and represents the data rate for a single voice channel. At the next level, 24 voice channels are multiplexed to form a DS1 channel. Further, 672 voice channels (or 28 DS1 channels) form a DS3 channel.

TABLE 4.2

North American Digital Hierarchy

DS Level	*Data Rate (Mbps)*	*Number of Voice Channels*
DS0	64 Kbps	1
DS1	1.544 Mbps	24
DS2	6.312 Mbps	96
DS3	44.736 Mbps	672
DS4	274.176 Mbps	4032

DS3 is a 44.736 Mbps interface. Figure 4.5 depicts the DS3 frame format. A DS3 frame is a multiframe of 4760 bits giving a data rate of 44.736 Mbps. Each multiframe contains seven M-

FIGURE 4.5

DS3 Frame Format

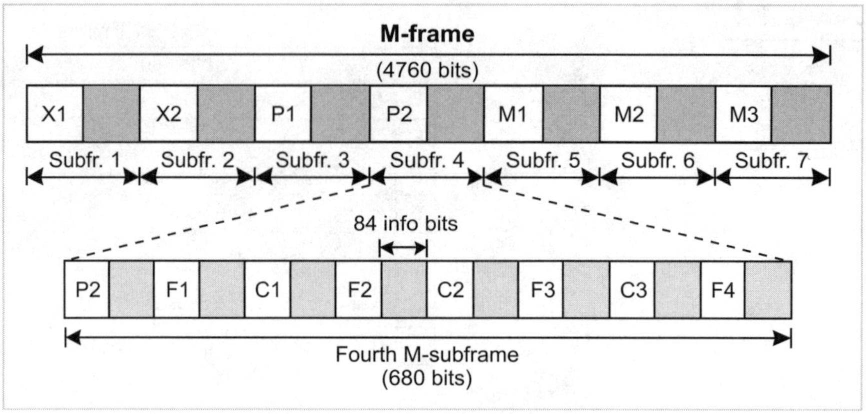

sub-frames, each with sub-frame containing 680 bits. An M-sub-frame is further divided into 8 blocks of 85 bits, 1 bit for overhead and the remaining 84 bits to carry payload. Explanations for some of the overhead fields are provided below (details of all overhead bits can be found in [ITU-T G.704]):

- **X-bits:** X1 and X2 bits indicate received errored multiframes to the remote end. Specifically, a binary value of 0 indicates error conditions like Loss of Signal (LOS), Out of-Frame (OOF), Alarm Indication Signal (AIS), or Slips.
- **P-bits:** P1 and P2 bits carry parity information.
- **M-bits:** M1, M2 and M3 bits are used to locate the sub-frames within a multiframe. A value of 010 (M1 = 0, M2 = 1 and M3 = 0) is used for this purpose.
- **F-bits:** F1, F2, F3 and F4 bits are used to locate the overhead bits within a sub-frame. A value of 1001 (F1 = 1, F2 = 0, F3 = 0 and F4 = 1) is used for this purpose.
- **C-bits:** These bits are used for specific purposes, the details of which are outside the scope of this book.

Recognizing the widespread use of DS3 systems, the ATM Forum has defined a means to carry ATM cells over them. The first standards related to ATM over DS3 were published in [ATMF UNI 3.1]. However, [ATMF DS3] supersedes these standards and defines the latest mechanism to transport ATM cells over DS3 transmission systems.

The ATM cells are mapped into the 84-bit space between two DS3 overhead bits (shown as '84 info bits' in Figure 4.5). These cells, however, are not directly mapped into the DS3 stream. Rather, this is a two-step process. The ATM cells are first mapped into a frame called the DS3 Physical Layer Convergence Protocol (PLCP) frame (see Figure 4.6). The PLCP frame is then carried in the DS3 frame.

The DS3 PLCP frame provides the means to transmit 12 ATM cells every 125 microseconds (i.e. 8000 times in a second) and provides an effective data rate of 40.704 Mbps (8 bits/byte * 53 bytes/cell * 12 cells/frame * 8000 frames/sec). A DS3 PLCP frame has 12 rows of 57 bytes each. Each row of a PLCP frame has four overhead bytes, followed by a 53-byte cell. A trailer of length 13 or 14 nibbles, follows the last row. It should be noted that there is no one-to-one

FIGURE 4.6

DS3 PLCP Frame Format

A1	A2	P11	Z6	First ATM cell	
A1	A2	P10	Z5	Second ATM cell	
A1	A2	P9	Z4	...	
A1	A2	P8	Z3	...	
A1	A2	P7	Z2	...	
A1	A2	P6	Z1	...	
A1	A2	P5	X	...	
A1	A2	P4	B1	...	
A1	A2	P3	G1	...	
A1	A2	P2	X	...	
A1	A2	P1	X	...	
A1	A2	P0	C1	Twelfth ATM cell	Trailer
PLCP framing (2 bytes)		POI (1 byt.)	POH (1 byt.)	PLCP Payload (53 bytes)	Trailer (13/14 nib.)

POI: Path Overhead Identifier POH: Path Overhead X: Unassigned

relationship between the beginning of the DS3 PLCP frame and that of the DS3 frame. The PLCP frame, however, is always nibble-aligned with the DS3 frame.

As shown in Figure 4.6, a number of fields are present in the PLCP apart from the ATM cells. The interpretation of various fields present in the PLCP frame is as follows:

- **A-bytes:** Framing bytes A1 and A2 are set to the values A1 = 0xF6 and A2 = 0x28 and used for frame alignment.
- **POI bytes:** The framing byte is followed by the Path Overhead Identifier (POI) byte. POI is used to identify the type of Path Overhead (POH) byte. In other words, the interpretation of the POH field is based on the value of the POI field. The various POH fields are explained below.
- **B1 bit interleaved parity byte:** Byte B1 is used for path error monitoring. It contains the Bit Interleaved Parity-8 (BIP-8) calculated over the 12×54 byte structure, consisting of the POH and the associated ATM cells of the previous PLCP frame.
- **G1 PLCP path status byte:** This byte carries the error monitoring information in the opposite direction of transmission, i.e. in the receiver-to-sender direction. Byte G1 consists of a 4-bit Far End Block Error (FEBE), a 1-bit Remote Alarm Indication (RAI) and three 'don't care' bits. The FEBE bits provide a count of the BIP-8 errors. The possible values are 0 through 8. If FEBE is not implemented, all bits are set to 1. Values other than those in the range 0–8, are treated as 0 errors. The RAI bit is used to indicate an occurrence of a failure along the path.
- **C1 cycle/stuff counter byte:** This byte determines if the PLCP frame ends with 13 or 14 nibbles. Nibble stuffing is done to align the PLCP frame to a DS3 frame. Nibble stuffing may be done on every third frame of a three-frame cycle spanning 375 msec. The C1 byte of the first frame in the three-frame cycle is coded to all 1's and indicates a trailer length

of 13 nibbles. The C1 byte of the second frame is coded as all 0's and indicates a trailer length of 14 nibbles. For the third frame in the cycle, if there is no nibble stuffing, the C1 byte is set to 01100110 and the trailer is of 13 nibbles. However, if nibble stuffing is done, byte C1 is set to 10011001 and the trailer is of 14 nibbles.

- **Zx growth bytes:** Z1-Z6 fields are reserved for future use.

This completes the explanation of the mapping of ATM cells in a DS3 frame. As far as the PMD functions are concerned, the cell delineation process is trivial, because the position of ATM cells is fixed in a PLCP frame. Once the PLCP frame is identified (by using framing bytes A1 and A2), the position of ATM cells is automatically determined. Thus, the cell delineation function, as described earlier, is not used. The HEC function remains the same as discussed earlier. As a configurable option, a self-synchronizing scrambler may be used.

4.7 E1 ATM Physical Layer Interface

Like North Americans who follow their own NADH multiplexing hierarchy, Europeans use their separate TDM hierarchy based on E1/E3 data rates. Table 4.3 shows the European digital hierarchy. At the lowest level of the hierarchy is E1, which supports 30 voice channels and operates at a rate of 2.048 Mbps. In addition to 30 voice channels, an E1 frame has two control channels, which are used for framing and other purposes. Figure 4.7 shows the E1 frame format. As shown in the figure, time slots 0 and 16 are reserved for control information and do not carry user data. The remaining 30 channels are available for user data.

Table 4.3

European Digital Hierarchy

Level	*Data Rate (Mbps)*	*Number of Voice Channels*
E1	2.048 Mbps	30
E2	8.448 Mbps	120
E3	34.368 Mbps	480
E4	139.260 Mbps	1920

Figure 4.7

E1 Frame Format

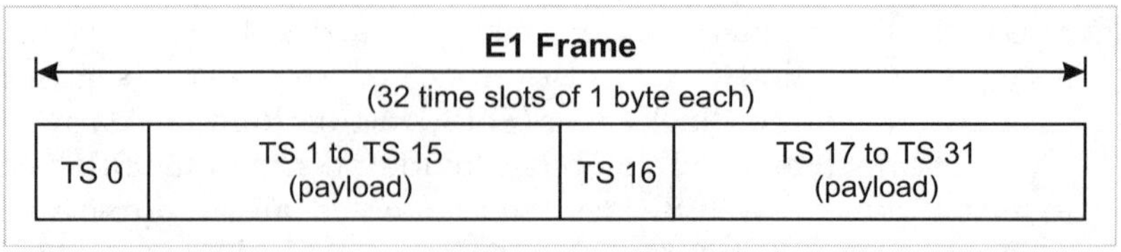

The procedure for mapping ATM cells in an E1 frame is detailed in [ITU-T G.804] and [ATMF E1]. Figure 4.8 shows how ATM cells are mapped in an E1 frame. Since the number of bytes available for user data in an E1 frame is 30, while the size of an ATM cell is 53 bytes, one cell spans multiple E1 frames. ATM cells are byte-aligned in the E1 frame and the

FIGURE 4.8

Mapping ATM Cells in E1 Frame

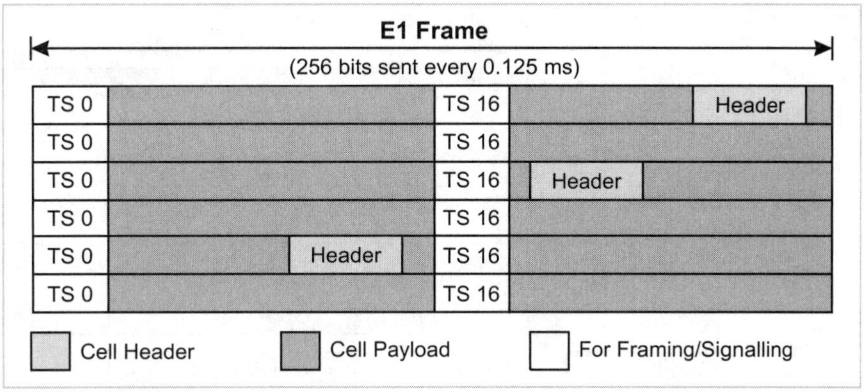

effective cell transfer rate is 2.048 * (30/32), i.e. 1.920 Mbps. The time slot 0 is used for framing and OAM functions (like RAI and CRC-4 procedures). Time slot 16 is reserved for signalling, the details of which are available in [ITU-T G.804]. As per [ATMF E1], TC functions like cell rate decoupling, cell delineation and payload scrambling are performed as described in [ITU-T I.432].

4.8 E3 ATM Physical Layer Interface

Besides E1, E3 links are very popular in Europe. E3 operates at 34.368 Kbps. Each E3 frame contains seven bytes of overhead and 530 bytes of payload capacity (58 + 59*8), and is transmitted every 125 ms.

[ATMF E3] defines a means to map ATM cells in an E3 frame. This mapping is shown in Figure 4.9. ATM cells are byte-aligned in the payload part. The explanation of seven overhead bytes is provided below:

- **Frame alignment (FAx) bytes:** The bytes FA1 and FA2 are set to the values A1 = 0xF6 and A2 = 0x28, and used for frame alignment.
- **Error monitoring (EM) byte:** EM byte is used for error monitoring. It contains the Bit Interleaved Parity-8 (BIP-8) calculated over all bits, including the overhead bits, of the previous 125 ms frame.
- **Trail trace (TR) byte:** This byte is used to repetitively transmit a trail access point identifier so that a trail receiving terminal can verify its continued connection to the intended transmitter.
- **Maintenance and adaptation (MA) byte:** This byte carries the error monitoring information in the opposite direction of transmission, i.e. in receiver-to-sender direction.
- **Network operator (NR) byte:** As the name suggests, this byte is allocated for maintenance purposes specific to individual network operators.
- **General purpose communication Channel (GC) byte:** This provides a channel for general purpose (e.g. to provide data/voice channel connection for maintenance purposes).

FIGURE 4.9

Mapping ATM cells in E3 frame

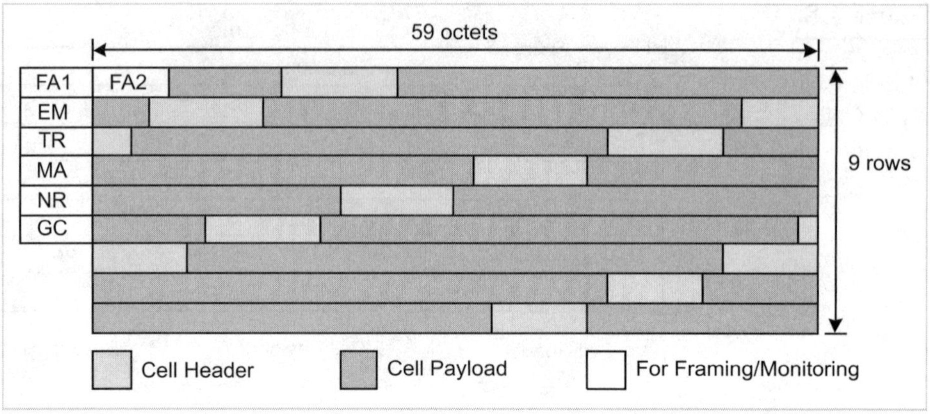

As per [ATMF E3], TC functions like cell rate decoupling, cell delineation and payload scrambling, are performed as described in [ITU-T I.432].

4.9 Fractional T1/E1 ATM Physical Layer Interface

Apart from defining the means to carry ATM cells on full T1 and E1 links, the ATM Forum specification [ATMF FRA E1/T1] also defines the means to carry ATM cells at rates of 64 Kbps or a multiple thereof (up to a maximum of 30, giving a data rate of 1.920 Mbps). This may be done over T1 or E1 interface or any other interface. For example, if the E1 interface is used, any of the 30 payload channels can be used for carriage giving data rate of N*64 Kbps. The signalling channels TS 0 and TS 16 are not used.

The mechanism for such a transport is similar to that discussed in Section 4.5 (ATM over DS1) and Section 4.7 (ATM over E1). Thus, the TC functions like cell rate decoupling, cell delineation and cell scrambling are performed as described in [ITU-T I.432].

4.10 SONET/SDH-based ATM Physical Layer Interface

Synchronous Optical Networks (SONET) and Synchronous Digital Hierarchy (SDH) are telecommunication standards for transmission based on optical fibers. In 1985, SONET was conceived by Bellcore, a research wing of the regional bell operating companies. Subsequently, around 1990, the American National Standards Institute (ANSI) formalized the SONET standard. Being an ANSI standard, SONET is primarily used in North America for carrying bit-rates based on the North American Digital Hierarchy (1.5/6/45 Mbps).

SDH, developed by ITU-T around 1990, is a standard similar to SONET. However, the major difference between SONET and SDH is in their domain of applicability. Since SDH is an international standard, it has a much wider reach. In fact, SDH is used in most parts of this world, barring North America and Japan, where SONET is prevalent.

Both SONET and SDH provide a rich set of features that include vendor-independent and sophisticated signalling structure, very high bandwidth support (~Gbps), and better performance monitoring. Because of their enhanced capability, SONET and SDH provide significant advantages over traditional telecommunication systems. Some of these advantages are explained below:

- **Interoperability:** SONET/SDH standards define a formal infrastructure for fiber-based transmission at the local, national and international levels. Most of the older transmission equipment had proprietary interfaces, which led to a lot of confusion. The standardization process aims to alleviate some of these problems and help in providing greater interoperability.
- **Unifying the American and European standards:** Prior to the formalization of SONET/SDH standards, the digital hierarchies used for multiplexing 64 Kbps voice channels in Europe and the US were incompatible. The Americans, along with the Japanese, followed the T1/T3 rates, whereas the Europeans followed the E1/E3 rates. The SDH and SONET hierarchies are an attempt to support both T1/T3 and E1/E3 digital hierarchies.
- **Reduction of operating costs:** The SONET/SDH standards help in cutting down the operating costs because of the reduced equipment requirements, increased network reliability and enhanced performance monitoring. The overhead bytes included in the SONET/SDH frames help in the tracking, isolation and resolution of problems in real-time.
- **Greater revenues:** The enhanced service capabilities and greater reliability of SONET/SDH-based transmission systems help in realizing greater revenues. Moreover, as the SONET/SDH standards define a very flexible architecture, accommodating future applications with a variety of transmission rates is relatively simple.

4.10.1 SONET Hierarchies

Like North American and European digital hierarchy, SONET/SDH standards define a digital hierarchy. At the lowest level of the SONET hierarchy is a 'base signal', referred to as the Synchronous Transport Signal Level-1 (STS-1). STS-1 operates at a transmission rate of 51.84 Mbps. All SONET signals are either carried as base signals (STS-1) or a multiple thereof. Table 4.4 depicts some of the STS-N levels, where N denotes the level, along with their data rates. This table also includes the optical counterpart of an STS-N level, given by Optical

TABLE 4.4

SONET/SDH Hierarchies

SONET		*SDH*	*Date rates (Mbps)*		*DS equivalent*	
STS level	*OC level*	*STM level*	*Gross*	*Effective*	*DS1*	*DS3*
STS-1	OC-1	N/a	51.840	49.536	28	1
STS-3	OC-3	STM-1	155.520	148.608	84	3
STS-12	OC-12	STM-4	622.080	594.432	336	12
STS-48	OC-48	STM-16	2488.320	2377.728	1344	48

Carrier Level-N (OC-N). As shown in the table, STS-N and OC-N are equivalent signals, the difference being that STS-N is an electrical signal, while OC-N is for optical signals. Two data rates, viz. gross data rates and effective payload data rates are also shown in the Table 4.4. The difference in the two values accounts for the overhead bytes included in a SONET frame.

The base signal for SDH is different from that of SONET and is referred to as Synchronous Transfer Mode Level-1 (STM-1). As is clear from the table, there is no SDH equivalent for STS-1. This is because the SDH starts at STS-3. Otherwise, the simple rule (STM-N ≡ STS-3N) defines the mapping between SONET transmission rates and SDH transmission rates. Both STS-N and STM-N signals can be sub-divided to carry lower speed channels. The last two columns in Table 4.4 give the DS1 and DS3 equivalents for STS-N and STM-N signals. This is an important aspect of SONET/SDH standards, because it helps in seamless interworking of networks following different hierarchies.

4.10.2 SONET Frame Format

As mentioned earlier, STS-1 forms the base signal for all transmission based on SONET standards. The STS-1 signal is carried in a 9 × 90 frame, which is transmitted periodically every 0.125 millisecond. This implies that 8000 frames are sent every second. Due to the synchronous nature of transmission, a frame is sent every 0.125 ms irrespective of whether there is data to be sent or not.

The basic SONET frame format is depicted in Figure 4.10. Each frame has 9 rows and 90 columns, with each row and column consisting of 1 byte. This gives a total data rate of

FIGURE 4.10

BASIC SONET Frame Format

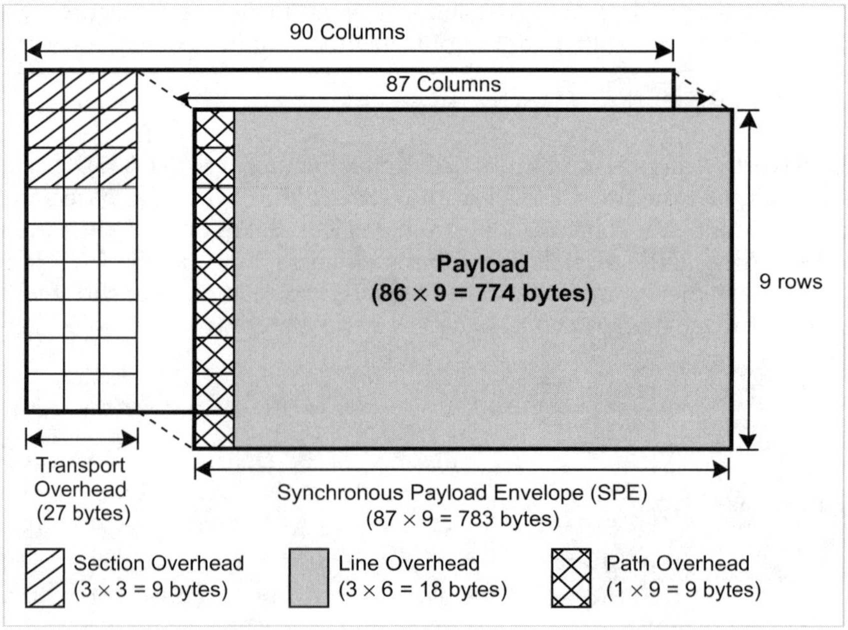

$9 \times 90 \times 8 \times 8000 = 51.84$ Mbps. However, all this is not available to the user for transmitting data because of the overheads involved. The first three columns of each frame are reserved for carrying the *transport overhead*. The transport overhead has two parts—the *section overhead* and *line overhead*. Section overhead occupies nine bytes of each frame and is used for various purposes like framing, local orderwire, performance monitoring (of STS-N signal), and carrying of datacommunication channels. The datacommunication channels within themselves carry Operations, Administrations, Maintenance and Provisioning (OAM & P) information. The 18-byte line overhead is used for performance monitoring (of individual STS-1 channels) and for express orderwire. The line overhead is also used to carry other information like the pointer to the start of the synchronous payload envelope, data channels for OAM & P, protection switching information, the line alarm indication signal and line far-end receive failure indication.

The remaining 87 columns form the Synchronous Payload Envelope (SPE). The SPE occupies 783 bytes and has 87 columns and 9 rows. Note that it is not necessary for an SPE to always start from the fourth column, first row. In fact, SPE can start from anywhere within the frame. The SPE pointer contained in the line overhead is used to identify the beginning of an SPE. The advantage of allowing an SPE to span more than one frame is to provide greater flexibility. Consider a scenario wherein a SONET frame is being filled and a payload arrives for transmission. If the aforementioned flexibility is not provided, the payload has to wait till the transmission of the next frame. This could mean a greater delay and a degraded performance. In order to obviate this possibility, SPE can span more than one frame, allowing for the payload to be filled as soon as it arrives.

The SPE has two parts—the path overhead and the user payload. Path overheads are used for performance monitoring (of SPE), and for carrying signal labels, path status and path trace. User payload, as the name suggests, is used to carry user data.

4.10.3 SONET Physical Layer Architecture

For the purpose of fault management, a SONET path is divided into distinct zones. These distinct zones are managed by different sub-layers, which are all a part of the SONET physical layer (see Figure 4.11). The lowest sub-layer of the SONET physical layer is the photonic sub-layer. The photonic sub-layer specifies the properties of the light and the fiber. The section sub-layer sits on top of the photonic sub-layer. The section sub-layer corresponds to a section and manages a single, point-to-point, fiber link. Sections are a continuous run of fiber without any intermediate network element. For example, the link between two repeaters is a section. Next in the hierarchy comes the line sub-layer which is concerned with multiplexing of tributaries on a single line and demultiplexing them at the other end. A line sub-layer corresponds to a line and manages links between two multiplexing devices. The line sub-layer is transparent to the presence or absence of repeaters. At the top of the physical layer hierarchy, is the path sub-layer. The path sub-layer manages end-to-end connections, spanning multiple multiplexers and regenerators.

From the above discussion, it follows that there is a great deal of similarity between the four sub-layers of the SONET physical and the four lower layers of the OSI reference model.

FIGURE 4.11

SONET Physical Layer Architecture

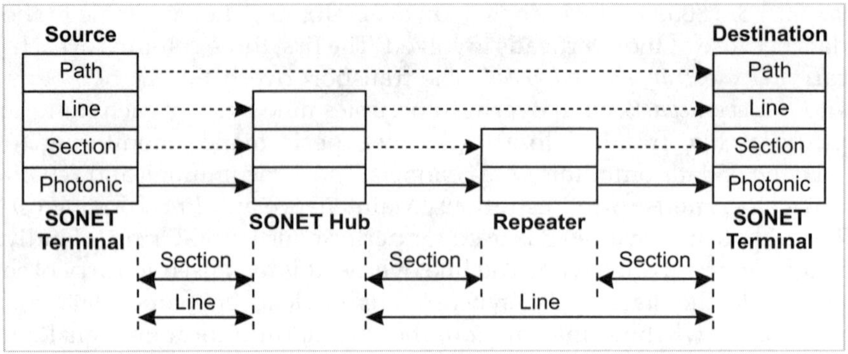

The photonic sub-layer can be compared to the physical layer, section sub-layer to the data link layer, line sub-layer to the network layer, and finally, the path sub-layer to the transport layer. This comparison is important because the knowledge and expertise built in understanding the OSI model can be used to analyse the issues of the four sub-layers of the SONET physical layer.

4.10.4 Mapping ATM Cells into STS-3c Frame

In the previous sub-sections, the SONET/SDH hierarchies and frame formats were explained. This section elaborates upon the information already provided and details the mechanisms of mapping ATM cells to STS-3c frame.

[ATMF UNI 3.1] defines a means to carry ATM cells in an STS-3c frame. STS-3c stream is multiplexed and switched through the network as a single entity. This is different from STS-3, which has three separable STS-1s that can be independently added or dropped. STS-3c has a single payload and STS-3c cannot add or drop a part of the payload. It should, however, be noted that both STS-3c and STS-3 have the same rate (155.52 Mbps) and the transport overhead. However, only one set of path overheads is needed for STS-3c payload envelope. This reduces the percentage overhead in an STS-3c frame. Thus, the SPE has a path overhead equal to that in one STS-1 frame i.e. 9 bytes and not 27 bytes. Therefore, the total space available for cell transfer in a frame is equal to 3 * 87 * 9 – 9 = 2340 bytes, giving a data rate of 149.760 Mbps (2340 * 8000 * 8). Since 2340 is not an integral multiple of 53, a cell may cross the boundary of SPE of one STS-3c frame to the other.

Figure 4.12 depicts the frame format of SONET STS-3c, as defined by [ATMF UNI 3.1]. As seen from the figure, there are three categories of overhead bytes, viz. the section overhead, line overhead and the path overhead. The section overhead and line overhead together form the transport overhead.

The section overheads applicable are as follows:

- **A-bytes:** A1 and A2 fields in the STS-3c are used for framing. The pattern present in fields A1 and A2 indicates the beginning of a frame. The pattern in byte A1 and A2 is set to 0xF628 for each of the STS-1s.

FIGURE 4.12

Mapping ATM Cell in a STS-3c Frame

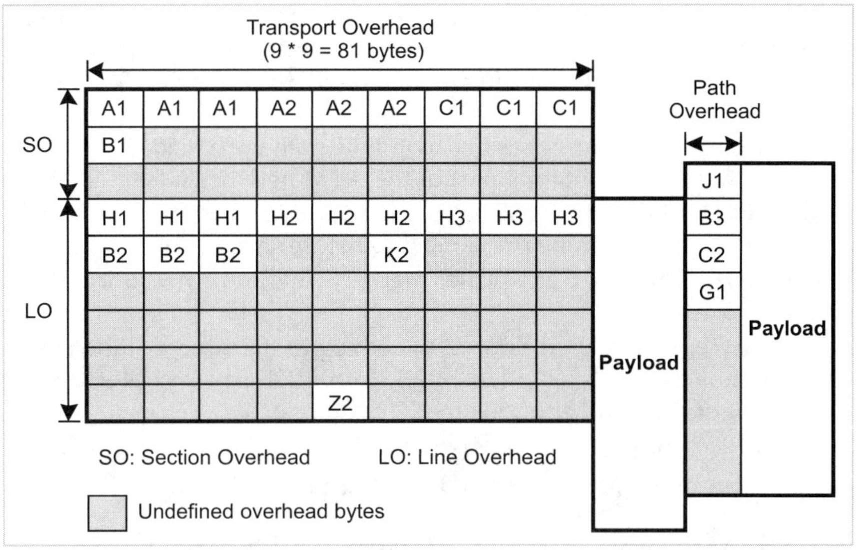

- **C1:** One byte C1 (STS-1 ID) field has a unique identifier to each of the STS-1 in the STS-3C.
- **B1 byte:** One byte B1 (BIP-8) is used for section error monitoring. The B1 byte of only the first STS-1 is used. The B1 bytes of the rest of the STS-1s are not defined. The *n*th bit of B1 provides even parity over the *n*th bits of each byte of previous STS-N frame after scrambling. The receiver recalculates the BIP-8 over the received data. If the value calculated by the receiver and B1 match, there is no error. If they do not match, an error in indicated.

The line overheads applicable are as follows:

- **H-bytes:** The two bytes H1 and H2 are used to indicate the offset (in bytes) the first byte of the beginning of the SPE. The beginning of SPE is also the beginning of the path overhead bytes. H1 and H2 byte of only the first STS-1 are required. The H1 and H2 byte of the remaining two STS-1s contain a concatenation indicator, which is 0x93FF, indicating a concatenation of more than one STS-1. The pointer byte H3 of each of the STS-1 is used for the purpose of frequency justification.
- **B2 byte:** The byte B2 (BIP-8) of each STS-1 is used for line error monitoring. The BIP-8 is calculated over all the bits of the line overhead and the SPE before the scrambling of the previous frame.
- **K2:** Automatic Protection Switching byte K2 is used for signalling between line level entities. Line level entity also uses this field to indicate any error condition to the peer and recovery from any such error. Only the bits numbered 6 through 8 of this field are defined. K2 byte of only the first STS-1 is defined.
- **Growth byte:** The growth byte Z2 is used for performance monitoring. It carries the results of the error monitoring information in the opposite direction of transmission, i.e. from the receiver to the sender.

The H1 and H2 mentioned above, indicate the position of the path overheads, which is followed by a stream of cells. Only one set of path overheads is required and is contained in the path overhead corresponding to the first STS-1 of the STS-3c. The applicable fields of path overhead are as follows:

- **J1 byte:** The receiver uses one byte state path trace field (J1) to verify the path connection. This field may contain a part of the ATM network address of the source.
- **B3 byte:** The byte B3 (BIP-8) is used for line error monitoring. The BIP-8 is calculated over all the bits of the previous SPE before scrambling.
- **C2 byte:** The byte C2 indicates mapping of the ATM within the payload field. It is set to 0×13.
- **G1 byte:** The byte G1 is used to convey to the source some information to enable it to monitor the total path. The bits 1 through 4, Far End Block Error (FEBE), indicate the count of the number of bits in B3 found in error.

As per [ATMF UNI 3.1], the TC functions like cell rate decoupling, cell delineation and payload scrambling, are performed as described in [ITU-T I.432].

4.10.5 Mapping ATM Cells into STS-48c Frame

Apart from [ATMF UNI 3.1] specification that provides the means to carry ATM cells into STS-3c frame, the ATM Forum has specified in [ATMF PHY 2.4G] the means to carry ATM cells at much higher rate (at 2.4 Gbps). For this, six different types of Physical Medium Dependent (PMD) sub-layers are defined that are based on single-mode fiber. As far as the frame format is concerned, it is a 4320×9 frame, with 144 columns of transport overhead and the rest carrying the Synchronous Payload Envelope (SPE). If this is compared with the STS-3c frame shown in Figure 4.12, then the 9 columns of transport overhead is replaced by 144, implying a 16 times increase. This is clear because STS-48c has 48 STS-1 in a concatenated fashion. In the SPE, while there is only one column of path overhead, there are fifteen bytes of fixed stuff in which ATM cells are not carried. For carriage of ATM cells, a total of $4160 \times 9 =$ 37440 bytes are available. This provides a cell rate of 2.396 Gbps. For details of cell format and overhead bytes, the reader is referred to [ATMF PHY 2.4G].

4.11 Universal Test and Operations PHY Interface for ATM (UTOPIA)

In the preceding sections, a number of physical layer standards for ATM were explained. These standards basically define the means to carry ATM cells over the underlying physical medium structure. However, these standards provide no guideline as to how the ATM layer and physical layer exchange data and control information. In other words, no standard interface has been defined between the ATM and the physical layer. Moreover, in an ATM system that has more than one physical layer interface, there is a need to define a standard

ATM-PHY interface for all physical layer interfaces. ATM Forum's UTOPIA is a step towards fulfilling this need.

Universal Test and Operations PHY Interface for ATM (UTOPIA) defines an interface between the physical layer (PHY) and the ATM layer. The aim of this specification is to define a common, standard interface between the ATM and PHY layers of ATM sub-systems. For this, various reference configurations are defined as shown in Figure 4.13. This allows an ATM layer to easily interface with the wide range of physical layer options available. Another motivation behind approving UTOPIA is to support data rates from sub-100 Mbps, up to 155 Mbps, with a common interface, requiring only an 8-bit wide data path.

The configuration in Figure 4.13(a) is captured in ATM Forum UTOPIA Level 1 specification (chip-to-chip interface) [ATMF UTOPIA L1]. Level 1 allows just one PHY to be interfaced with one ATM layer, using an 8-bit interface. The physical layer standards covered under this standard include 155.52 Mbps (SONET/OC-3c), 155.52 Mbps (8B/10B block coded), 100 Mbps (4B/5B TAXI), 44.736 Mbps (DS-3) and 51.84 Mbps (OC-1).

The configuration in Figure 4.13(b) is captured in ATM Forum UTOPIA Level 2 specification [ATMF UTOPIA L2]. Level 2 allows multiplexing of multiple PHYs over one UTOPIA interface. In addition to the physical layer standards supported by Level 1, the Level 2 standard specifies a 16-bit interface to support line rates at 622 Mbps.

FIGURE 4.13

UTOPIA Reference Configuration

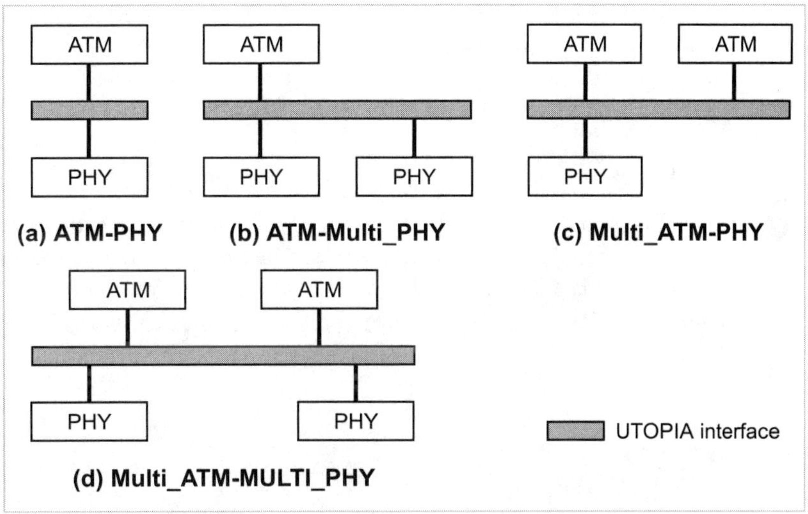

The ATM Forum UTOPIA Level 3 specification [ATMF UTOPIA L3] provides functionality similar to UTOPIA Level 1 and 2 but provides higher data rates. Towards this end, the data path is extended to support a 32-bit data path with speeds up to 3.2 Gbps. The data rates are further extended till 10 Gbps in ATM Forum UTOPIA Level 4 specification [ATMF UTOPIA L4].

For details on control signals and data exchange, the reader is referred to [ATMF UTOPIA L1], [ATMF UTOPIA L2], [ATMF UTOPIA L3] and [ATMF UTOPIA L4].

Conclusion

This chapter discussed the physical layer of the ATM Protocol Reference Model. At this layer, the main concern is to define means so that ATM cells can be carried over existing physical layer structures like T1/T3, E1/E3 and SONET/SDH without major modifications. For this purpose, ITU-T and ATM Forum have defined various specifications, as discussed earlier. Apart from the standards discussed in this chapter, ATM Forum has defined many other standards which are available at its website.

The physical transport, apart from the transport of ATM cells, must also provide network management facility so that various faults and management parameters can be reported. In this regard, SONET/SDH medium is quite advantageous, owing to its extensive network management support.

Review Questions

Q 1. What are the important issues in carrying ATM cells over a physical layer interface? Can ATM cells be carried over any physical layer? Give examples of the physical layers over which ATM cells can be carried highlighting the mechanism of doing so.

Q 2. Explain the following terms:
- *Header Error Control*
- *Cell Rate Decoupling*
- *Cell Scrambling*

Q 3. What are 'idle cells' and 'unassigned cells'? How do you distinguish between the two?

Q 4. What is the Cell Delineation procedure? In this procedure, explain clearly the safeguards against false misalignments due to bit-error and against false delineation during resynchronization? Comment on how the values of ALPHA and DELTA affect misalignments and resynchronization.

Q 5. How are ATM Cells mapped on the following physical layer interfaces:
- *DS1*
- *DS3*
- *E1*
- *E3*
- *Fractional E1/T1*
- *STS-3c*
- *STS-48c*

Q 6. What is Fractional E1/T1 and how do ATM cells transport over it? If in a fractional T1 line, discontinuous slots are available for ATM transport (say 2nd, 4th and 6th slots) how will the ATM cells be organized and carried over it?

Q 7. *What is UTOPIA? What is the motivation behind its standardization? What are the improvements in various UTOPIA levels?*

FURTHER READING

This chapter was based on different ITU-T and ATM Forum standards as listed in Table 4.1. Apart from this, few other books [ATM W. Goralski] and [ATM M. Prycker] also provide good coverage of the topic.

Chapter 5

ATM Layer

5.1 Introduction

ATM layer is the core layer of the ATM protocol stack. All the functionality essential to ATM is performed at this layer. Hence, the name of this layer. This layer multiplexes several user cell streams into a single stream and demultiplexes them at the destination. At the intermediate nodes, the ATM layer translates the connection identifiers (VPI/VCI) and performs cell switching. This layer also implements mechanisms for traffic policing and congestion indication.

The ATM layer receives from the upper layer a 48-byte payload, to which it appends a 5-byte cell header to from a 53-byte cell. The cell header contains—besides the VPI/VCI field—other fields like the cell priority bit, payload type field and header error check field. The format of the cell header at the User-Network Interface (UNI) is different from that at the Network-Network Interface (NNI). The difference is due to the fact that some functionality required at the UNI, is not required at the NNI. [ITU-T I.361] provides the cell header format at UNI and NNI. This chapter elaborates upon the frame formats at UNI and NNI and mentions the difference between the two formats. This chapter also mentions the functions of the ATM layer.

5.2 ATM Cell Header Structure at UNI

The structure of cell header at the UNI is shown in Figure 5.1. The 5-byte structure has the bits in the ascending order (from right to left) and the bytes in the ascending order (from top to bottom). The transmission of bits takes place in decreasing bit-order and increasing

FIGURE 5.1

ATM Cell Header Format at UNI ([ITU-T I.361])

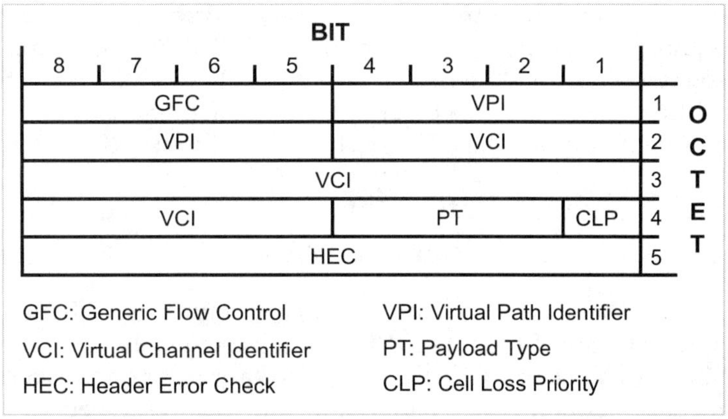

byte-order. That is, the eighth bit of the first byte is transmitted first and first bit of the fifty-third byte is transmitted last.

Before the cell header fields are detailed, note that some of the possible cell header values are reserved for cells generated at the physical layer. These physical layer cells have VCI = VPI = 0 and are identified by the first bit of the fourth byte, which is set to 1. Figure 5.2 shows the cell header values, which are reserved for the physical layer. As already explained in Chapter 4, the idle cell is used for cell rate decoupling. Physical layer OAM cells are used for management purposes and are identified by the fourth bit of the fourth byte, which is set to 1. The last row entry indicates a set of header values reserved for the physical layer, but which are yet to be defined. The bits marked 'P' indicate that these bits are available to the physical layer for various purposes.

FIGURE 5.2

Cell Header Values Reserved for Physical Layer

	Octet 1	Octet 2	Octet 3	Octet 4
Idle cell	00000000	00000000	00000000	00000001
Phy Layer OAM cells	00000000	00000000	00000000	00001001
Reserved for Phy Layer	PPPP0000	00000000	00000000	0000PPP1

5.2.1 Generic Flow Control (GFC) Field

This 4-bit Generic Flow Control (GFC) field is used for flow control between end-systems and the ATM network. This field is used only by the network to limit the amount of data injected by an end-system. There is no flow control in the reverse direction.

The scope of GFC depends upon the configuration of the end-system. An end-system can be either configured as controlled equipment or uncontrolled equipment. GFC procedures are not used for uncontrolled equipment. Further, each ATM connection from a controlled equipment may be controlled or uncontrolled. GFC mechanisms can be used only for the cells sent by a controlled terminal on a controlled ATM connection. All other cell flows are sent with this field set to 0000. Very few implementations support this functionality and hence, set this field to 0000. The GFC mechanism is discussed at length in Chapter 8.

5.2.2 Virtual Path Identifier (VPI) and Virtual Channel Identifer (VCI)

At the UNI, the cell header contains an 8-bit Virtual Path Identifier (VPI) field and a 16-bit Virtual Channel Identifier (VCI) field, which is used to relay/switch cells. However, not all values of VPI and VCI are available to carry user data. Some values are reserved for network management purposes, while some others are reserved for specific ATM layer functionality. The reserved values of VPI/VCI are called *pre-assigned* values. The pre-assigned VPI/VCI values are shown in Table 5.1 along with the function they perform.

TABLE 5.1

Pre-assigned VPI/VCI Values [ITU-T I.361]

Use	*VPI*	*VCI*
Unassigned Cell	0	0
Invalid	Any VPI value (except 0)	0
Meta-signalling	Any VPI value	1
General broadcast signalling	Any VPI value	2
Point-to-point signalling	Any VPI value	5
Segment OAM F4 flow cell	Any VPI value	3
End-to-end OAM F4 flow cell	Any VPI value	4
Segment OAM F5 flow cell (Note 1)	Any VPI value	Any VCI value (Note 2)
End-to-end OAM F5 flow cell (Note 1)	Any VPI value	Any VCI value (Note 2)
VP resource management cell	Any VPI value	6
VC resource management cell (Note 1)	Any VPI value	Any VCI value (Note 2)
Reserved for future VP functions	Any VPI value	7
Reserved for future VC functions (Note 1)	Any VPI value	Any VCI value (Note 2)
Reserved for future functions	Any VPI value	VCI = 8 to VCI = 31

Note 1: These cells are differentiated from user cells on the basis of the payload type field.
Note 2: Except VCI values 0, 3, 4, 6 and 7.

The values of VCI from 1 through 31 for any VPI are reserved and not used for data transfer. VPI = 0 and VCI = 0 is used for indicating *unassigned cells*. The ATM layer generates unassigned cells when it has nothing to send to the lower layer. If the lower layer expects a continuous stream of cells, then the ATM layer, in the absence of any user data, keeps on sending these cells. At the destination, these cells are not passed on to the higher layer. Thus, unassigned cells at the ATM layer are equivalent to idle cells at the physical layer.

VCI = 0 for any VPI value (except VPI = 0) is not used and is an invalid combination.

The VCI = 1 for any VPI is reserved for *meta-signalling*. The VCI = 2 is reserved for general *broadcast signalling* and VCI = 5 is reserved for *point-to-point signalling*. The signalling procedures are elaborated in Chapter 11.

The VCI = 3 and VCI = 4, for any value of VPI, are reserved for *segment* OAM F4 flow cells and end-to-end OAM F4 flow cells, respectively. These OAM cells are used for the management of the virtual paths. The segment F4 flow cells are used to carry maintenance information for a segment of a connection, while the end-to-end OAM F4 flow cells are used to carry maintenance information between the source and the destination. Similarly, segment OAM F5 flow cells and end-to-end OAM flow cells are defined for the virtual channels. Since they carry information about the VC, they are carried on the same VC over which they carry user information. These cells are identified by the PT field and explained in next section.

The value VCI=6 for any VPI is reserved for VP *resource management cells*. These cells carry information about the resources available with a node and other characteristics of the virtual path. The resource management cell is used for traffic management, the procedures of which are explained in Chapter 8. The VCI = 7 for all VPI is reserved for future VP functions. Similarly, for virtual channels, the resource management cells are defined for all values of VCI except VCI = 0, 3, 4, 6, 7. Also, provisions have been made to have reservations for future VC functions.

Note that the number of bits actually used for VPI and VCI fields are determined by negotiation between the user and the network. However, the bits cannot be allocated arbitrarily. It is required that the allocated bits of VPI and VCI be contiguous. Further, the bits allocated for both VPI and VCI should be the least significant bits, i.e. for VPI, the bits allocated should begin at bit 5 of second byte and for VCI, the bits allocated should begin at bit 5 of fourth byte in the cell header. All the unused bits are set to zero.

A user-network interface using all the available values of the VPI and VCI has to its disposal 256 virtual paths with 65,536 virtual channels for each virtual path. However, the number of reserved VPI/VCI fields reduces the number of connections available to a user.

5.2.3 Payload Type (PT) Field

This 3-bit field in the cell header identifies the type of the payload in the cell. The various values of the field Payload Type Identifier (PTI) are shown in Table 5.2. The possible combinations of the PTI field are classified in two categories.

The first category corresponds to user data cells and contains the top four entries of the Table 5.2. These entries correspond to those PTI values that have the most significant bit set to 0. For this category, the second most significant bit indicates whether congestion is being experienced or not. A value of 1 indicates that congestion is being experienced while a value of 0 indicates that there is no congestion. If an intermediate node receives a user data cell with this bit set to 1, it should initiate congestion control measures. Also, an intermediate node that is experiencing congestion sets this bit to 1 for all the outgoing cells. The least significant bit of the payload type field for a user data cell carries ATM-user-to-ATM-user indication. At the source end, this bit is passed to the ATM layer by the higher layer along with the cell payload. This bit is carried transparently through the network to the destination where it is delivered by the ATM layer to the upper layer along with the cell payload. One possible use of this bit is mentioned in Chapter 6 when the AAL5 layer is discussed.

TABLE 5.2

The Payload Type Identifier Coding [ITU-T I.361]

PTI coding	*Interpretation*
0 0 0	User data cell, congestion not experienced. ATM-user-to-ATM-user indication = 0
0 0 1	User data cell, congestion not experienced. ATM-user-to-ATM-user indication = 1
0 1 0	User data cell, congestion experienced. ATM-user-to-ATM-user indication = 0
0 1 1	User data cell, congestion experienced. ATM-user-to-ATM-User indication = 1
1 0 0	OAM F5 segment associated cell
1 0 1	OAM F5 end-to-end associated cell
1 1 0	VC resource management cell
1 1 1	Reserved for future VC functions

The second category corresponds to non-user data cells. This category contains PTI values that have most significant bit set to 1. For this category, the interpretation of the last two bits is independent of the previous category. As shown in the table, the values of 100 and 101 indicate OAM F5 segment flow and OAM F5 end-to-end flow respectively. The VC resource management cells have their payload type field set to 110. The VC field reserved for future use has payload type field set to 111.

5.2.4 Cell Loss Priority (CLP) Field

The single-bit Cell Loss Priority (CLP) field is used to assign a two-level priority to a cell. The cells with CLP bit set to 0, have a higher priority than the cells with their CLP bit set to 1. Thus, at the time of congestion, when there are excess cells, the cells with CLP bit set to 1 are subject to discard, prior to cells with the CLP bit set to 0. This field is not used by the cells generated by the physical layer. For the physical layer cells, this bit is always set to 1.

Similarly, for unassigned cells, this bit is always set to 0. It is through this bit that the physical layer idle cells and the ATM layer unassigned cells are distinguished from each other.

The ATM layer at the source end-system sets this bit to 0 or 1, depending upon the priority of the cell. However, if an intermediate node is facing congestion or if the source exceeds its allotted rate, the value of the CLP bit may be changed from 0 to 1. In this case, the priority of the cell is reduced. However, no intermediate node can change the CLP bit from 1 to 0, i.e. the priority of the cell can never be increased. The use of CLP field is further explained in Chapter 8.

5.2.5 Header Error Control (HEC) Field

The last byte of the cell header provides an error control mechanism for the remaining bytes of the cell header. The HEC is not applicable for the payload part of the cell. The error-handling mechanism is able to either correct single-bit errors or detect multiple-bit errors. The details of HEC procedures were discussed in Chapter 4.

5.3 ATM Cell Header Structure at NNI

The structure of cell header at the NNI is shown in Figure 5.3. The header is very similar to the cell header at the UNI except that the GFC field is done away with and the corresponding bits added to the VPI field. Thus, the VPI field at the NNI is of 12 bits length, whereas at UNI, the field is of 8 bits length only. The GFC field is omitted at the NNI because if the amount of data injected into the network is controlled, then within the network, there should be no need to enforce such a form of flow control. However, some services do require flow control within the network. But the primitive flow control provided by the GFC is not sufficient for such services and hence, the field has been done away with.

Figure 5.3

ATM Cell Header Format at NNI ([ITU-T I.361])

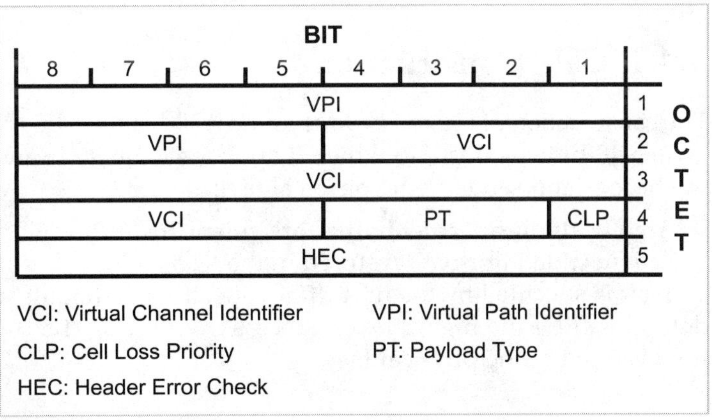

The reason for adding the resulting bits to the VPI field is straightforward. Within the ATM network, connections from several individual UNI connections are merged on a high-speed link (see Figure 5.4). Since the intermediate nodes must identify each connection uniquely, theoretically the maximum virtual connections between two ATM switches must be equal to (or greater than) the sum of individual connections at each UNI. In order to ensure this, the number of VPI bits at NNI is increased by 4. These additional bits allow connections from more than one UNI to be merged on to a single NNI and still allow each of the connections to be uniquely identified. The interpretation of the rest of the fields in the cell header for NNI, is the same as UNI.

FIGURE 5.4

Multiplexing of Several UNI Connections at NNI

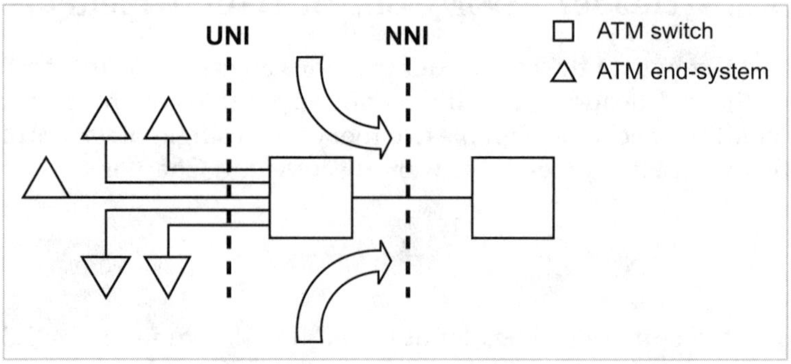

5.4 ATM LAYER FUNCTIONS

The functionality of the ATM layer is much more than just switching the values of the connection identifiers in the cell header. It ranges from simple functions like cell construction to far more complex ones like congestion control, traffic monitoring and providing QoS. The various ATM layer functions are discussed below.

5.4.1 Cell Construction

The construction of the cell is done at the ATM layer. The ATM layer receives from the upper layer a 48-byte payload, to which it appends a 5-byte header, fills in the necessary values in the header and sends to the physical layer.

Among the header fields that are filled, the VPI/VCI values are determined using the per-connection information stored in the translation table at the ATM layer. The value of the CLP bit is specified by the user. If not specified, a default value of 0 is used. The PTI field is determined by the higher layer process, requesting the transmission of payload. The HEC field is filled at the physical layer.

In order to appreciate how the ATM layer constructs an ATM cell, consider the following APIs provided by the ATM layer (as per [ITU-T I.361]):

- ATM-DATA request (ATM-SDU, submitted loss priority, congestion indication, ATM-user-to-ATM-user indication);
- ATM-DATA indication (ATM-SDU, congestion indication, ATM-user-to-ATM-user indication, received loss priority).

From the APIs given above, it is clear that the higher layer provides the necessary information to the ATM layer to fill the various fields of an ATM cell. Depending upon the higher layer protocol sitting above the ATM layer, one or more parameters are determined. For example, if AAL5 sits above the ATM layer, then the ATM-user-to-ATM-user indication is used to specify the continuation/end of AAL5 frame. For other AALs, this field is not used. Similarly, other fields (like submitted loss priority and congestion indication) may or may not be specified by the higher layer protocol. If they are not specified, the default values are used.

5.4.2 Cell Multiplexing and Demultiplexing

The ATM layer multiplexes cells from various logical connections using the VPI/VCI field onto the underlying physical medium. The allocation of bandwidth depends upon the connection parameters fixed at the time of the connection establishment. Any free bandwidth is distributed to various contending connections by using some fair algorithm. At the receiving end, the ATM layer demultiplexes cells and hands them over to the appropriate higher layer protocols.

5.4.3 Cell Relaying and Forwarding Using VPI/VCI

One of the basic functions of the ATM layer is to replace the VPI/VCI value of an incoming cell to a new VPI/VCI value. Depending upon the type of connection (virtual path connection or virtual channel connection), either only the VPI value or both VPI/VCI values are swapped. The mapping between incoming and outgoing VPI/VCI values is maintained within an ATM switch.

Apart from VPI/VCI swapping, the only other field that changes during the cell relaying process is the HEC field. The value of this field depends upon the contents of the cell header. Since with a change in VPI/VCI field, the value of HEC also changes, the HEC must be re-calculated. This recalculation of HEC is done at the physical layer and not at the ATM layer.

5.4.4 Interpretation of Pre-assigned Cell Header Values

As mentioned earlier, a number of pre-assigned header values are reserved for special purpose cells. This includes, among other types, the unassigned cells, the cells carrying signalling information, OAM flow cells and resource management cells. The ATM layer differentiates between these cells and takes appropriate action.

5.4.5 Support for Multiple QoS Classes

ATM provides multiple QoS classes to the users, which allows applications having varied requirements to be supported. The QoS required by the application is agreed upon at the time of connection establishment. The intermediate nodes from the source to the destination maintain, among other information, a mapping from the connection identifiers to the QoS parameters. The cells arriving on a link from different connections get a differential treatment, depending upon the QoS parameters associated with the connection to which the cell belong.

5.4.6 Usage Parameter Control

The ATM layer monitors each connection for the amount of cells injected in the network. This amount must remain below a certain threshold value, depending upon the parameters of the connection. If the end-system injects more cells than the value agreed upon, the ATM network takes appropriate measures. The measures include dropping of the cells sent in excess to the agreed value or tagging the cell by changing the CLP bit from 0 to 1. All these measures are within the purview of Usage Parameter Control (UPC) and are detailed in Chapter 8.

5.4.7 Explicit Forward Congestion Indication

In case of congestion, the network, apart from employing congestion control measures, also informs the end-users about the impending congestion. This enables the end-users to slow down the rate at which they inject the cells into the network, thereby alleviating the congestion situation. If congestion is detected, the ATM layer sets the second bit in the payload type field of the user cells to 1. This process is referred to as Explicit Forward Congestion Indication (EFCI).

When a cell is on its way to destination, no intermediate node can reset the congestion bit from 1 to 0. The indication is provided only in the direction in which the cells are travelling (i.e. the forward direction). It is the responsibility of the destination end-system to request the source to slow down the data rate.

Conclusion

This chapter provided an overview of the ATM layer. The primary concern of the ATM layer is efficient multiplexing of ATM cells from various connections, fair allocation of bandwidth to contending connections, and policing connections so that traffic contract is adhered to. Some of these aspects are revisited in Chapter 8 (traffic management) and Chapter 11 (signalling).

Review Questions

Q 1. Why are different header structures used at UNI and NNI interface?

Q 2. What are the pre-assigned VPI/VCI values? Explain with examples why they are used.

Q 3. Why do 'segment OAM F5 flow cells' and 'end-to-end OAM F5 flow cells' not have pre-assigned VPI/VCI values? How are these cells identified?

Q 4. What are the important functions of the ATM layer?

Q 5. Which fields of ATM cell header can be modified by an ATM switch? Under what condition does this modification take place? Explain in detail.

FURTHER READING

For details of the topics covered in this chapter, the reader is referred to [ITU-T I.361] which provides the definition of the ATM layer. Apart from this, [ATM W. Goralski] also provides good coverage of the topic.

Chapter 6

ATM Adaptation Layer

6.1 Introduction

The ATM Adaptation Layer (AAL) is used to map the requirements of the applications to the services provided by the ATM layer. At the source, the AAL layer takes data packets from the upper layer, converts them into 48-byte payload and hands them over to the ATM layer. At the destination, the reverse procedure takes place and the cells are assembled to form a higher layer data packet. This process of segmentation at the source and reassembly at the destination is the primary task of the AAL layer. However, different applications offer a different type of traffic to the network. In order to support such diverse applications, several variants of the adaptation layer are provided. Currently, five AALs are defined. They are AAL1, AAL2, AAL3/4, AAL5 and AAL0. This chapter elaborates upon the need, features and structure of these different types of AALs.

6.2 Service Classes and ATM Adaptation Layer

The organization of the AALs is based on the concept of service classes. A service class defines the characteristics of traffic offered by an application to the network. The various parameters used to characterize traffic are:

- **Bit Rate:** This is the rate of traffic offered by an application to the network. It may be either at a constant bit rate or variable bit rate.
- **Timing Relationship:** Some of the applications require a strict timing relationship between the source and the destination. Examples of such applications include voice and video. However, typical data applications (e.g. FTP) do not require any timing relationship.

- **Nature of Connection:** Applications may prefer the session between the sender and the receiver to be either connection-oriented or connectionless. The choice depends upon the inherent requirements of the applications. Generally, voice applications prefer connection-oriented session due to lower transit delays. On the other hand, FTP applications (which have vary small bursts of data transfer) may prefer connectionless session to avoid delays involved in establishing an end-to-end connection between the sender and the receiver.

Using the three parameters mentioned above, eight service classes are possible, out of which only five service classes are defined. Figure 6.1 shows the five service classes, the properties of each of them and the corresponding AAL used. All service classes, barring Class X, have well-defined properties. For Class X, all parameters are user-defined. This service class uses just the cell relay service of the ATM network to carry cells across the network.

FIGURE 6.1

Service Classes, Parameters and Relationship with AALs

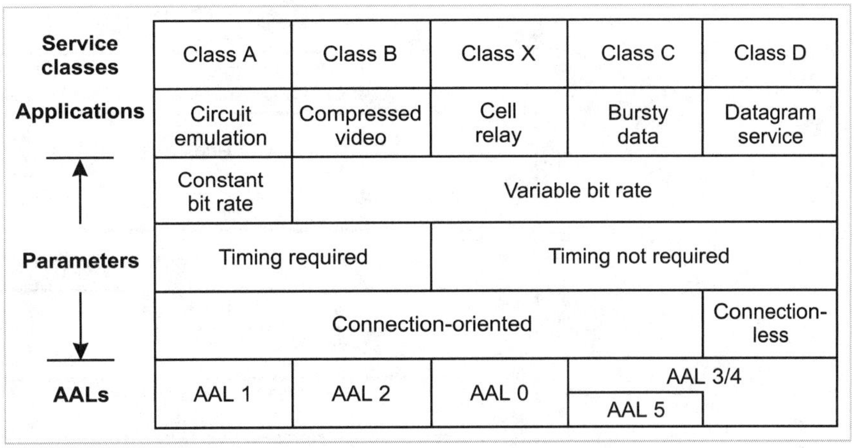

AAL1 carries time-sensitive, constant bit rate traffic on a connection established between end-systems. Therefore, the function of AAL1 is to take care of delay variations, to establish connection between peers and to maintain a timing relationship between them.

AAL2 provides function for the transfer of time-sensitive and variable bit rate traffic. AAL2 differs from the AAL1 in the sense that it carries variable bit rate instead of the constant bit rate. Moreover, AAL2 is more advanced as compared to AAL1, details of which are explained later in this chapter.

AAL3/4 provides for connectionless/connection-oriented, time-insensitive, variable bit rate data transfer. Apart from common functionality like segmentation and reassembly, AAL3/4 provides certain enhanced functionality like multiplexing of multiple end-to-end data flows on a single virtual circuit.

AAL5 provides functions for the transfer of Class C traffic. It is a very simple AAL with very limited functionality. The all-important function of AAL5 is segmentation and reassembly.

AAL0 does not perform any functions. For this AAL, data is received from the user in 48-byte payload and transferred to the ATM layer as such, without any processing. AAL0 is also called the null AAL. AAL0 is not discussed any further.

In order to perform the desired functionality, the AAL layer is divided into two sub-layers—viz. the Convergence Sub-layer (CS) and the Segmentation and Reassembly (SAR) sub-layer (see Figure 6.2). The CS interfaces directly with the user and accepts data from it. The CS optionally adds a header and/or a trailer to the user data, and passes it to the SAR sub-layer. The SAR sub-layer segments the data received from CS into smaller data units. The SAR sub-layer may also add a header and/or a trailer to these smaller data units to form a 48-byte SAR-PDU. This 48-byte SAR-PDUs forms the payload of the 53-byte cell at the ATM layer.

FIGURE 6.2

Structure of AALs [ITU-T I.363]

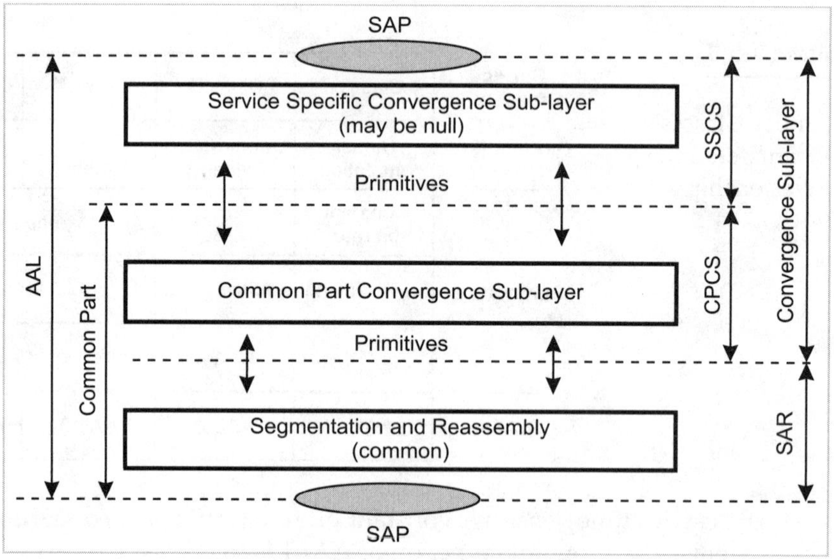

The CS itself is further divided into two sub-layers, viz. the Common Part Convergence Sub-layer (CPCS) and the Service Specific Convergence Sub-layer (SSCS). CPCS provides generic functionality common across applications and is fixed for an AAL. SSCS provides additional functionality required for specific applications.

Note that the above classification of the AAL layer into sub-layers is generic in nature and varies in functionality from AAL to AAL. In other words, the classification is not the same across all AALs. For example, the SAR sub-layer in AAL1 and AAL2 do not perform any segmentation and reassembly, because the data units accepted from higher layers are small in size and can fit into one or, at most, two cells. Thus, for AAL1 and AAL2, the use of the term SAR is not quite correct.

6.3 ATM Adaptation Layer 1 (AAL1)

The ATM Adaptation Layer 1 (AAL1) is used for constant bit rate applications, wherein data is transmitted and received at a constant rate. The ITU-T recommendation [ITU-T I.363.1] defines the following services, which are to be provided by the AAL1 layer to the AAL user:

- The transfer of service data units with a constant source bit rate and their delivery with the same bit rate;
- The transfer of timing information between the source and the destination;
- The transfer of structure information between the source and the destination;
- The indication of lost or errored information which is not recovered by AAL type 1, if needed.

The meaning of the first service is self-explanatory. The second service relates to clock recovery at the receiving side, while the third service defines the means to transfer block of information. The final service provides the means for error detection and correction. All these service definitions will be clearer within the course of this section.

In order to achieve the above-mentioned service objectives, the AAL1 is divided into two sub-layers, viz. the Segmentation and Reassembly (SAR) sub-layer and the convergence Sub-layer (CS). The functions of these sub-layers are elaborated below.

6.3.1 Segmentation and Reassembly (SAR) Sub-layer

The SAR sub-layer in AAL1 accepts a 47-byte blocks from the CS, and prepends a single byte header to it to form a 48-byte SAR-PDU. This 48-byte PDU, which forms the payload of a cell, is passed to the ATM layer. Thus, the SAR layer does not perform any segmentation and re-assembly functions, and is to an extent, a misnomer.

Figure 6.3 shows the format of the SAR PDU. The first four bits of the SAR-PDU header represent the Sequence Number (SN) field. This field consists of a Convergence Sub-layer Indication (CSI) bit and a 3-bit sequence number field that numbers cells from 0 to 7. The last four bits of the header contain the Sequence Number Protection (SNP) field, which is used to protect the SN field. The SNP field consists of a 3-bit CRC field and a one-bit parity check field.

The functions of the SAR sub-layer for AAL1 are summarized below:

- **Mapping function:** The SAR sub-layer provides a mapping between the CS and the ATM layer. As mentioned above, the SAR sub-layer prepends a one-byte header to the 47-byte packet received from the CS and passes this to the ATM layer. At the receiving end, the SAR sub-layer receives 48-byte data from the ATM layer, strips off the one-byte header and passes the 47-byte block to the CS.
- **Sequence numbering:** The SAR sub-layer provides a modulo-8 sequence-numbering mechanism, using the 3-bit SN field. The SAR sub-layer does not itself generate the sequence number. It is the CS that provides a sequence number along with the 47-byte

FIGURE 6.3

AAL1 SAR PDU [ITU-T I.363.1]

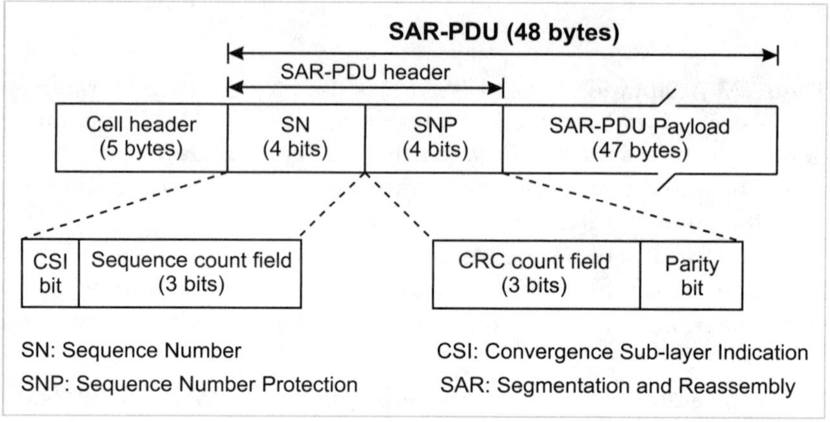

block to the SAR sub-layer. At the receiving end, the reverse process occurs, i.e. the sequence number contained in the SN field is passed to the CS. This sequence number is then used by the CS to detect lost or mis-inserted SAR-SDUs.

- **Synchronization:** This function is performed by using the CSI bit of the SN field. The use of this bit for synchronization is explained later in this section.
- **Error protection:** The SAR sub-layer also provides for error protection of the 4-bit SN field using the SNP field. While sending a SAR-PDU, the sender calculates the checksum and places it in the CRC field. Further, it calculates even parity for the first seven bits of the header and writes in the eighth bit.

The receiver has two modes of operation: *Correction Mode* and *Detection Mode* (see Figure 6.4). In the correction mode, single-bit errors are corrected. No error correction is done in the detection mode, only error checking is performed. The receiver starts in the correction mode.

FIGURE 6.4

Receiver Operatiol Modes [ITU-T I.36.1]

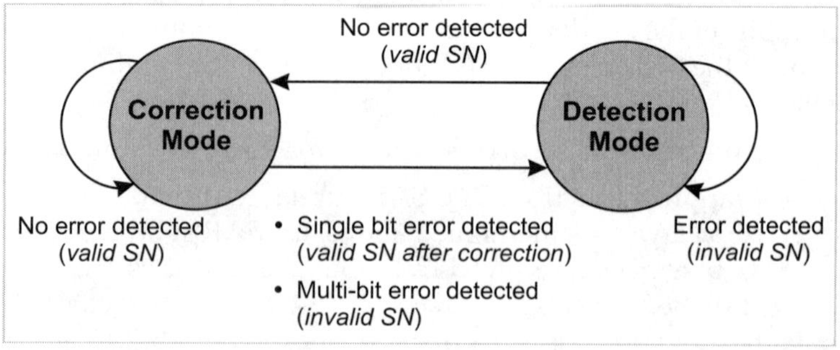

In the correction mode, the receiver examines the SAR-PDU header by checking the CRC bits and the even parity bit. If an error is detected, the receiver changes its mode to the detection mode. If the error in the header is a single-bit error, the error is corrected, otherwise the SN field is declared invalid. In the detection mode, the SAR-PDU header with an error is declared to have an invalid SN field. However, if a header is found without an error in the detection mode, the receiver changes its mode to correction mode. Tables 6.1 and 6.2 show the operations of the receiver in the correction mode and the detection mode respectively.

TABLE 6.1

Operations in Correction Mode [ITU-T I.363.1]

CRC Syndrome	*Parity*	*Action on current SN+SNP*	*Reaction for next SN+SNP*
Zero	No violation	No corrective action valid.	Continue in correction mode.
Non-zero	Violation	Single bit correction on syndrome. Declare SN valid.	Switch to detection mode.
Zero	Violation	Correct parity bit Declare SN valid.	Switch to detection mode.
Non-zero	No violation	No corrective action: multi-bit errors are uncorrectable. Declare SN invalid.	Switch to detection mode.

TABLE 6.2

Operations in Detection Mode [ITU-T I.363.1]

CRC Syndrome	*Parity*	*Action on current SN+SNP*	*Reaction for next SN+SNP*
Zero	No violation	No corrective action. Declare SN valid.	Switch to correction mode.
Non-Zero	Violation	No corrective action. Declare SN invalid.	Continue in detection mode.
Zero	Violation	No corrective action. Declare SN invalid.	Continue in detection mode.
Non-zero	No violation	No corrective action. Declare SN invalid.	Continue in detection mode.

6.3.2 Convergence Sub-layer (CS)

The primary function of the CS is to maintain a timing relationship between the sender and the receiver. CS does not add any header to the transferred data. Instead, it passes two fields—a sequence number and a CSI bit—to the SAR along with the 47-byte SAR-PDU payload. These two fields are filled by the SAR sub-layer in the SAR-PDU header (as discussed in the previous section).

The functions of CS for AAL1 are summarized as follows:

- **Handling cell delay variations:** The CS is responsible for handling cell delay variations occurring during cell transfer. Delay variations occur when the time taken by one cell to reach the destination is not equal to the time taken by other cells. This inequality results from unequal processing and queueing delays in intermediate switches. In order to handle cell delay variations, a buffer is maintained at the destination, whose size depends upon the average cell delay variation. All received cells are written to this buffer, from which they are read at a constant bit rate. However, it is possible that when the bits are read from the buffer, there are not enough bits in the buffer. In such a situation (called the buffer underflow), it may be necessary to insert appropriate number of dummy bits to maintain bit count integrity (i.e. the number of bytes be a multiple of 47). Similarly, in case of a buffer overflow, it may be necessary to drop the additional number of bits from the buffer.

- **Cell sequencing:** The sequence count value passed by the SAR sub-layer is processed at this sub-layer to detect lost or mis-inserted cells. This sub-layer compensates for the lost cells by inserting dummy SAR-PDU payloads and discards the mis-inserted cells.
- **Timing information transfer and clock recovery:** For maintaining timing relationship between the sender and the receiver, some information is required by the latter for it to be able to maintain the same frequency as the former. This layer provides mechanisms for the transfer of timing information between AAL1 peers. Mechanisms for clock recovery are detailed later in this section.
- **Partially-filled cells:** Under normal circumstances, until 47 bytes of the SAR-PDU payload are filled, a cell is not transmitted. For applications with very low data rates, this may result in excessive assembly delays. In order to circumvent this problem, AAL1 allows partially-filled cells to be transmitted. The number of valid bytes in a payload is determined at the time of subscription (for PVCs) or fixed via signalling (for SVCs).
- **Structured Data Transfer (SDT):** In order to support N × 64 Kbps circuits, a mechanism called the SDT is defined. SDT is discussed later in the section.

6.3.3 Source Clock Frequency Recovery

One of the important applications of AAL1 is circuit emulation. In circuit emulation, a logical ATM channel emulates the functionality of circuit-based technologies like T1/E1. In order to achieve this, the number of bits transmitted by the sender must be equal to the number of bits received by the receiver, and both, in turn, must be equal to the link rate of the circuit being emulated. For example, for emulating a T1 circuit, 1.544 megabits must be transmitted and the same number of bits must be received per second. Proper circuit emulation requires that there be a timing relationship between the sender and the receiver. In order to obtain this timing relationship, two methods are currently defined. These methods are the Synchronous Residual Time Stamp (SRTS) method and the Adaptive Clock method. Both these methods are detailed in the following sub-sections.

6.3.3.1 Synchronous Residual Time Stamp (SRTS) Method

This method assumes the existence of a common reference clock available to both the sender and the receiver. This method conveys to the receiver the difference between the common reference clock and the sender clock. The receiver then adjusts its clock on the basis of the time stamp received from the sender. A 4-bit time stamp is provided for a block of eight cells.

Another assumption in this method is that the frequency of the network clock is relatively lower than that of the sender clock. Usually, the sender clock frequency is a multiple of the network clock frequency. The number of sender clock cycles in one network clock cycle is known to both the sender and the receiver. This nominal number of clock cycles is the reference point of the sender and the receiver. In case of any variation in the sender clock, it is the difference between the obtained number of sender clock cycles in one network clock cycle and the calculated nominal value that is transmitted as the residual time stamp. The 4-bit time stamp provided for a duration of 3008-bit time (number of bits in 8 SAR-PDU payloads of 47 bytes each) is sufficient to meet the requirements of most of the applications having a constant bit rate traffic.

The 4-bit residual time stamp is transmitted in the CSI bit contained in the SAR-PDU header. The CS passes this bit along with 47-byte block and the sequence number to the SAR sub-layer. From the block of 8 SAR-PDUs, the PDUs numbered 1, 3, 5 and 7 carry the residual time stamp. The CSI bit in the even numbered PDUs (i.e. 0, 2, 4 and 6) carry information about the SDT and are not used for time stamping. The most significant bit of the residual time stamp is carried in the SAR-PDU header with sequence count of 1.

6.3.3.2 Adaptive Clock Method

The adaptive clock method does not require any 'reference' clock between the sender and the receiver. Nor does it require any recovery-related information to be carried from the sender to the receiver. This method is based on the fact that the amount of data received is a measure of the frequency of the sender clock and can be used to recover the same.

In this method, the receiver maintains a buffer, which has an associated fill level. The fill level indicates the amount of data in the buffer. The nominal sender's clock frequency is known to the receiver before the data transfer. In order to estimate the sender clock, the receiver performs the following steps, as also shown in Figure 6.5:

Step 1. The data is copied into a buffer and read using a locally-generated clock.

Step 2. The fill level of the buffer, which is the measure of the sender clock frequency, is used to control a Phase Locked Loop (PLL), which generates the local clock frequency proportional to the fill level.

Step 3. The generated clock is used to read the data in the buffer.

The mean fill level is kept at a chosen fixed level, depending upon the nominal sender clock frequency. Since the data is copied into the buffer in quantas (i.e. 47 bytes SAR-PDU), there is a bursty change in the fill level of the buffer even if data is arriving at a constant rate. This could lead to unnecessary changes in the frequency of the local clock even if it is at the same frequency as the sender. In order to avoid this, an upper fill level and a lower fill level are maintained instead of a single fill level. The upper and lower fill level are also chosen

FIGURE 6.5

Estimation of Receiver Clock

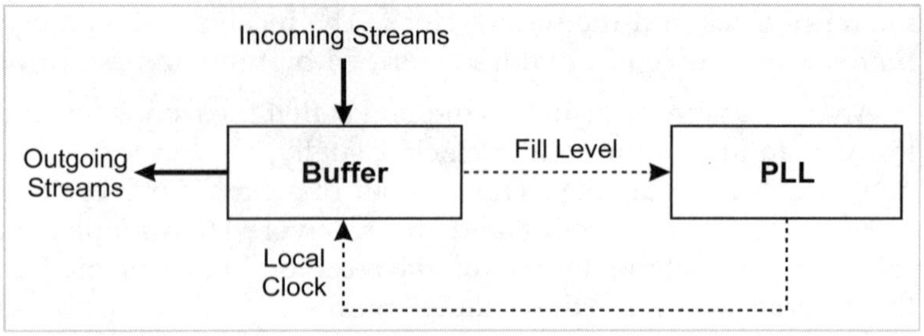

taking into consideration the delay variations in the data arrival procedure, which could cause fluctuations in the buffer fill level.

The local clock frequency is altered only if the fill level of the buffer goes either above the upper fill level or below the lower fill level. If the fill level is between the upper and lower fill level as in Case 1 (see Figure 6.6), the frequency of the clock is not changed. However, if the fill level exceeds the upper fill level as in Case 2, the frequency of the local clock is increased so that more data is read from the buffer and the fill level comes down. Similarly if the fill level drops below the lower fill level as in Case 3, the frequency of the local clock is decreased, so that less data is read from the buffer and the fill level comes back within the specified range.

FIGURE 6.6

Buffer Fill Level

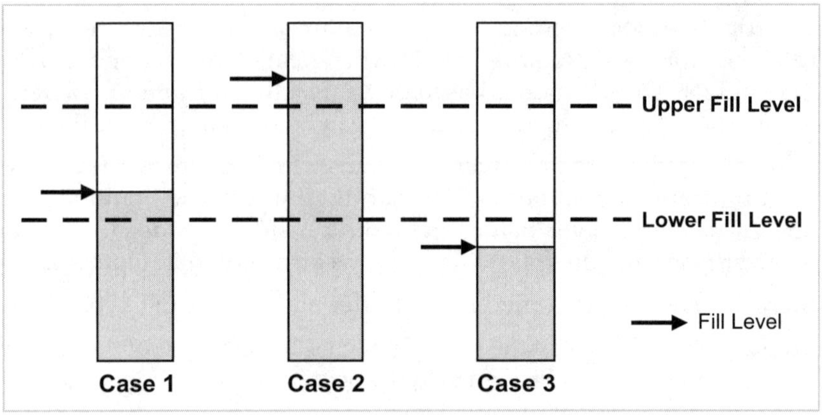

6.3.4 Structured Data Transfer (SDT)

Structured Data Transfer (SDT) is used to support Nx64 Kbps circuits. The unstructured mode is used for emulating a typical T1/E1 type of link. In contrast, SDT is used to support *fractional* T1/E1 links.

In order to support SDT, a SAR-PDU is classified as either a non-P format PDU or a P-format PDU (see Figure 6.7). The SAR-PDUs discussed so far are of non-P format and have their CSI bit set to 0. In contrast, information pertaining to SDT is transferred by using

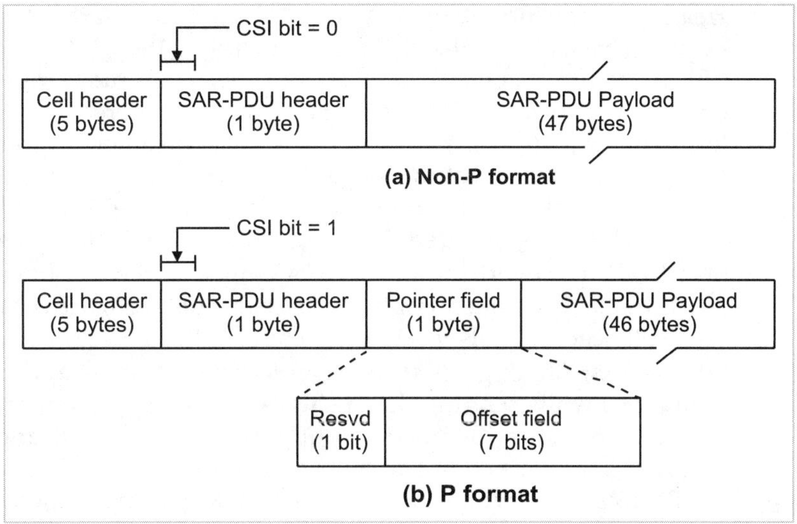

FIGURE 6.7

Format of SAR-PDU

P-format PDUs, which have the CSI bit set to 1. A P-format PDU can be only sent in an even number of PDUs (i.e. in PDUs numbered 0, 2, 4 and 6). In case SDT is not used, the CSI bits in the even-numbered PDUs are set to 0. Recall that the odd-numbered PDUs are reserved for sending SRTS information (in case the said mechanism is being used).

A P-format PDU has an additional one-byte field preceding the payload. This field (as shown in Figure 6.7) contains a 7-bit field referred to as the offset pointer. The offset pointer measures the number of bytes between the end of the pointer field and the first start of the structured block. Although the bit can have a maximum value of 127 (corresponding to $2^7 - 1$), the permitted values are 0 to 93, both inclusive. The value of 93 comes from the fact that SDT information can be sent every second byte (i.e. zeroeth, second, fourth, ...). Now, odd-numbered PDUs (in non-P format) have 47 bytes in payload, whereas even-numbered PDUs (in P-format) have 46 bytes in payload. Thus, two consecutive PDUs—one in P-format and other in non-P format—result in a total of 93 bytes.

6.3.5 Analysis of AAL1

The following points summarize various aspects of AAL1.

- **Characteristics:** AAL1 is designed to carry voice and to emulate the characteristics of a typical TDM connection. Hence, delay and jitter guarantees are of significant importance in an AAL1 connection.
- **Sub-layering:** Since AAL1 accepts a stream of incoming bytes, as against a data packet, there is no read to segment frames into cells or to reassemble cells from frames. Thus, SAR is actually a misnomer for AAL1. The two sub-layers CS and SAR can be combined to form a single layer. The only reason for dividing AAL1 into two sub-layers is to make the approach more modular.

- **Header/trailer format:** In AAL1, the only header added is the one-byte SAR-PDU header. This has four bits of SN field, including three bits of sequence count and one-bit CSI. Although sequence count is not used for frame retransmission, it is required to obtain timing information via the SRTS method. Moreover, sequence count ensures that lost or mis-inserted cells are easily detected. The other four bits provide error protection to the SN field, but not to the payload. Error correction for payload is not necessary because a voice circuit can tolerate occasional bit errors. This also implies that only a small SNP field is required (i.e. a big CRC for the entire payload is not necessary).
- **Overhead:** The overhead in AAL1 is limited to the one-byte header attached in every AAL1 frame. The header implies a 1.88 per cent (1/53) of overhead, which, in conjunction with cell header, totals to 11.3 per cent (6/53).
- **Drawbacks:** Although AAL1 supports delay-sensitive applications, there are inherent limitations while sending data using AAL1. One of the most important drawbacks of AAL1 is that it is ill-suited to support low-rate applications because of the delay involved in filling a complete cell. For example, for a 32 Kbps application, it takes 12 ms to fill a cell (48*8/32,000 sec). For a 4 Kbps application, the packetization delay increases to 96 ms, which is unacceptable. Moreover, the data in AAL1 is always carried in bundles of 64 Kbps, which reduces flexibility. While the partially-filled cells' method supports low-rate, delay-sensitive applications, this comes at the cost of wasted bandwidth due to excessive padding in the payload. Further, AAL1 does not define any compression or silence detection mechanisms, which could otherwise lead to better bandwidth utilization.

6.4 ATM Adaptation Layer 2 (AAL2)

AAL2 provides the means for bandwidth-efficient transmission of low-rate, short and variable length packets in delay-sensitive applications. In order to overcome the limitations of AAL1, AAL2 was introduced in 1997 (see [ITU-T I.363.2]), some four years after other AALs were standardized. The delay in standardization was due to the inherent complexity involved in defining a standard that could support the transmission of compressed time-sensitive data.

The characteristic features of AAL2 are summarized below.

- It provides the means for bandwidth-efficient transmission of low-rate, short and variable length packets in delay-sensitive applications.
- It provides for the transfer of time-sensitive constant bit rate traffic as well as variable bit rate traffic.
- It provides the means for compression and silence suppression techniques.
- It enables more than one AAL2 channel on a single ATM virtual circuit. This is done by multiplexing the data of multiple users on the same PDU that can be carried on a single ATM virtual circuit.

The AAL2 layer is divided into two parts, viz. the Common Part Sub-layer (CPS) and the Service-Specific Convergence Sub-layer (SSCS). The structure of AAL2 is shown in Figure 6.8. The CPS provides the basic functionality of AAL2, which includes packaging the variable payload into cells and providing error correction. The SSCS directly interacts with the AAL user. This sub-layer may be used to enhance the services provided by CPS or tailor the services provided by CPS to support specific AAL2 user services. This sub-layer may even be null and only provide a mapping between AAL user primitives and the CPS service primitives. Since the CPS forms the core of AAL2, only CPS is discussed in this chapter. The SSCS for AAL2 will be discussed in Chapter 20 when the techniques for carrying voice over ATM/AAL2 are explained.

FIGURE 6.8

Structure of AAL2 [ITU-T I.363.2]

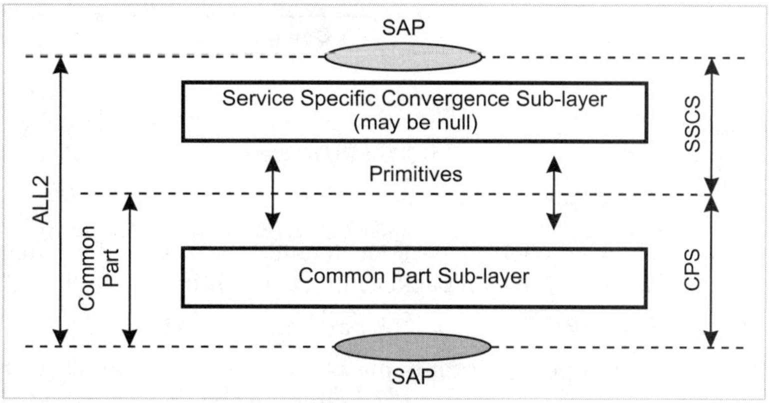

6.4.1 Common Part Sub-layer (CPS)

The CPS in AAL2 provides capabilities to transfer CPS-SDUs from one CPS user to another CPS user. The user of CPS can be either an SSCS entity or the layer management entity. In the first case, data received by (and sent to) the SSCS entity is application-specific data. In the second case, the data generated by layer management is used for the exchange of management-related information.

The CPS layer does a two-step operation on the data received from the CPS user. First, it adds a header to the received data and forms a CPS packet. In the second step, it breaks up the CPS packet, adds another header and gives it to the ATM layer as 48 bytes CPS-PDU, to be carried in the payload of the ATM cells. The detailed AAL2 operation, as a four-step process, is shown in Figure 6.9. The steps are as follows:

Step 1. Three user data frames—viz. A, B and C, each corresponding to a different AAL2 channel arrive for transfer.

Step 2. For each user data frame, a CPS packet is formed by prepending a CPS header.

FIGURE 6.9

AAL2 Operations

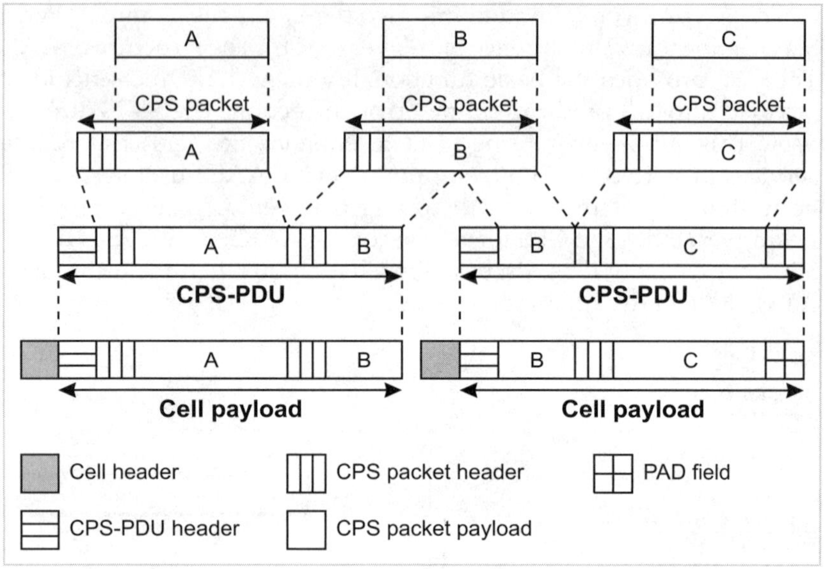

Step 3. The CPS packets are concatenated to form a CPS-PDU by pre-pending a CPS-PDU header. One CPS packet can span multiple CPS-PDU.

Step 4. The CPS-PDU forms the payload of an ATM cell.

Here, two important points may be noted. First, the use of two levels of headers allows different users of the same virtual circuit to be distinguished clearly. Second, the time that AAL2 waits before dispatching a CPS-PDU, is controlled by CU_timer (see [AAL2 C. Liu]). If sufficient bytes are not received within the expiry of this timer, CPS-PDU is sent with padding bytes. This bounds the maximum packetization time for a CPS packet. In essence, the use of CU_timer provides AAL2 the means to support delay-sensitive applications.

6.4.1.1 Structure of CPS Packet

The structure of the CPS packet is shown in Figure 6.10. The first field shown in the CPS packet is the Channel Identifier (CID) and identifies an AAL2 channel. This single byte field is used for multiplexing a number of AAL2 users on the same ATM virtual channel. Putting it differently, this field is used as a multiplexing identifier to identify the packets from different users. An AAL2 channel is a bi-directional channel and the same values of channel identifiers are used in both the directions. The value of 0 is not used, while 1 is reserved for layer management procedures. The values of 2 to 7 are also reserved. Permitted value of CID for CPS users is between 8 and 255, both inclusive.

Here, it should be noted that channel identifier values 8 to 255 can be used both by the SSCS entity as well as the layer management entity. In order to differentiate between the two classes of users, the User-to-User Indication (UUI) field is used. This field serves a two-fold

Figure 6.10

CPS Packet Format [ITU-T I.363.2]

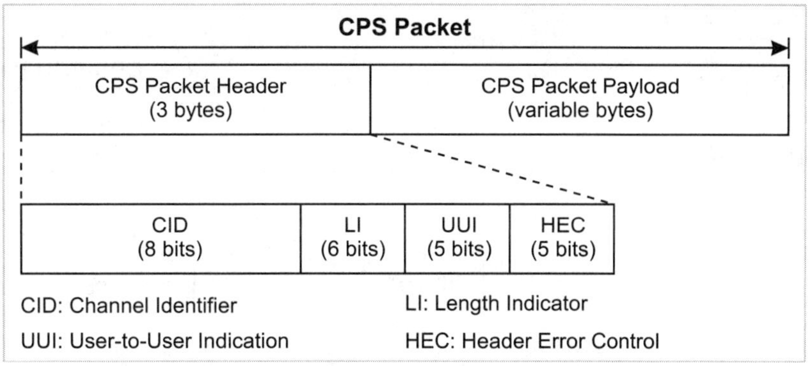

purpose. First, it is used to transparently transfer a 5-bit information between the CPS users. Second, it helps to distinguish between the two CPS users, i.e. the SSCF entity and layer management entity. A value of 0–27 is reserved for SSCS entity, whereas 30 and 31 for layer management entity. Values 28 and 29 are reserved for future use.

The CPS also provides mechanisms for error detection over the CPS header. The 5-bit HEC field is used for this purpose. These five bits contain the CRC checksum calculated over the first 19 bits of the CPS header (i.e. the HEC bits are excluded from the calculation).

The length of the CPS packet payload is carried in the Length Indicator (LI) field and can range between 1 to 45 or can be 64-bytes. The length stored in the LI field is one less than the actual length of the payload. This is done to accommodate a data length of 64 bytes, which otherwise requires 7 bits. Thus, an LI field—with all bits set to zero—indicates a payload packet length of 1, while an LI field with all bits set to 1 indicates a value of 64. Therefore, a CPS-packet may range from 4-bytes (3 bytes header + 1 byte payload) to 67-bytes (3-bytes header and 64-byte payload).

6.4.1.2 Structure of CPS-PDU

The CPS packets are carried inside the CPS-PDUs. A CPS-PDU consists of 48-bytes and forms the payload of the cell. The CPS-PDU has a one-byte header. The remaining 47-bytes are either filled by the CPS packets, or in the absence of any CPS packet, are occupied by padding bits. Each CPS-PDU may contain several CPS packets. The multiple CPS packets are placed in the CPS-PDUs one after the other without any gap between them. The last CPS packet in such a case may not fit in the CPS-PDU and have some overflowing bytes. In such a case, the overflowing bytes of the CPS packet are put in the next CPS-PDU.

The format of the CPS-PDU is shown in Figure 6.11. A 6-bit Offset Start Field (OSF) in the CPS-PDU header is used to mark the start of the first CPS packet in the PDU. It contains the number of bytes between the end of the header and the beginning of first CPS packet. The first CPS packet need not start immediately after the header as some bytes of a CPS packet in the previous CPS-PDU may overflow into this CPS-PDU. This scenario is shown in Figure 6.9. In the first CPS-PDU, the offset start field is set to zero as the start of the first CPS packet (A)

FIGURE 6.11

CPS-PDU Format [ITU-T I.363.2]

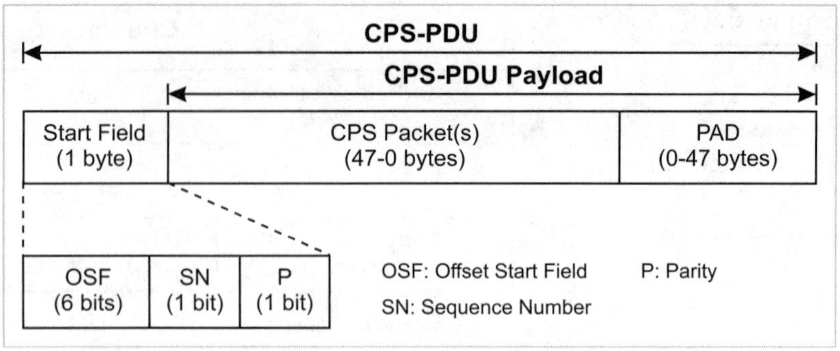

coincides with the beginning of the payload field of the CPS-PDU. Since the CPS packet B does not fit exactly into the first CPS-PDU, the remaining bytes are transferred as a part of the second CPS-PDU. The OSF field of the second CPS-PDU contains the offset of CPS packet C.

Within the CPS-PDU, the end of one CPS packet indicates the start of the next CPS packet. The length of each individual CPS packet is found by looking into the CPS headers. In order to differentiate between the beginning of a CPS packet and padding field, the first byte of the CPS packet is never set to zero. At the end of a CPS packet, a non-zero byte indicates the beginning of a new CPS packet and a zero value indicates the beginning of the padding field. This explains why only non-zero values of channel identifier are allowed.

With the above knowledge, the definition of offset field needs a slight modification. The offset start field actually defines the number of bytes between the end of the CPS-PDU header and the first beginning of the CPS packet or the padding field. The value of the first byte at the specified offset determines if the byte marks the beginning of a CPS packet or padding field (i.e. if the byte is non-zero, it indicates a new CPS packet, else it indicates padding).

Since the CPS packet size may be as large as 67 bytes, it is possible that neither the start nor the end of a CPS packet lies within a CPS packet. In such a case, the offset start field is set to 0.

The seventh bit in the CPS-PDU header is the Sequence Number (SN) bit. This field is used to number the stream of CPS-PDUs (modulo 2). In the outgoing CPS-PDU stream, this bit is set to 0 and 1 in alternate CPS-PDUs. At the receiver, the same value of sequence number bit in two successive PDUs indicates either an errored CPS-PDU header or a missing CPS-PDU between the two received CPS-PDUs.

However, it should be noted that this scheme assumes that the loss of more than one CPS-PDU is rare. This assumption stems from the fact that the loss of one CPS-PDU and the loss of any odd number of CPS-PDUs is not distinguishable. To distinguish between the lost and errored CPS-PDU header and to protect the remaining offset start field, a parity bit is provided. The eighth bit in the header is set to maintain an odd parity over the eighth bits in the header.

6.4.2 Analysis of AAL2

The following points summarize various aspects of AAL2.

- **Characteristics:** Like AAL1, AAL2 is designed to carry delay sensitive-applications like voice. However, it is different from AAL1 in many respects. First, as compared to AAL1, AAL2 is much more suited to carry low data rate applications and variable bit rate traffic, owing to the way it is structured. Second, unlike AAL1, it does not provide circuit emulation support. Hence, the clock recovery mechanism is missing in AAL2. What AAL2 does provide is timer-protected delivery using CU_timer. This timer provides an upper-bound on the packetization delay, thereby enabling AAL2 to support delay-sensitive applications.
- **Sub-layering:** The sub-layering in AAL2 is slightly different from that in AAL1. AAL2 is divided into two sub-layers—SSCS and CPS. SSCS provides enhanced functionality and is implementation-dependent. The CPS sub-layer first encapsulates the user payload in a CPS packet and then encapsulates the CPS packet in one or more CPS-PDU payloads. A CPS-PDU may even have multiple CPS packets. The first step is comparable to any header-based encapsulation scheme. The second encapsulation is necessary to delineate one CPS packet from another. Consider what happens if an ATM cell carrying multiple CPS packets is dropped. There is a possibility of mis-concatenation. In order to avoid this and to delineate CPS packet boundaries, a scheme similar to that used for ATM cell delineation, based on the CPS packet header HEC field, can be used. Since a CPS packet is of variable length, while ATM cells are of fixed length, this form of packet delineation is time-consuming and computationally very expensive. Thus, the alternative is to clearly mark the beginning of the first CPS packet in every cell so that even when previous cells are dropped, the next CPS packet is identified correctly. This is done by the offset field of the CPS-PDU packet header which is why another layer of encapsulation is necessary. The difference between AAL1 and AAL2 in this regard is that a cell for AAL2 connection can have multiple CPS packets, while for AAL1 connection, a cell can have only one SAR-PDU.
- **Header/trailer format:** In AAL2, there are two headers, the CPS packet header and the CPS-PDU header. For the CPS packet header, the 8-bit CID provides 255 simultaneous connections. For a DS1 link and a full-rate ATM connection (1.544 Mbps), almost 386 channels with 4 Kbps voice coding are possible. For a DS3 link, this number increases to 11200 (44.736/.004) possible channels. Thus, an 8-bit value is trade-off between higher packet overhead and lesser number of simultaneous channels. For a maximum packet length of 64 bytes, a 6-bit is necessary. The HEC, like in AAL1, provides error correction to the CPS packet header only.

  The CPS-PDU header is simple. The 6-bit OSF provides a pointer for the location of the next CPS packet or padding field. Since an ATM cell contains 48-bytes, a 6-bit length field is sufficient. The remaining two bits provide a preliminary form of sequencing and error protection.
- **Overhead:** On the face of it, overhead in AAL2 is more than that in AAL1. This is because in addition to the one byte CPS-PDU header, there is a 3-byte per CPS packet

overhead. However, if compared to the partial filled method of AAL1, AAL2 is much better. Also, techniques like voice compression, silence detection and suppression, and idle voice channel deletion make AAL2 much more efficient than AAL1. For more information on the efficiency of AAL2, the reader is referred to [AAL2 J. Baldwin] and [AAL2 GDC].

6.5 ATM Adaptation Layer 3/4 (AAL3/4)

The two AALs discussed earlier (i.e. AAL1 and AAL2) provide transmission of time-sensitive data. However, most data applications do not require such timing relationship between the sender and the receiver. The applications may require some upper bounds on the delay but are resilient to any variation in the delay themselves. Such data applications do not need the complexity of maintaining timing relationships as provided by AAL1 and AAL2.

In order to support such applications having time-insensitive, variable bit-rate traffic, that AAL3 and AAL4 were proposed. AAL3 provided connection-oriented services for the transfer of bursty traffic. This was made to allow certain connection-oriented services like frame relay over ATM. AAL4 provided a connectionless service for the transfer of bursty traffic. It provided options for assured data delivery along with flow control mechanisms, as well as unassured data delivery without any flow control mechanisms. The two AALs had a multitude of similarities between them. Both AAL3 and AAL4 accepted variable length frames (from variable bit-rate traffic sources) which were broken down to form the cell payload. Both of them did not require any timing relationship between the peers. In fact, both of them targeted the applications for which it did not matter whether AAL3 was used or AAL4 was used. As a result, these two AALs were combined to form a single AAL3/4. It combined several features of both the AALs.

Like AAL1, AAL3/4 is divided in two parts, viz. the Convergence Sub-layer (CS) and Segmentation and Reassembly (SAR) sub-layer. The CS itself is divided into two parts, viz. Common Part Convergence Sub-layer (CPCS) and Service Specific Convergence Sub-layer (SSCS). The structure of AAL3/4 is depicted in Figure 6.2. The SSCS may be defined to support certain added functionality required by some applications. It may also be null, in which case it just provides a mapping to the primitives of the underlying CPCS. Irrespective of the type of SSCS, the CPCS and the SAR sub-layer are always the same. Therefore, the CPCS and the SAR sub-layer are together termed as common part AAL. Here, only the SAR and the CPCS sub-layers are discussed.

6.5.1 Segmentation and Reassembly (SAR) Sub-layer

This layer in AAL3/4 breaks up the received data blocks (SAR-SDU) from the CPCS, adds its own header and trailer, and passes it to the ATM layer as blocks of 48-byte SAR-PDUs, which form the payload of a cell. The format of SAR-PDU is shown in Figure 6.12. AAL3/4 SAR-PDU has a 2-byte header, a 2-byte trailer and a 44-byte payload. This implies that AAL3/4

FIGURE 6.12

SAR-PDU Format [ITU-T I.363.3]

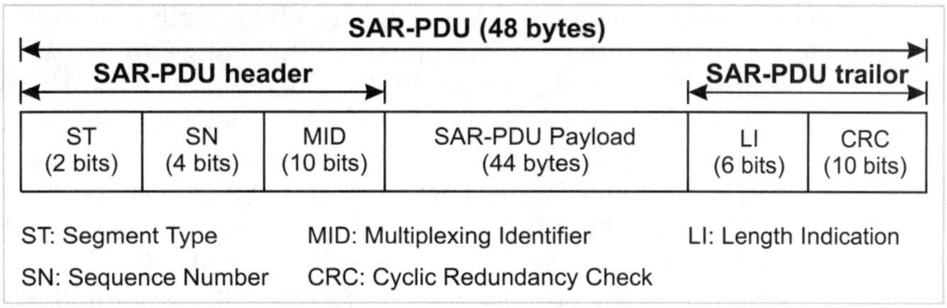

has four bytes of overhead per cell, which is the highest for any type of AAL. These overhead bits provide a number of functions not performed by any other AAL.

The functionality of SAR is as follows. The SAR-SDU received from the CPCS is broken into 44-byte blocks. These 44-byte blocks form the payload of the SAR-PDUs and are delivered to the ATM layer as 48-byte cell payload. At the receiving end, the payload part of various SAR-PDUs is assembled to form a SAR-SDU.

In order to delineate the SAR-SDU boundaries from a stream of incoming SAR-PDUs, the two-bit Segment Type (ST) field in the SAR-PDU header is used. The possible ST types and the values are given in Table 6.3. The interpretation of first three rows of the Table 6.3 is self-explanatory. As far as last row entry SSM is concerned, it is used when the SAR-SDU has length less than or equal to 44 bytes and the whole SDU is transferred in just one SAR-PDU.

TABLE 4.4

Coding of Segment Type (ST) Field [ITU-T I.363.3]

Segment type	*Encoding*	*Usage*
BOM	10	Beginning of message
COM	00	Continuation of message
EOM	01	End of message
SSM	11	Single segment message

The 4-bit Sequence Number (SN) field in the SAR PDU header is used to detect lost and mis-inserted PDUs. The SAR-PDUs generated by the SAR sub-layer are numbered modulo 16 with the sequence number of each SAR-PDU incremented by one relative to the previous SAR-PDU. At the time of reassembly, any lost or mis-inserted PDU is detected by using the Sequence Number field. Note that the sequence numbering is done for each SAR-PDU, derived from a SAR-SDU and corresponding to a particular multiplexed channel. As shall be shortly seen, AAL3/4 allows data belonging to multiple logical connections to be interleaved over the same ATM virtual connection, using the MID field (and hence, forming logically different connections at the SAR level). Thus, the SAR-PDUs are numbered sequentially for an SAR-SDU, corresponding to a particular value of MID.

All the bytes in the SAR-PDU may not contain useful data in the entire 44-byte payload field as in the case of SAR-PDUs, with segment type EOM or SSM. In order to indicate the number of bytes of useful data in SAR-PDU, the 6-bit Length Indication (LI) field in the trailer is used. The SAR-PDU with segment type set to BOM or COM has the length indication field set to 44 (because both of these SAR-PDUs always have useful data in all their 44-byte payload fields). The SAR-PDU with segment type set to SSM has valid values of length indication field as multiples of 4 (between 8 and 44). The lower limit is 8 because the smallest size of the CS PDU is 8 bytes. All the CS PDUs are 32-bit aligned, i.e. their lengths are multiples of 4. As a result, the payload field of SAR-PDU is also a multiple of 4. The SAR-PDU with segment type field set to EOM, has a length that is multiples of 4 (between 4 and 44). The lower limit is 4 once again, because the CS PDUs are 32-bit aligned. So, if there are any non-zero bytes overflowing from the last but one SAR-PDU, it will also be a multiple of 4. For SAR-PDU with segment type set to EOM, the value of 63 is used for abort indication. This provision allows the SAR sub-layer to abort the transmission of a partially transmitted SAR-SDU. Upon receiving such an SAR-PDU, the receives ignores the payload field. Moreover, the receiver also discards the partially received SAR-SDU. The SAR uses the abort functionality on a request from the CPCS.

SAR in AAL3/4 allows data belonging to multiple SAR-SDUs to be interleaved over the same ATM virtual connection, thus forming logically different connections at the SAR level. These logically different connections, which are merged into the same ATM virtual connections, are called *SAR connections*. A 10-bit Multiplexing Identifier (MID) field in the SAR-PDU header acts as an identifier for these SAR connections. The CPCS provides to the SAR sub-layer along with the SAR-SDU, the MID as well. The MID field is filled in each SAR-PDU formed from the SAR-SDU received from the CPCS. At the receiving end, the SAR-SDUs with different multiplexing identifiers are assembled separately. The receiver identifies to which of the SAR-SDU particular SAR-PDU belongs, using the MID field, and accordingly, concatenates the payload field of the SAR-PDU to the partially received SAR-SDU.

In SAR, means exist to detect and handle lost or mis-inserted SAR-PDUs, as well as errored SAR-PDUs. As mentioned earlier, the SN field is used to identify lost or mis-inserted SAR-PDUs. In order to detect errored SAR-PDUs, a 10-bit CRC field is used. The checksum is calculated over the SAR-PDU header, SAR-PDU payload and the length indication field. The SAR-SDUs received with bit errors or lost or mis-inserted SAR-PDUs, are either discarded or delivered to the CPCS sub-layer with an appropriate error indication.

6.5.2 Common Part Convergence Sub-layer (CPCS)

CPCS provides two modes of data transfer, namely message mode and streaming mode. In message mode, an entire CPCS-SDU is received from the upper layer and transferred to the SAR sub-layer. The SAR transmits this as a single SAR-SDU. In this mode, CPCS has whole data in place before it actually passed it to the lower layer.

In contrast, the streaming mode allows CPCS to start transferring data before it has received the complete CPCS-SDU from the upper layer. In such a case, CPCS transfers the partially-submitted CPCS-SDU to the SAR sub-layer. However, the whole CPCS-SDU is still transferred as a single CPCS-PDU, i.e. the CPCS header and trailer is not inserted to every partially submitted segment to the SAR sub-layer. Rather, the header is inserted before the first segment and the trailer inserted only after the last segment. Each segment is, however, transmitted as a different SAR-SDU. In streaming mode, CPCS also indicates to the SAR sub-layer whether any more segments will follow this one or not. Only the first SAR-PDU of the first segment carries the segment type as BOM. Also, the last SAR-PDU of the last segment carries the segment type as EOM. Rest of all the intermediate SAR-PDUs have their segment type field set to COM.

If the CPCS has partially transmitted a CPCS-SDU and subsequently, it detects some error, it requests its peer to abort the receiving process and discard the partially received CPCS-SDU. The mechanism to initiate the aborting process has been explained in the previous section.

CPCS inserts a header and a trailer for clear delineation of the CPCS-SDU. The format of CPCS-PDU is shown in Figure 6.13. It has a 4-byte header and 4-byte trailer, between which is the CPCS-PDU payload and a padding field. The length of the payload field is between 1 and 65535 data units.

FIGURE 6.13

CPCS-PDU Format [ITU-T I.363.3]

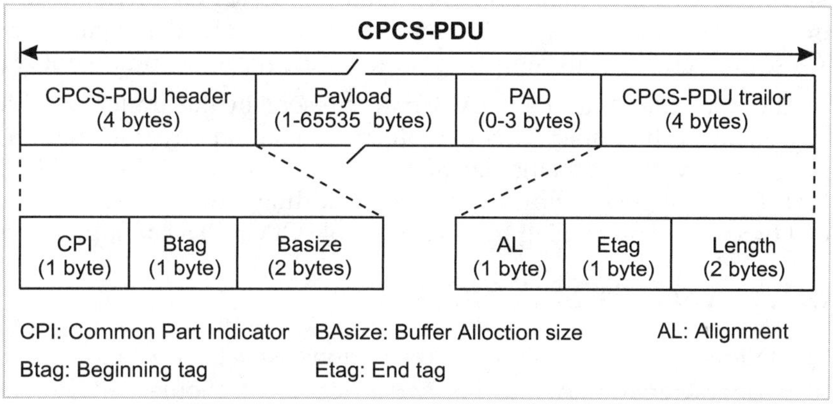

The first field is one-byte Common Part Indicator (CPI). It indicates how the field in the header and trailer of the CPCS-PDU are to be interpreted. Specifically, this field defines the counting units for the Buffer Allocation size (BAsize) field and the length field. Currently, only one possible value of CPI (0 ¥ 00) is defined. This value of CPI indicates that the counting units of BAsize and length are in bytes.

The second field in the CPCS header is a 1-byte Btag. The Btag field along with the Etag field, which is contained in the CPCS trailer, provides preliminary form of error protection. (Note that since SAR sub-layer provides extensive error handling mechanisms, the same functionality need not be duplicated at the CPCS.) The Btag and Etag fields collectively establish an association between CPCS-PDU header and trailer. The sender sends the same value in Etag and Btag for a given CPCS-PDU. At the receiver, if there is a mismatch between the Btag and the Etag field, an error is indicated.

The last field of CPCS header is a 2-byte Buffer Allocation size (BAsize) field. In message mode, this field is set to the length of the CPCS-PDU payload. In streaming mode, when the first segment of CPCS-SDU is transmitted and the CPCS-PDU header is inserted, the CPCS does not know how much more data will be sent by the upper layer. Under such circumstances, this field is used by the sender to indicate to the receiver the maximum buffering requirements for receiving a CPCS-SDU. At the receiving end, the space is allocated for the whole payload field when the first segment is received. If the receiver allocates a space less than the size of the CPCS-SDU, then at some point in time, the amount of received data will exceed the buffer size. In that case, a buffer of larger size will have to be allocated and the whole data already present in the smaller buffer will have to be copied into the larger buffer. This copying of buffers is an expensive operation. In order to avoid this, this field carries a value greater than or equal to the CPCS-PDU payload size. The exact length of the payload field is carried in the length field present in the trailer. This contains the exact length of the payload field for both the message mode and the streaming mode.

Between the CPCS-PDU payload and the trailer, a padding (PAD) field can be present. This field is used to make the total length of the CPCS-PDU, without counting header and trailer, a multiple of 4 bytes. The number of bytes added as a padding may range from 0 to 3. These padded bytes are simply ignored by the receiver and do not convey any information.

The one-byte Alignment (AL) field present in the trailer is added to achieve a 32-bit alignment of the trailer. This byte does not contain any information and is used for filling purpose only. Note that the size of header is also 4 bytes. The CPCS-PDU payload along with PAD field is also a multiple of 4 bytes. By adding the alignment field, the trailer also becomes of 4 bytes, thereby making the size of whole CPCS-PDU an integral multiple of 4 bytes.

6.5.3 Analysis of AAL3/4

The following points summarize the various aspects of AAL3/4:

- **Characteristics:** AAL3/4 provides both connectionless and connection-oriented services for the transfer of time-insensitive, variable bit-rate traffic.
- **Sub-layering:** AAL3/4 is divided into two sub-layers, SAR and CS. The primary function of the SAR sub-layer is the segmentation and reassembly of cells. In addition, SAR also provides the means to multiplex many logical channels on a single virtual channel. The CS sub-layer is divided into CPCS and SSCF. The CPCS provides two modes of data transfer, viz. message mode and streaming mode. For this, the CPCS uses a header-cum-trailer-based CPCS-PDU format.
- **Header/trailer format:** The header format in AAL3/4 is rather complicated. AAL3/4 user data is encapsulated in CPCS-PDU, which has a header and a trailer. The CPCD-PDU header provides many fields, which serve no concrete purpose. The CPI field defines only one way of interpreting other header/trailer fields, and is thus, of no use. However, in future, this field may prove to be an effective means to support larger-size CPCS-PDUs. The Btag and Etag provide preliminary form of error correction, and can be done away with. The BAsize is useful only in the streaming mode and that too, when the source can correctly specify the accurate buffer requirements.

The SAR-PDU has a two-byte header and a two-byte trailer. The header provides the means to sequence SAR-PDUs (using SN), multiplex many logical channels on a single virtual channel (using MID) and delineate the SAR-SDU boundary (using ST). A one-bit ST field can provide similar functionality as that provided by the two-bit field, albeit with loss of some functionality. The use of BOM segment-type provides additional information about the beginning of SAR-SDU, which is, otherwise, not available using a one-bit field. The 10-bit CRC is large enough to provide error correction for entire SAR-PDU.

- **Overhead:** Considering only SAR-PDU overheads, as large as 4 bytes are wasted per cell. This amounts to a total overhead of 17 per cent (9/53), which is the maximum for any AAL. If the overhead corresponding to CPCS-PDU header/trailer is added, this percentage increases further.

6.6 ATM Adaptation Layer 5 (AAL5)

Although AAL3/4 provides the means for transfer of time-insensitive data, the overhead associated with AAL3/4 (which is roughly 17%) is unacceptable to a majority of applications. Many data applications can do without error handling mechanisms provided by AAL3/4. This assumption stems from the fact that underlying links are generally reliable and that error-handling mechanisms are also present in the higher layer transport protocols.

AAL5 aims at providing an extremely simple mode of data transfer with minimal overheads. Like AAL3/4, AAL5 also provides the means for the transfer of time-insensitive data. However, it does not support most of the services supported by AAL3/4. The sequence numbers provided in the SAR-PDUs of AAL3/4 is removed in AAL5. Moreover, no multiplexing and demultiplexing are provided at the SAR sub-layer level. The segment type field used to delineate the boundaries of the AAL3/4 SAR-SDUs in a stream of SAR-PDUs is replaced by a much simpler implementation. This implementation requires just one bit, which too is placed in the cell header and not in the SAR-PDU. Thus, SAR-PDU in AAL5 contains 48 bytes of data with no byte for overhead. This is the reason why AAL5 is also sometimes called the Simple and Efficient Adaptation Layer (SEAL).

The basic structure of AAL5 is same as that of AAL3/4. Just as in AAL3/4, AAL5 is divided into CS and SAR sub-layers. Further, CS is divided into CPCS and SSCS sub-layers. Here again, only the common part of AAL (i.e. CPCS and SAR sub-layers) is discussed.

6.6.1 Segmentation and Reassembly (SAR) Sub-layer

The segmentation and reassembly sub-layer in AAL5 is very simple as compared to the other AALs. This layer does not add any header or trailer to the SAR-SDU. It just breaks down the SAR-SDU into 48-bytes SAR-PDUs, which in turn form the payload of the cells. The beginning and end of the SAR-SDU is indicated through the ATM-User-to-ATM-User indication bit, which is the rightmost bit among the 3 bits in the Payload Type Identifier (PTI) field of the cell header. A value of 1 of this bit indicates the end of a SAR-SDU. On the other hand, a value of 0 indicates the beginning or continuation of an SAR-PDU. This value is passed to the ATM layer along with the SAR-PDU.

6.6.2 Common Part Convergence Sub-layer (CPCS)

The CPCS of AAL5 is quite similar to that of AAL3/4. Just as in AAL3/4, AAL5 also provide two modes of data transfer, namely message mode and streaming mode. In order to repeat what was said about these two modes in AAL3/4, "In message mode of data transfer, an entire CPCS-SDU is received from the upper layer and only then transferred to the SAR sub-layer. On the other hand, streaming mode allows CPCS to start transferring data before it has received the complete CPCS-SDU from the upper layer. In such a case, CPCS transfers the partially submitted CPCS-SDU to the SAR sub-layer for transfer."

The operation in message mode is same as in AAL3/4. The operation in the streaming mode by SAR sub-layer is, however, somewhat different. In streaming mode, only the last SAR-PDU of the last segment has its ATM-User-to-ATM-User indication bit set to 1. All the remaining SAR-PDUs of the CPCS-SDU have the above-mentioned bit set to 0.

Figure 6.14 shows the format of CPCS-PDU for AAL5. Unlike AAL3/4 CPCS-PDU, there is no header in the AAL5 CPCS-PDU, only a 4-byte trailer. The first field in the trailer is the one-byte field CPCS User-to-User Indication (CPCS-UU) field. This field is transferred as such between CPCS end-users.

FIGURE 6.14

AAL5 CPCS-PDU Format [ITU-T I.363.5]

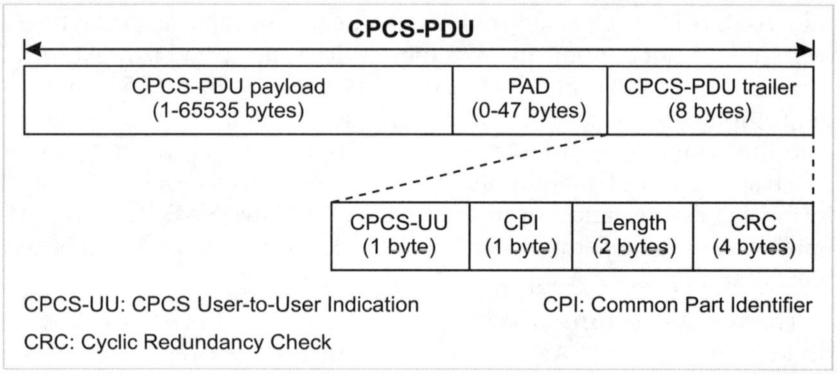

The 2-byte length field carries the number of bytes of the CPCS-PDU payload. The permissible value of the length is between 1 and 65535. A value of zero indicates an abort function. This abort function is similar to that defined for AAL3/4.

Just as in AAL3/4, there is a Common Part Indicator (CPI) field in the CPCS-PDU trailer. However, its functions are not the same as that of AAL3/4. Here, it is used only for the purpose of 64-bit alignment of the trailer. Other functions of the CPI field are intended for further study.

The 4-byte CRC field has the checksum calculated over the whole of CPCS-PDU, including the payload, pad field and the first four bytes of the CPCS-PDU trailer.

Between the CPCS-PDU payload and the trailer, a padding (PAD) field is present. This field is used to make the total length of the CPCS-PDU including the trailer an integral multiple of 48 bytes. The number of bytes added as padding may range from 0 to 47. These padded bytes are ignored by the receiver and do not convey any information. The pad field is required so that the whole CPCS-PDU may be exactly filled in the cells and so that no cell carries less than 48 bytes of information. Note that there is no field in individual cells to specify the number of useful bytes in the cell.

6.6.3 Analysis of AAL5

The following points summarize various aspects of AAL5.

- **Characteristics:** AAL5 is the simplest of AALs and provides an efficient means to transfer data. It is presumed that in the long run, most of the applications (barring certain delay-sensitive applications) will use AAL5.
- **Sub-layering:** The sub-layering in AAL5 is similar to that in AAL3/4, but the structure of SAR-PDU and CPCS differs greatly. This results in loss of functionality in AAL5. For example, multiplexing multiple logical connections in a single virtual channel is supported in AAL3/4 but is not supported in AAL5.
- **Header/trailer format:** The header format in AAL5 is extremely simple. The SAR layer does not add any header or trailer, thereby removing the overheads associated with it. The CPCS-PDU adds a trailer, which is of eight bytes. Adding a trailer instead of a header has its advantages and disadvantages. A trailer-based scheme can support streaming mode, which is not possible in a header-based scheme because the length of the frame is known only when the complete PDU has arrived. The disadvantage is that even if there is an error in the trailer, the buffer for payload has to be allocated. This is not the case in a header-based scheme, where the packet is discarded as soon as a severe error in the header is detected.

  The rationale for choosing various trailer fields and for determining their appropriate sizes is simple to understand, and hence, left unexplained. The only point to note is that since the complete CPCS-PDU is protected by CRC, a very large CRC is required.
- **Overhead:** Overhead in AAL5 is the least among all AALs. For a CPCS-PDU of maximum size (65,536 bytes), the overhead is 0.01 per cent. Even for a modest CPCS-PDU of 1500 bytes, the overhead is 0.5 per cent, which is quite nominal.

Conclusion

This chapter discussed the four types of AAL, namely AAL1, AAL2, AAL3/4 and AAL5. Different layers are required to support different types of application. Thus, at the AAL layer, the difference between application level services comes to the fore. Because of the number of AALs specified, it can be said that the integrated platform provided by ATM exists only till the ATM layer. The difference between voice and data applications gets visible at the AAL very clearly.

Apart from the AALs explained in this chapter, there is Signalling ATM Adaptation Layer (SAAL) that is used for signalling and is detailed in Chapter 11. In that chapter, the SSCS sub-layer is revisited. In Chapter 13, the need and mechanism for AAL2 signalling is explained. In Part V of this book, applications of AAL1 (e.g. circuit emulation), AAL2 (e.g. and voice over-ATM) and AAL5 (e.g. data transport) are detailed.

Review Questions

Q 1. What is the need for different AALs in ATM? On what basis are the different AALs categorized? Explain the different AALs in brief.

Q 2. Wirth respect to AAL1, write notes on the following:
- *Correction and detection modes*
- *Clock recovery*
- *Structured Data Transfer (SDT)*

Q 3. What are the main differences between AAL1 and AAL2?

Q 4. What are the drawbacks in AAL1? How does AAL2 remove these drawbacks?

Q 5. Differentiate between AAL3/4 and AAL5 with regard to data transfer. Why is AAL5 called the Simple and Efficient Adaptation Layer (SEAL)?

Q 6. Calculate the bandwidth required to carry six 32 Kbps voice channels over AAL2 with a packet fill delay of 6 ms. In general, what is the relation between packet fill delay and bandwidth requirement? In this question, what is the bandwidth required if the packet fill delay is increased to 8 ms? What would it be if the packet fill delay is decreased to 4 ms? How does silence suppression and idle channel removal change the calculations? (Note: For calculations, find the bandwidth requirements for ATM cell. Ignore any physical layer overheads).

[Ans. 283 Kbps, 265 Kbps, 318 Kpbs]

Further Reading

For details of the topics covered in this chapter, the reader is referred to ITU-T specifications for AAL. [ITU-T I.363] provides a general description, while [ITU-T I.363.1], [ITU-T I.363.2], [ITU-T I.363.3], and [ITU-T I.363.5] describe AAL1, AAL2, AAL3/4 and AAL5 respectively. Apart from this, few other books like [ATM W. Goralski] also provide good coverage of the topic. The paper [GDC AAL2] provides good comparison between AAL1 and AAL2. The reader is also referred to the 'Further Reading' section of Chapter 20.

PART 3

Core Concepts of ATM

CHAPTER 13
AAL2 Signalling

CHAPTER 14
ATM Network Management

CHAPTER 15
ATM Security

The previous part elaborated the ATM protocol reference model and explained the three layers, namely the physical layer, the ATM layer and the ATM adaptation layer. The layered model is one way to understand a technology. Another approach is based on functional capabilities that provide a viable alternative to understand the way technology works. This part looks at the core concepts of ATM in terms of its functional capabilities.

To start with, Chapter 7 of this part discusses the parameterization of traffic and services in ATM networks. This leads to two sets of parameters defined for ATM. The first set of parameters, called traffic parameters (or traffic descriptors), is presented from the user's perspective, and provides quantitative measurements of the traffic characteristic. Traffic description includes parameters like Peak Cell Rate (PCR), Sustainable Cell Rate (SCR), Maximum Burst Size (MBS), Burst Tolerance (BT) and Minimum Cell Rate (MCR). The other set of parameters, called service parameters (or service descriptors), is from the network's perspective, and includes parameters that describe the quality with which the network is carrying user traffic. Service descriptions include Cell Loss Ratio (CLR), Cell Transfer Delay (CTD), Cell Delay Variation (CDV) and Cell Error Ratio (CER) among others. The chapter also looks at the concept of service classes.

Chapter 8 describes the six key elements of traffic management. These elements are Traffic Contract Management (including connection admission control), Traffic Shaping (using leaky bucket techniques), Traffic Policing (including usage parameter control and network parameter control), Priority Control (using a bit in ATM cell header), Flow Control (including ABR flow control) and Congestion Control (including cell discard techniques and congestion notification).

Chapter 9 discusses switch design and architecture. The chapter first explains the buffering techniques (including input buffering, output buffering, and shared buffering). This chapter also discusses important switching architectures like shared-memory architecture, shared-medium architecture and space-division architecture. Thereafter, the functionality of ATM switch along with a number of ATM switch architectures including Prelude, PARIS, Knockout, Starlite and Moonshine are discussed.

In ATM, the addresses are also referred to as ATM End System Address (AESA). An AESA identifies the location of one or more ATM interfaces. AESA is based on the generic Network Service Access Point (NSAP) format. As per this format, an NSAP address is divided into two parts—the Initial Domain Part (IDP) and the Domain Specific Part (DSP). AESA format and other aspects of ATM addressing are described in detail in Chapter 10.

Chapter 11 talks about signaling protocols in ATM and covers various ATM signalling standards (like Q.2931, Q.2971, and UNI 4.0). The presentation includes both point-to-point and point-to-multipoint signalling. The various software components of the ATM signalling stack including Signalling ATM Adaptation Layer (SAAL) are also explained in this chapter.

While signalling is used to establish switched virtual connections, the purpose of routing in ATM is to route the signalling messages to the destination. In order to provide the means to route signalling messages, the ATM Forum has come up with Private Network-to-Network Interface (PNNI) protocol. ATM Routing using PNNI is described in detail in Chapter 12.

Apart from ATM signalling, another important signalling protocol in ATM is the AAL2 signalling protocol. Note that ATM signalling provides the means to establish/release ATM Virtual Channel Connection (VCC) or Virtual Path Connection (VPC). In the context of AAL2, the AAL2 signalling protocol encompasses the control plane functions to establish, clear and maintain AAL2 connections. AAL2 signalling and the means to manage AAL2 connections are described in detail in Chapter 13.

Chapter 14 focuses on ATM network management. In ATM, the focus is on fault management and performance management through the use of special cells, termed as Operations and Management (OAM) cells. Apart from OAM procedures based on OAM cells, there is the Integrated Layer Management Interface (ILMI) defined by ATM Forum which specifies how the Simple Network Management and ATM Interface MIB are used to provide status and configuration information to any ATM device. These protocols are described in detail in Chapter 14.

Chapter 7

ATM Traffic and Service Parameterization

7.1 Introduction

It is well-known fact that ATM networks provide Quality of Service (QoS). But what are the parameters that define the QoS? How does one define the service quality in ATM? Are all users of ATM networks provided with similar quality? Or are there means to differentiate between different users?

This chapter tries to address these questions; it elaborates upon the quantitative aspects of QoS. In order to do so, two sets of parameters are defined. The first set of parameters, called *traffic parameters* (or *traffic descriptors*), is from the users perspective and provides quantitative measurements of the traffic characteristic. The other set of parameters, called *service parameters* (or *service descriptors*), is from the network perspective and includes parameters that describe the quality with which the network is carrying user traffic (see Table 7.1). For an overview of traffic and service parameters, refer [Gen S. Kasera].

This chapter first defines the traffic and service descriptors of ATM. It then describes the six service categories in ATM. These *service categories* essentially relate the traffic characteristics and QoS requirements to network behaviour. In other words, the network behaves differently for different service categories; the network behaviour for different ATM service categories is briefly explained in this chapter. The detailed network behaviour, as manifested in the tools and techniques for traffic management, is discussed in next chapter.

7.2 ATM Traffic Parameters

In ATM, several traffic parameters are defined. Some of these traffic parameters are grouped together to form what is referred to as source traffic descriptor. As the name suggests, a source

TABLE 7.1

Basics of Traffic and Service Parameters [Gen S. Kasera]

Traffic Parameters	Definition	Traffic parameters specify various aspects and characteristics of traffic. They measure two important aspects of traffic, viz. the amount of data carried and the rate at which it is carried.
	Elements	• **Average Data Rate (ADR):** It is the average data rate for the duration of the data transfer. • **Maximum Allowable Data Rate (MADR):** This is the maximum data rate that a user is allowed at any instant of time. • **Burst Duration (BD):** It is the time duration for which a user is allowed to send data at the maximum allowable rate.
Service Parameters	Definition	Service parameters specify how the offered load is serviced by the network.
	Elements	• **Delay:** It is the measure of time taken by a data unit to reach one boundary from another. • **Delay variation (or jitter):** Jitter is the variation in delay over time. • **Loss ratio:** It is the ratio of total data lost to the total data transmitted. • **Error ratio:** It is the ratio of total erroneous data received to the total data transmitted. • **Throughput:** It is the traffic injected into the network, minus the lost/corrupted data. • **Goodput:** It is the ratio of useful data received and the actual amount of data injected into the network.

traffic descriptor is a set of traffic parameters belonging to the ATM source. The source traffic descriptor is used to capture the intrinsic traffic characteristics of the connection requested by a particular source. The source traffic descriptor includes the following parameters:

- **Peak Cell Rate (PCR):** This is the maximum rate at which a user is allowed to inject data into the network. Specifically, PCR defines an upper bound on the traffic that can be submitted by an ATM source. The inverse of PCR gives the 'minimum inter-arrival time' of cells for a given connection. The maximum value of PCR is bounded by the link rate.
- **Sustainable Cell Rate (SCR):** This is the measure of the long-term average of user traffic. Specifically, SCR is an upper bound on the long-term average of conforming cells for an ATM connection. An ATM source may, at times, send at a rate greater than SCR. However, the average rate of the connection must remain less than SCR.

  Once PCR and SCR are defined, it is important to ask, "For how long can a user send cells at PCR and still ensure that the long-term average is less than or equal to SCR?" Putting it differently, how much data can a user send at its peak cell rate without overshooting the average rate of the SCR? This is defined by the MBS and BT as explained below.

- **Maximum Burst Size (MBS):** This is the amount of data that an ATM source can send at its peak cell rate. Specifically, MBS is the number of cells an ATM source can send at the PCR. If MBS and PCR are known, then the maximum burst duration is determined using the relation T_{MBS} = MBS/PCR, where T_{MBS} is the duration for which a user can send data at the peak cell rate.
- **Burst Tolerance (BT):** BT is a measure of the interval between consecutive bursts during which cells are sent at PCR. In other words, BT is the time interval after which an ATM source can again send data at PCR without violating the long-term average of SCR.

The value of BT is given by the formula: $BT = (MBS - 1)/\left[\left(1/SCR\right) - \left(1/PCR\right)\right]$. This formula can be derived by considering a cell stream that arrives either at PCR or at SCR (see Figure 7.1). A cell stream arriving at PCR will have cells recurring at every 1/PCR second. Similarly, a cell stream arriving at SCR will have cells recurring at 1/SCR second. Now, by definition, BT is the time interval for which an ATM source must wait before sending cells at PCR again. Since MBS cells can be sent at PCR, the time required to send (MBS – 1) cells at PCR is (MBS – 1)/PCR, while the time required to send at SCR is (MBS – 1)/SCR. Note that (MBS – 1) cells are taken because at first, the cell for both streams arrive simultaneously and are not considered. Now, after sending MBS cells at PCR, the ATM source must wait till the long-term average again reaches SCR, before sending cells at PCR again. This waiting time is {(MBS – 1)/SCR – (MBS – 1)/PCR} which is equal to $(MBS - 1)/\left[\left(1/SCR\right) - \left(1/PCR\right)\right]$.

- **Minimum Cell Rate (MCR):** This parameter is defined for low-priority applications involving best-effort transfer. Specifically, MCR is the minimum cell rate that the network must provide to a connection. Its value can even be zero.
- **Maximum Frame Size (MFS):** This parameter specifies the maximum size of AAL PDU for the Guaranteed Frame Rate (GFR) service category. GFR is detailed in a Section 7.5.6.

FIGURE 7.1

Diagrammatical Representation of Burst Tolerance (BT)

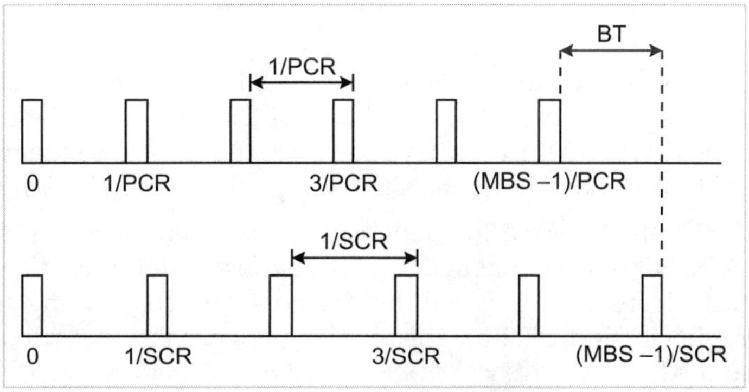

All the afore-mentioned parameters collectively define the source traffic descriptor. PCR, SCR and MCR are expressed in terms of cells per second. MBS and MFS are expressed in terms of cells and BT in terms of seconds. It will be clear in subsequent sections that different service categories specify different parameters. No category includes all the above parameters. All the parameters except BT are either configured (for PVC), or negotiated at the time of signalling (for SVC).

Besides the source traffic descriptor, another term called the connection traffic descriptor is defined by ATM traffic management specification [ATMF TM 4.1]. Connection traffic descriptor specifies the traffic characteristics of an ATM connection and includes the source traffic descriptor, the Cell Delay Variation Tolerance (CDVT) and the conformance definition. While conformance definition for various service classes is discussed in next chapter, the relevance of CDVT is explained below.

A cell stream arriving at the user network interface experiences some cell delay variation while entering the network. The reasons for this include multiplexing of several incoming cell streams or due to the variable processing delays in switches at the network ingress point. The delay variation may also result from insertion of physical layer overhead cells or management cells (OAM cells). This delay variation or randomness affects the inter-arrival time (i.e. 1/PCR) between consecutive cells of a connection as monitored at the UNI. Due to this, a cell stream injected at PCR may suffer some delay variation, resulting in the rate exceeding PCR. Cell delay variation tolerance or CDVT represents an upper-bound on the delay variation (i.e. jitter) of cells at a UNI. Thus, the measurement of cell rates (PCR/SCR) at the network ingress point is done keeping in account the CDVT for that interface.

7.2.1 An Illustrative Example

In this sub-section, three scenarios are presented that explain the relationship between some of the traffic descriptors defined above and also the circumstances under which a user cell is declared non-conforming. Figure 7.2 is referred in the process.

Figure 7.2(a) illustrates the case when a user starts sending data at low rates, steadily increasing the rate till the rate exceeds the PCR. All the cells received in excess of the peak cell rate are declared non-conforming, shown by the shaded region in Figure 7.2(a).

Figure 7.2(b) depicts another case where the user starts sending the data in bursts, increasing the rate till the rate reaches PCR, after which the rate is held steady. In this case, the user is allowed to send data at PCR only for T_{MBS} time, where T_{MBS} = MBS/PCR. After time t*, when MBS cells have been sent at PCR, subsequent cells are declared non-conforming till time t**, shown by the shaded region in Figure 7.2(b). Here, (t** – t*) is the BT.

Finally, Figure 7.2(c) depicts the case when the user sends data at a rate slightly less than PCR, but never equal to it. In this case also, after time t*, the average data rate will exceed the SCR. Here also, cells will be declared non-conforming till time t**, shown by the shaded region in Figure 7.2(b). The relation between t* and t** depends upon the values of PCR, SCR and BT.

FIGURE 7.2

Scenarios to Explain Relationship Between Traffic Descriptors

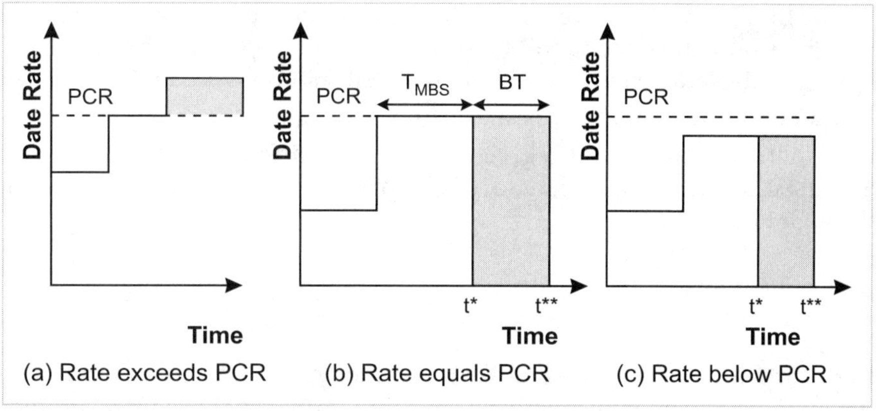

7.3 ATM Service Parameters

In ATM, the service parameters are commonly referred to as Quality of Service (QoS) parameters. ATM Forum recommendation [ATMF TM 4.1] specifies six QoS parameters. Depending upon the capability of the network and the requirements of individual connections, a network may offer one or more QoS parameters. The six QoS parameters defined are:

- Cell Loss Ratio (CLR)
- Maximum Cell Transfer Delay (maxCTD)
- Peak-to-Peak Cell Delay Variation (peak-to-peak CDV or P2P-CDV)
- Cell Error Ratio (CER)
- Cell Mis-insertion Ratio (CMR)
- Severely Errored Cell Block Ratio (SECBR)

Out of these six parameters mentioned above, the first three are negotiable, while the last three are not. Here, negotiable implies whether the value of that particular QoS parameter can be decided by a user or not. For example, the value of CLR can be specified by a user in the signalling message (for SVCs). The network can then decide whether the specified value is acceptable or not. The provision for specifying desirable values exists only for those parameters that are negotiable (that is, only for CLR, maxCTD and peak-to-peak CDV).

All six QoS parameters cover the following three broad categories of assessment:

- **Speed:** It is important for an ATM network to specify how fast it can deliver cells at the destination. CTD and CDV measure the speed of ATM networks in delivering cells. While CTD measures the transit delay, CDV measures the variation in these delays.
- **Accuracy:** After speed comes the accuracy with which cells are delivered. For this, three parameters—CER, CMR and SECBR—are specified. All these parameters specify the ratio of error in the cells delivered.
- **Dependability:** The question of measuring accuracy arises only when the cells are delivered. To measure whether a cell is actually delivered or not, CLR is specified. Thus,

CLR measures the dependability of a connection. There is a subtle difference between accuracy and dependability. While the former specifies the degree of error in a delivered cell, the latter measures the ratio of cells not delivered within a specified time.

Table 7.2 lists each of the service parameters and the formula to evaluate it. Typical values used in the network are also given in the table. Most of the parameters are barring P2P-CDV and maxCTD are self-explanatory. The interpretation of these two parameters is provided through Figure 7.3.

7.4 Factors Affecting QoS Parameters

In the previous section, various QoS parameters, as defined by the ATM Forum, were introduced. This section explains the factors that influence these QoS parameters.

- **Propagation Delay:** This represents the delay in transporting bits over the physical media. Propagation delay depends upon the distance between the source and the destination, as well the speed of the links between them. This factor affects only the CTD.
- **Media Error Statistics:** This represents the error characteristics introduced by the physical media. Errors introduced by the media may be single-bit or burst error. Preferably, fiber is the underlying physical media because it is less susceptible to external interference and because the errors in fiber seldom occur in bursts. This factor affects all parameters, barring CTD and CDV.
- **Switch Architecture:** Switch architectures have a significant impact on the performance measures of ATM connection. The next chapter discusses the correlation between QoS parameters and switch architecture. Notably, buffering and scheduling strategy are the two most important design issues that affect QoS parameters like CTD, CDV, and CLR.
- **Buffer Capacity:** Unlike other factors discussed above which require an unbounded optimization, buffer requires a bounded optimization. Having very large and very small buffer sizes may have their own disadvantages. A very small buffer size leads to frequent buffer overflows during load conditions while a very large buffer leads to high delays. The latter may not be acceptable to time/delay sensitive applications. This factor affects CTD, CDV, CLR and SECBR.
- **Traffic Load:** Traffic load on the network has a significant affect on the critical QoS parameters. A fluctuating traffic load implies variable amount of buffering from the source to the destination. A high traffic load may also cause occasional buffer overflows. As the traffic load increases, CLR, CTD and CDV value also increases.
- **Number of Intermediate Nodes:** This affects all of the QoS parameters. The number of intermediate nodes affect delays in the form of additional processing at each of the intermediate node. The loss probability due to buffer overflow at an intermediate node also increases with the increase in the number of nodes.

TABLE 7.2 QoS Parameters in ATM

QoS parameter	*Definition*	*Criteria of assessment*	*Formula*	*Approximate range*
CLR	It is the fraction of cells that are either not delivered to the destination or delivered after a pre- specified time.	DEPENDA-BILITY	((Cells lost)/(Total cells transmitted))	1 × 10 –7
maxCTD	It is the (1 - a) quantile of CTD	SPEED	See Figure 7- 3	1-1000 msec
P2P-CDV	It is the difference between (1 - a) quantile of the CTD and the fixed CTD that could be experienced by any delivered cell during the entire connection holding time.	SPEED	See Figure 7- 3	1-130 msec
CER	It is the ratio of the total number of cells delivered with error to the total number of cells delivered.	ACCURACY	((Errored cells delivered)/(Total cells delivered))	1×10^{-6}
CMR	It is the number of cells, meant for some other destination, inserted per second.	ACCURACY	((Mis-inserted cells in time T)/T)	1 cell/day
SECBR	It is the ratio of severely errored cell blocks to the total transmitted cell blocks. (A severely errored cell block outcome occurs when more than M error cells, lost cells or mis-inserted cell outcomes are observed in a received cell block of N cells.)	ACCURACY	((Severely errored cell blocks)/(Total transmitted cell blocks))	1 ×10 –4

FIGURE 7.3

Cell Transfer Delay Probability Density Model

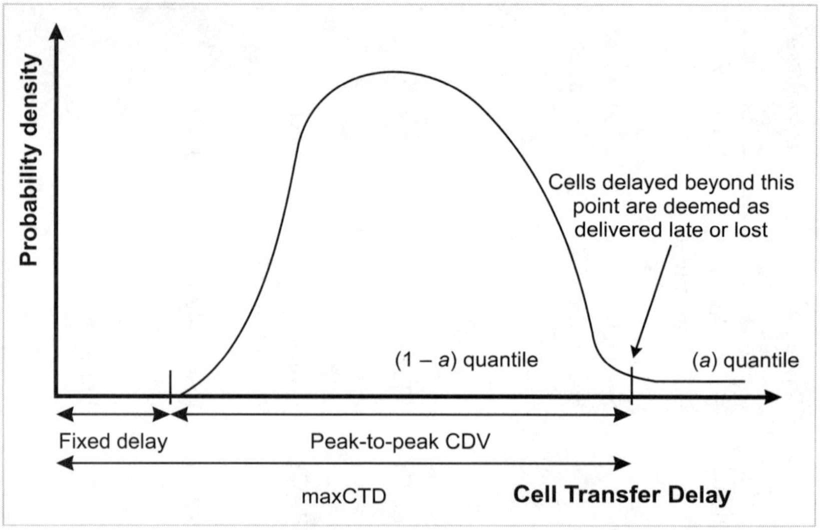

- **Resource Allocation:** This is also one of the significant factors affecting three negotiable QoS parameters. Resource allocation on a per connection basis in the form of buffers in the intermediate routes manifests itself in the form of lower cell loss during conditions of high load. Both bandwidth and buffer allocation have impact on both the cell transfer delay and cell delay variation.
- **Failures:** A failure in the form of link breakdown, switch crashing or a port failure affects cell loss. A malfunctioning switch also corrupts the outgoing cells, thereby affecting CER and SECBR.

Table 7.3 summarizes the various factors as well as the parameters on which the factors have an effect.

TABLE 7.3

Factors Affecting QoS Parameters ([ATMF TM 4.1], Copyrights 1999 the ATM forum)

Attributes\QoS Parameter	*CER*	*SECBR*	*CLR*	*CMR*	*CTD*	*CDV*
Propagation Delay					X	
Media Error Characteristics	X	X	X	X		
Switch Architecture			X		X	X
Buffer Capacity		X	X		X	X
Number of Intermediate Nodes	X	X	X	X	X	X
Traffic Load			X	X	X	X
Failures	X	X	X			
Resource Allocation			X		X	X

7.5 ATM Service Categories

Depending upon a number of attributes, ATM service offerings are classified into the following categories:

1. **CBR:** Constant Bit Rate
2. **rt-VBR:** Real-Time Variable Bit Rate
3. **nrt-VBR:** Non-Real-Time Variable Bit Rate
4. **UBR:** Unspecified Bit Rate
5. **ABR:** Available Bit Rate
6. **GFR:** Guaranteed Frame Rate

The classification is based on one or more of the following attributes:

- **Time Sensitivity:** Some of the categories require strict bounds on delay and delay variation (e.g. voice/video). This is the reason why CTD and CDV parameters are specified for these categories. CBR and rt-VBR are examples of such service categories.
- **Nature of Traffic Flows:** Some applications, like voice for examples, require constant bit rate. For such applications, a CBR service category is appropriate. In contrast, the traffic offered by data applications is generally bursty. Thus, it is preferable that instead of allocating a fixed bandwidth for such applications, bandwidth is statistically shared among a number of similar applications. Service categories that cater to such class of applications are nrt-VBR, ABR, UBR or GFR. Now, depending upon the service category, different parameters are specified. For example, SCR is specified for nrt-VBR, MCR for ABR and MFS for GFR.
- **Provisioning:** To meet the requirements of certain applications, the network may have to reserve some resources. For example, in order to handle CBR applications, the network has to provision resources in advance. Obviously, such provisioning can also get translated into higher costs. On the other hand, the network may offer leftover bandwidth at a lower price (e.g. UBR). UBR kind of service is useful for low-priority applications that can tolerate excessive delays and even some cell loss.

Table 7.4 shows the differences in each of the service categories. It lists down the characteristics of each of the category with respect to timing sensitivity, priority and also gives relevant traffic descriptors and service descriptors. A brief description of each of the service categories is provided in the following sub-sections.

7.5.1 Constant Bit Rate (CBR) Service Category

CBR service category is used by applications that require a constant bandwidth allocated to them throughout the life of the connection. This service category places strict upper bounds on cell transfer delay and cell delay variation. Applications using CBR are voice, video and circuit emulation. The bandwidth requirement of the CBR traffic is characterized by PCR only. This is because the average rate and peak cell rate for CBR connections is equal, and so,

Table 7.4

ATM Service Categories

Service Category	*Traffic Descriptors*	*Nature of Traffic*	*Negotiable Service Descriptors*	*Time-sensitive*	*Applications*
CBR	PCR, CDVT,	Stable	maxCTD, P2P-CDV, CLR	Yes	Uncompressed Video, Voice, Circuit Emulation
rt-VBR	PCR, CDVT, SCR, MBS	Bursty	maxCTD, P2P-CDV,	Yes	Compressed Video and Voice
nrt-VBR	PCR, CDVT, SCR, MBS	Bursty	CLR	No	Data-transfers, Banking transactions, multimedia e-mail
UBR	PCR, CDVT, SCR, MBS	Bursty	—	No	E-mail and FTP
ABR	PCR, CDVT, MCR	Bursty	CLR is low (not quantifiable)	No	E-mail and FTP
GFR	PCR, CDVT, MCR, MBS, MFS	Bursty	CLR is low (not quantifiable)	No	E-mail and FTP

only one of them is specified. As delay and jitter are critical, CDV and CTD parameters are specified. Loss may also be an important criterion and hence, CLR is also specified.

7.5.2 Real-time Variable Bit Rate (rt-VBR) Service Category

rt-VBR service category is intended for applications that have bursty traffic and require strict bounds over CTD and CDV. For rt-VBR traffic, the bandwidth requirements are specified by SCR and PCR. Again, since real-time applications may be sensitive to loss, delay and jitter, CLR, CTD, and CDV are specified for rt-VBR traffic. Applications using VBR are compressed video, voice and other real-time applications with bursty traffic.

7.5.3 Non Real-time Variable Bit Rate (nrt-VBR) Service Category

nrt-VBR service category is for applications having bursty traffic, but which are not sensitive to cell delay and delay variation. For the same reasons as in rt-VBR, bandwidth requirement of this service category is characterized by PCR, SCR and MBS. For non-real time applications, since delay and jitter are not important, CDV and CLR are not specified. However, CLR is specified for the nrt-VBR traffic. Typical applications of nrt-VBR are off-line transfer of video, multimedia e-mail and bank transactions.

7.5.4 Unspecified Bit Rate (UBR) Service Category

UBR service category is intended for applications having no requirement, whatsoever, with regards to cell transfer delay, cell delay variation and cell loss. This is the reason why this category does not have any negotiable parameters associated with it. A UBR connection is characterized by PCR. However, the PCR value has little significance because the network never rejects a UBR connection due to unavailability of resources. If the network is unable to carry the offered load, the load belonging to this category is dropped. In essence, this service category is intended to consume the leftover capacity of a network. Example applications of this service category include simple file transfer and e-mail.

7.5.4 Available Bit Rate (ABR) Service Category

ABR service category is used by applications that can adapt to closed-loop feedback given by the network. The feedback information contains information about the rate at which the user can transmit data. The flow control procedure for ABR service category are explained in the next chapter.

The ABR service category places no strict bounds on cell delay, cell delay variation, or cell loss. A source adjusting to data rates according to the feedback given by the network is expected to experience low cell loss and receive a fair share of available bandwidth. In other words, although this category does not have any QoS parameters associated with it, compliant users are assured of a low cell loss rate. The traffic is characterized by PCR and MCR. The applications may or may not specify MCR (that is, specifying MCR is optional). A zero MCR can also be specified. In case MCR is specified and non-zero, the network is guaranteed to provide MCR rate throughout the life of the connection. Example applications of ABR service category include data transfer and e-mail.

Like UBR, this service category tries to utilize the available bandwidth of the network. However, the important difference between the two is that while a minimum value (MCR) is specified in ABR connections that the network must always support, the network is not committed to support the PCR specified for UBR connections. Moreover, unlike ABR, there is no feedback mechanism in UBR that monitors the rate of user data.

7.5.6 Guaranteed Frame Rate (GFR) Service Category

The GFR service category, like ABR, attempts to dynamically utilize the residual bandwidth in the network. However, unlike ABR, there is no complex flow control mechanism. GFR service category is defined to carry data belonging to non-real time applications.

A GFR connection is characterized by PCR, MCR, MBS and MFS. The interpretation of these parameters and nature of commitment provided by the network is similar to that in an ABR connection. The only difference is that GFR guarantees are based upon complete AAL5

PDUs rather than individual cells. This is why an additional parameter MFS is used which specifies the maximum number of cells that can be contained in a AAL5 PDU. The MFS is also used to determine the maximum size of a buffer that an intermediate ATM switch must support.

7.6 QoS and QoS Classes

The previous sections explained how traffic and service parameters define a service category. The number of permutations and combinations that is possible does not make life any easier. Hence, it cannot be assumed that the user of ATM services has complete knowledge of such traffic and QoS parameters. (It is like asking a telephone subscriber, how much delay he or she wants during a conversation. All he or she desires is a comfortable talk.) Hence, specifying QoS parameters is not a pre-requisite in establishing an ATM connection. To make this an optional feature for not too intelligent users, few QoS classes are specified. These are mapped to a set of default QoS values. Each class broadly maps to one of the service categories. There are two categories of QoS classes:

- **Specified QoS class:** Currently, four specified classes are defined by the ATM Forum
  - **Class 1:** It specifies the QoS parameters typically required for circuit emulation and constant bit rate voice and video.
  - **Class 2:** It specifies QoS parameters typically required for variable bit rate voice and video and other multimedia traffic.
  - **Class 3:** It specifies QoS parameters for interoperation of connection-oriented protocols, such as frame Relay.
  - **Class 4:** It specifies QoS parameters for connectionless services such as LAN emulation, IP, or SMDS services.
- **Unspecified QoS class:** This class does not specify any of the QoS parameters. Hence, there is no explicit commitment by the network. The service offered by the network is just "best effort". This is the QoS class typically used by the UBR traffic.

Conclusion

This chapter explained the different ATM traffic parameters and the relation between them. The service parameters and the factors affecting them were detailed next. Thereafter, the different service categories of ATM were described. The next chapter will continue on some of the concepts introduced in this chapter and will explain the different tools for managing traffic in ATM networks.

Review Questions

Q 1. What are the different traffic parameters in ATM? Describe these parameters in brief. What is the need for so many parameters?

Q 2. *Derive the formula BT = (MBS – 1)/(1/SCR) – (1/PCR))*

Q 3. *What are the basic service parameters in any network? How do they map to the service parameters in ATM networks?*

Q 4. *Explain the difference between maxCTD and P2P–CDV with the help of the transfer delay probability density model.*

Q 5. *What is the need for so many service categories in ATM? Explain with the help of the relationship between traffic parameters, service parameters and service categories. Why do different services categories are defined by different parameters?*

Q 6. *What are the factors that affect the ATM QoS parameters? Discuss in brief the impact of these factors on various ATM QoS parameters.*

Further Reading

For details of the topics covered in this chapter, the reader is referred to [ATMF TM 4.1] and [ITU-T I.371]. The ATM Forum specification [ATMF TM 4.1] has good normative and informative text to augment an understanding of various traffic and service descriptors in ATM.

Chapter 8

ATM Traffic Management

8.1 Introduction

It is imperative for a network to ensure that the traffic flowing through it is managed properly. This is necessary to ensure that the network resources are utilized optimally and the service guarantees of the network users are met. The tools and techniques used by the network to do so come under the purview of traffic management. Typically, there can be six basic elements of traffic management—traffic contract management, traffic shaping, traffic policing, priority control, flow control and congestion control (see Table 8.1). For details of these elements of traffic management, refer [Gen S. Kasera].

Table 8.1 Basics of Traffic Management [Gen S. Kasera]

Definition	Traffic management is the act of managing network traffic, providing service guarantees to user connections and ensuring optimal utilization of network resources.
Elements	• **Traffic Contract Management:** It provides the means to allocate network resources to contending user connections and to manage network resources efficiently (i.e. deciding whether a new connection request can be accepted or not). • **Traffic Shaping:** It provides the means to alter the traffic characteristics of the connection, and deliver a more predictable and conforming traffic to the network. • **Traffic Policing:** It provides the means to ensure that a connection adheres to the parameters of its traffic contract and that non-conforming connections are penalized.

Contd.

Table 8.1

Contd.

• **Priority Control:** It refers to treating packets of unequal importance unequally (i.e. packets of higher priority are given a preferential treatment as compared to packets of lower priority). • **Flow Control:** It provides the means to synchronize the sender's sending capacity and the receiver's receiving capacity. • **Congestion Control:** It provides the means to ensure that total traffic load on the network remains below the carrying capacity of the network.

This chapter discusses the tools and techniques used in ATM for traffic management. The ATM traffic management specifications [ITU-T I.371] and [ATMF TM 4.1] mention these six elements in detail. Table 8.2 provides a bird's eye view of the six elements of ATM traffic management. In the following sections, the basic description as provided in the table is expanded, and a more detailed description provided.

8.2 ATM Traffic Contract Management

The traffic contract management procedures in ATM are very extensive. A traffic contract is defined by two sets of parameters: the connection traffic descriptor and QoS parameters. The first of these parameters includes three things: the source traffic descriptor, the CDVT and the conformance definition. While the source traffic descriptor and CDVT have been explained in the previous chapter, the conformance definition is provided in this section. In essence, the conformance definition is used to distinguish between a conforming cell and a non-conforming cell of a connection. The second set of parameters (i.e. QoS parameters) includes service descriptors like CLR, maxCTD and peak-to-peak CDV.

The contract parameters are specified either explicitly or implicitly. The parameters are explicitly specified when the network uses the values specified by the user. The values are implicitly defined if they are assigned by a network using a default rule (in absence of explicitly-defined traffic parameters contract).

During traffic contract management, the Connection Admission Control (CAC) mechanisms decide whether a call request is to be accepted or not. CAC procedures determine if sufficient resources are available in the network to support the requested call. CAC procedures also ensure that the performance of existing connections is not degraded by accepting the new request. If sufficient resources are not present or the acceptance of connections may endanger the QoS guarantees of the existing connections, the network rejects the call. Thus, CAC ensures that resources are not allocated in excess to the existing capacity. An unchecked admission may lead to a situation where the demand for the resources far exceeds the availability, thereby causing congestion. In this sense, the CAC mechanism acts as a *preventive congestion control* mechanism.

TABLE 8.2

Elements of ATM Traffic Management [ITU-T I.371] and [ATMF TM 4.1]

Name	*Mechanism*	*Description*
Traffic Contract Management	Connection Admission Control (CAC)	Set of actions taken by the network at the call establishment phase (or during call re-negotiation phase) in order to establish whether a VCC or VPC can be accepted or rejected.
Traffic Shaping	Leaky Bucket (other mechanisms also possible)	Mechanism that alters the traffic characteristics of a stream of cells on a VCC or a VPC to achieve a desired traffic characteristics, in order to ensure conformance or to achieve better network efficiency whilst meeting the QoS objectives.
Traffic Policing	Usage Parameter Control (UPC)/Network Parameter Control (NPC)	Set of actions taken by the network to monitor and ensure that the traffic contract is respected in terms of traffic offered, at the user access and the network access, respectively.
Priority Control	Selective Cell Discard	Refers to the action of selectively discarding cells with CLP bit set when buffer overflow occurs within the network.
Flow Control	Generic Flow Control	Provides a link-level flow control mechanism to control data offered at a UNI.
	ABR Flow Control	A fast resource management mechanism to utilize any leftover bandwidth in the network for connections belonging to ABR service category.
Congestion Control	Selective Cell Discard	As defined above.
	Frame Discard	A mechanism whereby instead of discarding cells, complete higher layer PDUs are discarded.
	Explicit Forward Congestion Notification	A congestion notification mechanism used to assist the network in avoidance of and recovery from a congested state.

8.2.1 Classification of CAC Algorithms

There are many variants of CAC algorithms, but all of them fall under the following two basic categories:

- **Non-statistical or peak bandwidth allocation:** If the CAC algorithm allocates resources at the peak cell rate of the connections, the algorithm is said to be non-statistical in nature. For example, consider a connection that has a peak cell rate requirement of 5 Mbps and an average cell rate requirement of 1 Mbps. Now, a non-statistical CAC algorithm will allocate resources for 5 Mbps of data, despite the fact that the data will not always arrive at the peak rate. Although such algorithms lead to gross underutilization of network resources, they are simple to implement.
- **Statistical CAC algorithm:** As is obvious from the previous example, non-statistical CAC algorithms lead to underutilization because connections do not always operate at their peak rates. Another class of algorithms, called the statistical CAC algorithm, provides better resource utilization by taking into account the statistical nature of connections. Thus, using statistical allocation, the sum of peak rates of all connections can exceed the link capacity. A parameter, called the *statistical gain*, measures the advantage of using the statistical algorithm over the non-statistical algorithm. The statistical gain is defined as the ratio of connections admitted with statistical multiplexing to that of connections admitted with peak rate allocation.

As compared to non-statistical algorithms, statistical algorithms are performance-efficient but difficult to implement. This is because statistical algorithms require accurate modelling of source traffic. Unless there are accurate models that can predict source traffic requirements, statistical algorithms will lead to degradation of throughput or sub-optimal allocation. If representative models are unavailable, peak bandwidth allocation is preferable. Even if there are models that can capture source behaviour, for proper CAC implementation, a user must be aware of such models and must provide the accurate parameters used in the modelling—such intelligence from the user cannot be assumed.

Another hurdle in statistical allocation is the computational requirements in terms of CPU usage. Arriving at a statistical solution is generally not trivial and can require solving complicated numerical problems. However, many statistical algorithms have been proposed, discussion of which is beyond the scope of the book. The reader is referred to [TM H.G. Peros] and [TM K. Shiomoto] and the references contained therein for an elaborate survey of state-of-the-art CAC algorithms.

8.2.2 Requirements of a CAC Algorithm

There are many factors that are required from a good CAC algorithm. Amongst these factors, the three important ones are:

- **Minimum processing complexity:** Ideally, the CAC algorithm should have minimum processing complexity. This is because an ATM switch operates in real-time and requires that connection establishment and release requests be served with minimum latency. Under such circumstances, a CAC algorithm that requires lot of processing time and power, may reduce the number of SVCs that can be handled by the switch. Since the number of SVCs that can be established/released by a switch is an important performance benchmark, CAC algorithms should be relatively simple to implement.

- **Model independence:** As mentioned before, accurate modelling of source traffic is a non-trivial task. Thus, it is better that the CAC algorithm be model-independent. Model independence here implies that the CAC algorithm should be applicable even if the model used to characterize the source behaviour is altered.
- **Balance between cell loss and link utilization:** While cell loss is a performance measure that concerns the user, link utilization is a factor that is important from the network's point of view. Lower link utilization and greater bandwidth allocation leads to lower cell losses and vice-versa. Thus, the CAC algorithm should strive to achieve a balance between the two orthogonal measures.

Apart from the factors mentioned above, a CAC algorithm should be *scalable*, i.e. the computation required should be independent of the number of connections.

8.2.3 CAC Algorithms for ATM Service Categories

This section discusses the type of CAC algorithms that can be applied to various ATM service categories. A CBR connection, for example, is characterized by its PCR only (i.e. it does have an SCR). In this case, the peak bandwidth allocation scheme is applicable. This is because cells are transmitted at a constant rate and there is no statistical gain to be achieved. In general, the greater is the burstiness of a connection, the more is the advantage of statistical multiplexing. Figure 8.1 presents a simple CAC algorithm based on the PCR value of CBR connections. The connection is rejected if the PCR value exceeds a certain threshold value. This threshold value can be a percentage of the link rate. Such a clause ensures that no connection occupies a significant portion of the link capacity. The connection request is also rejected if the number of CBR connections exceeds a pre-specified maximum value. The second '*if*' condition determines if sufficient bandwidth is available for the new connection. Since a switch services the requests of different service categories, the factor %_RESVD_FOR_CBR_CONN measures the part of bandwidth that is reserved specifically for CBR connections.

The peak bandwidth allocation scheme for CBR connections, as mentioned above, does not take into account the QoS parameters (i.e. CLR and CDV) of the request. There are two methods that take into account CDV, viz. negligible and non-negligible CDV methods, details of which can be found in [TM J. Roberts].

For VBR connections, the effectiveness of the CAC algorithm depends upon the burstiness of the connection, where burstiness is the ratio of SCR and PCR (i.e. burstiness = SCR/PCR). If the burstiness tends towards 1, then the peak bandwidth allocation is a suitable option. However, as the burstiness value reduces, the link utilization achieved using peak allocation also falls. In the extreme case where SCR/PCR << 1, link utilization can be very low. If instead of PCR, SCR is used to allocate resources, there is a possibility that simultaneous bursts cause severe cell loss. Thus, the preferred option is to use a value between SCR and PCR, where the exact value could be chosen by monitoring the behaviour in an actual ATM network.

FIGURE 8.1

A CAC Algorithm

```
simple_cac_algorithm
n: the number of pre-existing connections
PCRn+1: Peak Cell Rate of the incoming call
if((MAX_ALLOWED_PCR < PCRn+1) or (MAX_NO_OF_CBR_CONNECTIONS < (n+1)))
    Reject the call
if (PCRn+1 < (%_RESVD_FOR_CBR_CONN * LINK_RATE - Σ PCRi_n PCR_i ))
                                                 i=1 to N
    Accept the call
    Increment n
else
    Reject the call
```


The CAC procedure for ABR and GFR connections is simple because the only parameter of any significance is the MCR. A peak bandwidth allocation using MCR is a reasonable scheme. CAC for UBR is the simplest, because the network is not obliged to guarantee anything. However, an upper limit on the number of UBR connections may be maintained to provide some bandwidth to UBR connections.

8.2.4 Cell Conformance and Connection Compliance

Closely associated with the concept of traffic contract is the notion of conformance for a cell and compliance for a connection. Conformance of a cell carried over an ATM connection is defined in relation to the traffic parameters specified in the conformance definition of the traffic contract. Each cell is either declared conforming or non-conforming, depending upon the adherence of that cell to the conformance definition. For each service category, a conformance definition is explicitly defined in the specifications [ITU-T I.371] and [ATMF TM 4.1].

User traffic may occasionally have a cell or two that are non-conforming. In such a case, the network is not suppose to give up its commitment of providing QoS guarantees for the connection. For such connections that have occasional non-conforming cells, a term 'connection compliance' is defined. A connection is compliant if it has not 'too many' non-conforming cells. The exact definition of 'too many' is network-specific. For a compliant connection, the network provides QoS guarantee for at least the number of cells that are conforming. For a non-compliant connection, the network is not obliged to provide QoS guarantees to any cell on that connection.

In order to define conformance, two distinct categories of cell flows are defined. These categories are as follows:

- **CLP transparent (CLP = 0+1):** This category refers to the cell flows in which the CLP bit is ignored by the network. For this category, cells with CLP = 1 and cells with CLP = 0 are given identical treatment. Hence, the tagging option (i.e. setting the CLP bit from 0 to 1) is not applicable to CLP transparent cell flows.
- **CLP significant (CLP = 0):** This category refers to the cell flows in which the CLP bit is *not* ignored by the network. For this category, CLR objectives apply to cells with CLP = 0. That is, even if cells carrying CLP = 1 are dropped by the network, this is not counted against the performance guarantees provided by the network. Tagging is allowed but is an optional feature.

Now that the two distinct cell flows are defined, the conformance definition of various service guarantees is explained. For each service category, depending upon the type of cell flow, one or more conformance definition is provided. For example, three definitions exist of VBR category—viz. VBR.1, VBR.2 and VBR.3.

Table 8.3 lists the conformance definition of all the service categories, viz. CBR, rt-VBR, nrt-VBR, ABR, GFR and UBR. For certain service categories (e.g. VBR), there is more than one conformance definition to cater to different possibilities. The second column of the table shows the names of conformance definitions associated with different service categories, while the third column lists the traffic descriptors required for conformance definition. A subscript of 0 + 1 implies that the particular parameter is CLP-transparent. Similarly, the subscript of 0 implies that the parameter is CLP significant. As an example, consider the CBR service category, which is CLP transparent. The concept of CLP significance and tagging is used to allow the connection to use the available bandwidth for rates between PCR and SCR (or between PCR and MCR for the ABR category). Since, for the CBR connection, only PCR is specified (SCR or MCR is not specified) the tagging option is not applicable and PCR is referred to as PCR0+1. The implication of the fourth column will be clear when the GCRA algorithm is discussed. In the fifth column, the tagging option specifies whether non-conforming cells should be tagged or discarded. In case the tagging option is applicable, it is the user's prerogative to specify whether tagging is wanted or not. Thus, even if the tagging option is applicable, its use for certain categories is not mandatory. In the last column, the CLR objective specifies whether CLR is computed over the aggregate CLP = 0 + 1 flow or CLP = 0 flow.

8.2.5 ATM Service Level Agreement (SLA)

Before moving over to the next topic, one important concept related to traffic contract needs some elaboration. This concept, referred to as Service Level Agreement (SLA), is rather generic and applies to anyone who provides a service. In simple terms, an SLA can be viewed as a set of parameters that define the quantitative benchmarks for the service provided. For example, a very common SLA is 99.999 per cent availability (also called five 9's) of a network service.

Table 8.3 Conformance Definition for Various Service Categories

Service Category	*Definition Name*	*Traffic Descriptor*	*Conformance Definition Option*	*Tagging Objective*	*CLR*
CBR	CBR.1 CDVT	PCR_{0+1},	GCRA(1/PCR_{0+1}, CDVT)	n/a (Note 1)	CLR_{0+1}
rt-VBR/ nrt-VBR	VBR.1	PCR_{0+1}, SCR_{0+1}, CDVT	GCRA(1/PCR_{0+1}, CDVT) GCRA(1/SCR_{0+1}, BT_{0+1} + CDVT)	n/a (Note 1)	CLR_{0+1}
VBR.2	PCR_{0+1}, SCR_0, CDVT		GCRA(1/PCR_{0+1}, CDVT) GCRA(1/SCR_0, BT_0 + CDVT)	n/a (Note 1)	CLR_0
VBR.3	PCR_{0+1}, SCR_0, CDVT		GCRA(1/PCR_{0+1}, CDVT) GCRA(1/SCR_0, BT_0 + CDVT)	Applicable	CLR_0
ABR	ABR.1 MCR_0	PCR_0	See [ATMF TM 4.1]	n/a (Note 1)	CLR_0
GFR	GFR.1	PCR_{0+1}, CDVT MCR_0	1. GCRA(1/PCR_{0+1}, CDVT) 2. Cells in a frame have same CLP value 3. MFS not exceeded	n/a (Note 1)	CLR_0
	GFR.2	PCR_{0+1}, CDVT MCR_0	1. GCRA(1/PCR_{0+1}, CDVT) 2. Cells in a frame have same CLP value 3. MFS not exceeded	Applicable	CLR_0
UBR	UBR.1	PCR_{0+1}, CDVT	Network specific	n/a (Note 1)	n/a
	UBR.2	PCR_{0+1}, CDVT	Network specific	Applicable	n/a

n/a: not applicable.

Note 1: In this case, the non-conforming cells are discarded.

In the context of ATM, an SLA specifies the reliability and performance of an ATM VP or VC provided by a service provider. The important parameters of any ATM SLA are the latency of the network, the delay variation and the loss ratio. Apart from this, there are certain common elements of an SLA like availability of the network, complaint response time, and time to repair a problem. Table 8.4 shows the typical SLA attributes provided by different service providers. A range in the table refers to the quotes made by different service providers.

TABLE 8.4

Typical SLA Attributes for ATM

Attribute	*Value/Range*
Access Speed	OC1, OC3, DS1, DS3
Service Category	CBR, rt-VBR, nrt-VBR
Traffic Policing	Cell Drop
Latency	25 to 50 ms
Delay Variation (Jitter)	0.6 to 2 ms
Loss	0.01% to 1.0%
Availability	99.999%
Complaint Response Time	15 minutes to few hours
Mean Time to Repair	2 to 4 hours

8.3 ATM Traffic Shaping

Traffic shaping is typically implemented using the leaky bucket algorithm. In this scheme, bursty data generated by a source is stored in the buffer and sent out at a lower rate, thereby transforming bursty traffic with a high PCR to a more stable traffic with lower PCR. This mechanism is referred to as peak cell rate reduction. Traffic shaping makes the traffic characteristics more predictable and hence, the network can adequately satisfy the requirement of the resources of such a source.

Figure 8.2 shows how traffic shaping reduces the PCR. Since traffic shaping delays cell transmission, it is generally not suitable for real-time applications, which are sensitive to delays and delay variations.

8.4 ATM Traffic Policing: Usage Parameter Control

In ATM, the mechanism to monitor traffic and enforce the traffic contract is termed as Usage Parameter Control (UPC). Depending upon whether traffic monitoring is performed at UNI or NNI, the monitoring functions are referred to as Usage Parameter Control (UPC) or Network Parameter Control (NPC) respectively. However, UPC is considered to be a generic concept and includes NPC as well.

FIGURE 8.2

Peak Cell Rate Reduction

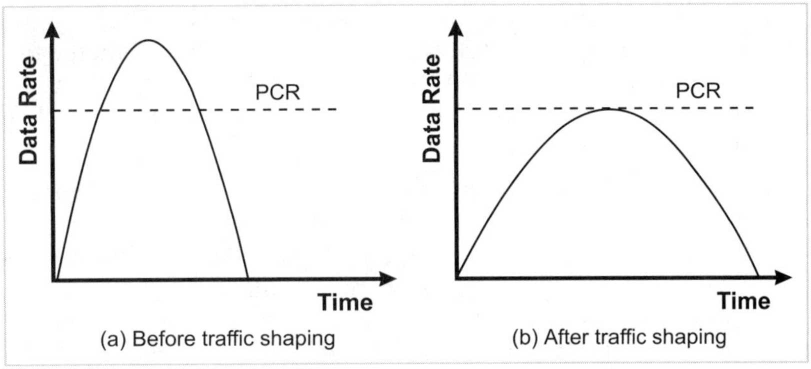

There is a minor difference in opinion between the ITU-T and the ATM Forum over the use of UPC/NPC functions. While the ATM Forum makes both UPC as well as NPC a network-dependent option, ITU-T recommends the use of UPC while leaving NPC as a network option.

UPC applies to both user connections as well as signalling channels. This implies that along with controlling the amount of user cells, the UPC function also monitors the amount of signalling information entering the network. Methods for monitoring meta-signalling and OAM flows are for further study.

Several algorithms are defined to monitor traffic and to enforce traffic contract. The most commonly used algorithm is the Generic Cell Rate Algorithm or GCRA. GCRA can also be viewed as a 'Leaky bucket' algorithm. The next section elaborates upon the GCRA algorithm.

8.4.1 Generic Cell Rate Algorithm (GCRA)

The Generic Cell Rate Algorithm or GCRA is used to define conformance of the cells with respect to the negotiated traffic contract. GCRA may be used by the UPC function to enforce conformance. However, the network is not obligated to use this function and may use any other algorithm. The algorithm checks every cell for its conformance. The GCRA is used to define conformance with respect to PCR and CDVT, as well as SCR and BT.

GCRA is defined in terms of two parameters: The increment (I) and limit (L), and is denoted as GCRA(I,L). The increment value specifies when the next cell is expected, while the limit specifies the permissible deviation from the expected arrival time. For example, for CBR applications, I is equal to 1/PCR and L equal to CDVT. This form of GCRA is represented as GCRA(1/PCR, CDVT). Similarly, for VBR applications, GCRA is defined in terms of SCR, CDVT and BT.

GCRA checks conformance to the traffic contract, using the twin concepts of expected and actual cell arrival times. For an incoming cell, the expected arrival time is referred to as the Theoretical Arrival Time (TAT). The conformance of a cell is decided on the basis of its TAT. The cell is said to be conforming if it arrives at time (TAT-L) or after. All cells arriving before time (TAT-L) are marked as non-conforming cells.

Figure 8.3 shows two equivalent versions of the GCRA algorithm. In Figure 8.3(a), the Virtual Scheduling Algorithm is depicted. In this algorithm, the first incoming cell initializes the algorithm and sets the TAT of the first cell to ta(1). For the kth cell, the conformance is decided on the basis of TATk, according to the following rules (here, TATk is the theoretical arrival time of the kth cell):

1. If the kth cell arrives after time TATk, the cell is declared conforming and the TAT for the next cell (i.e. TATk + 1) is set to (ta(k) + I) and k is incremented.
2. If the cell arrives between time TATk and TATk - L, the cell is still declared conforming. However, the TAT of the next cell is still set to (TATk + I) and not to (ta(k) + I). This prevents a smart user from obtaining a data rate of 1/(I – L), instead of 1/(I). Moreover, the value of k is incremented.

3. If the cell arrives before (TATk – L), the cell is declared non-conforming and the TAT for next cell is again set to TATk. (Note that a cell will not be declared non-conforming if it

FIGURE 8.3

Equivalent Versions of the Generic Cell Rate Algorithm

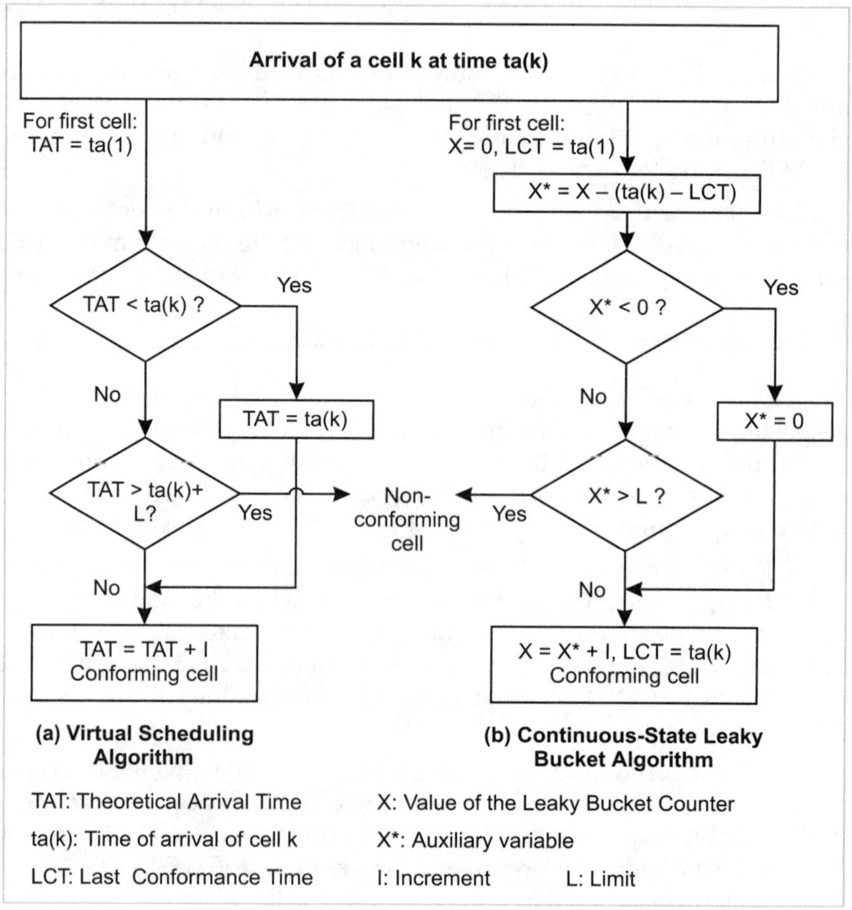

arrives after its TAT. Thus, even if the TAT of a cell is set to a value much less than the actual arrival time, conformity of the cell towards traffic contract will remain unaffected).

The three steps mentioned above ensure that cells transmitted above the allotted rate are declared as non-conforming. Consider what happens if a malicious user starts sending cells at a rate slightly greater than the allotted rate (PCR); i.e. at a spacing slightly less than I, say (I – δ), where δ < L. The first arriving cell initializes the algorithm and sets TAT2 to I. Let the expected arrival times of subsequent cells be as shown in upper part of Figure 8.4 (expected behaviour). The second cell arrives at instant I-d and is still conforming. The TAT3 is set to TAT2 + I by rule 2. The third cell now arrives at 2(I – δ). This cell is conforming if 2*δ < L. However, going by the figure, the third cell is also conforming. It is the fourth cell which is declared as non-conforming because (3*δ) > L. In a general case, a user can only send n cells at a rate greater than allotted rate, where n is the largest number such that (n*δ) < L. For example, if δ = 0.4 L, the user is able to send only two cells, before the third one is declared as non-conforming.

FIGURE 8.4

The Cell Arrival Process

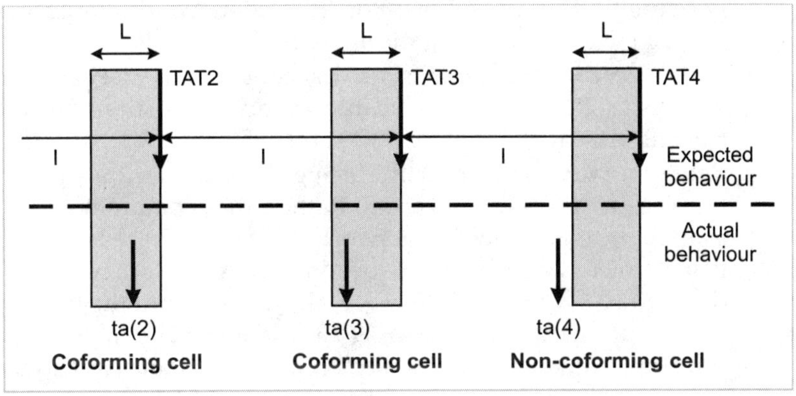

Figure 8.3(b) shows the *continuous-state leaky bucket* algorithm, an equivalent algorithm to the one explained above. The continuous-state leaky bucket algorithm can be considered as a bucket with a capacity of (I + L), where each cell pours in I data-units of fluid into the bucket and where the bucket is leaking at a rate of 1 data-unit/time-unit. The capacity of the leaky bucket at any instant of time is maintained by the variable X. An incoming cell is declared conforming if, at the time of its arrival, the fluid in the bucket is less than or equal to L. Else, the cell is declared non-conforming.

The algorithm maintains a Last Conformance Time (LCT) which stores the last time a conforming cell was accepted. On the arrival of a cell, the amount of fluid drained since LCT is subtracted from the bucket capacity. This is depicted in the figure by the expression: X* = X – (ta(k) – LCT). In this expression, (ta(k) – LCT) gives the amount of liquid drained since the last conforming cell arrived (note that in a unit time, 1 data unit is drained, so, in (ta(k) – LCT) time, equal amount of data units are drained). Thus, the right-hand side of the expression gives the capacity of the leaky bucket at time ta(k).

Once the capacity of the leaky bucket is determined, the conformity of a cell depends upon whether there are I data units available in the bucket. This check is performed by the two decision boxes shown in the figure. If the amount of fluid is less than L, the cell is declared conforming and LCT is set to the new cell arrival time and the amount of fluid is increased by I. Else, the cell is declared non-conforming and the value of variables X and LCT are left unchanged.

8.4.2 Traffic Policing of ATM Service Categories

GCRA is a generic method used to monitor and enforce the traffic contract. The exact mechanism depends upon the service category being policed. The following sub-section elaborates upon how the traffic of various service categories is policed to ensure conformance to the traffic contract.

- **Policing CBR connections:** For CBR connections, conformance is ensured using the GCRA(1/PCR, CDVT) algorithm.
- **Policing VBR connections:** For VBR connections, a cell needs to conform to both PCR as well as SCR. This is why conformance is ensured by using a dual leaky bucket algorithm GCRA(1/PCR, CDVT) and GCRA(1/SCR, CDVT + BT).
- **Policing ABR connections:** Policing ABR connections is slightly difficult because the traffic descriptors are not fixed. Rather, these parameters are adjusted according to the available capacity in the network. Consider a scenario where a backward resource management cell is sent, requesting an ATM source to reduce its data rate. In such cases, the GCRA algorithm must take into account the time the ATM source will take to modify its data rate. Owing to the complexity involved in accurately measuring conformance, the precise definition of a compliant ABR connection is network-specific. The ATM Forum recommendation [ATMF TM 4.1] specifies a dynamic-GCRA (D-GCRA) algorithm to ensure conformance to an ABR connection. This algorithm measures conformance against the explicit rate feedback provided to the ATM source. An important difference between GCRA and D-GCRA is that the increment value in the latter changes with time, which is not the case in the former.
- **Policing UBR connections:** Policing UBR connection is not a critical issue in ATM networks, because no service guarantees are provided to this type of connection. However, UBR connections can be policed using the GCRA(1/PCR, CDVT) algorithm, the decision of which is network-specific.
- **Policing GFR connections:** GFR connections are policed using mechanisms similar to that applied for CBR connections (i.e. using the GCRA(1/PCR, CDVT) algorithm). In addition to this, the policing function may also discard cells that are in excess of the Maximum Frame Size (MFS).

8.4.3 Policing Actions

Once it is determined through the GCRA algorithms that a particular cell is non-conforming, appropriate actions are taken against the offending cell. The UPC or the policing function has the liberty to decide upon the particular course of action. Besides other possibilities, the following action may be taken against the non-conforming cells:

- **Cell passing:** The simplest action is to simply allow the cell to pass through. In case the network is lightly loaded, this may not be a very bad option, because it will result in better utilization of network resource. However, during overload conditions, this action is not recommended.
- **Cell tagging:** This refers to the marking of the CLP bit in the ATM header as 1. Cell tagging applies to CLP = 0 cells only.
- **Cell discarding:** This refers to the process of dumping non-conforming cells.

The UPC function may choose to discard any non-conforming cell. It may even decide to tag these cells and then allow them to enter the network. The tagged cells are treated differently from rest of the cells. During congestion, tagged cells are the first ones to get dropped. The network may also choose to implement a mix-and-match of the two policies, i.e. cell discard and cell tagging. If the connection is non-conforming within a certain threshold, the network may choose call tagging. However, if this exceeds certain limits, the network may start discarding the cells at the boundary of the network right away.

8.5 ATM Priority Control

Although attempts have been made to incorporate priority control features in ATM network, not much headway has been made in this direction. Knowing well the advantages of priority control in any network, the critics are quick to point this lacuna in ATM. Incorporating priority in ATM is rather difficult. Complex priority control algorithms will further overload ATM layer functionality. Processing every cell according to its priority will mean additional delays, which might not be acceptable to network users.

Currently, selective cell discard is the only priority control function specified in ATM. This function is based on the CLP bit of the ATM header. The cells with CLP bit set to 0 are considered as high-priority cells, while the cells with CLP bit set to 1 are considered as low priority cells. If the network is forced to drop some cells, the network selectively drops the cells with lower priority.

The cells can be either marked by the user or by the UPC function at the network ingress point. In the first case, the user classifies cells into two priority levels and sends high-priority cells with CLP bit 0 and low-priority cells with CLP bit 1. If there is congestion along the destined path, the network first discards the cells with CLP bit 1. This scheme ensures that high-priority cells are given preferential treatment. As an example, the CLP bit can provide two priority levels for applications like the layered encoder. In a layered video encoder two types of data streams are produced, an *essential* layer and an *enhancement* layer. The essential layer is sufficient to produce an average quality picture, while the enhancement layer improves the picture quality. The layered encoder is an example where cells belonging to the same application have unequal priority. For such applications, during overload conditions, the cells of the enhancement layer can be selectively discarded.

8.6 ATM Flow Control

Flow control in ATM networks has a slightly different role to play as compared to conventional data networks, for several reasons. First, ATM is designed to be a high-speed technology. Thus, traditional hop-by-hop flow controls for error checking (as used in X.25) cannot be used in ATM. Moreover, interactive multimedia applications, which communicate in real-time, preclude the use of flow control because a delayed cell is as good as a lost cell.

Currently, flow control in ATM is used only to optimize bandwidth utilization by having feedback mechanisms. Two such flow control mechanisms are specified. One of the mechanisms is Generic Flow Control that provides a link-level flow control at the UNI. The other mechanism is ABR flow control which is used to utilize any leftover bandwidth in the network for connections belonging to the ABR service category.

8.6.1 Generic Flow Control (GFC)

Generic Flow Control is a link-level flow control scheme defined in ITU-T recommendation [ITU-T I.361] to control the amount of data flowing at the UNI. This mechanism is applicable only to data flowing in the forward direction (i.e. user to network direction), not in the backward direction. The assumption here being that if the data entering the network is controlled from both sides of a connection, a uni-directional flow control at one side is sufficient.

Recall that in the UNI cell header, there are four bits for Generic Flow Control (GFC bits). The 4 bits in the cell header were defined keeping in mind the flow control requirement across UNI and access control within the Customer Premises Network (CPN). However, as per the current standards, GFC flow control mechanism is used to control flow of traffic from various ATM connections and to limit the transmission capacity at the UNI interface.

The scope of GFC depends upon the configuration of the end-system. An end-system can be either configured as uncontrolled equipment or configured as controlled equipment. GFC procedures are not applicable for uncontrolled equipment. Connections from a controlled equipment fall into two different categories, which are:

- **Controlled connections:** This refers to connections whose cell emission rate is controlled by GFC signals. Typically, no resource is reserved in advance for controlled connections. Hence, connections belonging to ABR, UBR, GFR and nrt-VBR service categories are examples of connections that can be controlled.

- **Uncontrolled connections:** This refers to connections whose cell emission rate is not controlled by GFC signals (except for the HALT signal, explanation of which follows). Typically, resource is reserved in advance for uncontrolled connections. Connections belonging to CBR and rt-VBR service categories are generally uncontrolled.

It follows from the above that GFC mechanisms apply only to cells sent by a controlled terminal on a controlled ATM connection (with an exception). All other cells flows are sent with the GFC field set to 0000. Table 8.5 gives the possible values of the GFC bits in cells moving in user-to-network direction.

TABLE 8.5

Interpretation of GFC Bits in User-to-Network Direction [ITU-T I.361]

Bit values	*Interpretation*
0 0 0 0	Terminal is uncontrolled. Cell is assigned or on an uncontrolled ATM connection
0 0 0 1	Terminal is controlled. Cell is unassigned or on an uncontrolled ATM connection.
0 1 0 1	Terminal is controlled. Cell on a controlled ATM connection Group A.
0 0 1 1	Terminal is controlled. Cell on a controlled ATM connection Group B.
All other values are ignored.	

As shown in the table, controlled connection is divided into two categories: Class A and Class B. This categorization depends upon the output queue for which the cells of that connection are meant. Therefore, cells arriving at the UNI can be queued into one of the three queues: One for uncontrolled traffic and two for controlled traffic (corresponding to Class A and Class B). Note that the presence of two queues is optional, because by default, only one queue is assumed. The GFC procedures for the default or 1-q model (i.e. only a single controlled queue is maintained) are elaborated below. For the optional two-queue or 2-q model, the reader is referred to [ITU-T I.361].

The GFC procedure assumes a master-slave relationship between the network and the user. Hence, the network element at the UNI is referred to as controlling equipment, while the end-system is referred to as the controlled equipment.

Figure 8.5 shows interpretation of GFC bits in user-to-network direction (i.e. controlling equipment to controlled equipment). As seen from the figure, there are two important GFC signals: HALT and SET. The explanation of the various GFC signals are as follows:

- **HALT:** This signal is used to stop transmission of assigned ATM layer cells for both controlled and uncontrolled ATM connections. This signal is sent to limit the amount of cells generated by an ATM end-system at a UNI. For example, if HALT and NO_HALT signals are sent alternatively, the transmission capacity of the link gets reduced by half. Therefore, by appropriately mixing HALT and NO_HALT signals, the transmission capacity of the physical link can be controlled.

FIGURE 8.5

Interpretation of GFC Bits in Network to User Direction

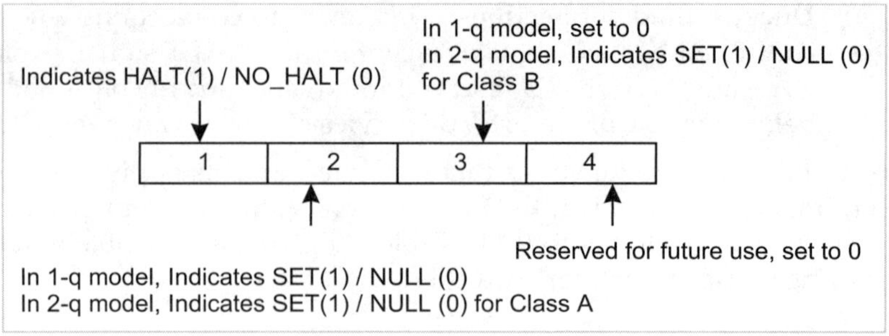

- **SET:** The controlling device issues SET signal to set the credit counter to a specified integer value. This credit counter specifies the maximum data the controlled device can send until the next SET signal is received.
- **NULL:** The controlling device may at times send NULL signal, indicating that no action has to be taken.

At the start-up, a controlled device behaves as an uncontrolled one. It receives a HALT or a SET signal from the controlling device and thereafter, performs controlled GFC procedures. After either of these signals is received, the controlled equipment remains in the controlled mode for a given period of time (implemented by a timer whose default value is 5 seconds). After the timer expires, the controlled equipment reverts back to uncontrolled mode.

8.6.2 ABR Flow Control

For ABR traffic, the source controls its data injection rates according to the loading conditions of the network. At any given time, the current load of the network is conveyed to the source through Resource Management (RM) cells. These RM cells carry various pieces of information about the current state of the network. The information includes, among other things, the bandwidth currently available, the state of congestion and the impending congestion.

Although the ABR service does not provide any QoS guarantee, it ensures an efficient means to access the bandwidth available in the network. In order to achieve this goal, a flow control feedback loop is used. In the early stages of standardization, two flow control approaches were considered viz. credit-based flow control and rate-based flow control. Following is an explanation of these flow control approaches.

- **Credit-based flow control:** The credit-based flow control was proposed to be a hop-by-hop feedback flow control mechanism. In this mechanism, the source periodically receives credit from the next hop, using which it sends data. The credit specifies the amount of data the source can send. For every cell sent, the sender decrements the credit available. On receiving credits, the source sends data at any rate, provided it has the sufficient credits to do so. Once the credit count reaches zero, the sender is prohibited to send any more data. Each node in the path maintains a separate queue for each VC. A receiver monitors the individual VC queue lengths, the buffers available, and the total amount of credit still not consumed by the network. The unconsumed credit includes the cells received but not yet processed, the cells in transit and the credit still left with the sender. The credit generated by the receiver depends upon the resources available. The sender also sends control cells periodically, specifying the amount of credit still left. This helps in synchronizing the sender and the receiver, and helps in accounting for any lost cells.

- **Rate-based flow control:** The rate-based congestion control in ATM was proposed to be an end-to-end feedback flow control mechanism. There is one feedback loop from the source to the destination. The destination sends back control information towards the source. Any intermediate node can also change the control information and forward it towards the source. The source adjusts its rate according to the control information received.

Both credit-based and rate-based schemes have their own advantages and disadvantages. The credit-based flow control scheme is desirable because of its greater responsiveness to transient changes in the network. This is due to the fact that in case there is a temporary overload condition, all a node has to do is to stop sending credits to the previous hop. This flexibility is not available in rate-based flow control, because rate-based is an end-to-end flow control mechanism and there is a significant delay before the control information reaches the source.

However, credit-based flow control has a serious drawback in that it requires per-link per-VC monitoring. This form of monitoring complicates switch design and hampers scalability. In this regard, rate-based flow control is simple, scalable and does not require state information of a number of VCs to be stored in intermediate switches. After carefully weighing the pros and cons of both the schemes, the ATM Forum has adopted the rate-based flow control approach.

It follows from above that ABR flow control is rate-based end-to-end flow control mechanism. In this mechanism, the source generates a Resource Management (RM) cell (called the forward RM cell), which is looped back by the destination and sent back to the source (looped back cell is called the backward RM cell). This end-to-end flow control model is depicted in Figure 8.6.

FIGURE 8.6

End-to-end ABR Flow Control Model

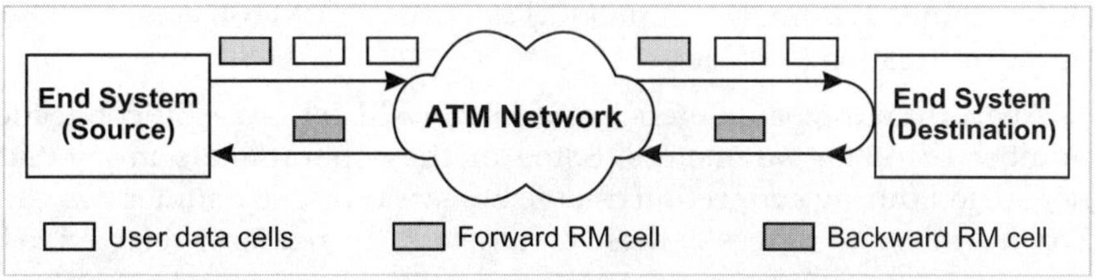

An intermediate node has three options with regard to conveying feedback information to the source. As the simplest of options, a node can directly insert information in the forward or backward RM cells. Alternatively, a node can set the congestion bit of the ATM cell header to convey congestion notification. The destination in this case then updates the backward RM cell according to the received information. Lastly, a node can directly generate a backward RM cell. In this case, the communication no more remains end-to-end. This option is advantageous, in the sense that it alleviates the latency involved in end-to-end feedback.

For ABR flow control, RM cells play a very critical role because they are used to carry feedback information. RM cells are defined for the management of individual VCs as well as VPs. For a VP, RM cells are sent over VCI = 6 of that VP with PTI bits in the ATM header set to 110. For a VC, RM cells are sent over the same VPI/VCI value as that of the VC connection with PTI bits set to 110. Note that even though the PTI value is same in both the cases, a VP RM cell is distinguished from a VC RM cell through the VCI.

Although detailed description of the RM cell format is beyond the scope of this book, explanation of some of the fields that form a part of the RM cell is provided below. The information an RM cell carries includes the Minimum Cell Rate (MCR) acceptable to that connection and the Current Cell Rate (CCR) of the source. The RM cell also has a bit, No Increase (NI) bit, which specifies whether the source can increment its rate or not. If this bit is 1, the source cannot increase its data rate. This bit may be set by any intermediate node to indicate an impending congestion. If this bit is 0, the source can increase the rate by the Rate Increase Factor (RIF) negotiated during the connection set-up.

The RM cell may arrive at the source with a Congestion Indication (CI) bit set. This indicates a possible congestion somewhere between the source and the destination. If an RM cell arrives with CI bit 1, the source decreases its data rate by the Rate Decrease Factor (RDF), a factor also negotiated at the time of connection establishment. Note that although the NI bit and CI can both be set during congestion, they serve different purposes.

RM cells also have a field that specifies the Explicit Rate (ER) at which the source is meant to transfer data. This cell rate is filled in by the destination, representing the rate at which it can receive data. On the way back to the source, this field can be decreased or left unchanged by the intermediate nodes (but never increased).

A single-bit DIR field indicates whether the RM cell is travelling from source to destination or is looped back. RM cells arriving at the source may either be the ones that looped back after reaching the destination or they might have been generated by an intermediate switch to notify the source about the congestion. A single-bit field Backward Notification (BN), differentiates a source-generated RM cell from an RM cell generated by an intermediate node or the destination.

Apart from the parameters carried in an RM cell, an ABR connection is specified using a number of other parameters. Some of these parameters are constant, while others are negotiated during connection establishment. Table 8.6 enumerates various ABR parameters. For the units and ranges of these parameters, the reader is referred to [ATMF TM 4.1].

8.7 ATM Congestion Control

In ATM, both reactive and preventive mechanisms exist to tackle the problem of congestion. CAC is an example of preventive mechanism, where resources are allocated in such a manner

TABLE 8.6

ABR Parameter Description (Copyrights 1999 the ATM Forum)

Label	*Name*	*Description*
PCR	Peak Cell Rate	Cell rate that the source may never exceed.
MCR	Minimum Cell Rate	Rate at which source is always allowed to send.
ICR	Initial Cell Rate	Rate at which a source should send initially and after an idle period of time.
RIF	Rate Increase Factor	Controls the amount by which the cell transmission rate may increase upon receipt of a RM cell.
Nrm	number of cells/RM cell	Maximum number of cells a source may send for each forward RM cell.
Mrm	—	Controls allocation of bandwidth between RM cells, backward RM cells and data cells.
RDF	Rate Decrease Factor	Controls the decrease in the cell transmission rate.
ACR	Allowed Cell Rate	Current rate at which a source is allowed to send.
CRM	Missing RM-cell Count	Limits the number of forward RM cells count that may be sent in the absence of received backward RM cells.
ADTF	ACR Decrease Time Factor	Time permitted between sending RM-cells before the rate is decreased to ICR.
Trm	Time between RM cell	Upper-bound on the time between forward RM-cells for an active source.
FRTT	Fixed Round-Trip Time	Sum of fixed propagation delays from the source to a destination and back.
TBE	Transient Buffer Exposure	Negotiated number of cells that the network would like to limit the source to sending during start-up periods, before the first RM cells returns.
CDF	Cut-off Decrease Factor	Controls the decrease in ACR associated with CRM.
TCR	Tagged Cell Rate	Limits the rate at which a source may send out-of-rate forward RM cells.

that congestion remains a distant possibility. Among the reactive congestion control mechanisms, three techniques, viz. selective cell discard, frame discard and explicit forward congestion indication—are discussed below.

8.7.1 Selective Cell Discard

During congestion, a network may decide to discard cells with the CLP set to 1 and protect the cells with the CLP bit set to 0. This mechanism is referred to as selective cell discard. The basic idea behind this mechanism is to protect high-priority cells while dropping the low-priority ones. Also note that one of the means of traffic policing is tagging the non-conforming cells. Tagging works on the assumption that in case of congestion, tagged cells are the first ones to be discarded.

8.7.2 Frame Discard

In ATM networks, a large IP packet is segmented and carried in a number of cells as the 48-byte payload. Thus, during congestion, if one of the cells belonging to an IP packet is dropped, the remaining cells carried by ATM network become dead cells because these cells cannot be properly reassembled at the receiving end. This is because there is no retransmission at ATM or AAL level. Carriage of dead cells can aggravate congestion to a great extent.

Depending upon how the dropped cells are bunched, the throughput of the network varies. In the worst-case scenario where a cell from each packet is discarded, the throughput can be nearly zero. With the same level of cell discard, the best-case packet discard strategy discards cells that belong to the same packet giving a much better throughput as compared to distributed cell discard strategy discussed earlier (see Figure 8.7). Thus, it is preferable that if one cell of a packet is discarded, the rest of the cells (called dead cells) of that packet are also discarded. In other words, rather than discarding cells on an individual basis, it is better to discard cells on a frame basis.

FIGURE 8.7

Worst and Best Case Packet Discard

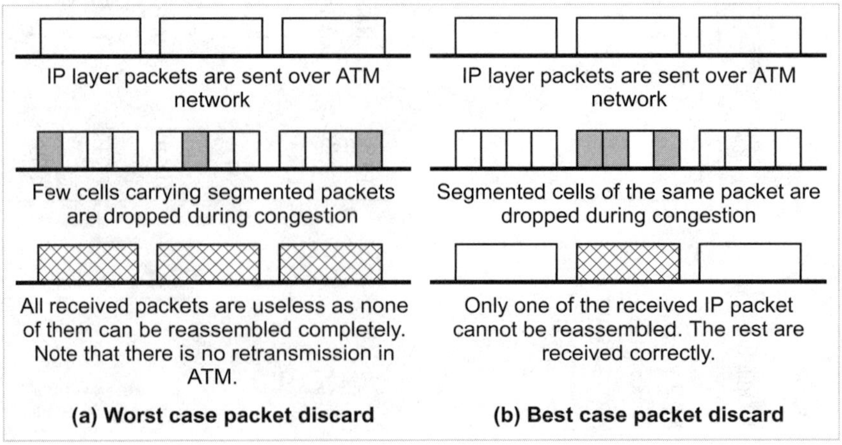

To tackle the packet discard issue, various schemes are currently defined in which packets are discarded at frame level rather than cell level. These schemes are as follows:

- Partial Packet Discard (PPD)
- Early Packet Discard (EPD)
- Random Early Discard (RED)

Each of the schemes is discussed in the following sections.

8.7.2.1 Partial Packet Discard (PPD)

One of the simplest approaches is to drop a cell without any consideration for cell boundaries. This scheme, referred to as Drop Tail, though the simplest to implement, leads to transmission of corrupted or dead cells. Thus, some sort of intelligent discard is desirable.

A relatively simple packet discard strategy is Partial Packet Discard (PPD). In this scheme, once a cell of a frame is dropped, all the remaining cells (barring the end-of-frame cell) of this frame are also dropped. The cells of the frame prior to the one that is dropped are still carried till the destination. For AAL5, implementing this mechanism is straightforward because means exist to identify AAL5 frame boundaries at ATM layer. Recall that the third bit in the PTI of an AAL5 PDU is set to '1' for the last cell and '0' for the remaining cells. Hence, this bit is used to identify the start and end of an AAL5 PDU. Once a cell of an AAL5 PDU is dropped, all the successive cells are also dropped till an end-of-frame cell is not received. The end-of-frame cells are not dropped because they provide frame delineation. At the receiving end, since the cells cannot be reassembled properly, the received cells are discarded. It is then up to the transport layer protocol to retransmit the packet, which in turn would lead to retransmission of cells corresponding to the packet.

To understand why end-of-frame cells are not dropped, consider the situation depicted in Figure 8.8. If cells marked gray are discarded, then the receiver will merge cells shaded cells to form an AAL5 PDU, which is actually incorrect. Thus, to provide unambiguous frame delineation, end-of-frame cells are never discarded.

FIGURE 8.8

Example of How Dropping Last Cell Leads to Wrong Reassembly

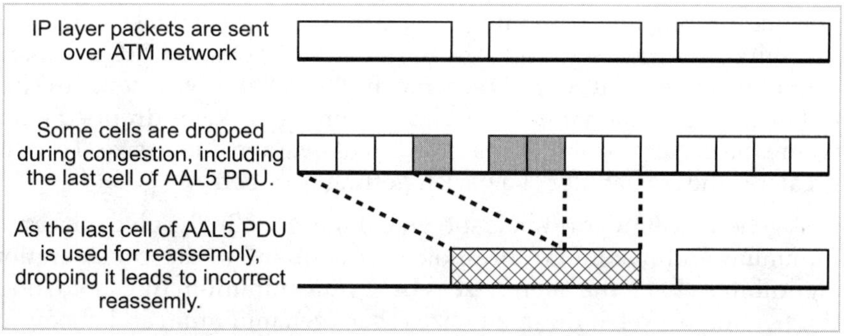

The PPD is a relative simple scheme that fares well compared to the Drop Tail scheme. On an average, PPD scheme drops only half the cells and thus, dead cells are still carried till the destination. Thus, PPD is a congestion control technique (reacting to congestion), rather than congestion avoidance technique.

8.7.2.2 Early Packet Discard (EPD)

One of the drawbacks of PPD scheme is that some of the dead cells are still carried till the destination. The reason for this is that the decision to drop cells is taken instantly rather than with foresight. The problem can be avoided if such a situation can be predicted a priori whereby carrying half of the dead cells can be avoided.

To avoid this drawback, Early Packet Discard (EPD) is used. In this strategy, an entire incoming frame is discarded to prevent any possibility of buffer overflow. The problem of carrying dead cells (which are at the beginning of a frame) or carrying the end-of-frame cell also gets automatically eliminated. To discard frames using EPD, buffers are monitored. This is done to ensure that the occupancy of the cells in queuing buffer of a switch remains within a threshold level. Once the occupancy level crosses a given threshold of buffer, a complete frame is discarded. This is followed till the buffers fall below the threshold level. Even if the queue crosses threshold, the cells belonging to a partially transmitted frame are not dropped and admitted provided there is buffer space available.

The merit of this scheme over the PPD scheme is that it does not carry some of the dead cells that are carried in PPD scheme. However, one of the demerits of the EPD scheme is that it is unfair to low bandwidth flows. In particular, if the buffer occupancy is hovering around the threshold value, then high rate flows are likely to have more opportunities at getting their packet accepted. Once a packet is accepted, it pushes the buffer occupancy above the threshold and a subsequent packet from a low bandwidth flow is then likely to be dropped.

8.7.2.3 Random Early Discard (RED)

Given that EPD penalizes the low bandwidth flows, it is desirable to have fairness in the packet discard scheme whereby packets from non-conforming flows can be adequately penalized. One solution is to perform per flow buffer accounting. This scheme can penalize non-conforming flows but impedes scalability of the switches. Thus, it is desirable to perform accounting at a coarser level. The Random Early Discard (RED) achieves fairness while still providing the benefits of EPD scheme. In this scheme, whenever the queue size in the switch buffer exceeds certain threshold, each incoming packet is dropped with a certain probability, where the exact probability is a function of average queue size. Thus, greater the queue size, greater is the probability of a packet getting dropped.

A general RED algorithm is shown in Figure 8.9. In the given figure, min^{th} and max^{th} are the minimum and maximum threshold of the queue under consideration. The essence of the algorithm is that if the queue size is below the minimum threshold, no action is taken against the arriving packet. If the queue size is between minimum and maximum threshold, packet is

Figure 8.9

General RED Algorithm

For each packet arrival,
    Calculate the average queue size *avg*
    If $min_{th} < avg < max_{th}$
    Calculate probability p_a
    With probability p_a
    Drop the arriving packet
Else if $max_{th} < avg$
    Drop the arriving packet

marked (or dropped) with certain probability. However, if the queue size exceeds maximum threshold, no new packet is accepted.

The RED scheme penalizes flows in proportion to the amount of traffic they contribute and thus prevents any of the flows from getting a disproportionate amount of resources. However, one drawback of RED is that it is unfair to low speed TCP flows. This is because RED drops packets when the maximum threshold is crossed, and it is possible that one of the packets belongs to a flwo that is currently using less than its fair share of bandwidth. Since TCP reacts strongly to packet loss, the lost packet will force reduction in congestion window resulting in an even lower rate.

8.7.2 Explicit Forward Congestion Indication (EFCI)

In ATM, congestion notification is sent to the destination using a mechanism known as the Explicit Forward Congestion Indication (EFCI). In this mechanism, the congestion bit in the ATM header (second PTI bit) is set to 1 to indicate an impending congestion in the network. Upon receiving an EFCI notification, the destination should somehow communicate to the source to slow down the transmission rate because congestion is prevalent in the downstream direction (in source to destination direction). How the destination actually conveys this information depends upon the protocols running at the end-system. No standard procedures exist in this regard. Moreover, since EFCI is optional for all of the service categories (except ABR), it is recommended that the network should not rely on this congestion control mechanism.

8.8 Dynamics of TCP Traffic Over the ATM Networks

Today, ATM is merely viewed as a transport technology. There is still some time before native ATM applications become as ubiquitous as, say, applications like the World Wide Web (WWW), TELNET or FTP. Till then, the means to carry TCP/IP traffic over the ATM networks occupies significant importance. This is why today 'Performance measures of TCP traffic over ATM networks' is a important topic in the area of research.

Allyn and Sally [TM Allyn Romanow], in their analysis, have found out that in carrying TCP/IP traffic, ATM networks perform poorly as compared to other packet-based networks (like frame relay). The following reasons are cited:

- **Delivery of inactive cells or dead cells:** The primary reason of low throughput of TCP over ATM is the delivery of inactive cells or dead cells. Since a packet is fragmented into many cells, a cell dropped from a packet renders the remaining cells of the same packet as dead or inactive. In the absence of the means to drop the dead cells, throughput of TCP over ATM drops drastically. This problem is exacerbated by factors like: large TCP packets, small buffers, large TCP window-size and large number of active connections. If one considers a large TCP packet, even a single dropped cell results in appreciable number of dead cells.

- **Link idle time:** Throughput also suffers when the congested link is occasionally idle. Links are idle when the TCP connections (i.e. the TCP window) try to achieve equilibrium using slow-start mechanisms. When a TCP connection is established, or restarted after a period of inactivity, a slow-start scheme is recommended to avoid congestion. Therefore, it is quite possible that when the connections go through slow-start together (a phenomenon referred to as *synchronization*), the link may be idle for some time.

To improve the performance of TCP over ATM, two strategies, viz. early packet discard and partial packet discard—have been suggested. In both the strategies, attempts are made to minimize the possibility of the network carrying dead cells.

Conclusion

The success of ATM depends upon its ability to provide its users an integrated solution to their requirements and that too, under severe constraints. In this quest for success, effective traffic management becomes very important. Traffic management techniques in ATM are different from other networks, as the latter usually caters to a small spectrum of users only. ATM offers itself as a single, multi-service network, and thus, requires dealing with traffic characteristics in which the experience obtained from other technologies is of little use. In ATM, several policies have been brought together to obtain a workable traffic management solution. This includes CAC, UPC, frame Discard and several congestion control mechanisms, among others. Traffic management in ATM is still far from being stable and is still an actively-pursued research area.

Review Questions

Q 1. *Why is the design of a good and efficient CAC algorithm so difficult in ATM?*

Q 2. *What is the difference between peak-bandwidth CAC algorithm and statistical CAC algorithm?*

Q 3. *What is the connection between CAC algorithm and the routing protocol PNNI? (Hint: You may refer to Chapter 12).*

Q 4. *What is the difference between the conformance of a cell and compliance of a connection? Is it possible that all cells of a connection are conforming yet the connection is non-compliant?*

Q 5. *Explain the GCRA algorithm. What is the significance of 'Limit' field for the GCRA algorithm? In ATM, which parameter gives the value of Limit field?*

Q 6. *Prove that virtual scheduling algorithm and continuous-state leaky bucket are equivalent.*

Q 7. *Why do VBR connections require dual-rate GCRA algorithm? Explain with the help of an example.*

Q 8. *Why was rate-based flow control chosen over credit-based flow control in ATM? Explain how ABR flow control works with reference to some of the important parameters of the Resource Management cell.*

Q 9. *Why are packet discard strategies important when carrying IP traffic over ATM?*

Q 10. *What are the different packet discard schemes? Compare between PPD, EPD and RED schemes. (Hint: Refer [TM M. Labrador]).*

Further Reading

For details of the topics covered in this chapter, the reader is referred to [ATMF TM 4.1] and [ITU-T I.371]. The ATM Forum specification [ATMF TM 4.1] has good normative and informative text to augment an understanding of various traffic and service descriptors in ATM. For specific topics, a number of survey papers are available for reference. For example, [TM K. Shiomoto] and [TM H. G. Peros] provide good survey on CAC while [TM M. Labrador] provides a good overview on various packet discard techniques. Few papers authored by Raj Jain including [TM Raj Jain] and [TM Raj Jain 2] are good references for congestion control techniques.

Chapter 9

ATM Switching

9.1 Introduction

A switch can be viewed as a network element that transfers packets from input ports to the appropriate output ports. Here, a port refers to a point-of-attachment in the switch. Accordingly, switching is the process of transfer of packets from input ports to output ports. This transfer is also referred to as internal routing of packets. Switching or internal routing forms the core functionality of the switch. Table 9.1 provides generic information on switching. For details on these topics, refer [Gen S. Kasera]. As some of these topics need clarity, they are further elaborated in this chapter.

Table 9.1

Basics of Switching [Gen S. Kasera]

Definition	Switching is the process of transfer of packets from input ports to output ports.
Components	• **Input module:** Input module accepts packets arriving at the input ports and processes the packet. • **Switching fabric:** This provides the means to route packets from input ports to output ports. • **Output module:** Processes the packet just before transmission on the outgoing link.
Performance Measures	• **Packet loss probability:** This parameter refers to the probability of a packet being dropped within the switch before it can be successfully transferred to its destined outlet.

Contd.

TABLE 9.1

Contd.

	• **Delay:** This refers to the time required to switch a packet, right from the time it enters the input module to the time it leaves the output module. • **Throughput:** Ratio of output traffic rate and input ratio rate. • **Design complexity:** Refers to the complexity involved in designing the switch. • **Scalability:** Scalability refers to the ease with which a switch can be expanded. • **Cost:** Refers to the commercial viability of the switch.
Issues	• **Conflict and Contention:** Conflict is a phenomenon occurring during the switching process, whereby the switching process is hindered. • **Blocking:** Depending upon whether there are contentions/conflicts in the switching process, a switch can be classified as *blocking* or *non-blocking*. • **Buffering:** Processing of storing excess packets, includes *input buffering, shared buffering* and *output buffering*. • **Memory Look-up:** Memory in switches is used for two purposes—to store connection-related information (like outgoing labels and connection priority) and to buffer packets. • **Multicasting:** One-to-many transmission technique, in which instead of sending a copy of data to each recipient, data is sent to a group of receivers using a multicast address.
Architecture	• **Shared Memory Architecture:** In this architecture, a common pool of memory is used to buffer packets for all inlets and outlets. • **Shared Medium Architecture:** In this architecture, a broadcast medium is used to switch packets to their destined outlets. The broadcasting is done typically through time division multiplexing. • **Space Division Architecture:** In this architecture, small switching elements are used to form a switch.

Apart from a general discussion on switching concepts, this chapter discusses switching in ATM. With the phenomenal growth in the speeds of optical transmission systems like SONET/SDH, there is severe pressure on ATM switch designers to develop switches that match the transmission capability of optical fiber. By current estimates, in terms of speed, switches will lag the transmission systems for many decades to come. It is said that without high-speed switches, it is like building a ten-lane highway interspersed by a single-lane tunnel. Against this backdrop, it is imperative to design switches that are capable of forwarding packets speedily, are highly scalable and easy to implement. This chapter discusses issues related to high-speed switching in ATM. A number of case studies are presented for understanding the topic.

9.2 Components of a Typical Switch

A typical switch has three basic components: *input modules*, *switching fabric* and *output modules*, as shown in Figure 9.1. These components are detailed below.

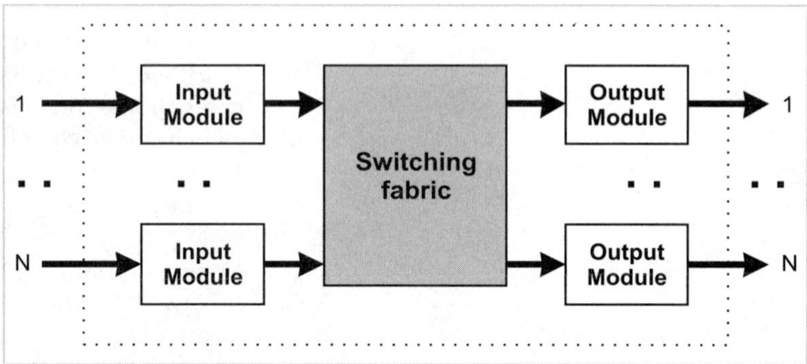

Figure 9.1

A Typical Switch

- **Input module (IM):** The input module accepts packets arriving at the input ports. The input module then extracts the virtual circuit number (i.e. the logical address or label) from the packet header of the arrived packets. The extracted label is used to index a virtual circuit table and to determine the outgoing label and the output port of the arrived packet. This function may also be performed by the output module. Other functions of the input module include deframing and unscrambling, synchronization, classification of the packet into various categories (like signalling packets, management packets, etc.), error checking and internal tagging (i.e. associating route information with packets so that they can be forwarded to their destined output). Other functions of the input module are dependent on the transmission media and the underlying technology.
- **Switching fabric:** The switching fabric provides the means to route packets from input ports to output ports. The switching fabric consists of transmission links and switching elements. A transmission link is a passive entity (without any decision-making intelligence), which merely carries packets. A switching element, on the other hand, performs functions like internal routing. The switching fabric is also referred to as the *interconnection structure*.

- **Output module:** The functionality of the output module is tightly linked to the transmission media and the underlying technology. Depending upon the overall design of the switch, it can perform functions like label swapping, error control and data filtering.

9.3 Performance Measures in Switch Design

Before discussing various switching issues and switching architectures, it is important to define parameters based on which different approaches are appraised. These parameters (referred to as *performance measures*) can be classified either as QoS-related parameters or non-QoS parameters. QoS-related parameters include *packet loss probability, switching delay* and *throughput*. Non-QoS parameters include *design complexity, scalability* and *cost*. These parameters are defined below.

- **Packet loss probability:** This parameter refers to the probability of a packet being dropped within the switch before it can be successfully transferred to its destined outlet. Overload conditions, internal design and shortage of buffer space are the three main reasons for packet loss in a switch. The desirable values of packet loss probability range between 10^{-8} and 10^{-12}.
- **Delay:** This refers to the time required to switch a packet, right from the time it enters the input module to the time it leaves the output module. The delay includes the time required to process a packet at every intermediate switching element. Thus, the greater is the number of switching elements, the greater is the switching delay. The delay parameter also includes the waiting time of a packet in the switch buffers.
- **Throughput:** There are two ways in which people define the throughput of the switch. According to one school of thought, throughput is equal to the port speed of an input link. According to others, throughput is the sum total of link rates of all input links. However, both the measures may not reflect the maximum data rates achievable under different traffic conditions. A better measure is the ratio of output traffic rate and input ratio rate. This ratio takes into account the cases when the internal design of a switch does not permit 100 per cent throughput under unity load.
- **Design complexity:** Switch designers are generally wary of making complicated switch designs. A complex switch design invariably translates into either higher costs or limited switching capability. For this reason, a simple design with slightly lesser capability is preferred over complex designs.
- **Scalability:** Scalability, in the present context, refers to the ease with which a switch can be expanded. There are two aspects associated with scalable switch design. First, a scalable switch must be capable of supporting additional inlets without any deterioration in other performance measures. Secondly, it must also be capable of supporting any increase in the port speed of its inlets.
- **Cost:** Needless to say, cost is a significant factor in deciding the success or failure of a switch design. Quite often, performance-efficient switch designs prove to be a

commercially unviable option, thereby relegating the design to a place in research papers only. Thus, the cost aspects related to the commercial production of switches must always be borne in mind while proposing new switch designs.

9.4 Switching Issues

Most of the issues in switch design pertain to two important aspects: *buffering* and *internal routing*. The first aspect relates to the placement of buffers in the switch, while the second relates to the design of the switching fabric (specifically speaking, the design of the interconnection structure). These two are, however, not independent concepts, because fixing one of them imposes constraints on the other. In fact, there are two ways of designing a switch. Either one can choose a particular buffering technique and then decide from one of the interconnection structures possible. Alternatively, one can choose a particular interconnection structure and then decide upon the buffering technique that suits it best. In this section, the first approach is adopted.

First, the need for buffering is explained and then various buffering techniques and associated interconnection structures are discussed. A few other switching issues in this section are also discussed. In the following sections, three different interconnection structures are explained and the applicability of different buffering techniques is discussed.

9.4.1 Conflicts and Contentions

One of the fundamental challenges in designing a high-speed switch is tackling the problem of *conflicts*. Conflict is a phenomenon occurring during the switching process, whereby the switching process is hindered. Conflicts are of two types: *external conflicts* and *internal conflicts*. In order to understand external conflict, consider the following case: '*N*' packets simultaneously arrive at '*N*' different inlets, all destined for the same outlet (see Fig. 9.2). Here, all *N* packets cannot be switched onto the same outlet in a single time slot. In cases like

Figure 9.2

External Conflict

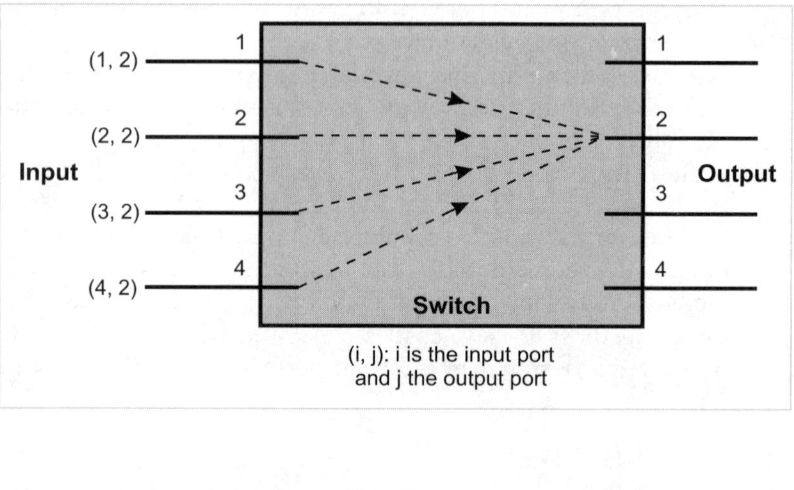

this, there is a conflict (or contention) between different inlets for an outlet. This phenomenon is termed as an external conflict. The most common way to handle external conflict is to buffer the excess packets. In the scenario presented above, N–1 packets have to be buffered, while one packet is allowed to go through.

In contrast to external conflicts, internal conflicts are the result of the way the interconnection structure is designed. Internal conflict occurs when a switch is incapable of forwarding arrived packets even when an external conflict has not occurred; i.e. a conflict occurs even when 'N' arrived packets are destined for 'N' different outlets. Generally, any internal resource of a switch that is shared by packets coming from multiple paths can lead to internal conflicts. For example, in a multi-stage switching architecture, the internal switching elements are potential candidates for internal conflicts. The problem of internal conflicts for multi-stage switches is analysed in Section 9.8 of this chapter.

9.4.2 Blocking and Buffering

Depending upon whether there are contentions/conflicts in the switching process, a switch can be classified as *blocking* or *non-blocking*. In a non-blocking switch design, as long as an outlet is free, packets coming from any inlet can be switched onto the free outlet. In contrast, a blocking switch design leads to conflicts, either external or internal. The external conflicts cause external blocking and internal conflicts cause internal blocking.

There are mainly three conditions under which blocking (internal and external) occurs. These are:

1. Packets from multiple inlets arrive simultaneously, all destined for the same outlet.
2. There are more than one contenders for a shared resource like memory, system bus and internal switching elements.
3. Packets arrive at a rate faster than the maximum rate at which they can be switched.

Many alternatives exist for tackling the problem of blocking. The simplest solution is to *discard packets* that are in excess of that which can be switched. Although this is a simple solution, in case of traffic bursts, this would lead to huge packet losses, thereby disrupting the service guarantees of a connection. Thus, this method is usually not adopted.

The next alternative is to have *dedicated switched paths* within the switch. This method is used in a majority of telecommunication switching systems. This option, however, would preclude the use of statistical multiplexing. This is because in statistical multiplexing, instead of allocating dedicated switched paths, resources are shared among multiple connections and a path is found only when packets arrive. Thus, dedicated internal switched paths cannot be used in technologies that employ statistical multiplexing (e.g. ATM).

The third option to avoid blocking is to use a *back-pressure* mechanism. In this technique, new inputs are allowed only if it can be ensured that this will not lead to packet loss. For applying back-pressure, feedback signals are sent to the previous stage for throttling the input rate. Instead of solving the problem, back-pressure merely shifts the problem to a preceding stage. This shifting process may aggravate the problem to such an extent that packets may have to be queued up at the input of the switch. As one shall observe shortly, input buffering is not the most preferred form of buffering. Thus, though switches using the back-pressure mechanism and giving optimal throughput have been designed, this technique is not very popular.

Finally, the best option is to *buffer* excess packets that cannot be switched at one go. These buffered packets are then served, as and when the resources are available. Indeed, buffering or queueing is a key issue in all switch designs. The most important design considerations are where to buffer and what buffer size to maintain. The first consideration (i.e. *where to buffer*) influences the throughput and the scalability of the switch. There are basically three approaches to the placement of buffers, viz. *input buffering*, *output buffering* and *central buffering*. All three approaches are detailed in the following sub-sections.

The second consideration (i.e. *what buffer size to maintain*) also influences the throughput of the switch. Along with throughput, the buffer size also affects two important performance measures, viz. *loss* and *delay*. Choosing the optimal value of buffer size is an important design decision. This decision is essentially a trade-off between loss and throughput on one side, and delay and cost on the other. A large value of buffer size means that even during heavy loads, there is very little loss, thereby giving a better throughput. However, a large value also means that the mean waiting time of a packet in the buffers increases, thereby increasing the average delay of the packet. Moreover, keeping large buffers also increases the cost of the switch. The inverse logic applies to a small buffer size.

9.4.2.1 Input Buffering

In switch designs employing *input buffering*, the blocking problem is solved by buffering the packets at each inlet of the switch. The packets are buffered until the control logic allows the packets to cross the switching fabric. In this scheme, each inlet maintains its separate buffers. Figure 9.3 shows a schematic diagram for input buffering.

FIGURE 9.3

Input Buffering

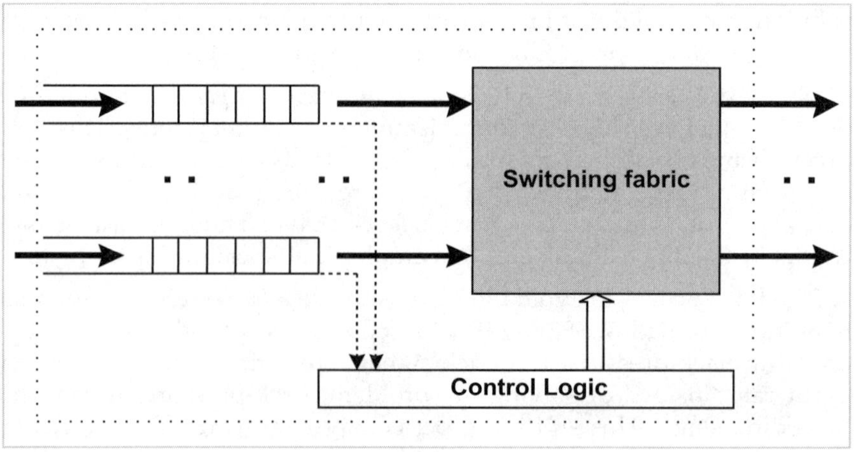

The performance analysis of input buffering is tightly coupled with the *scheduling strategy* used to serve the buffered packets. Some of the packet-scheduling approaches are discussed below.

One of the simplest packet scheduling strategies for input buffering is the *First-in-first-out (FIFO)* approach. In this approach, the first packet to enter the queue is also the first one to leave. This makes the design of the input queues very simple. However, the FIFO approach suffers from a serious problem called the *'head-of-line' (HOL) blocking*. Consider a case in which the first packet (say X) in one of the input queues is waiting for its destined outlet to be free. Now, packets queued up behind packet X cannot be switched even if their respective outlets are free. In such cases, it is said that the first packet is causing a head-of-line blocking. Due to this head-of-line blocking problem, a significant portion of the switch capacity is wasted. Thus, the maximum throughput is in the range of 55–70 per cent of the switching capacity only. Table 9.2 shows the maximum throughput achievable by using the FIFO approach for different number of input ports, (N). For as less as two inputs, the saturation throughput is only 75 per cent. This highlights the impact that HOL blocking has on the throughput of an input queued switch. Figure 9.4 shows the mean waiting time of packets as a function of the offered load. Offered load is the percentage of time that an input port receives packets for forwarding. It is observed from Fig. 9.4 that as the offered load approaches the 60 per cent mark, the waiting time approaches infinity. Thus, the throughput does not exceed the 60 per cent mark due to the HOL problem mentioned above.

TABLE 9.2

The Maximum Throughput Achievable Using Input Buffering with the FIFO Approach ([Swi M.G. Hluchyj] © 1988 IEEE)

N	*Saturation throughput*
1	1.0000
2	0.7500
3	0.6825
4	0.6533
5	0.6399
6	0.6302
7	0.6234
8	0.6184
μ	0.5858

Another issue associated with the FIFO approach needs to be considered here (in fact, this issue applies for the next scheduling approach as well). In case two or more packets at the head of different queues contend for the same outlet, one packet has to be selected out of the many contending packets. This selection can be done on a fixed priority basis; i.e. each queue has a static priority associated with it and the output contention is solved by choosing the packet of the highest priority. The other option is to assign priorities to queues on a dynamic basis, thereby precluding the possibility of starvation packets of low-priority queues. These two approaches constitute a trade-off between design complexity and fairness.

FIGURE 9.4

Mean Waiting Time for Input Buffering with the FIFO Approach, for N = ∝

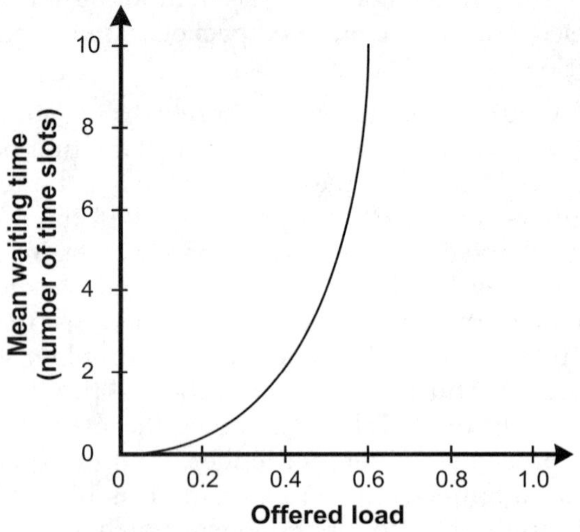

Source: ([Swi M.G. Hluchy] © 1988 IEEE)

Another approach used for scheduling is the, *multiple* queueing approach. In this approach, instead of maintaining a single queue, multiple queues are maintained at each inlet. In fact, to completely solve head-of-line blocking, the number of queues maintained at each inlet is equal to the number of outlets. By keeping multiple queues, significant improvements are observed in the throughput of the switch with near-optimal capacity utilization. This improvement, however, comes with a problem. The buffer requirements immediately shoot up by an order of magnitude. For a 256 × 256 switch, this would mean 65,536 queues and a total of 65536*B buffers, where 'B' is the number of buffer elements in each queue.

The third strategy used is the *Maximum matching approach* that tends to maximize the number of matches between inputs and outputs (i.e. finding the largest set of packets arriving at input ports that do not have any contention for the output ports), thereby yielding the best throughput achievable. There are, however, a couple of problems associated with this approach. First, it leads to starvation under certain traffic conditions. For example, if there is a packet lying in buffer 1 destined for outlet 2, and packets continuously arrive at buffer, 1 and 2, destined for outlets 1 and 2, respectively. Under this scenario, to maximize throughput, while the newly arrived packets will be allowed to pass through, the buffered packet will wait indefinitely. This will lead to starvation. Secondly, the computation involved in determining the maximum match can be very time-consuming at times, rendering the approach unsuitable for high-speed switching. Thus, the preferred approach is to go for heuristic, which provides a near-optimal solution (i.e. maximal solution), but not the exact solution. *Parallel Iterative Matching (PIM), iterative round-robin matching with slip (SLIP)* and *Least Recently Used (LRU)* are examples of maximal matching approach. Interested readers are referred to [Swi Nick McKeown] for details.

9.4.2.2 Output Buffering

In the *output buffering* approach, the blocking problem is solved by buffering the packets at each outlet of the switch. Note that the placement of buffers is always defined in relation to the switching fabric. Thus, 'buffering at the output' implies that the buffered packets have crossed the switching fabric and are waiting for their destined outlets to be free, that is, the decision of 'where to switch' has already been made, and only then have the packets been buffered. Figure 9.5 shows a schematic diagram for output buffering.

FIGURE 9.5

Output Buffering

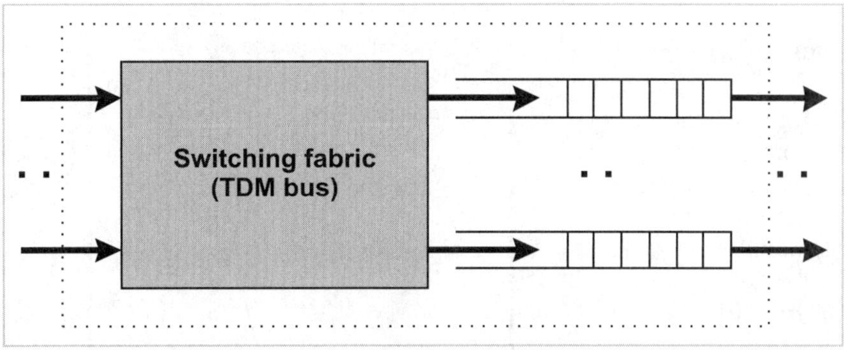

Output buffering is preferred over input buffering because of its better performance in terms of throughput and delay, for the reasons explained below. However, this technique mandates that the switches be capable of internally switching multiple packets in a single time slot. This pre-condition is necessary to avoid input buffering. In order to understand why this condition exists, reconsider the case when 'N' packets arrive at 'N' different inlets, all destined for the same outlet. If only one cell is switched out in a single time slot and the switch is not employing input buffering, where and how are the remaining N-1 packets be buffered? If packets are to be buffered at the output, the N-1 packets must have to cross the switching fabric. Now, to do this, the switching fabric must operate at a speed that is N times the port speed. In other words, the switch must be capable of switching internally multiple packets in a single time slot. This is the reason why a *speed-up factor* is required in switches employing the output buffering technique.

The packets in output queues are scheduled, typically using a FIFO approach. Since there is no HOL problem similar to the one that exists in input buffering, the throughput in the output-buffered switch is near-optimal. Figure 9.6(a) shows the mean waiting time of packets as a function of the offered load. The figure establishes the fact that with output buffering, a near 100 per cent throughput is achievable. At first look, the graph depicted in Figure 9.6(a) does not seem to be very clear. This is because by increasing the buffer size, there is an associated increase in the mean waiting time of packets. For a buffer of infinite size, the mean waiting time approaches infinity. But if one compares this graph with that for input buffering, one will observe that in the latter case, the mean waiting time reaches infinity near the 60 per cent mark.

Thus, the maximum throughput in an input-buffered switch cannot exceed 60 per cent. In contrast, the maximum throughput in an output-buffered switch can reach nearly 100 per cent. As far as the mean waiting time is concerned, the increase in this value is attributed to the increase in buffer size. The rationale for keeping a larger buffer can be appreciated by looking at Figure 9.6(b). A larger buffer results in a lower packet loss. For achieving a given packet loss, (say, 10^{-10}) at a given load (say, $p = 0.90$), a very large buffer is needed. In other words, the switch must have ample buffers to guarantee a packet loss of 10^{-10} under heavy loads ($p = 0.90$). The two graphs explain why the buffer size is a compromise between throughput and packet loss, on one side, and delay (mean waiting time), on the other.

FIGURE 9.6

Performance of Output Buffering (a) Mean Waiting Time as a Function of the Offered load p, for N = ∞. (b) Packet loss Probability as a Function of Buffer size B, for N = ∞

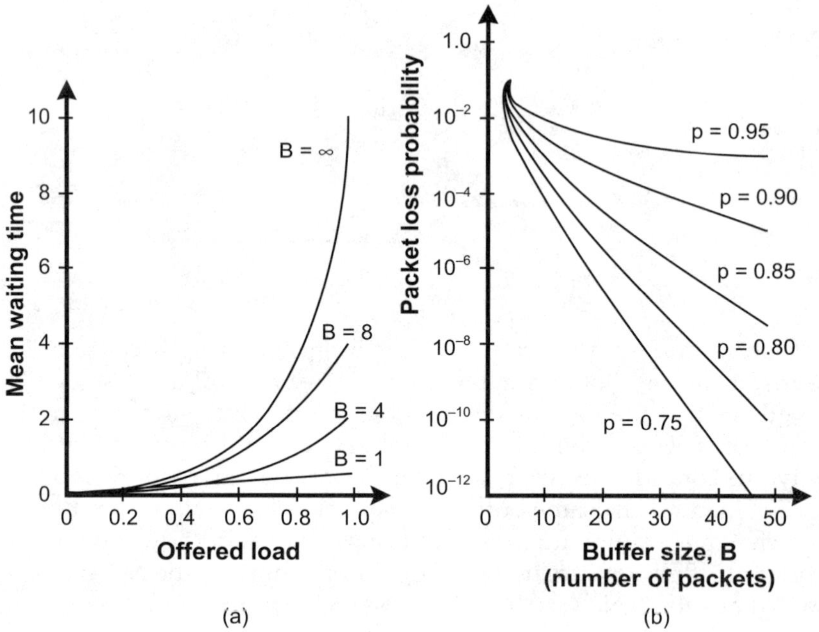

Source: ([Swi M.G. Hluchy] © 1988 IEEE)

Although output buffering is performance-efficient, the speed-up factor leads to severe scalability problems. In order to solve this problem, instead of having a single buffer, multiple buffers are kept at each outlet, one for each inlet. After doing this, it is no larger necessary for the memory to operate at 'N' times the port speed. This case, however, is not very different from the multiple input buffering approach discussed earlier. Thus, output buffering also leads to scalability problems because of excessive buffer requirements.

9.4.2.3 Shared Buffering

One of the drawbacks of both the input buffering and output buffering techniques is that buffers in both these approaches have to be tailored to handle the most severe of traffic bursts.

Thus, the size of each buffer must be above a given threshold, wherein the threshold value depends upon the loss probability desired from the switching fabric. This leads to poor memory utilization of the buffers because the average traffic behaviour is much less severe. The low memory utilization because of keeping dedicated buffers for each output has prompted the switch designers to look for low-cost alternatives. The *shared buffering approach* is an excellent option in this regard. In this approach, a large pool of memory is shared between all inlets and outlets. This leads to substantial cost reduction because of the sharing of memory buffers. Figure 9.7 shows a schematic diagram for shared buffering.

FIGURE 9.7

Shared Buffering

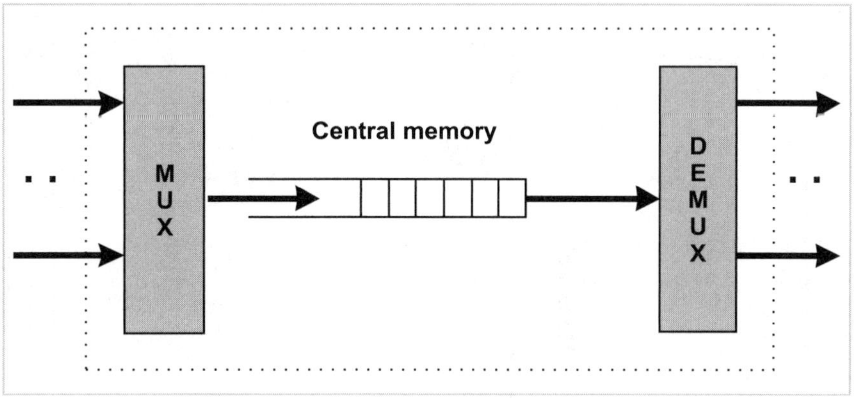

Although it is cost-effective, the shared-memory approach leads to some design complications. First, the memory management required for a shared buffering system is much more complicated as compared to the memory management requirements for other buffering approaches. This is because though a single memory is used for maintaining all queues, distinct logical queues for each outlet must still be maintained. Two alternatives exist for maintaining logical queues. The first option is to have a fully shared memory, in which case the whole memory is available for the queue of a single outlet. The other option is to maintain an upper limit for the queue of each outlet (partitioning). The two approaches constitute a trade-off between lower packet loss and greater fairness.

Second, the memory of the shared buffer must be equal to at least the total bandwidth of all the inlets, and moreover, it must operate at N times the port speed. All these factors, especially the memory management complexity, make the shared-memory switch difficult to design.

Keeping the design complications aside, the performance of shared buffering is almost as good as that of output buffering. If all the parameters are kept the same, the packet loss probability in the former is much better than in the latter. Figure 9.8(a) shows the mean waiting time of packets as a function of the offered load. Figure 9.6 again establishes the fact that with infinite buffering, a near 100 per cent throughput is achievable. If one compares Figure 9.6(b) with Figure 9.8(b) for $p = 0.9$, then shared buffering fares much better. In order to achieve a given packet loss, the number of buffers required in shared buffering is much less than that required in output buffering. This result is not surprising because, in the case of the former, buffers are not dedicated for each outlet as is the case in the latter. Thus, even if one of the buffer queues is sufficiently loaded, the spare capacity of other queues can be used to buffer packets. Note that Figure 9.8(b) applies for a complete sharing approach. If buffer partitions are used, then obviously, the result will tend towards that of output buffering. In the extreme case of shared buffering, when a distinct partition is maintained for each output queue, the result will converge with that of output buffering.

FIGURE 9.8

Performance of Shared Buffering (a) Mean Waiting Time as a Function of the Offered Load, for $N = \infty$ (b) Packet Loss Probability as a Function of Buffer Size B, for p = 0.90

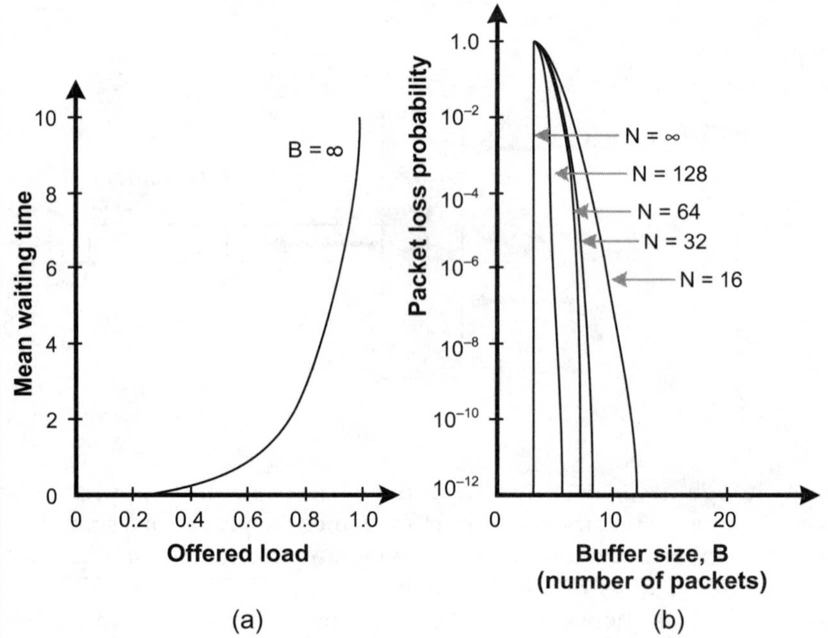

Source: ([Swi M.G. Hluchy] © 1988 IEEE)

9.4.3 Memory Look-up

Memory in switches is used for two purposes—to *store connection-related information* (like outgoing labels and connection priority), and to *buffer packets*. Thus, for each packet, memory is accessed for retrieving connection-related information and in some cases, for buffering. Thus, memory management and efficient memory look-up have important implications for switch design.

One of the techniques that is quite commonly used to speed up the access time for information retrieval is to employ Content Addressable Memory (CAM). For application areas that require some form of searching or translation, CAMs prove to be an attractive alternative to software solutions like searching and hashing. The trade-off is between the speed of the CAM (albeit at a higher cost) and the simplicity, reliability and the low-cost of software options. Earlier, because of the prohibitive costs of CAMs, people used to opt for the low-cost software option. However, with a fall in prices, there has been a phenomenal growth in the sales of CAMs. Now CAMs can be seen in switches, bridges and routers. Some of the application areas of CAMs are as follows:

- MAC address look-up in layer 2 bridges/switches
- VPI/VCI translation in ATM switches
- DLCI translation in frame relay switches
- Other address translation and protocol conversion in gateways
- Longest prefix matching for network layer protocols like IP4
- ARP cache in network servers

CAM is significantly different from the way data is stored and accessed in Random Access Memory (RAM). For reading data from a RAM, the specific memory address is supplied. Again, for writing something to the memory, the memory location must be supplied. In contrast, reading data from CAM does not require the memory address of any location. The read operation is similar to the way one recalls the names of the people whom one has met. The picture of the person is compared with a number of visual images stored in the brain. When a match is found, the information associated with the image is fetched and one recalls the name, age, and other details of the person. Similarly, for reading something from the CAM, some information is placed in a Special Purpose Register (SPR). The information here can be a virtual circuit identifier, an MAC address or any other piece of data. This information is then compared with the information stored in all the entries of CAM. Note that information stored in the SPR is compared with only a part of every CAM entry (that is, one compares a picture with a picture only, not with the name). Once a match is found, a particular flag is set to inform the user that the given information was located in the CAM. It is also possible that more than one match is found. In that case, some priority is accorded to sort the matching entries. The entry with the highest priority is then returned to the user.

9.4.4 Multi-casting

Traditional communication techniques normally involve the exchange of data between two entities only. Today, however, new applications like audio/video conferencing and collaborative computing demand simultaneous data exchange between more than two entities. This has led to new communication paradigms, one of which is *multi-casting*. Multi-casting is an attractive alternative to multiple, point-to-point connections, because it helps in reducing the network traffic significantly. Multi-casting is essentially a one-to-many transmission technique, in which instead of sending a copy of data to each recipient, data is sent to a group of receivers using a multi-cast address. For multi-casting data, intermediate network elements (like switches) interpret the multi-cast addresses and make multiple copies of data only when required.

Depending upon the way a switch is designed, multi-casting can be inherently natural or can involve additional design complexity. As a general rule, it is much easier to multi-cast in those switch designs that broadcast packets, as compared to other switch designs.

The manner in which multi-casting is handled is explained when different switching architectures are discussed.

9.5 Switching Architectures

Different writers tend to classify switching architectures differently. Most of them, however, categorize all switching architectures under three broad categories, viz. *shared-memory architecture, shared-medium architecture* and *space-division architecture.* The following sections discuss each of the three architectures. In each section, first the *basic aspects* of the architecture are covered, followed by a discussion on issues pertaining to *buffering, multi-casting* and *scalability*.

9.6 Shared-Memory Architecture

A shared-memory switch uses a common pool of memory to buffer packets for all inlets and outlets. The shared memory is typically a dual port memory. A dual port memory is advantageous because it permits simultaneous read and write operations, thereby reducing the time required to perform a given set of operations by half. However, it occupies more space than a single-port memory. Figure 9.9 shows a schematic diagram of the shared-memory architecture. Examples of switches employing the shared-memory approach include CNET's Prelude switch and Hitachi's shared buffer switch.

Figure 9.9

Shared-memory Architecture

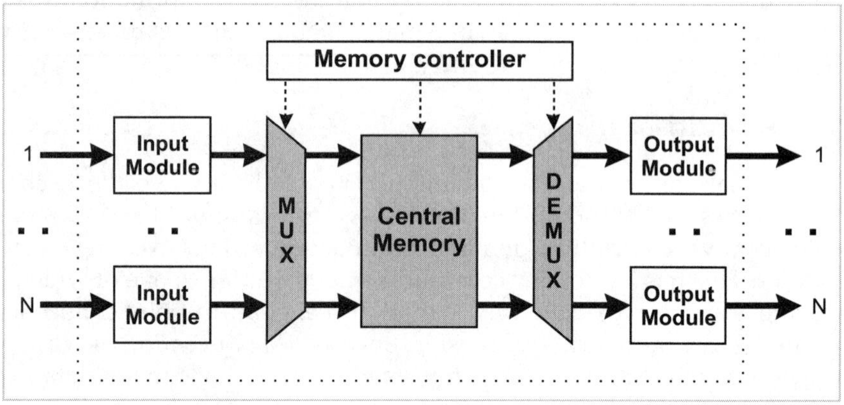

The operation of a shared-memory switch is relatively straightforward. Packets from all the inlets are multiplexed into a single stream and written to the dual port memory. At the output end, the packets are read out sequentially and demultiplexed on the basis of their outgoing port identifier. It is the task of the memory controller to decide upon the order in which the packets are written and read out. Some aspects of shared-memory architecture are detailed below.

9.6.1 Buffering

As is obvious from its name, a shared-memory switch uses the shared buffering approach. Thus, a shared-memory switch has all the advantages and disadvantages associated with shared buffering. The most important advantage is the significant cost reduction because of lower memory requirements. The most important disadvantage is that complicated memory management is required.

9.6.2 Multi-casting

Shared-memory switches are not very well suited for broadcasting/multi-casting. This is because of the centralized approach adopted. Still, there exist techniques to incorporate multi-casting in a shared-memory switch. This, however, requires additional logic in the switching fabric. Two of these techniques are:

- **Copy networks:** One of the simple (but inefficient) methods to implement multi-casting, without complicating the internal design, is to employ copy networks. A copy network makes multiple copies of a multi-cast packet before the packet enters the switch. After the copies are made, each packet is treated as a unicast packet. Note that a copy is made for each outlet for which the original multi-cast packet was destined, with each copy having its distinct label. Thus, though copy networks simplify the internal design of the switch, complicated logic is now required at the beginning of the switch. Moreover, copy networks result in unnecessary wastage of buffers and the internal capacity of the switch. After all, multi-casting is useful only if the packets are duplicated as late in the switching process as possible.
- **Address buffering:** The drawbacks of a copy network can be removed by making multiple copies of the packet addresses, rather than copies of the packets themselves. In this approach, when a multi-cast packet arrives, a pointer entry is copied in each of the logical queues of the outlet for which the multi-cast packet is destined. Since, instead of the whole packet, only a small address is copied, considerable memory can be saved. This copying, however, requires that the shared memory must now operate at N^2 times the port speed. This is because in the worst case, if N multi-cast packets arrive, each destined for all the outlets, the central memory must do N^2 write operations in a single time slot. An $O(N^2)$ speed-up factor can cause severe scalability problems. One of the ways to handle this problem is to have a multi-bank memory (with each output queue having its own memory). By keeping multi-bank memories, the speed-up factor decreases by an order of N. This, however, increases the total cost of the system.

9.6.3 Scalability

The main problem in scaling shared-memory switches is the memory look-up time. For completely non-blocking behaviour, both the read and write operations must each take place at N times the port speed. Thus, for an $N \times N$ switch having links operating at a speed V, and bus width w, the memory speeds required for single-port and dual-ported memory are

(2NV/w) Hz and (NV/w) Hz, respectively. Implementing multi-casting further aggravates the problem. Overall, shared-memory switches do not scale very well.

9.7 Shared-Medium Architecture

In a shared-medium architecture, a broadcast medium is used to switch packets to their destined outlets. The broadcasting is done typically through time division multiplexing (using either a bus or a ring interconnect). Figure 9.10 shows a schematic diagram of the shared-medium architecture. FORE System's AS-X200 and NEC's ATOM are examples of switches using the shared-medium approach.

Figure 9.10

Shared-Medium Architecture

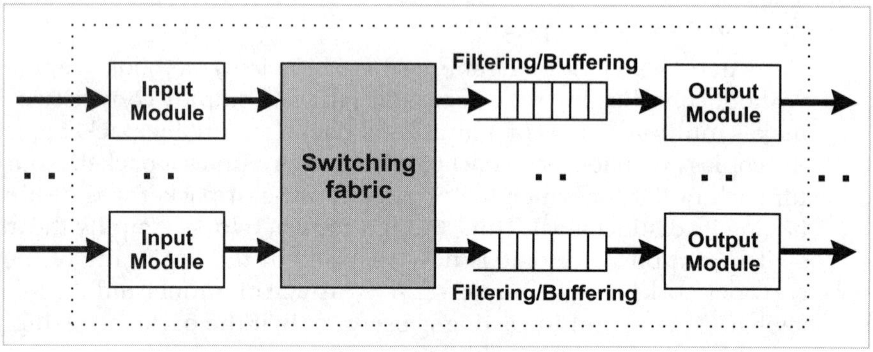

The operation of a shared medium switch can be explained as follows:

As a packet arrives, the input module determines the outgoing port of the packet, using the packet's virtual circuit number. The outgoing port number, along with other information like the multi-cast identifier, packet priority, etc. is attached to the packet in the form of a routing tag. This tag is used for subsequent processing of the packet. The packet is then broadcast to each outlet, using the broadcast medium. At the output side, the address filters decode the outgoing port address of each packet. If the packet is destined for that outlet, the packet is copied to the output buffer. Otherwise, the packet is discarded. The buffered packets are served as and when the outgoing transmission link is free.

Some aspects of shared-medium architecture are discussed below.

9.7.1 Buffering

In order to avoid external blocking, output buffering is used in shared-medium architecture. Thus, this switch architecture is optimal in terms of throughput. It also follows from the type of queuing that the shared medium (i.e. the bus or the ring) must operate at N times the port speed. Each filtering and buffering module also requires this speed-up factor of N.

9.7.2 Multi-casting

As mentioned earlier, multi-casting is inherently natural for switching fabrics (like shared-medium) that use the broadcast mechanism for transferring packets. In order to broadcast a packet to all outlets (or to a subset of it), the only requirement is that each address filter must be capable of recognizing a multi-cast address. This can be easily done through a bit-map vector, wherein a set bit in one of the fields implies that the outlet corresponding to that bit must accept the packet.

9.7.3 Scalability

The speed-up factor required for the address filters, memory buffers and the broadcast medium hamper the scalability of a shared medium switch. For an N × N switch having links operating at a speed V, the bus must operate at (NV/w) hz, where w is the width of the bus. This means that a 16-port switch with OC-3 (155 Mbps) links, the operating speed of the bus must be 80 Mhz and 40 Mhz for 32 bits and 64 bits, respectively. As the number of ports is increased to 256, the new speed requirements turn out to be 1.28 Ghz and 0.64 Ghz for the corresponding bus width. This multiplicative increase in speed requirements has rendered the shared-medium approach unsuitable for large switching systems. Nonetheless, switches up to few Gb/s, employing the shared-medium approach, are not uncommon.

9.8 Space-Division Architecture

Due to the speed-up factor involved in both the shared-medium approach (the bus/ring must operate at N time the port speed) and shared-memory approach (read/write operations must be at N times the port speed), neither of the two architectures scale very well. The main reason for this is the use of a resource (like TDM bus or central memory) that must be capable of handling all the packets from all the inlets. Viewing at the problem differently, one realizes that neither of the two architectures exploits the availability of parallel switching paths.

Space-division architecture helps in solving problems associated with both these architectures. In this sub-section, two categories of space-division switches, viz. crossbar switches and banyan-based switches, are discussed.

9.8.1 Crossbar Switching

Crossbar switches have been used since ages. They are still pervasively used in telecommunication networks. The design of a crossbar switch is relatively straightforward. It consists of an N × N matrix of *crosspoints*, each of which can either be enabled or be disabled. When a crosspoint is enabled, a dedicated path is set up between an inlet and outlet. For the proper functioning of the switch, only one crosspoint corresponding to an output link can be enabled. The converse, however, is not mandatory, because an input can be simultaneously connected to more than one output link (provided no other input is connected to the given set of outputs). Figure 9.11 provides a schematic diagram of a crossbar switch.

FIGURE 9.11

Crossbar Switch

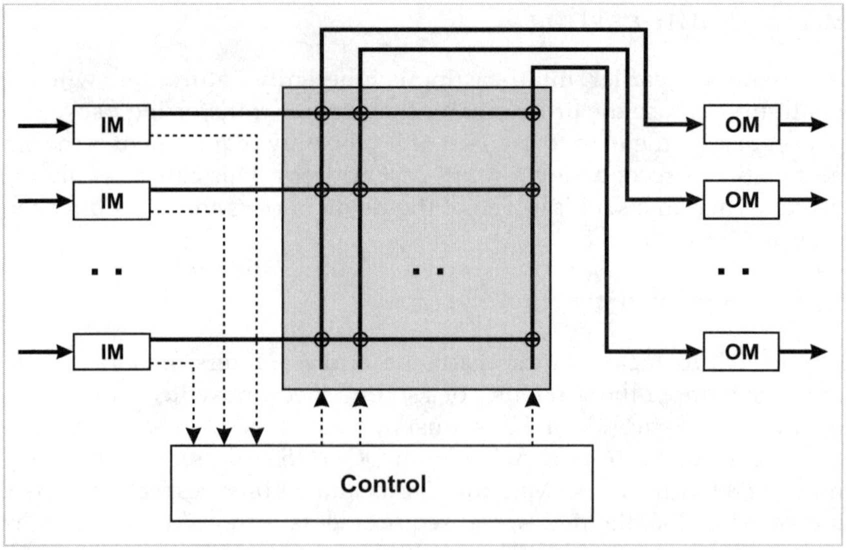

Crosspoints can be implemented by using transmission gates, which can exist in two states. In a *cross state*, there is no connection between horizontal and vertical lines, i.e. the vertical input is connected to the vertical output and the horizontal input is connected to the horizontal output. When idle, a crosspoint remains in this state. In *bar state*, the vertical input is connected to the horizontal output and horizontal input is connected to the vertical output. Now, it can be verified that for connecting input line i to output line j, it is sufficient to keep bar switches (i, k), $k = 1, 2, \ldots (j-1)$ and (m, j), $m = (i+1), \ldots, N$ in the cross state and the bar switch (i, j) in the bar state.

9.8.1.1 Buffering

In a crossbar switch, buffering can be done in several ways. The simplest option is to have FIFO queues at each inlet (i.e. input buffering). This scheme is depicted in Fig. 9.12(a). Since this leads to HOL blocking, variants of input buffering can be used instead. The relative merits and demerits of various forms of input buffering have been discussed earlier. For input buffering, besides tackling the HOL blocking, there must also exist means to arbitrate HOL packets in different queues destined for the same outlet (i.e. to solve the external blocking problem). The solution to this problem leads to two different approaches for the design of the switching fabric. One approach is to have a centralized controller. The task of the centralized controller is to find a matrix configuration that maximizes the number of packets being switched in a single time slot. The requirement is that not only must the solution be optimal, the computation must also be done in real-time. In order to achieve this aim and to solve the problem of external blocking, the controller should have global knowledge of the packets present in the input queues and the destined outlets of each packet. This requirement complicates the design of the controller, thereby increasing the cost of the switch.

Another approach is to opt for a distributed controller. In this approach, each output maintains its separate arbiter. The arbiter inspects packets coming from all the inputs and chooses one of them by some fair algorithm. Control signals are then sent to all but one inlet, requesting to block the transmission of packets. For details, readers are referred to [Swi Fouad A. Tobagi].

The next buffering option for the crossbar switch is to use output buffering. Output buffering mandates that the crossbar must operate at speeds greater than the port speed. This is required to prevent either input buffering or packet loss (due to packet discard). The actual speed-up factor required depends upon the traffic characteristic. For bursty traffic with peak rates that are small as compared to port speed, a speed-up factor of 2 is sufficient. As the ratio of the peak rate to the port speed is increased, the speed-up factor required also increases.

Another option is to use, what is termed as, *crosspoint buffering* (see Figure 9.12(b)). This option is attractive because it alleviates problems associated with both input buffering (HOL blocking) and output buffering (the speed-up factor). In this approach, packets are buffered at each crosspoint. Crosspoint buffering, however, leads to severe scalability problems. This is because an N × N switch already has N^2 crosspoints. With crosspoint buffering, the total number of buffers required goes up to N^2B. This enormous memory requirement and associated design complexity renders the crosspoint buffering approach unsuitable except for very small systems.

FIGURE 9.12

Buffering in Crossbar Switches

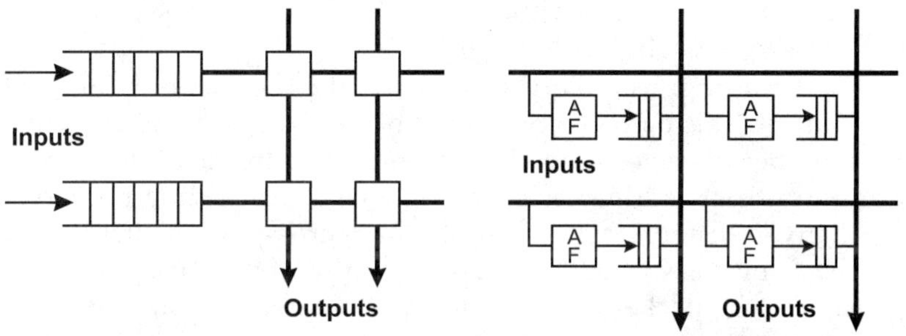

(a) Crossbar switch with input buffering.

(b) Crossbar switch with crosspoint buffering

Source: ([Swi H. Ahmadi] © 1989 IEEE)

9.8.1.2 Multi-casting

Multi-casting in crossbar switches is much more difficult than it may seem to be in the first place. The following will make this point more clear. Multi-casting seems easy because each inlet is connected to all the outlets, and hence, theoretically, a packet can easily be broadcast to all the outlets. However, consider what happens if input buffering (with FIFO) is used and one or more destined outlets of the multi-cast packet is not available. If the multi-cast packet is held till all the destined outlets are free, the packet can get excessively delayed (which will disrupt its delay and jitter bounds). Moreover, the FIFO approach in input buffering will lead to the HOL blocking problem. If the multi-cast packet is forwarded to one or more of the free outlets (without waiting for all the outlets to be free), the control logic must keep track of the outlets to which the packet has already been sent, and the outlets to which the packet is due.

Keeping track of all this is not a simple task. If output buffering is used instead of input buffering, then in the worst case, the crossbar must operate at N times the port speed. This worst case scenario results in the crossbar switch losing all the advantage it had over the shared-medium approach.

9.8.2 Banyan-based Switching

One of the major drawbacks of crossbar switching is that the number of crosspoints required for an $N \times N$ switch is N^2. This drawback prompted designers to look for alternatives that would require fewer numbers of crosspoints than N^2. *Banyan-based multi-stage switches* are one such alternative. These switches derive their name because of their tree topology. It is customary to refer to a multi-stage Banyan switch as a *Banyan network*. Banyan networks are characterized by the following attributes:

- **Multi-stage network:** All Banyan networks are composed of $b \times b$ crosspoint switches. The total number of stages in a Banyan network is $\log_b N$ with N/b crosspoint switches in each stage. This gives a total number of $b^2 * (N/b) * \log_b N$ crosspoints. For a 32-input/32-output Banyan network composed of 2×2 crosspoint switches, this comes to 320 crosspoints. This is almost one-third of the crosspoints required for a single-stage crossbar switch ($32 \times 32 = 1024$).
- **Self-routing:** In a Banyan network, there exists only one path between an input and output. This path is represented by a k-bit output address ($a_1, a_2, \ldots, a_k$), where k is the number of stages in the Banyan network. In each intermediate stage j, the jth bit of the output port address is used to make a forwarding decision. For example, consider a Banyan network composed of binary crosspoint switches. In this network, if $a_j = 0$, the packet is forwarded to the upper output of the crosspoint switch (at the jth stage), and if $a_j = 1$, the packet is forwarded to the lower output. Since, once the port address of a packet is decided, the packet is automatically routed to the specified output, Banyan networks are said to exhibit *self-routing behaviour*. Figure 9.13 shows how a packet with output address 010 is routed to the destined output port. Note that the leftmost bit is used in the first stage, the second bit from the left in the second stage, and so on.

The above characteristics impose the following limitations on a Banyan network:

- **Internal and external blocking:** One of the most severe limitations of Banyan networks is *internal* and *external blocking*. Internal blocking occurs when the inputs of an intermediate crosspoint switch contend for the same output link (see Figure 9.14(a)). This problem can be solved by a number of means like buffering conflicting packets, internal speed-up and sorting. External blocking, on the other hand, occurs when two or more packets are destined for the same output port in a single time slot (see Figure 9.14(b)). Buffering is used to resolve the problem of external blocking.

FIGURE 9.13

Self-Routing in a Banyan Network

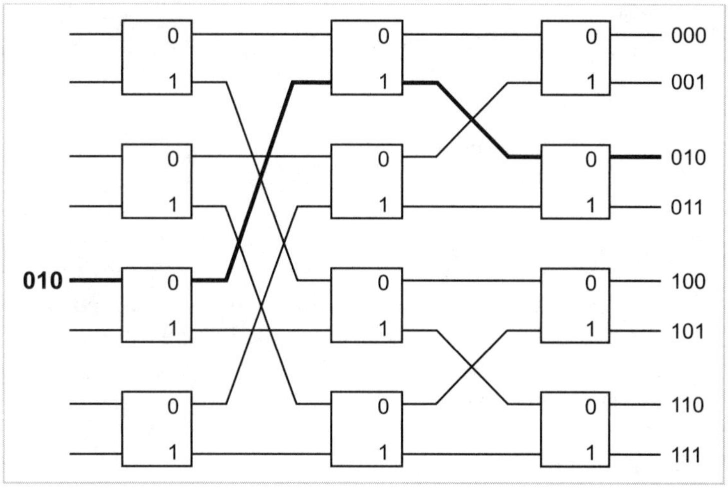

- **Non-realizable permutations:** The reduced requirement of crosspoints in a Banyan networks has an assistant performance limitation. It is that not all permutations are now realizable. Recall that in an N × N crossbar matrix, without external blocking, there can be N simultaneous paths for data transfer and a total of N! possible permutations. In contrast, a banyan network allows only $2^{kN/2}$ permutations, where $k = \log_2 N$, number of stages in the switch. This factor can be calculated by knowing that each crosspoint switch can be in one of the two states and that there are a total of $(N/2)*\log_2 N$ crosspoint switches. In essence, this factor means that there can be cases in which there can be a non-realizable permutation without external blocking.

9.8.2.1 Blocking Solutions in Banyan Networks

In Banyan networks, the problem of blocking can be solved by using one of the following options: *buffering*, *internal speed-up*, and *sorting*. In this section, buffering and sorting techniques are discussed.

- **Buffering:** Both internal and external blocking problems can be solved by buffering conflicting packets. In Banyan networks, three types of buffering are possible—*input buffering*, *output buffering* and *internal buffering*. Internal buffering is again classified into input buffering, output buffering and shared-memory buffering. The general issues pertaining to each of these techniques have already been discussed. Nonetheless, a few comments on internal buffering are in order. With internal buffering, one of the two conflicting packets in a 2 × 2 switch is allowed to pass through, while the second one is buffered. In the next time slot, for the buffered packet to move forward, either the buffer of the next stage should be empty or the buffered packet of the next stage should move forward. In case neither these conditions holds true, some mechanism is required to inform the packet of the preceding stage to remain in its buffer. This is achieved by using upstream control signals. Indeed, internal buffering has its own share of problems. As the number of intermediate stages increases, internal buffering disrupts the delay and jitter guarantees. Moreover, determining the optimal buffer size is not easy, because this tends to be a function of traffic behaviour. For traffic distributed uniformly across all the links, a fewer number of buffer elements are required than when traffic is concentrated across a few links.

- **Sorting:** Banyan networks have an interesting property in that when the packets at inputs are sorted according to their output port addresses, there is no internal blocking. Going back to Figure 9.14, one will observe that a packet with higher output address (011) is above the packet with a lower output address (010). For the same inputs, Fig. 9.15 shows how the internal blocking is removed by introducing a *Batcher-Bitonic sort network* before the Banyan network. By introducing a sorter before the Banyan network, the network becomes internally non-blocking. The problem of external blocking, however, still remains.

9.8.2.2 Multi-casting

Due to the connection-oriented approach of Banyan networks, it is difficult to incorporate multi-casting capability. One way to incorporate multi-casting in Banyan networks is to introduce a copy network before the Banyan network. In this model, the switch consists of a copy network, followed by a distribution network and a routing network. The copy network makes multiple copies of multi-cast packets. Then, the distribution network, which essentially is a buffered Banyan network, distributes packets evenly across all inputs of the routing network. Between the copy network and distribution network are the Broadcast and Group Translators (BGTs). The purpose of a BGT is to translate the destination address of the copies of packets to the appropriate destination address. The routing network forwards the packets to the appropriate output port.

9.8.2.3 Scalability

Banyan networks have many desirable features that make their implementation attractive. First, they can be constructed modularly from smaller switches. Second, their regularity and interconnection pattern make their Very Large Scale Integration (VLSI) implementation a good proposition. Still, internal and external blocking severely hamper the throughput of large banyan networks. In fact, the maximum achievable throughput for a moderate size, single-buffered banyan network (N = 32) is as low as 0.4 [Swi Fouad A. Tobagi]. This value falls to 0.26 for larger networks (N = 1024). The throughput can be improved by increasing the buffer size. Similar performance gains are also observed by introducing distribution networks before the routing network and by decreasing the number of intermediate stages.

9.9 Switching in ATM

ATM has emerged as the preferred transfer mode for high-speed transmission of data. This development is the result of sustained research in the field of switching over a period of one-

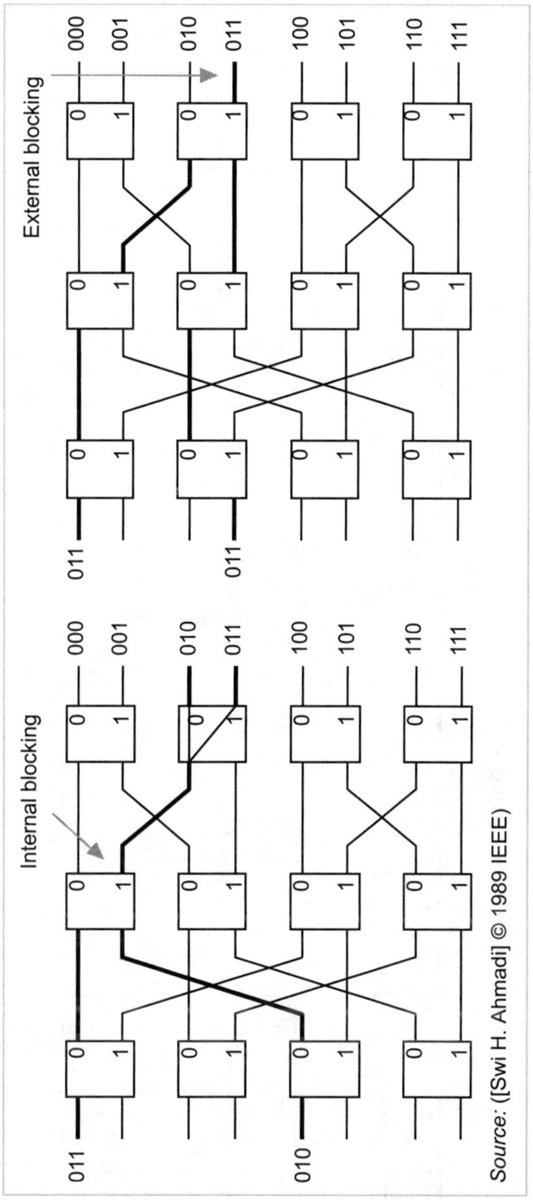

Source: ([Swi H. Ahmadi] © 1989 IEEE)

FIGURE 9.14 Internal and External Blocking in Banyan Networks

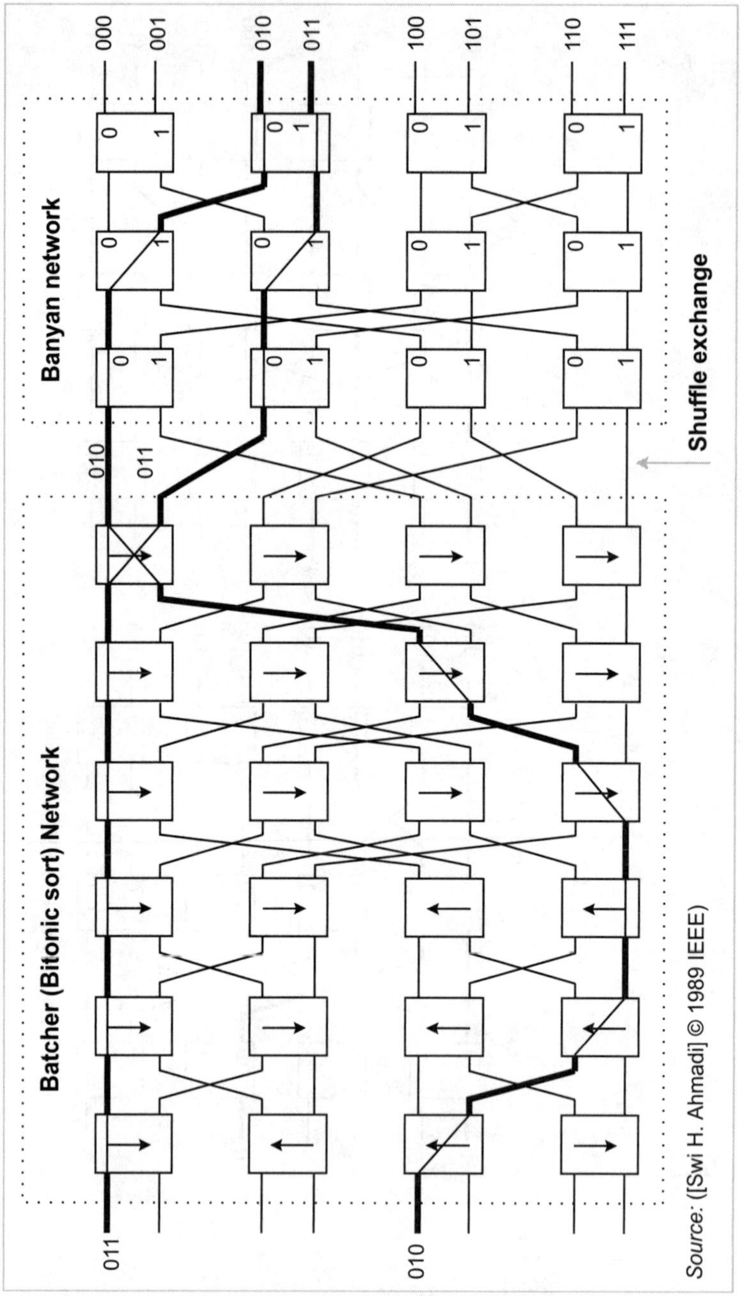

FIGURE 9.15
Batcher-Banyan Network

and-a-half decades (1985-2000). The endeavour is not only to develop switching fabrics capable of switching millions of cells per second, but also to guarantee the QoS promises of ATM technology. After all, efficient switching operations and efficient traffic management of cells hold the key to delivering the QoS promises of ATM.

The fact that ATM is a hybrid of traditional circuit switching (for voice transfer) and packet switching (for data transfer) has an important bearing on ATM switching architectures. However, neither packet-switching architectures nor circuit-switching architectures are directly applicable to ATM switches. The following paragraphs explain why.

Although ATM is connection-oriented, it is significantly different from circuit switching. In circuit switching, dedicated switched paths are setup once a voice channel is established. In contrast, ATM uses a store-and-forward switching that is based on statistical multiplexing. This implies that ATM cells are buffered until the destined output ports are free. While buffering is a key design issue in ATM switches, voice frames are never buffered in circuit switches. Moreover, the speeds at which ATM switches operate are many times higher than the conventional circuit switches. Further, while voice is carried in fixed voice channels (usually 64 kbps), ATM switches are capable of supporting multiple data rates (from as low as few kbps to as high as few Gbps). Finally, ATM switches do Usage Parameter Control (UPC) and Network Parameter Control (NPC) at the ingress point of the network. They also support traffic management features like prioritized transfer (based on multiple classes of traffic), congestion notification and flow control. All these features are not required in conventional circuit switches.

Even compared against packet-switching technologies like X.25 and frame relay, ATM switching is different. This is despite the fact that ATM technology is inspired from frame relay. Speed is again a crucial difference. The early packet switches operated at very low speeds (~64 kbps), and thus, software-based processing was possible. Compare this with 150 Mbps ATM switches that make hardware-based processing a prerequisite. Moreover, the low speeds and protocol complexity of X.25 makes it a poor candidate for carrying real-time data like voice and video. Although, frame relay is much simpler as compared to X.25, but like X.25, it also supports transmission of variable-sized frames. Variable size frames introduces unpredictable delay in the switch and hence, increases the jitter. Thus, both X.25 and frame relay are not well suited for real-time applications. In contrast, ATM provides services like CBR and rt-VBR, which require bounded values of end-to-end transit delay and cell delay variation. This makes buffering of cells a very critical issue.

In essence, switching in ATM is a complicated process because of the following reasons:

- The high-speeds at which ATM switches operate (few Mbps to few hundred Gbps),
- The large number of inputs (~1000) that ATM switches must support,
- The variety of data-rates that ATM switches support,
- The traffic management features that ATM switches support

The following sections explain the functions of a typical ATM switch. An ATM switch is more than just an interconnection structure that buffers and routes cells. On the contrary, the control plane functions and traffic management functions of an ATM switch pose significant design complexity. Although this chapter is not an attempt to explore the design issues associated in performing the control plane and traffic management functions, there is a mention of the related functions that an ATM switch must perform.

The function of an ATM switch is divided according to the three planes of the three-dimensional reference model. Recall that the ATM reference model defines three planes—the user plane, the control plane and the management plane. The following sections elaborate upon the functions of an ATM switch in the three different planes.

9.10 ATM User Plane Functions

In the user plane, the primary function of an ATM switch is to transfer user cells from incoming links to outgoing links. The ATM switch, however, remains transparent to the user information carried in the 48-byte payloads of every cell. The only part of the cell inspected and processed is the 5-byte cell header. The cell processing includes VPI/VCI translation, HEC verification, cell tagging, and explicit forward congestion indication.

In the context of user plane, the three core modules (input module, switching fabric and output module) perform three distinct sets of functions. The functions of each of the module are discussed below.

9.10.1 Functions of the Input Module

The Input Module (IM) of an ATM switch performs several functions. The most important among these is the VPI/VCI translation. The following steps are associated with the table look-up and VPI/VCI translation:

1. As a cell arrives, the VPI field of the cell header is extracted. This value is used for locating an entry in the translation table associated with each link. (Note that the VPI/VCI values are not necessarily unique across different interfaces. Hence, in case they are not unique, an ATM switch must have means to distinguish between cells received from one link with cells received from another. One of the ways to ensure this is to use separate translation table for each link and use the link identifier along with VPI/VCI). The table look-up may be through hashing, search trees, or through content addressable memories. In order to minimize the table look-up time, only a subset of the total VPI/VCI values is allocated. For example, only 12 bits for VCI and 8 bits for VPI may be used. This reduces the search space, and hence the search time of table look-up.
2. The table entry is then used to determine whether the particular VPI value corresponds to a VPC or a VCC. In other words, whether the VPI is a part of a VPC or part of a VCC. Note that a VPC is switched, based on VPI value only, whereas a VCC is switched using the VPI/VCI pair.
3. If the VPI value corresponds to a VPC, then relevant information is fetched from the translation table entry. This information includes the new VPI value and the outgoing

link identifier. Additionally, other information like cell priority, delay bound and jitter bound may also be fetched from the table.

4. If the VPI value corresponds to a VCC, another level of search is made using the VCI value. Again, the appropriate information is fetched from the translation table.
5. The old values of the VPI/VCI are then replaced by the new values (in case of step 3, only VPI value is replaced). A routing tag is then attached to the cell. This tag contains the information that was fetched from the translation table. Apart from the parameters already mentioned, a multi-cast cell may also carry a bit-map corresponding to the set of outlets for which the cell is destined. The cell is then forwarded to the switching fabric.

Apart from the above steps, the input module also performs the following functions:

- It verifies the HEC value of every cell and discards erroneous cells.
- It discards empty cells.
- It identifies the cell boundaries (using the cell delineation procedures).
- It converts optical signals into electrical signals (applicable for SONET/SDH based interfaces only).
- It separates the user cells from signalling cells and OAM cells. The signalling cells are forwarded to the CAC module and OAM cells to the management module.
- It does usage parameter control and network parameter control (UPC/NPC) at the ingress point of the network.

9.10.2 Functions of the Switching Fabric

The primary function of an ATM switching fabric is to transfer cells from input modules to output modules. Another important function of the switching fabric is to buffer excess cells. Besides routing and buffering, other functions of switching fabric include the following:

- Multicasting/broadcasting cells
- Providing support for fault tolerance
- Traffic concentration and multiplexing
- Traffic expansion and de-multiplexing
- Congestion notification by setting EFCI bit

9.10.3 Functions of the Output Module

The output module is the direct opposite of the input module. Thus, the functions of the output modules can easily be deduced from these of the input module. To summarize, the output module performs the following functions:

- It generates the new HEC value and adds it to the cell header.
- It generates empty cells, as and when required, and sends it (cell rate decoupling).
- It multiplexes user cells with signalling cells and OAM cells.
- It converts electrical signals into optical signals (applicable for SONET/SDH-based interfaces only.

9.11 ATM Control Plane Functions

As discussed earlier, the control plane in the ATM reference model is responsible for the establishment, monitoring, and release of switched virtual circuits. All these control plane functions are performed by exchanging signalling messages. As far as signalling is concerned, an ATM switch forwards all the signalling cells received over the reserved VPI/VCI pair (0, 5) to the CAC module. The CAC module processes every connection request, and depending upon the availability of resources, determines whether new connections can be accepted or not. Detailed discussion of signalling procedures is provided in Chapter 11. For the purpose of signalling, PNNI routing is used to determine the optimal routes. A detailed discussion of PNNI-based routing procedures is provided in Chapter 12.

9.12 ATM Management Plane Functions

The management plane functions of an ATM switch are broadly categorized into two important categories, namely fault management and performance management. These management procedures are performed by using the OAM cells. OAM procedures are covered in detail in Chapter 14. There also exist other mechanisms for management, some of which are detailed in Chapter 14.

9.13 Case Study of ATM Switching Architectures

The architectural issues related to ATM switch architectures are not very different from those discussed in the previous part of the chapter, and is not repeated. Nonetheless, specific implementations of ATM switches are elaborated, and thus, major switching architectures are covered once again. As mentioned already, there are three classes of switching architectures, viz. shared-memory architecture, shared-medium architecture and space-division architecture. Thus, specific implementations of each of these architectures are looked at. The following switch implementations are discussed:

- **Prelude:** This is based on shared-memory architecture (see Section 9.14).
- **Packetized Automated Routing Integrated System (PARIS):** This is based on shared-medium architecture (see Section 9.15).
- **Knockout:** This is based on space-division architecture (see Section 9.16)
- **Starlite:** This is a multistage switch (see Section 9.17)
- **Moonshine:** This is again a multi-stage switch with some improvement over Starlite design (see Section 9.18)

9.14 Prelude Switch

Prelude switch was designed by CNET in France. The aim was to provide high-throughput and flexible switching capability for integrated broadband networks. The throughput is required to support high bandwidth applications like video. The flexibility is required to provide a multi-service environment that is capable of supporting different applications including voice, video and data. The Prelude switch is described in detail in [Swi M. Devault].

The prototype built by CNET supported a packet size of 16 bytes, with 15 bytes of payload and 1-byte header. Prelude switch design assumes that the number of input ports and output ports is equal to the number of bytes in a packet (i.e. 16). The need for this assumption will soon be clear. The figures in this section however assume a packet size of four bytes (three bytes of payload and one-byte header).

As illustrated in Figure 9.16, there are four distinct stages in a Prelude switch. In the first stage, the packets are converted from serial to parallel form. Thus, the state delivers a cell stream byte-by-byte without any phase relationship from one output to another. Since the clocks are derived from remote clock, the output ports operate based on these remote clocks.

Figure 9.16

Overall Structure of Prelude Switch

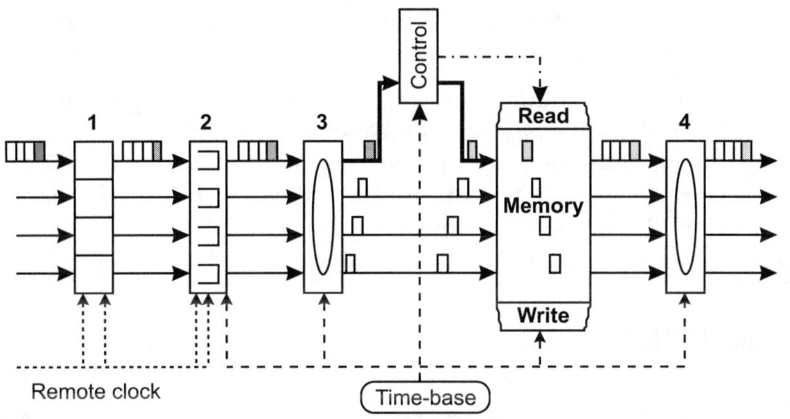

1: Framing and parallelization 2: Clock adaptation and phase alignment
3: Super-multiplexing 4: De-multiplexing and serialization

Source: ([Swi M. Devault] © 1987 IEEE)

This is followed by clock adaptation and phase alignment stage (stage 2). This state locks the clock and phase of each cell carrier on all outputs to a central clock. This stage time-shifts the packets by one byte from one link to the next (see Figure 9.17), resulting in a diagonal alignment. This alignment ensures that if there is header of a cell on output link in one cycle, then in the next cycle the header is on the next output link. As is clear from Figure 9.17, this stage ensures that there is at most one header on the output links.

FIGURE 9.17

Diagonal Alignment of Packets in Prelude Switch

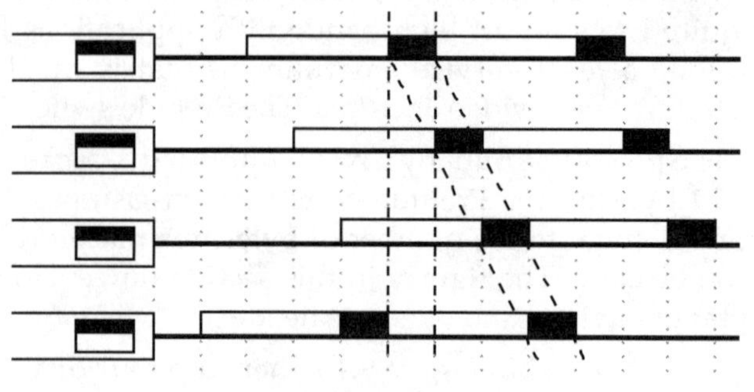

Source: ([Swi M. Devault] © 1987 IEEE)

The diagonally-aligned packets are then fed into the super-multiplexing stage (stage 3). The super-multiplexing stage is essentially a rotative space division switch that assumes cyclically N (here, 16) different switching patterns. The purpose of this stage is to extract the header information from all the incoming packets and put it on the first output that leads to the controller. Figure 9.18 shows a super-multiplexing stage for 4-byte packets.

As is clear from Figure 9.18, the headers of the four input lines are transferred to the first output line, while the payload of the packets are multiplexed onto the remaining output lines. Note that the multiplexing of payload is not done arbitrarily. The second output line always carries octet 1 of each payload. Similarly, the third output line carries octet 2 of each payload, and so on. Thus, the 15 output lines carry the 15 octets of payload, one octet each. This explains why the number of lines is equal to the number of bytes in a packet. Another

FIGURE 9.18

Parallel-Diagonal Super-Multiplexing of Packets Using a Rotative Space Division Switch in Prelude

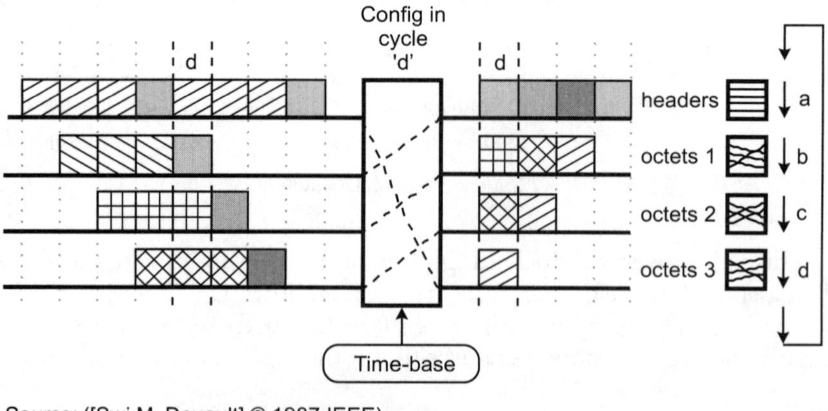

Source: ([Swi M. Devault] © 1987 IEEE)

important outcome of this form of multiplexing is the parallelism achieved. This allows the use of N memory modules, each module storing a byte.

The headers carried by the topmost output line are fed into the controller. The controller processes the headers and determines the output port for the packet using the address translation table. It also replaces the header information with that stored in the address translation table. The new header, along with the payload octet, is stored in the shared memory. Additionally, the address of the packet header is placed in the output queue for which the packet is destined. At each byte time, the controller delivers the address of a packet (i.e. the address of the header), which to be extracted and forwarded. Octets of the payload are extracted from other memory banks. The transmission of the extracted packets is done by stage 4, using techniques similar to the one used in stage 3. Stage 4, called the demultiplexing and serialization stage, re-assembles the octets from different lines to form a 16-byte packet. Figure 9.19 shows the reassembly of a 4-byte packet.

FIGURE 9.19

Extraction of Packets from Memory of Prelude

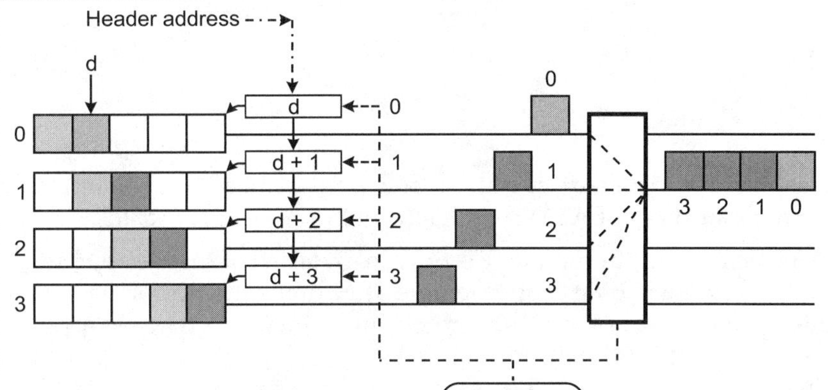

Source: ([Swi M. Devault] © 1987 IEEE)

9.15 Packetized Automated Routing Integrated System (PARIS) Switch

Packetized Automated Routing Integrated System (PARIS) switch is a typical, shared-medium switch designed to carry voice, video and data, all in packetized form (see Figure 9.20). The switch is capable of handling variable size packets (32 bits to 8 kilobits). To avoid packet loss due to bus arbitration delays, an input buffer capable of storing a maximum of four packets is maintained. With a round-robin bus arbitration scheme, an input buffer with four elements is sufficient to ensure that there is no packet loss. At the output side, two buffers are maintained for two priority levels, with each buffer having two elements. The PARIS switch is described in detail in [Swi I. S. Gopal].

FIGURE 9.20

PARIS Switch Architecture

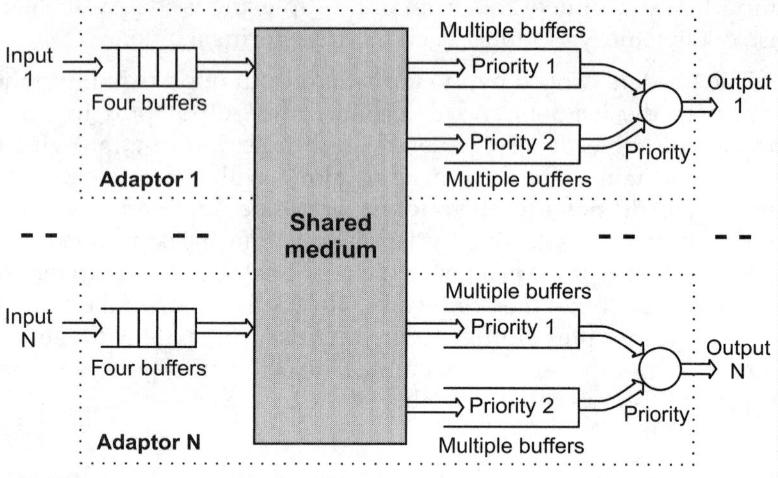

Source: ([I. S. Gopal] © 1987 IEEE)

9.16 Knockout Switch

The previous section discussed the functions of an ATM switch. This section describes one of the popular types of ATM switch referred to as *knockout switch.*

Typically, output buffering is optimal in terms of throughput, it requires that the memory buffers and switching fabric operate at N times the port speed. Knockout switches reduce this speed-up factor by deliberately introducing some loss probability in the switching fabric. To understand the fundamentals of knockout switching, let us get down to switching basics.

The speed-up factor of N is required because it is quite possible that cells arriving from all the N inlets are destined for the same outlet. Now, if it is decided that only L cells out of the maximum N are buffered in a cell-time, then the speed-up factor is reduced to L. This decision, however, introduces a finite probability of some cells (N–L in the worst case) getting dropped. If, somehow, the cell loss ratio of a connection is maintained within permissible limits, then this approach can easily be adopted. Interestingly, [Swi Y.S. Yeh] have shown that to achieve a cell loss probability of 10^{-10}, a value of L = 12 is sufficient (this result applies for 90 per cent load and for arbitrarily large values of N). Thus, even for a very large value of N, and even for high load, the speed factor is more or less constant at L.

Besides requiring a very low speed-up factor, knockout switches offer many other desirable features like *self-routing*, *low latency* and *non-blocking transfer*. Moreover, the buffer requirement has a linear growth with N. This is in sharp contrast to many other switching fabrics with output buffering that have quadratic growth with N. All this makes a knockout switch an excellent option for large switching systems.

9.16.1 Components of Knockout Switch

Figure 9.21 depicts a schematic diagram of a knockout switch. Cells arriving at an input are broadcast to all the output ports, using a passive broadcast bus. The broadcast cells are intercepted by a bus interface that precedes each output port. Indeed, the core of the knockout switch resides in the bus interface (see Figure 9.21). The bus interface first filters the cells destined for its output port and discards the remaining cells. It then queues up cells that are not discarded in a shared buffer. The buffered cells are then served on a first-in-first-out basis.

FIGURE 9.21

PARIS Switch Architecture

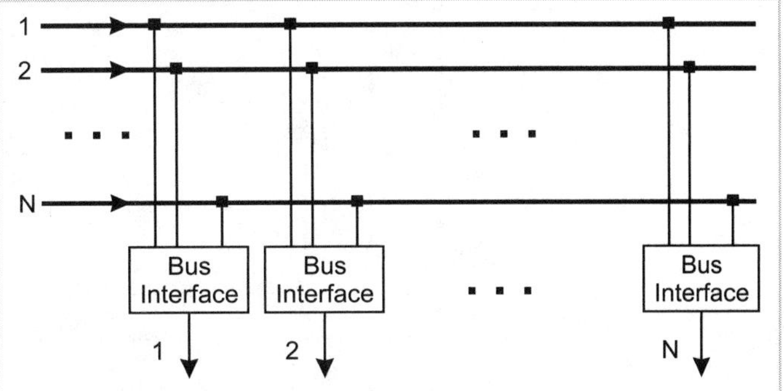

Source: ([Swi H. Ahmadi] © 1987 IEEE)

As shown in Figure 9.22, a bus interface has three major components: *cell filter*, *concentrator* and *shared buffer*. Each of these components is discussed below.

9.16.1.1 Cell Filters

Each bus interface has N cell filters, one for each input line. Each filter examines the output port address of the received cell. Irrespective of the input line on which the cell is received, a cell filter in the bus interface *i* allows only cells destined for output port *i* to pass through. All the remaining cells are blocked.

The implementation of a cell filter is straightforward. A cell carries the output port address as a binary number (with $\log_2 N$ bits). This number is compared bit-by-bit against the output address of the bus interface. If all the bits match, the cell is allowed to go through, otherwise as soon as a compare operation fails, the forward path is blocked for that particular time-slot. If some part of the cell goes through before the path is blocked, then the garbage bits are ignored by the subsequent stages of the bus interface.

9.16.1.2 Concentrator

The cells that successfully cross the cell filter enter the concentrator. The concentrator is an N-input, L-output element that concentrates a number of input lines (N) onto a given number of output lines (L). Specifically, if there are more than L cells for an output port in a particular

FIGURE 9.22

Basic Structure of a Knockout Switch

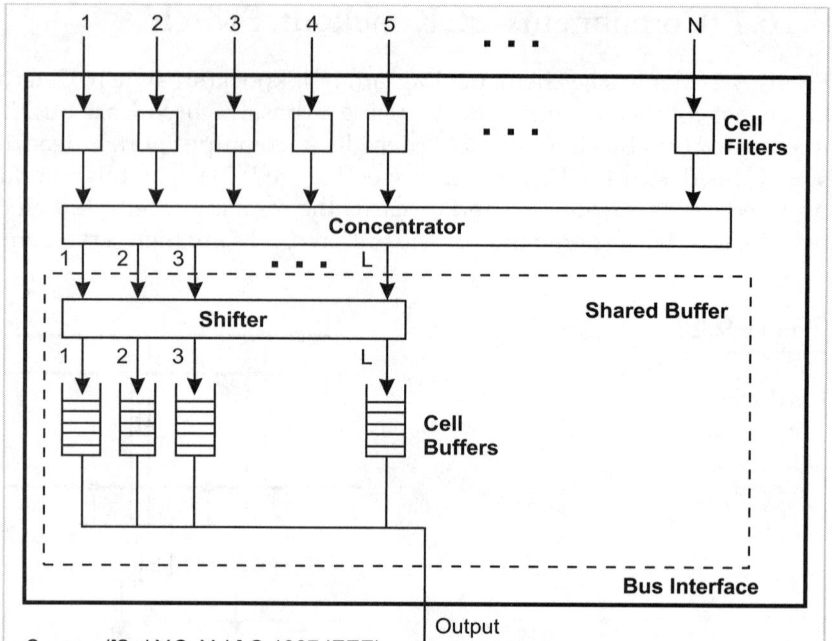

Source: ([Swi Y.S. Yeh] © 1987 IEEE)

time slot, only L cells are allowed to pass through, while the remaining cells are dropped. Thus, the concentrator introduces some cell loss probability in the switching process. It can be argued that since cell loss is unavoidable in any practical communication network (due to link failures, transmission errors, buffer overflows, etc.), a switch also has the right to induce some. However, the main issue is whether the Cell Loss Ratio (CLR) guarantees of ATM connection are maintained despite the induced cell loss. If yes, then the logic given for cell loss in knockout switch is justifiable. Now, the typical values of CLR are $\sim 10^{-10}$. It is experimentally proven that even for 100 per cent loading, a value of L = 12 (for arbitrarily large value of N) is sufficient to achieve a cell loss of 10^{-10}. Thus, the cell loss of a knockout switch is well with the permissible limits of CLR for ATM connections. It can be further proven that each additional output line of the concentrator (beyond L = 8) leads to a decrease in the loss probability by a factor of 10. This is a significant result, because even if the CLR guarantees were to reduce beyond 10^{-10}, with very little modification, the desired result can be obtained. This result also means that the buffer requirement is now a function of N*L (N output ports with L queues each) rather than N^2. This explains why knockout switches have linear growth with N.

The concentrator consists of *contention switches* and one-bit *delay elements*. A contention switch is a 2×2 switching element, which chooses a *winner* and a *loser*. When cell arrives at only one input port, the arrived cell is automatically declared as the winner. If cells arrive at both the input ports, the winner may be decided on the basis of a static priority or on an alternating priority basis. In case of the former, the input towards left is given a higher priority. The one-bit delay elements are single-input/single-output delay elements that ensure that the cells leave the concentrator simultaneously.

The peculiar design of the concentrator gives the knockout switch its name. This is because the cells are concentrated by using a *knockout competition*. As stated in previous paragraphs, the winner in a competition may be decided on the basis of a static priority or on an alternating priority basis. The knockout tournament is played across several sections, with one section for each concentrator output. In the tournament, each cell is allowed to compete (and lose) in at most L sections. As soon as a cell loses a match, it is eliminated from the competition for that section. This cell then moves towards the section on its right. There, it competes with other loser cells. A winner cell on the other hand, stays in the competition and fights with other winner cells in the next round. The tournament continues until L winners are chosen and K–L cells discarded. Here, K is the number of cells arriving simultaneously at the inputs of a concentrator. Note that a cell is discarded only if more than L cells simultaneously arrive at the input (that is, K > L). For all other cases, there is no cell discard.

Figure 9.23 depicts a 8-input, 4-output concentrator. The eight inputs are fed in pairs in four contention switches. The four winners remain in Section 1, while the four losers move towards section two. These four losers are said to be 'knocked out' of the competition for

FIGURE 9.23

The 8-input/ 4-output Concentrator

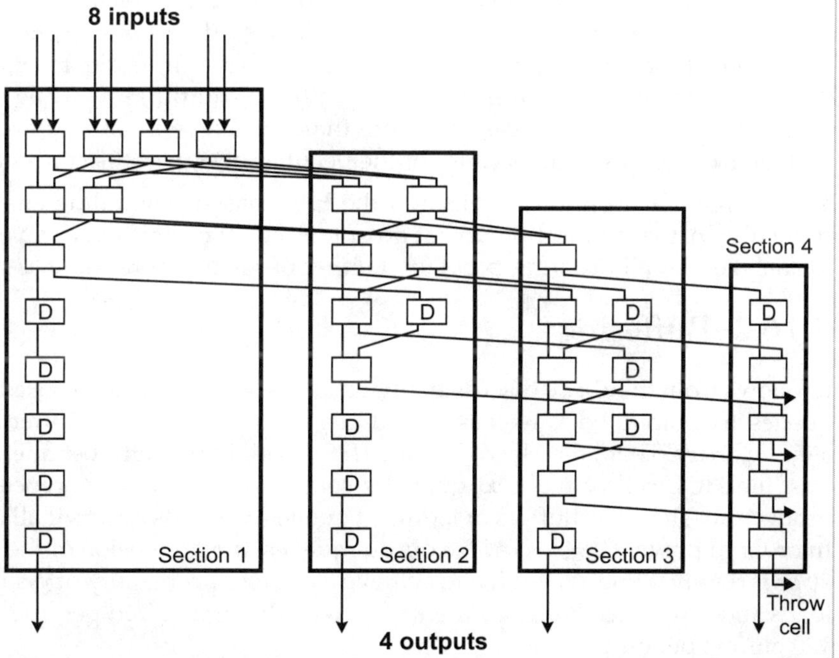

Source: ([Swi Y.S. Yeh] © 1987 IEEE)

section 1. However, they are still in the fray for the remaining three sections. The four winners compete in pairs to leave two winners. The remaining two then compete against one another to give the winner of Section 1. Similar competition is held in all the remaining sections for choosing a total of L winners (four in this case). The remaining N–L are discarded (again four cells in this case are discarded).

9.16.1.3 Shared Buffer

The cells winning a concentrator output enter the shared buffer, the third component of the bus interface. Note that the speed-up factor requirement for output buffering is reduced significantly because of the N by L concentrator. However, L cells for buffering still requires a speed-up factor of L. To minimize this requirement, L separate FIFO buffers are used.

In the design of the shared buffer, one issue needs to be considered. It is that the cells tend to emerge from the concentrator in left-to-right direction (i.e. the cells first emerge from the leftmost section, then from the second section, and so on). Now, this will lead to overloading of buffers towards the left and underloading of buffers towards the right. In order to prevent this phenomenon, a *shifter* is used. The role of the shifter is to distribute cells in a cyclic fashion such that the L buffers are filled equally. The shifter is designed in a manner such that the total number of cells in each buffer does not differ by more than one cell. This is maintained, irrespective of the number of cells arriving in a particular time slot.

For example, consider the operation of an 8-input shifter. In the first time slot, five cells arrive at inputs numbered 1-5. These cells are passed without any shifting to outputs 1-5. In the second time slot, suppose four new cells arrive. In order to ensure that the cells are distributed equally, the four new cells are shifted to outputs 6, 7, 8 and 1. This cyclic operation of the shifter continues endlessly. Note that the cells will always arrive contiguously, and that too from left to right, because of the peculiar design of the concentrator.

The cyclic distribution of cells and the FIFO-based scheduling ensures that the cells of a particular virtual connection do not go out of sequence. It is left as an exercise for the readers to analyze the cell-ordering problem if either of the two conditions is violated.

9.16.2 Buffering

It follows from the above discussion that a knockout switch uses shared output buffering. The queues are both filled as well as served in a cyclic fashion. The shifter first fills the queues in a cyclic order. Then, the Head-Of-Line (HOL) cell in the leftmost queue is first served. After this, the HOL cell of the next queue is served. This process is repeated cyclically. All the inputs fully share the buffers of L output queues. Cell loss occurs if all the L queues are filled. In order to prevent cell loss at the output queues, the dimension of the queue must be chosen appropriately. Experimentally, to achieve a cell loss probability of nearly 10^{-10} for 90 per cent loads and L = 12, ten buffer elements (i.e. B = 10) are required per queue. This gives a total of 120 buffers per output port.

9.16.3 Multi-casting

Like most other broadcast-based switching fabrics, a knockout switch can realize multi-casting easily. All that is required now is additional control logic in the cell filters. Recall that a cell filter makes a bit-by-bit compare operation with the received output port address to filter cells. To incorporate multi-casting, this scheme has to be changed. One of the solutions is to use a bit-map vector, one bit for each output port. Each filter then makes a check on the corresponding bit to find out whether it is set or not. If it is set, then the cell is allowed to pass through, otherwise it is blocked.

9.16.4 Scalability

A knockout switch can be easily expandable to large values of N (~1000) with port speeds of 50 Mbps. This ease of scalability is the result of modular design and linear growth with N. In fact, the knockout switch can grow modularly from N × N to JN × JN, where $J = 2, 3, \ldots$.

For example, four N × N knockout switches can be used to realize a 2N × 2N switch with minor modifications. In the new approach, each concentrator has L additional inputs, with the inputs being received from the concentrator outputs of the previous stage. The operations of the remaining elements remain the same.

For details of various aspects of knockout switch, refer to [Swi Y.S. Yeh].

9.17 Starlite Switch

Starlite switch, proposed by Huang and Knauer [Swi A. Huang], is one of the earliest implementations of the Batcher-Banyan-based switch design. The switch is designed to handle fixed length cells. Each cell is routed by using a routing tag.

Starlite switch consists of a concentrator, a sorter, a trap network and an expander. Figure 9.24 depicts the Starlite switch architecture. The key challenge in the design of this switch is handling the external blocking problem. Recall that by using a sorter before the Banyan network, the problem of internal blocking is solved. To solve the problem of external blocking, Starlite design uses a trap network followed by recirculating buffers. Consider an example for understanding the operations of the Starlite switch. A four-input Starlite input switch receives the following cells in a time-slot: (1, 2), (2, 4), (3, 2) and (4, 1), where, in the tuple (i, j), i is the input port address and j the output port address. Figure 9.25 depicts this scenario. After the cells are sorted, a cell is said to lose contention for an output port, if the cell on the preceding output line has the same output port address. In the given example, (3, 2) loses contention for the output port 2 as cell (1, 2) is also destined for the same output port 2. The cells that lose contention are then fed back into the input side of the sort network for re-entry in the next time-slot. These cells, however, are permitted to enter idle input ports only.

FIGURE 9.24

Starlite Switch Architecture

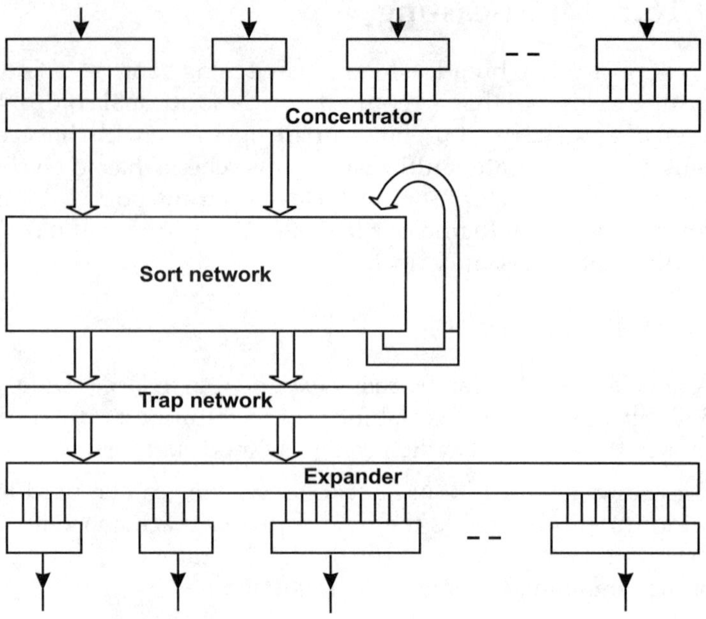

Source: ([Swi A. Huang] © 1987 IEEE)

Since, at any time, the number of recycled cells can be greater than the number of idle input ports, an intermediate buffering stage is introduced. This explains why buffers are placed after the sort network (see Figure 9.25).

FIGURE 9.25

Re-entry in a Starlite Switch

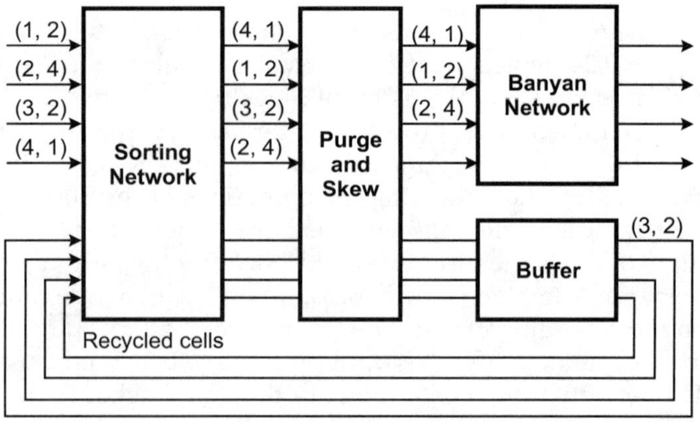

Source: ([Swi A. Huang] © 1987 IEEE)

Recycling cells leads to another design problem. If cells are simply allowed to re-circulate, there is a strong possibility of the cells getting out of sequence. This problem is solved by associating a priority with each cell, where priority is dependent on the age of the cell. This aging process ensures that cells in the buffers are given a higher priority over cells arriving at a subsequent point of time.

9.18 Moonshine Switch

A few years after the Starlite switch was proposed, Hui [Swi J.Y. Hui] proposed a new, 3-phase algorithm to handle the external blocking problem. In this work, the following drawbacks of Starlite design were pointed out:

- The trap network introduced cell loss in the switch.
- Despite using age-based priority scheme, cells could still get out of sequence.
- Almost half of the inputs are dedicated for the recycled cells.

The concentration network that was required to skew the outputs of the sort network resulted in *additional requirement* for chip sets and sub-system designs. Note that the output of the sort network contains cells with duplicate addresses. Removing some cells leaves holes in the output lines, and thus, cells need to be skewed to the top lines of the Banyan network.

Moonshine design also relies on sort network to solve the problem of internal blocking. In addition, sorting also helps to identify the number of packets destined for the same output port. This is subsequently used to tackle external blocking. How the external blocking problem is actually solved is explained by considering the following three phases in the Moonlight switch design:

- **Phase 1 (Send and resolve request):** In the first phase, instead of sorting the complete cells, the cells are sorted according to their output port address (see Figure 9.26(a)). Each input port i sends a short request packet which is merely a source-destination pair (i, j_i), where j_i is the output port for cell arriving on the ith inlet. These request packets are then sorted by the sorter in a non-decreasing order. The request is granted to the input port i only if j_i is different from the output port address of the request above it. This phase is called the *arbitration phase.* By using request packets instead of the complete cells, re-circulating buffers are no more required.
- **Phase 2 (Acknowledge winning port):** After the arbitration phase is over, the result of the contention is informed to all the requesting inputs (see Figure 9.26(b)). This is done in the second phase of the algorithm. In this phase, all request packets that had won the contention send an acknowledgement to their respective inputs. This acknowledgement is sent by using a fixed connection from the kth output of the sort network to the kth input.
- **Phase 3 (Send cell):** In the final phase, the inputs receiving acknowledgements transmit the complete cell. The remaining inputs, which did not receive any acknowledgement, buffer cells to retry in the next time slot.

FIGURE 9.26

Phases in the Moonshine Switch

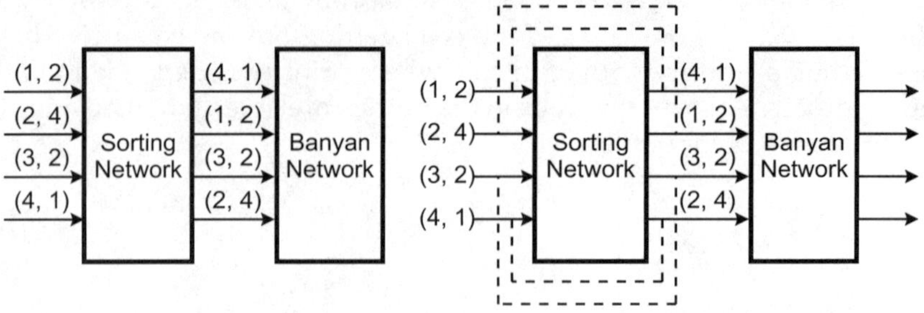

(a) Phase-1 (send and resolve request)

(b) Phase-2 (acknowledge winning port)

Source: ([Swi J.Y. Hui] © 1987 IEEE)

The three-phase algorithms remove most of the drawbacks of the Starlite design. Since the cells of a particular connection are buffered at the inputs, there is no way the cells can get out of sequence. Neither is there any possibility of cell loss, unless the buffers at the inputs overflow. Moreover, as there are no recirculating buffers, no inputs need to be dedicated for the recycled cells.

Moonshine design employs input buffering to buffer contending cells. As discussed earlier, input buffering leads to HOL blocking. One of the ways to solve this problem is to modify the original design to associate a priority with each cell. Further, instead of having a single round of phases 1 and 2, multiple rounds of each of these phases are held. After the first round of phase 1 and 2 is over, the cells next in line to HOL cells are given a chance to contend for idle outputs. One thing that must be ensured is that the winning cells of round one do not lose in the next round. To ensure this, the priorities of the winning cells are increased suitably to ensure that they do not lose in the next round. Phases 1 and 2 are repeated for a fixed number of times, after which the inputs can transmit winning cells.

The number of rounds to be held is a design issue. With each additional round, the speed-up factor of the switch must be increased appropriately. This is essentially a trade-off between increased throughput and speed-up overheads. Note that even without using the multiple-round approach, a speed-up factor is required for phases 1 and 2; this is because in these phases, no cell forwarding is done, merely additional processing is done.

CONCLUSION

Switching along with traffic management occupies a significant role in delivering the promises of ATM technology. This is because the technological strengths of ATM, viz. high-speed switching and the ability to provide QoS guarantees is inextricably linked with a proper switch design. The key challenges are achieving high-throughput, while still maintaining the delay and jitter guarantees for real-time application. Other modules in ATM switch like the CAC module and the management module pose additional challenges in switch design. Since both signalling and management functionality in ATM are quite complex and elaborate, the complexity reflects upon in ATM switch design.

Review Questions

Q 1. *Explain the merits and demerits of input buffering, output buffering and shared buffering?*
Q 2. *What is the difference between shared-memory and shared-medium architecture? Give examples of ATM switch designs based on these architectures.*
Q 3. *What is a Banyan network? What are the key problems in a Banyan network and how are they solved?*
Q 4. *Mention the key design challenges in ATM switch design.*
Q 5. *Describe the user plane, control plane and management plane functions of ATM switch?*
Q 6. *Describe the VPI/VCI handling function of an ATM switch.*
Q 7. *Why is knockout switch so named? Explain in the context of the functions of concentrator. Also highlight how the induced cell loss does not cause the performance of the switch to deteriorate.*
Q 8. *What is the role of shifter in the knockout switch? What happens to cell sequencing if the shifter distributes cells randomly?*
Q 9. *How is the external blocking problem solved in Starlite switch? What new problems does it cause?*
Q 10. *How does Moonshine switch solve the problems associated with Starlite switch? What new design challenges does it create?*

Further Reading

For further references on switching and the issues raised in this chapter, a number of good papers are available. [Swi M.G. Hluchyj] provides a good description on issues related to queueing. The switching architectures are discussed in [Swi Fouad A. Tobagi] and [Swi H. Ahmadi]. For more on ATM switching, the reader is referred to [ATM M. Prycker]. For the case study of ATM switch architectures, the reader is referred to [Swi M. Devault] for Prelude switch, [Swi I.S. Gopal] for PARIS switch, to [Swi Y.S. Yeh] for Knockout switch, [Swi A. Huang] for Starlite switch and to [Swi J.Y. Hui] for Moonshine switch. Apart from these papers, each of the above provides a number of other good references that readers can read and benefit from.

Chapter 10

ATM Addressing

10.1 Introduction

The primary goal of a network is to provide the means for two or more entities to communicate with each other. To communicate, some identifier is required so that one entity can uniquely and unambiguously identify another entity. This requirement is fulfilled by giving an address to the network entities. Addressing refers to the process of assigning unique numerical identifiers to network entities for the purpose of locating or identifying these entities. There can be various ways in which addresses can be viewed and classified. Table 10.1 provides some of the ways in which addresses can be classified; the table also provides few important addressing structures. For details on these topics, refer [Gen S. Kasera].

This chapter discusses the addressing structure applicable to the ATM end-systems. The addresses so used are called ATM End System Address (AESA). The AESA are based on NSAP format, details of which are provided in this chapter. Here, it may be noted that unlike Internet addressing where the end-system addresses are used for identification as well as message routing, the addresses in ATM are used only for establishment of virtual circuits (i.e. VPC/VCC). Once the VPC/VCC are established, the VPI and VCI are used for cell switching. The ATM addresses then have no role to play.

10.2 ATM End System Address (AESA) Format

The ATM address is also referred to as ATM End System Address (AESA). An AESA identifies the location of one or more ATM interfaces. Information on an ATM addressing and AESA is available in two documents published by the ATM Forum—the ATM Forum Addressing: Reference Guide [ATMF Addr RG 1.0] and the ATM Forum Addressing: User Guide [ATMF

TABLE 10.1

Basics of Addressing [Gen S. Kasera]

Definition	Addressing refers to the process of assigning unique numerical identifiers to entities for the purpose of locating or identifying these entities
Classification	• **Scope-based addressing:** Refers to the classification based on the scope or the geographical extent in which a given address is applicable. This leads to two class of addresses, viz. globally unique addresses and locally unique addresses. • **Utility-based addressing:** Refers to the classification based on the utility or the purpose that an address serves, which leads to two classes of addresses, viz. locators and identifiers. • **Layer-based addressing:** Refers to the classification based on the OSI layers, which leads to two classes of addresses, viz. link-layer addresses and network layer addresses. • **Functionality-based addressing:** Refers to the classification based on the number of entities being addressed, and leads to the four classes of addresses, viz. unicast, multicast, broadcast, and anycast addresses.
Examples	• **E.164 Addressing:** Defined by ITU-T, a number of up to 15 digits is used in this scheme to route ISDN/telephony calls. An E.164 number comprises of Country Code (CC), National Destination Code (NDC) and Subscriber Number (SN). • **Internet Addressing:** Defined by IETF, a 32-bit identifier is used in this scheme to identify the network part and the host part of an Internet host. • **Network Service Access Point (NSAP) Addressing:** Defined by ISO, this scheme is based on the concept of Service Access Points (SAP). An NSAP address comprises of an Initial Domain Part (IDP) and the Domain Specific Part (DSP).

Addr UG 1.0]. Both the documents are similar in flavour and content, although the former is more detailed in nature. This section briefly summarizes the content of the afore-mentioned documents, as applicable to the AESA structure.

AESA is based on the generic Network Service Access Point (NSAP) format. As per this format, an NSAP address is divided into two parts—the Initial Domain Part (IDP) and the Domain Specific Part (DSP). The IDP identifies a particular network-addressing domain that is part of the global network-addressing domain. The IDP field is itself divided into two parts—Authority and Format Identifier (AFI) and Initial Domain Identifier (IDI). The AFI field specifies the authority controlling the IDI and the format of the IDI. For example, AFI = 45 implies that the IDI is based on the E.164 address format. The IDI identifies the authority controlling the assignment of the Domain Specific Part (DSP). The meaning and format of DSP is controlled by the authority that is specified by the IDI field.

In ATM, the generic NSAP format is of 20 octets in length. Four different AESA formats are specified by the ATM Forum, viz. the Data Country Code (DCC) AESA format, the International Code Designator (ICD) AESA format, the E.164 AESA format and the Local AESA format. The type of format applicable is specified by the AFI field (i.e. different formats have different values of AFI). Thus, DCC format has AFI = 0×39, ICD format has AFI = 0×47, E.164 format has AFI = 0×45 and Local AESA has AFI = 0×49. Recall that the AFI field specifies the authority controlling the IDI and the format of the IDI.

The four different AESA formats are depicted in Figure 10.1. In the first three formats, the IDI field is represented by DCC, ICD and E.164 values, respectively. In the local AESA format, there is no IDI field. The DCC is a three-digit field that specifies the country to which the address is registered. The ICD is a four-digit (two-octet) field that identifies an authority responsible for the allocation and assignment of values of the Domain Specific Part (DSP). In E.164 AESA format, the E.164 specifies an ISDN number and is up to 15 digits in length (numbers with less than 15 digits are padded with leading semi-octets 0000). As the E.164 address is embedded in an NSAP structure, the address is called embedded E.164 address. In contrast, a native E.164 address is one that follows the structure provided in [ITU-T E.164] specification. The last format, Local AESA format, is used within a private network. Such addressing is useful when interconnection with other networks is not required.

Common to all the formats are the two basic parts—IDP and DSP. As stated earlier, the IDP consists of two fields: the Authority and Format Identifier (AFI) and the Initial Domain Identifier (IDI). The DSP consists of three fields—the High-Order Domain Specific Part (HO-DSP), the End System Identifier (ESI) and the Selector (SEL). The purpose of IDP, DSP, AFI and IDI fields are explained earlier. The components of DSP field are explained below.

The coding rules of HO-DSP field are specified by the authority that is identified by the IDP. The authority controls how addresses are assigned and interpreted within that domain. The authority may break the domain into sub-domains and use a portion of HO-DSP to identify individual domains. The division of domains into sub-domains provides means for hierarchical routing. Thus, the construction of HO-DSP should be such that it facilitates hierarchical routing. Note that the length of HO-DSP varies depending upon the value of IDP.

The ESI field identifies an end-system and is of 6 octets in length. One of the reasons to have this field of six octets is to allow the ESI field to be specified by an IEEE MAC address. For a particular value of IDP + HO-DSP, the ESI field is unique.

The selector field is a 1-octet field and is not used for the purpose of routing. This field can be used to identify a protocol within an end-system.

10.3 ATM Group Address

Apart from the individual address structure discussed in the previous section, the group address is also defined by ATM Forum. The ATM Group Address identifies a service (rather

FIGURE 10.1

ATM Forum Defined ATM End-system Address Formats

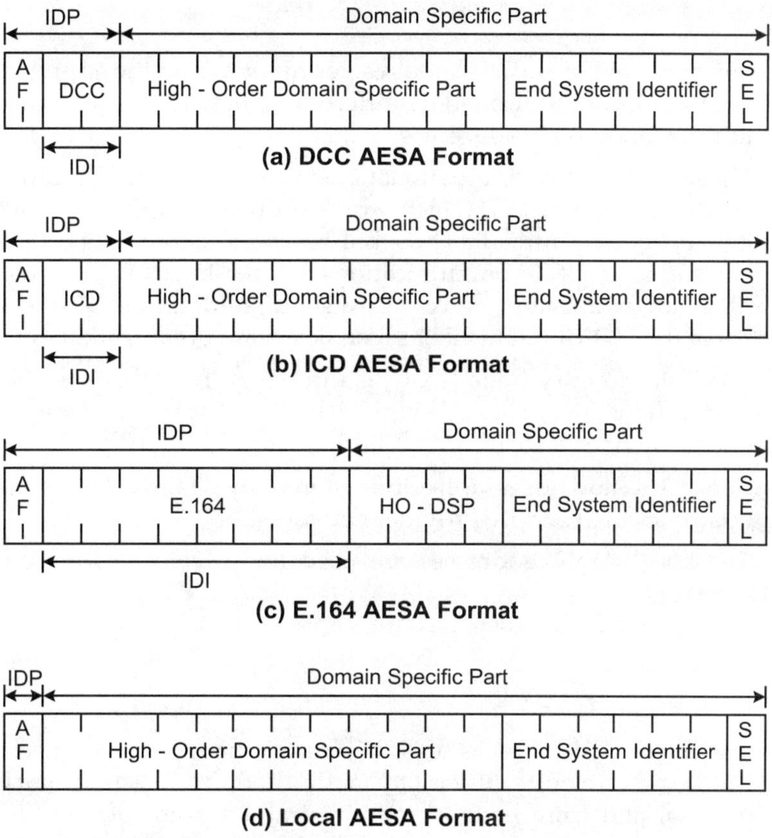

AFI: Authority and Format Identifier
IDP: Initial Domain Part
AESA: ATM End System Address
DCC: Data Country Code
ICD: International Code Designator
IDI: Initial Domain Identifier
HO-DSP: High-Order Domain Specific Part
SEL: Selector

Source: ([ATMF Addr RG 1.0], © 1996 the ATM Forum)

than an end-system). For example, these address can be used for Anycast service. The group addresses have different AFIs than that used for individual addresses. For example, a well-defined group address is given to LAN Emulation Configuration Server (LECS) service.

An ATM group is formed by one or more ATM end-systems. An ATM end-system can enter or leave a group anytime using the ILMI registration and deregistration procedures. Once a group is formed with one or more addresses, and an end-system wishes to avail a particular service, it uses the group address as Called Party Number. Seeing this group address, the intermediate ATM switch routes the call to the nearest server providing that service.

10.4 Acquiring ATM Address

Acquiring an ATM address requires permission from the appropriate organization. Different AESA formats are controlled by different bodies. This section discusses briefly the means to acquire different ATM addresses.

The assignment of International Code Designator (ICD) address is controlled by British Standards Institute (BSI), which works on behalf of the ISO. Acquiring ICD addresses for ATM has become difficult. Thus, BSI has reserved an ICD code 0×0124 for Identifiers for Organizations for Telecommunications Addressing (IOTA). Thus, for AFI = 0×39 and ICD = 0×0124, an organisation can obtain a six-digit identifier. This identifier forms the first three octets of the HO-DSP. The next seven octets are organizationally assigned.

The Data Country Code (DCC) is three-digit field that specifies the country to which the address is registered. The code is given by ISO. The national member body of the ISO controls the assignment of the DCC based addresses.

The E.164 addresses are administered by ITU-T. These numbers generally identify interfaces to ATM Service Provider networks.

The Local AESA addresses are used for private networks; thus its assignment is not monitored.

10.5 ATM Name System (ANS)

One of the basic drawbacks of ATM or any other form of numeric addressing scheme is that the addresses are quite difficult to handle, both from user's point of view (in memorizing the addresses) and from network administrator's point of view (in allocating/updating the addresses). If one takes the case of the Internet, it is accepted fact that the Domain Name System (DNS) is more popular than the 32-bit Internet addresses. Thus, machines are generally given lexicographic names. These names, suffixed with the domain names of the companies, form the host name. The DNS server then translates the names to IP addresses keeping users transparent from the actual address used for the hosts. Changing the addresses of network hosts is also easier. Taking cue from the success of domain names in Internet, the ATM Forum has come up with a name system referred to as ATM Name System (ANS), specified in [ATMF Addr ANS 2.0]. The ANS is a system that maintains mapping between names to ATM addresses and ATM addresses to names. Apart from this, ANS also facilitates discovering the location of services in a Switched Virtual Circuit (SVC) environment. The ANS is based on the IETF DNS, which is specified in [RFC 1034]. While IETF DNS is very popular, it remains to be seen whether ANS used in ATM networks can become popular. ATM Forum continues to bring the popular features on Internet in the purview of ATM standardization; however, the users will ultimately judge the efficacy of these features.

Conclusion

This chapter provided a brief overview of ATM End System Address (AESA) based on NSAP format. The different variations of this format were also discussed. Other related topics like group addressing, acquisition of ATM addresses, and ATM Name System (ANS) were also touched upon.

Review Questions

Q 1. *Which basic format is the ATM End System Address (AESA) based upon? Describe the high-level structure and components of this format.*

Q 2. *Why are different structures defined for AESA? What are these structures? Describe them in brief.*

Q 3. *Briefly describe the structure and use of Local AESA format.*

Q 4. *What is the use of name systems?*

Further Reading

Information on ATM addressing and AESA is available in two documents published by the ATM Forum—the *ATM Forum Addressing: Reference Guide* [ATMF Addr RG 1.0] and the *ATM Forum Addressing: User Guide* [ATMF Addr UG 1.0]. Apart from this, *ATM Forum Bi-Level Addressing* [ATMF Addr BA 1.0] is an informative document that provides 'bi-level' addressing concepts and examples. These concepts are useful for specifying addressing rules and recommendations for ATM Service Provider and private networks. Apart from this, the signalling protocols like UNI, PNNI and BICI provide an insight into the addressing used in ATM networks. For the references of these protocols, refer to the 'Further Reading' section of Chapter 16. The concept of name systems was also briefly introduced in this chapter. For details of ANS, the reader is referred to [ATMF Addr ANS 2.0].

Chapter 11

ATM Signalling

11.1 Introduction

Signalling is an act whereby signals are used to exchange information and/or to convey instructions. In the context of communication network, signalling is used between the user and the network, or between two network elements, to exchange various control information. In particular, signalling is used to dynamically establish, monitor and release connections. Besides connection establishment/release, signalling is used to exchange call-related information between the communicating entities. Call-related information refers to the parameters that define a connection. In a telephone call, the called party number and calling party numbers are call-related information. For dynamic virtual connections, the service parameters and the traffic descriptors are call-related information. Table 11.1 provides generic information on signalling. For details on these topics, refer [Gen S. Kasera].

This chapter discusses signalling in ATM. ATM standards provides elaborate signalling support for dynamic establishment and release of Switched Virtual Circuit (SVC). Signalling assumes a greater significance in ATM owing to the extensive traffic contracting and CAC mechanisms involved during the connection establishment phase. The information necessary for traffic contracting is exchanged between the user and the network through these signalling messages. The signalling information includes QoS parameters (like cell delay variation and cell loss ratio) and traffic descriptors (like PCR, SCR and MBS). Further, the VPI and VCI values that identify a connection in ATM are indicated by the network to the user through signalling messages.

TABLE 11.1

Basics of Signalling [Gen S. Kasera]

Definition	Signalling is used to dynamically establish, monitor and release connections
Taxonomy	• **Inband signalling:** Refers to using the same virtual channel to carry signalling information as that used to carry data. It also refers to using the same voice frequency band to carry signalling information as that used to carry voice (i.e. 300–3400 Hz). • **Outband signalling:** Refers to the technique in which the signalling information and data are carried on different channels. It also refers to using frequencies above the voice band (but below the upper threshold of 4000 Hz) to carry signalling information. • **Inchannel signalling:** Refers to the technique in which the same physical channel carries signalling information as well as voice and data. • **Common channel signalling:** Refers to the technique in which a separate channel is used solely to carry signalling information for a number of connections. • **Associated signalling:** Refers to the technique in which the signalling channels and the data paths pass through the same network elements. • **Non-associated signalling:** Refers to the technique in which there is no correspondence between signalling channels and data paths.
Issues	• **Acknowledgement:** Used by the receiver to inform the sender of the receipt of a signalling message. • **Timer protection:** Used by the sender to avoid inordinate delays in case the signalling messages or the acknowledgement gets lost or corrupted. • **Handshaking:** Refers to the two-way or three-way handshaking protocol for deciding the connection parameters through the exchange of signalling information. • **Call Identification:** Refers to the means to identify a particular signalling connection when more than one connection exists between two signalling entities. • **Finite state modelling:** Refers to the model through which states, events and outcomes of signalling are depicted.
Models	• **Point-to-point signalling:** Refers to the signalling for establishment and release of connection between two entities. • **Point-to-multipoint signalling:** Refers to the signalling for establishment and release of connections between a root and multiple leaves. • **Multipoint-to-multipoint signalling:** Refers to the signalling for establishment and release of connections between multiple entities.

11.2 ATM Signalling Protocol Stack

The signalling procedures for ATM are defined for both User-Network Interface (UNI) as well as Network-Network Interface (NNI). Figure 11.1 shows the protocol stack used for ATM signalling. The user initiating a call uses signalling procedures applicable at UNI, namely procedures mentioned in [ITU-T Q.2931] or ATM Forum's UNI Signalling Specification [ATMF UNI 4.0]. Within the network, the signalling information is exchanged using procedures of PNNI specification. Again, at the destination interface, the UNI signalling procedures are used to exchange signalling information.

Figure 11.1

ATM Signaling Protocol Stack

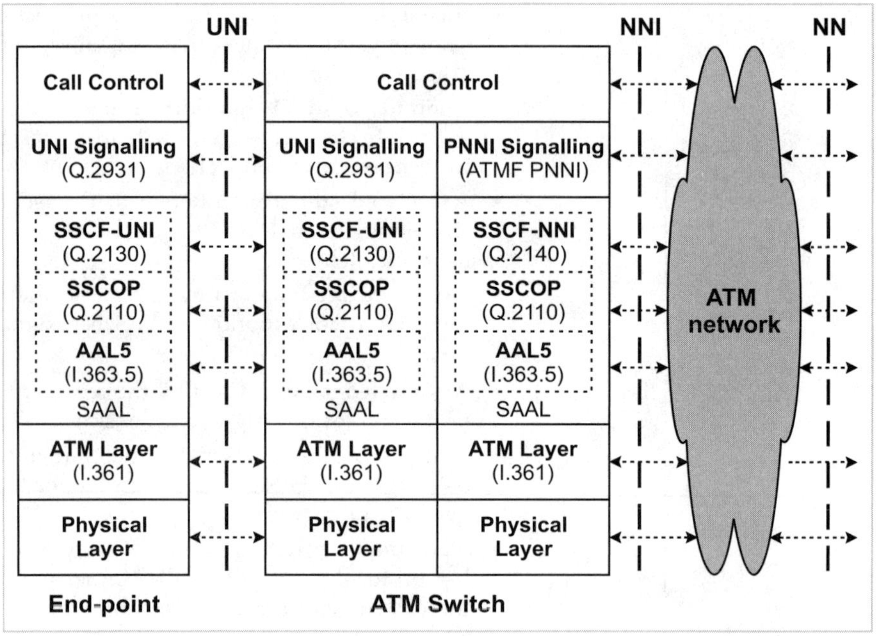

As is evident from the figure, there are three main components of ATM signalling protocol stack, viz. SAAL, UNI Signalling and PNNI Signalling. This chapter covers SAAL and UNI Signalling. The PNNI protocol defined in [ATMF PNNI 1.1] includes signalling as well as routing procedures. PNNI is discussed in Chapter 12. In the figure, the 'call control' refers to the proprietary application controlling the behaviour of UNI/PNNI signalling. The details of call control are outside the scope of this book.

11.3 Signalling ATM Adaptation Layer (SAAL)

The UNI Signalling layer sits on top of the Signalling ATM Adaptation Layer (SAAL). The SAAL can be viewed as a special AAL used for the purpose of signalling. SAAL is divided into two parts—Segmentation and Reassembly (SAR) and Convergence Sub-layer (CS). The CS is further divided into Common Part Convergence Sub-layer (CPCS) and Service Specific Convergence Sub-layer (SSCS). Further, SSCS is divided into Service Specific Connection-Oriented Part (SSCOP) and Service-specific Co-ordination Function (SSCF). In Figure 11.1, the SAR and CPCS layers, which are usually implemented in hardware, are merged together and shown as AAL5. The functions of both SAR and CPCS have been already discussed in Chapter 6.

SSCOP is the core sub-layer of SAAL and provides connection-oriented transfer of PDUs. It may be noted that among different layers of ATM, SSCOP is the only layer where retransmission is done. This is because signalling messages require reliability, as against normal user cells which may or may not require it. The SSCOP is detailed in Section 11.3.1.

SSCF provides a mapping between the primitives provided by SSCOP to that required by specific-user applications. The SSCF is detailed in section 11.3.2 for NNI and Section 11.3.3 for UNI.

In essence, SAAL provides an assured delivery of signalling messages on an otherwise unreliable ATM layer. AAL5 forms a part of SAAL and thus, all nodes carrying the signalling message from source to destination must support AAL5. Intermediate switches primarily concerned with switching of cells should support at least AAL5 in order to support signalling operations. However, it is quite possible that the intermediate switches do not support any other AAL, except AAL5.

It is customary to refer to the ATM signalling layer as layer 3. This is because of a direct mapping between the functionality of SAAL and that of layer 2 of the OSI reference model. Thus, SAAL is commonly referred to as layer 2 and the ATM signalling layer as layer 3.

11.3.1 Service-Specific Connection-Oriented Part (SSCOP)

As discussed in Chapter 6, there are various adaptation layers in ATM referred to as ATM Adaptation Layers (AALs). Prominent among these are AAL2 and AAL5. AAL2 is the layer used to carry voice while AAL5 is used to carry data. Apart from AAL for voice and data, there is another AAL that is used to carry signalling information. This AAL is called the Signalling ATM Adaptation Layer (SAAL). The peculiar characteristic that distinguishes it from AAL5 is that SAAL provides reliable delivery, which is not a feature of AAL5.

As AAL5 does not provide reliable delivery, a service specific layer is required that provides assured delivery. For this purpose, the Service Specific Connection Oriented Part (SSCOP) layer is used above AAL5. SSCOP is detailed in [ITU-T Q.2110].

The functions of SSCOP are similar to the functions of any acknowledged-mode data link layer (e.g. providing sequence numbering, flow control and error control). The important functions of SSCOP as follows:

- **Transfer of user data:** This refers to transfer of user data between SSCOP users. SSCOP supports both assured and unassured data transfer.

- **Sequence integrity:** SSCOP ensures that the order of SSCOP delivered to the SSCOP user at the receiving side is same as that delivered by the SSCOP user at the transmitting side.
- **Error correction by selective retransmission:** By delivering packets with sequence numbers, missing PDUs are detected and are selectively retransmitted.
- **Flow control:** This refers to the ability of SSCOP receiver to control the rate at which the SSCOP transmitting entity may send information.
- **Keep alive:** In case there are prolonged periods in which no data is exchanged between SSCOP peers, there is a keep alive mechanism to ensure that the connection is still alive.
- **Local data retrieval:** This refers to the mechanism whereby the local SSCOP user is allowed to retrieve the in-sequence SDUs that have not yet been released by SSCOP entity.
- **Connection control:** This refers to establishment, release and re-synchronization of SSCOP connection.

11.3.2 Service-Specific Co-ordination Function for NNI (SSCF-NNI)

The SSCF-NNI is a lightweight protocol. Its primary function is to map the particular requirements of the layer 3 protocol to the services provided by SSCOP. SSCF for NNI is used as defined in ITU-T standard [ITU-T Q.2140].

The SSCF-NNI provides the SAAL primitives to SAAL user. The SAAL user sees the SAAL as a single entity while internally, the SAAL is divided in three sub-layers (i.e. SSCF, SSCOP and AAL5). The functions of SAAL are represented by the functions of SSCOP that has already been discussed. Suffice it to say that the functions of SSCF-NNI is to map the services provided by SSCOP to PNNI signalling.

11.3.3 Service-Specific Co-ordination Function for UNI (SSCF-UNI)

Like SSCF-NNI, the primary function of SSCF-UNI is to map the particular requirements of the UNI layer 3 protocol (i.e. UNI signalling) to the services provided by SSCOP. SSCF for UNI is used as defined in ITU-T standard [ITU-T Q.2130].

The SSCF provides the SAAL primitives to SAAL user (i.e. UNI signalling). The following functions are provided by SSCF at the UNI:

- **Transfer of data:** Two modes of data transfer are supported by SSCF, namely the unacknowledged transfer and assured transfer of data. The data is assumed to be octet aligned with a maximum of 4096 octets. Note that the assured data transfer service of SSCF-UNI is provided in conjunction with SSCOP because SSCF-UNI uses the services of SSCOP.

- **Establishment and release of SAAL connections:** For assured transfer, SSCF provides means to establish and release SAAL connections. The SAAL connections are used to exchange assured data between SAAL peers.
- **Transparency of transferred information:** The SSCF provides transparent transfer of information in the sense that there is no restriction on the content, format or coding of information. Further, the information is transparent to SAAL in the sense that SAAL cannot interpret the information transferred.

11.4 UNI Signalling

UNI Signalling refers to the core signalling procedures for establishment and release of an ATM SVC. The UNI Signalling protocol has evolved in various ATM Forum specifications including [AMTF UNI 3.1], [AMTF UNI 4.0] and [AMTF UNI 4.1]. This chapter is based on [AMTF UNI 4.0]. One of the key changes in UNI 4.1 is the deletion of Leaf Initiated Join (LIJ) capability from UNI 4.0. Readers may note that even though LIJ is deleted from the latest ATM Forum specification, it is retained in this chapter for the purpose of elaborating the concept.

Subsequent sections of this chapter focuses on the UNI signalling messages and procedures. In this section, some of the important aspects of UNI signalling are summarized.

11.4.1 Non-associated Signalling

ATM provides support for both channel-associated signalling as well as channel, non-associated signalling. In channel-associated signalling, all the signalling messages for each VP is exchanged on VCI = 5 of that virtual path. In channel non-associated signalling, all the signalling messages of all the virtual paths are exchanged on VPI = 0 and VCI = 5. Currently, the standards support only the channel non-associated signalling mode. Thus, all signalling messages are carried on VPI = 0 and VCI = 5.

11.4.2 Point-to-Point and Point-to-Multipoint Signalling

The ATM UNI signalling standards exist for both point-to-point as well as point-to-multipoint signalling models. Sections 11.5 and 11.6 elaborate upon these models respectively. The basic point-to-point signalling procedure at the User-Network Interface (UNI) is laid out in the ITU-T recommendation [ITU-T Q.2931]. The Q.2931 point-to-point signalling procedures defined for ATM is derived from ITU-T's recommendation [ITU-T Q.931]. Q.931 lays down signalling specifications for ISDN at the UNI.

The point-to-multipoint signalling procedure at the UNI is detailed in [ITU-T Q.2971]. The ATM Forum has also published recommendations for signalling at the UNI. Both point-to-point and point-to-multipoint signalling procedures are laid out in ATM Forum's UNI Signalling Specification Version 4.0 [AMTF UNI 4.0]. The ATM Forum's UNI 4.0 directly references Q.2931 and Q.2971 for point-to-point and point-to-multipoint procedures, and suggests modifications that suit the interests of the business community. The changes between Q.2931 and UNI 4.0, and Q.2971 and UNI 4.0 are briefly summarized in Sections 11.5.7 and 11.6.4 of this chapter.

11.4.3 Supplementary Services

Apart from the point-to-point as well as point-to-multipoint signalling capabilities, ATM signalling procedure provides means for various supplementary services. A very popular set of such service is Line Identification supplementary services. *Line Identification* provides features that can present or restrict the identity of the parties involved in the call. The line identity services group consists of four related supplementary services, namely Calling Line Identification Presentation (CLIP), Calling Line Identification Restriction (CLIR), Connected Line Identification Presentation (COLP) and Connected Line Identification Restriction (COLR).

The Calling Line Identification Presentation (CLIP) supplementary service enables the subscriber to receive the calling party number along with the call setup request. The CLIP service can be used by the subscriber to screen incoming calls before accepting them. The

Calling Line Identification Restriction (CLIR) supplementary service prevents the calling subscriber's identity from being presented to the called subscriber, by not providing the former's number to the called subscriber at call setup time. Further, in case the calling subscriber has CLIR subscription and the called subscriber has CLIP subscription, the CLIR service over-rides the CLIP service, and the called subscriber is not presented the identity of the calling subscriber. An exception exists to this rule though, whereby national regulations may permit a category of users (e.g. police), to over-ride the CLIR service with the CLIP service.

The Calling Line Identification services (CLIP and CLIR) can allow/restrict the presentation of the calling number to the called subscriber. On the other hand, the Connected Line Identification services (COLP and COLR) provide the facility to allow/restrict the presentation of the connected subscriber number to the calling subscriber. If the calling subscriber has subscribed to the Connected Line Identification Presentation (COLP) service, then the connected party number is provided to the calling subscriber. This is especially useful in cases where the call may be diverted/forwarded by the called subscriber to some other destination. The COLR service is similar to the CLIR service. The Connected Line Identification Restriction (COLR) service prevents the identity of the connected party to be provided to the calling party.

Apart from these, there are other services related to User-to-User Signalling (UUS) (which allows a subscriber to send/receive a limited amount of subscriber generated information to/from another user in the call that is carried transparently through the network without any modification of the contents), Sub-addressing and Multiple addressing among others.

11.4.4 Anycast Capability

The ATM Anycast capability allows a user to request a point-to-point connection to a single ATM end system that is part of an ATM group. An ATM group is identified by an ATM group address, which is a specific type of ATM address.

To use the anycast capability, the calling user initiates the connection request with the ATM group address in the called party number. The network upon receiving such a request establishes a point-to-point connection with any member of the group. When one of the members receives such a request, it responds with its own address in the connected party number. Thereafter, an anycast connection does not differ much from a point-to-point connection.

11.4.5 Soft PVC

As discussed till now, there are two types of connections in ATM, Switched Virtual Connection (SVC) that is established through signalling and Permanent Virtual Connection (PVC) that is established through an administrator. Between these two extremes is the Soft PVC in which the establishment of the connection within the network is done by signalling. By configuration the switching system at one end of the soft PVC initiates the signalling for this. The network management system provisions one end of the soft PVC with the address identifying the egress interface from the network. The calling end has the responsibility for establishing, releasing, and (optionally) re-establishing the call.

11.5 ATM Point-to-Point Signalling

ATM point-to-point signalling procedures are defined in ITU-T specification Q.2931 and ATM Forum specification UNI 4.0. These specifications define the procedures for establishment and call release of ATM SVCs. Establishment of a point-to-point SVC in ATM involves a three-way handshake. In contrast, the call release procedure involves a two-way handshake only. This difference is attributed to the fact that a three-way handshake is required for proper parameter negotiation. And since the traffic-contracting phase in ATM warrants proper parameter negotiation, a three-way handshake for call establishment is justified. For call release, the case is different for two reasons. First, releasing a call requires no negotiation, and hence, two-way handshake is optimal. Second, it is mandatory on the part of an end-system to clear a call on receiving a call-clearing message. Thus, there are very few options available to an ATM end-system than to accept a release request and clear the call.

For establishing a call, an ATM end system (also called the user side of UNI or U-UNI) sends a SETUP message to its nearest network element (i.e. a switch) on a UNI (see Figure 11.2). The SETUP message contains all the mandatory information (and some optional information) required to establish a call. The mandatory elements include the called party number, Broadband bearer capability and ATM traffic descriptor parameters. Broadband high layer information, Broadband low layer information, and End-to-end transit delay are some of the optional elements.

On receiving a SETUP message, the network side of the UNI (also referred to as N-UNI) checks the parameters received in the SETUP message. If any mandatory element is missing, the request is rejected outright, and a RELEASE COMPLETE message is sent to the calling user. Otherwise, it performs Connection Admission Control (CAC) on the received parameters

FIGURE 11.2

Point-to-Point Call Establishment Procedure

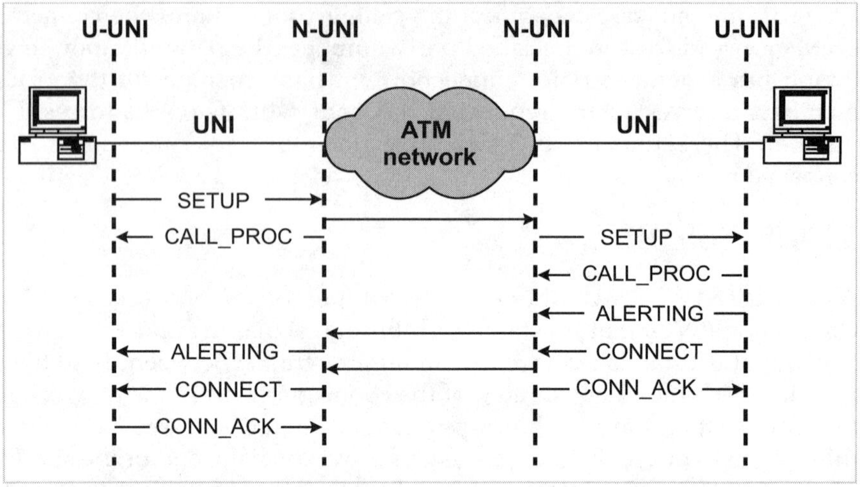

to ensure that it can support this new call establishment request. If adequate resources are available, switch at N-UNI allocates a VPI and VCI for that connection, and sends it towards the calling user in the CALL PROCEEDING message. The CALL PROCEEDING message indicates to the calling user that all the information necessary for establishing a call has been received and that no more information will be entertained.

Here, two things must be noted. First, sending of the CALL PROCEEDING message is optional, and hence, the VPI and VCI can be indicated to the calling user by the network in either the ALERTING message or the CONNECT message. Second, the calling user even has the right to specify its choice of VPI and VCI in the SETUP message. In this case, the N-UNI decides whether it can accept the specified VPI/VCI or not. If it cannot, then it sends a RELEASE COMPLETE message mentioning the appropriate cause. Else, it includes the specified VPI/VCI in the first message it sends towards the calling user.

The received SETUP message is forwarded through the network to the called user. (The process of forwarding is called routing and is achieved through PNNI protocol, which is discussed in next chapter). For the purpose of this discussion, the ATM network is treated as a cloud, which forwards signalling messages transparently. The only point to note is that resources are reserved in the path that is used to reach the called user.

When the SETUP message reaches the destination UNI, a VPI/VCI pair is allocated by the N-UNI and included in the SETUP message sent to the called user (U-UNI). The called user optionally sends a CALL PROCEEDING and an ALERTING message towards the network side. A CALL PROCEEDING message indicates that the user has received all the information required for establishing a call and is currently processing the request. However, this message has a local significance only, and is not forwarded to the calling user. In contrast, an ALERTING message has a global significance, and indicates to the calling user that the called user has been alerted. An ALERTING message is analogous to the ringing sound that one hears while making a telephone call.

If the called user is able to support the requested parameters, it sends a CONNECT message towards the calling party. Optionally, the called party may also modify the connection parameters, and send the modified parameters in the CONNECT message. The N-UNI then verifies this message. If it finds that the modified (or unmodified parameters) are correct, it sends the CONNECT message towards the calling party and a CONNECT ACKNOWLEDGE message towards the called party. Sending of the CONNECT ACKNOWLEDGE message completes the three-way handshake. When the calling side receives the CONNECT message, it sends a CONNECT ACKNOWLEDGE message to its N-UNI, thereby completing a symmetric three-way handshake. The connection is now ready for data transfer.

Here, a clarification on symmetrical handshaking must be provided. The CONNECT ACKNOWLEDGE message is required only at the destination side, not at the call-originating side. The CONNECT message sent by the called user (at the destination side) is verified by the N-UNI. If it is found to be acceptable, only then is the CONNECT message forwarded towards the calling user. Otherwise, the N-UNI at the destination side initiates call-clearing procedures. At the originating-side, the CONNECT ACKNOWLEDGE message sent by the calling party towards the N-UNI serves no useful purpose, as the message has already been verified. Here, it may be argued that the calling party should have the means to reject the parameters of the CONNECT message (that have been modified by the called party) if found unacceptable. But then, the called party cannot arbitrarily modify the parameters. This is permitted only within a given range and the range is specified by the calling party. Thus, either the called party rejects the message, or it modifies in such a manner that the parameters remain within the given constraints. In either case, the parameters will be acceptable to the called party. Thus, CONNECT ACKNOWLEDGE message is sent by the calling party only to keep the call-establishment procedures symmetrical at both the originating side and the destination side.

Till now, it is assumed that the connection request is not rejected anywhere. Now, consider the case when the connection establishment request is unacceptable either to the called user or to the network. Generally, a request is rejected if the mandatory elements are not specified; or the connection parameters cannot be supported by the entity rejecting the request (this may be due to resource unavailability or even due to incorrectly specified parameters).

In such cases, the call is released by sending a RELEASE COMPLETE message towards the calling user. This RELEASE COMPLETE message may either be generated by the N-UNI or the called user, depending upon who is rejecting the request. On receiving a RELEASE COMPLETE message, the calling user frees up any resource that it may have allocated for the call.

After the data transfer phase is over, either side (called user or calling user) can initiate the call clearing procedure. Releasing a connection entails freeing-up all resources allocated for the call, including the buffers and the VPI-VCI values along the path. If the calling user intends to release the connection, it does so by sending a RELEASE message toward the called user (see Figure 11.3). The N-UNI side on receiving the RELEASE message frees-up all of its resources and sends back a RELEASE COMPLETE message towards the calling user. It also forwards the RELEASE message towards the called user, so that resources are freed-up by each intermediate switch that was servicing the connection.

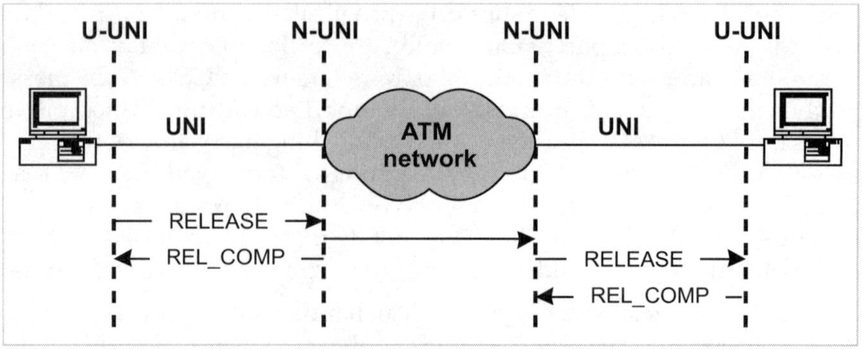

FIGURE 11.3 Point-to-Point Call Release Procedure

The calling user, on receiving the RELEASE COMPLETE message, assumes that the connection has been freed, and frees any resource still allocated for the call. The called party, on receiving the RELEASE message, also frees-up its resources and sends a RELEASE COMPLETE message back towards the N-UNI. The N-UNI, on receiving the RELEASE COMPLETE message, frees-up its resources too, and the finally the whole connection gets cleared, thereby completing the symmetric two-way handshake.

In order to realize different signalling flows as discussed above, different call states are maintained at the user side and the network side. For example, when the user initiates call establishment procedure, it moves to the Call Initiated state. Similarly, when the network receives such a request, it also moves to the Call Initiated state. ITU-T Q.2931 specification lists the different call states at the user side and the network side. These call states are used to implement the Finite State Machine for realization of the ATM signalling protocol. Details of these call states is provided in Section 11.5.3.

11.5.1 Signalling Messages

The messages used for ATM point-to-point signalling are defined in [ITU-T Q.2931]. All Q.2931 signalling messages are encoded using a TLV (type-length-value) format. The type field specifies how the contents of the message are to be decoded. The length field specifies the total length of the message and the value contains the actual contents of the message. The TLV format is the most common mechanism used to encode and decode protocol messages.

Based on TLV format, Q.2931 signalling messages have a common header of 9-byte. The format of the 9-byte header is shown in Figure 11.4. The first octet contains the protocol identifier field. It is set to '0×80' for all Q.2931 messages. The four most significant bits of the second octet is spare, and are set to '0'. The four least significant bits contain the length of the Call Reference Value (CRV), which is fixed and set to 3. The length of the CRV is followed by 3-bytes that indicate the actual CRV. Since an ATM end-system can have more than one call being established/released at a time, it needs to distinguish between different calls. Since the ATM end-system addresses alone are not suffice, CRVs are used. The CRVs have a local significance; i.e. they are unique per interface only. Just like VPI/VCI are swapped per-

FIGURE 11.4

Q.2931
Message Header

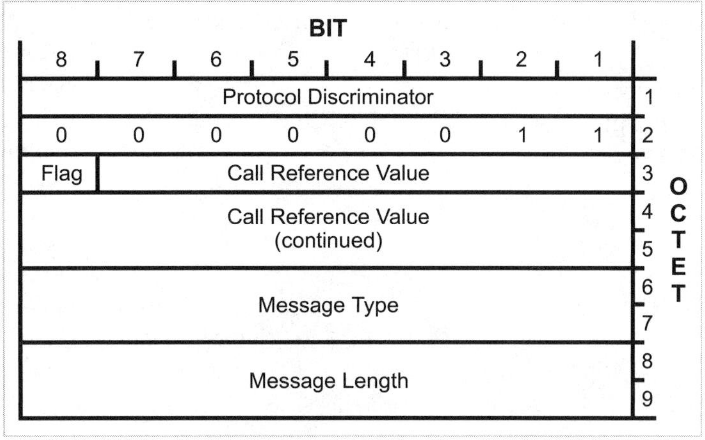

interface, CRV values are also swapped, the only exception being the global CRV (0×000000). This global CRV has global significance and is used for a specific purpose. The details of global CRV follow later in this section.

One bit of the CRV, called the CRV flag, is used to identify whether a call was initiated by the peer or initiated locally. When a call is initiated locally, a unique, 23-bit number is allocated, and the 24th bit set to '0' to form a 3-byte CRV value. The peer, while responding to this call, sets this flag to '1' in all its responses. Consider the scenario where two calls are simultaneously initiated by the two end-systems with the same CRV value. Also, both of them set the flag to 0 (the call being a locally-initiated one) and transmit their respective SETUP messages. Each of the side will receive the SETUP message with flag in CRV set to 0. The received SETUP will not be considered as a response by the peer for an already existing call, as both the sides are expecting messages with the same CRV and the flag set to 1. Thus, any possibility of the same CRV being allocated for two calls, is eliminated by using the flag in the CRV.

The sixth octet of the message contains the message type. The seventh octet contains spare bits, the extension flag, and other fields for error handling. Octets eight and nine contain the total length of the message, excluding the 9-byte header. The remaining part of the message contains various Information Elements (IEs) specific to a message. IEs are small units of information carried in signalling messages. The different types of IEs are discussed in the next sub-section.

All point-to-point messages are listed in Table 11.2. While discussing handshaking, the use of call establishment messages (like SETUP, CALL PROCEEDING, CONNECT, etc.) and call clearing messages (like RELEASE and RELEASE COMPLETE) had been elaborated.

STATUS and STATUS ENQUIRY messages are used for error recovery. A STATUS ENQUIRY message is sent to inquire the call-state of the peer. This is required when, due to some error condition, the call-state between peers becomes incompatible. The recipient of

TABLE 11.2

Point-to-Point Signalling Messages

Call Establishment Message	
SETUP	Used to initiate the procedure for call establishment.
CALL PROCEEDING	Used to indicate the calling user that all the information necessary for establishing a call has been received.
ALERTING	Used to indicate to the calling user that the called user has been alerted.
CONNECT	Used to indicate that the called user is willing to accept the call.
CONNECT ACKNOWLEDGE	Used to complete the symmetrical three-way handshake.
Call Release Message	
RELEASE	Used to initiate the procedure for call release.
RELEASE COMPLETE	Used to acknowledge RELEASE message or initiate call release.
Miscellaneous Messages	
STATUS ENQUIRY	Used to inquire the call-state of the peer.
STATUS	Used to respond with own call state during status enquiry.
RESTART	Used to restart (i.e. bring to idle condition) the requested virtual circuits.
RESTART ACKNOWLEDGE	Used to acknowledge that the requested restart procedure is complete.

the STATUS ENQUIRY message responds with a STATUS message containing its own call state. On receiving a STATUS message, the sender of the STATUS ENQUIRY message checks if the call-state received in the STATUS message is the same as its own call-state (i.e. call-states are compatible). If the call-states are found to be incompatible, then either the call clearing procedures are initiated or some other appropriate action (protocol-dependent or implementation-dependent) is taken.

Messages like RESTART and RESTART ACKNOWLEDGE are used to recover virtual circuits from temporary failures. A RESTART message is sent to request the recipient of the message to restart (i.e. bring to idle condition) the requested virtual circuits. While sending a RESTART message, the sender can request one of the following to be restarted: A particular virtual channel; or all virtual channels in the indicated virtual path; or all virtual channels controlled by the signalling entity. The recipient of the RESTART message responds with a RESTART ACKNOWLEDGE message to acknowledge the receipt of the RESTART message and to indicate that the requested restart procedure is complete. Both the RESTART and RESTART ACKNOWLEDGE message are sent using the global CRV (0×000000), because these messages do not refer to a particular call.

11.5.2 Information Elements

The Q.2931 signalling messages comprise of a number of Information Elements (IEs). The IEs present in different Q.2931 messages are shown in Table 11.3. These IEs are used for both point-to-point signalling and point-to-multipoint signalling.

TABLE 11.3

List of Information Elements Used in ATM Signalling [ITU-T Q.2931]

AAL parameters	It describes the AAL parameters for the call.
ABR additional parameters	It specifies additional ABR parameters for call establishment.
ABR setup parameters	It specifies a set of ABR parameters for call establishment.
Alternative ATM traffic descriptor	It specifies an alternative ATM traffic descriptor for the negotiation of traffic parameters during call setup.
ATM traffic descriptor	It describes the traffic parameters for the call.
Broadband bearer capability	It specifies the broadband bearer services to be provided by the network.
Broadband high layer information	It specifies information about higher layers which may be used for compatibility checking by an addressed entity.
Broadband low-layer information	It specifies information about lower layers which may be used for compatibility checking by an addressed entity.
Broadband repeat indicator	It indicates how repeated information elements shall be interpreted.
Call state	It indicates the state of a call or of a global interface.
Called party number	It identifies the called party of the call.
Called party sub-address	It identifies the sub-address of the called party of the call.
Calling party number	It identifies the calling party of the call.
Calling party sub-address	It identifies the sub-address of the calling party of the call.
Cause	It indicates the reason for generating certain messages.
Connection identifier	It indicates the local ATM connection resources on the interface.
Connection scope selection	It indicates to the network that the call shall be processed within the selected routing range.
Endpoint reference	It uniquely identifies an end-point of a point-to-multipoint call.

Contd.

TABLE 11.3

Contd.

Endpoint state	It indicates the state of a party in a point-to-multipoint call.
End-to-end transit delay	It indicates the maximum acceptable and the cumulative end-to-end delay for a call.
Extended QoS parameters	It specifies the various QoS parameters.
Generic identifier transport	It allows to carry identifiers unchanged between a call originating entity and the addressed entity.
Leaf sequence number	It is used to associate the response message with the corresponding LEAF SETUP REQUEST message.
LIJ call identifier	It uniquely identifies a point-to-multipoint call at a root's interface.
LIJ parameters	It allows Root to associate options with the call when the call is created (e.g. Network LIJ).
Minimum acceptable traffic	It specifies the minimum acceptable ATM traffic parameter in the negotiation of traffic descriptor parameters during call setup.
Notification indicator	It indicates information pertaining to a call.
QoS parameters	It specifies the QoS classes for the call.
Restart indicator	It identifies the class of facility to be restarted, i.e. a single VC, all VCs in a VP or all VCs on an interface.
Transit network selection	It identifies one requested transit network on the path of the call from the calling party to the called party.

An IE in a particular message can either be mandatory, or conditionally mandatory, or can be optional. An IE is said to be mandatory for a message, if that particular message is incomplete without it. For example, ATM traffic descriptor is a mandatory IE of the SETUP message. Thus, a SETUP message is not accepted unless it has this IE.

An IE is conditionally mandatory when its presence is required when certain conditions are satisfied. For example, in a RESTART message, the restart indicator IE indicates one out of the three possible options discussed in the previous section. Now, if option 1 is chosen (which corresponds to the restart request for a particular virtual channel), then the connection identifier IE carries the requested virtual channel number. Thus, connection identifier IE in this case becomes a conditionally mandatory IE.

An IE is optional when its presence or absence makes no difference to the handshaking procedure. Optional IEs are mainly used to carry precise or additional information. For example, the Alternate traffic descriptor is an optional IE of the SETUP message. It is included in the SETUP message to provide the intermediate ATM nodes an alternative in case the parameters in ATM traffic descriptor IE are unacceptable.

The Q.2931 procedure also allows multiple occurrences of some of the IEs to be present in the same message. For example, there can be multiple instances of Transit Network Selection IE in the same SETUP message. Multiple instances of the same IE provides the network the option of choosing one out of the many possible transit networks. Some other IEs like Cause IE and Called Party Sub-address IE can also have multiple instances, details of which are beyond the scope of this book.

Some of the important IEs listed in Table 11.3 are further explained below:

- **Broadband Bearer Capability IE:** This IE is a mandatory part of the SETUP message and indicates the type of broadband connection-oriented service requested by the user (see Figure 11.5). This IE contains four fields, viz. the bearer class, ATM transfer capability, susceptibility to clipping and user-plane connection configuration. The bearer class specifies the type of service requested by the user. Four bearer classes are defined by UNI 4.0, which include BCOB-A, BCOB-C, BCOB-X and Transparent VP. Among these, BCOB-X and Transparent VP provide ATM cell-relay service only, while BCOB-A and BCOB-C provide more than ATM only service (i.e. the network may look into the AAL information element to provide interworking based upon its contents). The second field, i.e. ATM transfer capability, is an optional field and indicates whether the requested connection is CBR, rt-VBR, nrt-VBR or ABR type. The third field, that is susceptibility to clipping, specifies whether the connection is susceptible to clipping in the initial part of the connection or not. Finally, the user plane connection configuration field specifies whether the type of connection (i.e. point-to-point or point-to-multipoint).

233

FIGURE 11.5

Broadband Bearer Capability IE

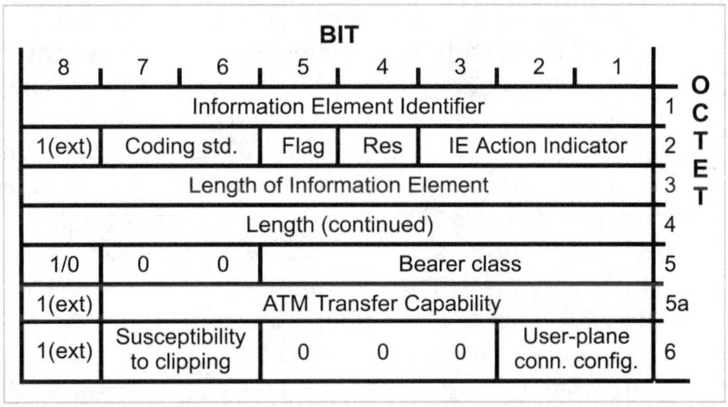

- **ATM Traffic Descriptor IE:** This IE specifies the desired traffic parameters of the connection request. It contains the parameters for both forward and backward directions. Each parameter is defined for CLP = 0 and CLP = 0 + 1 cell flows. The parameters for traffic descriptor include the Peak Cell Rate (PCR), the Sustainable Cell Rate (SCR) and the Maximum Burst Size (MBS). For ABR connections, the Minimum Cell Rate (MCR) is included. Note that the type of connection is determined by the contents of Broadband Bearer Capability. Thus, permissible values of ATM traffic descriptor are determined in conjunction with the contents of Broadband Bearer Capability.

Apart from the aforementioned parameters, the traffic descriptor also includes best-effort flag, frame discard field and tagging field. The presence of the best effort flag indicates that only a best effort service is requested. This flag is used for specifying UBR connections. The frame discard field indicates whether frame discarding is allowed in forward and backward direction or not. Depending upon the values of this field, the AAL5 packets may or may not be discarded by the network using the Early Packet Discard (EPD) or Partial Packet Discard (PPD). The tagging field in this IE indicates whether the intermediate nodes are allowed to tag a cell (i.e. set the CLP bit to '1') or not.

- **Alternative ATM Traffic Descriptor IE:** Alternative ATM traffic descriptor IE contains the same information as contained in ATM traffic descriptor. This IE is included in the SETUP message to provide an alternative set of traffic parameters. The called user may choose one of the two sets of traffic parameters, and indicate to the calling user the same by including it in CONNECT message.

- **Minimum Acceptable Traffic Descriptor IE:** As an option to allow user to negotiate the exact values of traffic descriptor, this IE is specified. This IE contains the minimum cell rates that are acceptable to the calling user. The called user can choose cell rates using the value specified in ATM Traffic Descriptor IE or Minimum Acceptable Traffic Descriptor IE or can choose intermediate values. Note, however, that among Minimum Acceptable Traffic Descriptor and Alternative ATM traffic descriptor, only one IE is permitted in SETUP message.
- **AAL Parameters IE:** This IE indicates the AAL type, and the associated parameters to be used for the call. The parameters used in this IE are not significant to the intermediate nodes and have only an end-to-end significance. The first field is the AAL type, on the basis of which the remaining fields are interpreted. For example, if the AAL type is specified as AAL1, the remaining fields will have information on clock recovery method, CBR rate, partially filled cells method, structured data transfer block size, etc. If the type is AAL3/4 this IE will contain information about the Multiplexing Identifier (MID) and the maximum CPCS-SDU size exchanged between peers.
- **QoS Parameters:** This IE contains the requested QoS class for forward and backward direction. The networks that cannot support the requested QoS set the QoS class to Unspecified QoS class (QoS class 0).
- **Extended QoS Parameters:** The QoS parameter IE only specifies the QoS class. The individual parameters are specified in the Extended QoS Parameters IE. The IE contains the maximum forward and backward Cell Delay Variation (CDV) and Cell Loss Ratio (CLR) acceptable for the connection. It also has cumulative cell delay variation field.

Although both QoS and Extended QoS IE can be included in SETUP message, presence of at least one of them is mandatory.

11.5.3 Call States

For point-to-point signalling based on Q.2931, different call states are maintained at the user side and the network side. Interpretation of these call states at the user side and at the network side of the UNI is different. The call states are defined in terms of incoming calls and outgoing calls. Q.2931 defines incoming and outgoing calls as viewed from the user side of a UNI. Thus, at the UNI of the calling party, the call is outgoing both for the U-UNI as well as the N-UNI. At the calling party end, the scenario is reversed and the call is incoming, both at the U-UNI as well as the N-UNI.

At the network side, the following 11 call states are maintained ([ITU-T Q.2931]):

1. **Null (N0):** No call exists.
2. **Call Initiated (N1):** This state exists for an outgoing call when the network has received a call establishment request, but has not yet responded.
3. **Outgoing Call Proceeding (N3):** This state exists for an outgoing call when the network has sent acknowledgement that the network has received all call information necessary to effect call.
4. **Call Delivered (N4):** This state exists for an outgoing call when the network has indicated that the remote user alerting has been initiated.
5. **Call Present (N6):** This state exists for an incoming call when the network has sent the call establishment request, but has not yet received a satisfactory response.
6. **Call Received (N7):** This state exists for an incoming call when the network has received an indication that the user is alerting, but has not yet received an answer.
7. **Connect Request (N8):** This state exists for an incoming call when the network has received an answer, but the network has not yet awarded the call.
8. **Incoming Call Proceeding (N9):** This state exists for an incoming call when the network has received acknowledgement that the user has received all call information necessary to effect call establishment.
9. **Active (N10):** This state exists for an incoming call when the network has awarded the call to the called user. This state exists for an outgoing call when the network has indicated that the remote user has answered the call.
10. **Release Request (N11):** This state exists when the network has requested from the user to clear the end-to-end connection (if any).
11. **Release Indication (N12):** This state exists when the network disconnected the end-to-end connection (if any) and has sent an invitation to disconnect the user-network connection.

At the user side too, 11 call states are maintained ([ITU-T Q.2931]):

1. **Null (U0):** No call exists
2. **Call Initiated (U1):** This state exists for an outgoing call, when the user requests call establishment from the network.

3. **Outgoing Call Proceeding (U3):** This state exists for an outgoing call when the user has received acknowledgement that the network has received all call information necessary to effect call establishment.
4. **Call Delivered (U4):** This state exists for an outgoing, when the calling, user has received an indication that remote user alerting has been initiated.
5. **Call Present (U6):** This state exists for an incoming call when the user has received a call establishment request, but has not yet responded.
6. **Call Received (U7):** This state exists for an incoming call when the user has indicated alerting, but has not yet answered.
7. **Connect Request (U8):** This state exists for an incoming call when the user has answered the call and is waiting to be awarded the call.
8. **Incoming Call Proceeding (U9):** This state exists for an incoming call when the user has sent acknowledgement that the user has received all call information necessary to effect call establishment.
9. **Active (U10):** This state exists for an incoming call when the user has received an acknowledgement from the network that the user has been awarded the call. This state exists for an outgoing call, when the user has received an indication that the remote user has answered the call.
10. **Release Request (U11):** This state exists when the user has requested the network to clear the end-to-end connection (if any) and is waiting for a response.
11. **Release Indication (U12):** This state exists when the user has received an invitation to disconnect because the network has disconnected the end-to-end connection (if any).

The state common to incoming and outgoing calls are the null state and the active state. The commonality stems from the fact that only these two states are stable states, and hence, exist for both incoming and outgoing calls. During the call establishment phase, excluding these two stable states, the number of states applicable for an outgoing call is three (U1, U3 and U4), while those for incoming call is four (U6, U7, U8 and U9). This difference in the number of states is because the Connect Request state is applicable for an incoming call only, not for an outgoing call. For an incoming call, after sending a CONNECT message (by the U-UNI of the destination side) or after receiving a CONNECT message (by the N-UNI of the destination side), the call state is changed to Connect Request. This is because the sender of the CONNECT message waits for a CONNECT ACKNOWLEDGE, and hence stays in the Connect Request state, while the receiver of the CONNECT message processes the message received, and thus again stays in the Connect Request state.

The case for an outgoing call at the originating UNI is slightly different. After sending a CONNECT message, the N-UNI side immediately goes into Active state. Thus, irrespective of whether the user sends a CONNECT ACKNOWLEDGE message or not, the N-UNI side stays in the *Active* state. The reason the CONNECT ACKNOWLEDGE message is sent by the-UNI at the very first place, is to allow symmetrical signalling procedures. Upon receiving the CONNECT message, the U-UNI side immediately sends a CONNECT ACKNOWLEDGE message to the N-UNI, and changes its state to Active. Thus, at the source UNI side, Connect Request state is not applicable.

11.5.4 Timer Protection

Q.2931 defines different sets of timers for the user side and the network side of a UNI (see Table 11.4). The table provides the list of timers maintained, the cause for normal stop, and the actions taken after the first and second expiry. Although not explicitly mentioned in the table, the number of timers on network and user sides are different. This is because of the inherent asymmetry in the Q.2931 signalling procedure that has already been mentioned in the previous section. The timer T313 that is started after sending a CONNECT message, is present only at the user side, not at the network side. The reason for this should now be clear. At the called UNI, the user expects a CONNECT ACKNOWLEDGE message after sending a CONNECT message and hence, starts the T313 timer. At the calling UNI, the N-UNI already changes its state to *active* and hence, there is no need to start a timer. Even if the calling user does not send the CONNECT ACKNOWLEDGE message, or the same message is lost, the connection is already in the *active* (data transfer) phase.

11.5.5 Signalling Example: Call Establishment

Having discussed the Q.2931 signalling messages, the information elements present in them, the various call states and protocol timers, the basic flow diagram is revisited and the steps involved in establishing and clearing an ATM virtual connection explained. The following steps are involved in call establishment (see Figure 11.6):

Step 1. To establish a call, a user forms a SETUP message with all the mandatory information elements and sends it to the nearest network element (N-UNI). It then changes its call-state from Null (U0) to Call Initiated (U1) and starts the timer T303.

Step 2. The N-UNI on receiving this SETUP message changes its state to Call Initiated (N1). It also sends a CALL PROCEEDING message to the calling user after which it changes it's state to Outgoing Call Proceeding (N3). Thus, N1 is used as a transient state only.

Step 3. The calling user on receiving this CALL PROCEEDING message stops the timer T303, starts timer T310, and changes its state to Outgoing Call Proceeding (U3).

Step 4. The intermediate network elements forward the SETUP message towards the called user.

Step 5. Upon receiving the SETUP message, the N-UNI at the destination side starts timer T303, changes its state from Null (N0) to Call Present (N6), and sends the SETUP message towards the called user (U-UNI).

Step 6. The called user on receiving the SETUP message, changes its state from Null (U0) to Call Present (U6). It then sends a CALL PROCEEDING message towards the N-UNI, and changes its state to Incoming Call Proceeding (U9). Here again, the U6 is a transient state only.

Table 11.4

Q.2931 Protocol Timers [ITU-T Q.2931]

Timer	*Cause for start*	*Cause for stop*	*Action at first expiry*	*Action at second expiry*
T301	ALERTING received	CONNECT received	Clear call	N.A.
T303	SETUP sent	ALERTING, CONNECT, RELEASE COMPLETE or CALL PROC received	Resend SETUP; Restart T303	Clear internal connection; Enter NULL state
T308	RELEASE sent	RELLEASE or RELEASE COMPLETE received	RELEASE sent; Restart T308 Enter NULL state	Release call reference value
T309	SAAL disconnected	SAAL reconnected	Clear internal connection	N.A.
T310	CALL PROC received	ALERTING, CONNECT, or RELEASE received	RELEASE sent	N.A.
T313	CONNECT sent	CONNECT ACK received	RELEASE sent	N.A.
T316	RESTART sent	RESTART ACK received	RESTART may be resent several times	RESTART may be resent several times
T317	RESTART received	Internal clearing of call reference	Maintenance notification	N.A.
T322	STATUS ENQUIRY sent	STATUS. RELEASE or RELEASE COMPLETE received	STAT ENQ may be resent several times.	STAT ENQ may be resent several times

Note: N.A. implies not applicable.

FIGURE 11.6

Point-to-Point Call Establishment

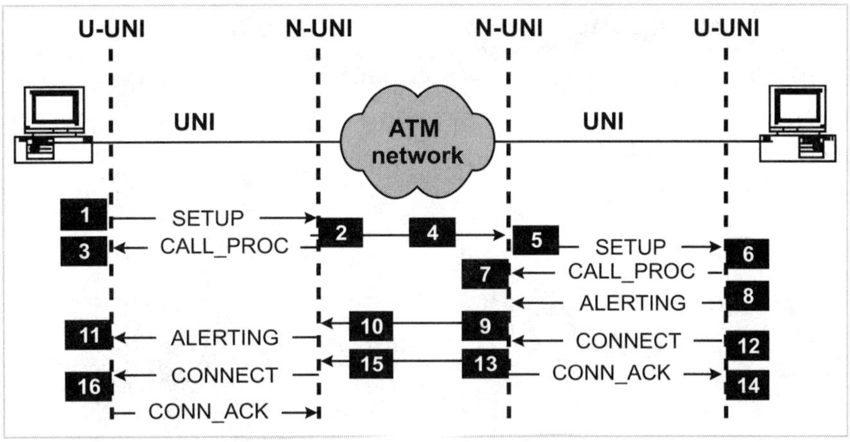

Step 7. The N-UNI on receiving this CALL PROCEEDING message stops timer T303, starts timer T310, and changes its state to Incoming Call Proceeding (N9).

Step 8. The called user then sends an ALERTING message towards the source, and changes its state to Call Received (U7).

Step 9. The N-UNI on receiving this ALERTING message stops timer T310, starts T301, and changes its state to Call Received (N7).

Step 10. The N-UNI at the calling side on receiving this ALERTING message sends it towards the calling party, and also changes its state to Call Received (N4).

Step 11. The calling U-UNI on receiving this ALERTING message stops the timer T310, starts T301, and changes its state to Call delivered (U4).

Step 12. The called U-UNI sends a CONNECT message, starts timer T313 and changes its state to Connect Request (U8).

Step 13. The N-UNI on receiving this CONNECT message stops timer T301, and changes it state to Connect Request (N8). It then sends a CONNECT ACKNOWLEDGE message towards the called user, changes its state to Active (N10) and also forwards the received CONNECT message towards the calling user.

Step 14. The called user on receiving the CONNECT ACKNOWLEDGE message stops timer T313 and changes it state to Active (U10).

Step 15. The N-UNI at the calling side on receiving the CONNECT message, sends it towards the calling U-UNI, and changes its state to Active (N10).

Step 16. Similarly, the calling user on receiving the CONNECT message changes its state to Active (U10) and stops the timer T301. The connection between the calling user and the called user is now ready for data transfer.

11.5.6 Signalling Example: Call Release

Both U-UNI as well as N-UNI at either the calling side or the called side can initiate the call release procedures. The following steps illustrate call release procedures initiated by the calling user (see Figure 11.7).

FIGURE 11.7

Point-to-Point Call Release

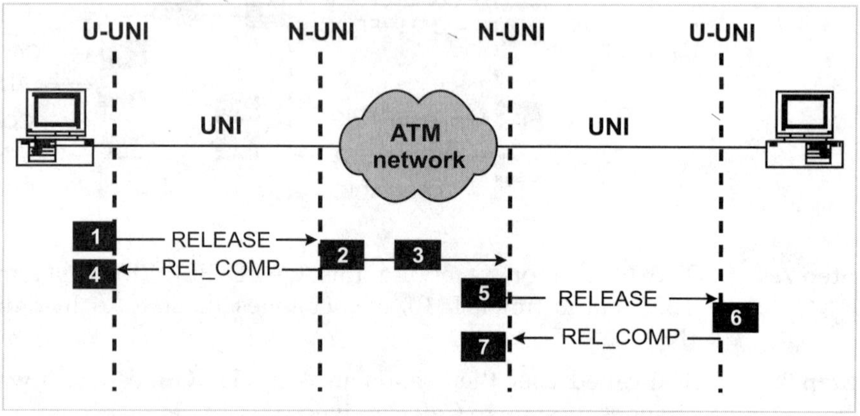

Step 1. For releasing the call, the calling user sends a RELEASE message towards the N-UNI, starts timer T306 and changes its state to Release Request (U11).

Step 2. The N-UNI on receiving this RELASE message changes its state to Release Request (N11). After freeing up the resources, the N-UNI changes its state to Null (N0) and sends a RELEASE COMPLETE message towards the calling user.

Step 3. The N-UNI also forwards the same RELEASE message towards the called user.

Step 4. On receiving the RELEASE COMPELTE message, the calling user stops timer T306 and changes its state to Null (U0).

Step 5. The N-UNI at the destination side starts timer T306, changes its state to Release Indication (N12) and sends the RELEASE message towards the called user.

Step 6. The called user on receiving this RELEASE message changes its state to Release Indication (U12). After freeing up any resources, the called U-UNI changes its state to Null (U0) and sends a RELEASE COMPLETE message towards the N-UNI.

Step 7. The N-UNI on receiving this RELEASE COMPLETE message stops timer T306 and changes its state to Null (N0).

11.5.7 Comparison between Q.2931 and UNI 4.0

As mentioned, both ITU-T Q.2931 and ATM Forum's UNI Signalling Specification 4.0 give specifications for point-to-point signalling procedures. However, these are not two different specifications. ATM Forum's UNI Signalling Specification refers to Q.2931 for point-to-point signalling specification, however, with some alterations. These alterations are as follows:

- Connection characteristic negotiation during connection establishment phase,
- Support for ABR capability, and
- Anycast capability.

For supporting the connection characteristic negotiation during the call establishment phase, two new information elements, viz. Alternative ATM Traffic Descriptor and Minimum Acceptable ATM Traffic Descriptor are introduced. In a call establishment request, user can include one of the two afore-mentioned IEs in the SETUP message. By specifying Alternative ATM Traffic Descriptor, the calling user can allow the called user to choose one of the two traffic descriptors. If the calling user accepts the traffic descriptors specified in Alternative ATM traffic descriptors, it will include an ATM traffic descriptor IE in the CONNECT message with the parameters specified in the Alternative ATM traffic descriptor. If the calling user includes Minimum ATM traffic descriptor in the SETUP message, the called user can select any value of traffic parameters between the ones specified in the ATM traffic descriptor and the Minimum ATM traffic descriptor IE. The user then includes the ATM traffic descriptors in the CONNECT message with its parameters as the parameters selected.

For supporting the ABR traffic capability, apart from the introduction of MCR for forward and backward direction in ATM Traffic Descriptor IE, two new IEs are specified. These IEs are ABR additional parameters and ABR setup parameters.

Another enhancement that ATM Forum's UNI 4.0 makes over ITU-T Q.2931 is the introduction of 'Anycast' capability. Through this service, a user may request for a point-to-point connection to one of the ATM end system from a group of end-systems. An additional IE called Connection Scope Selection is specified to support the ATM Anycast capability.

ATM Forum's UNI 4.0 also specifies four additional QoS classes. It also introduces the support for negotiation of individual QoS parameters through Extended QoS parameters IE.

11.6 ATM Point-to-Multipoint Signalling

In point-to-multipoint calls, as specified by ITU-T Q.2971, one ATM end-system (termed as root) establishes a unidirectional connection with a number of other ATM end-systems (termed as leaves). The procedures specified in Q.2971 allow only the root to initiate procedures to add a leaf to a point-to-multipoint call. ATM Forum's UNI 4.0 additionally specifies the Leaf Initiated Join (LIJ) capability. Using LIJ capability, leaves can initiate procedures for establishing a new point-to-multipoint call, or can even initiate procedures for adding themselves to an already existing connection, with or without the intervention from the root. The LIJ procedures are discussed towards the end of this chapter.

For point-to-multipoint calls, the following handshaking procedures are applicable:

- A point-to-multipoint call is established between two end-systems, using a three-way handshake. This is because normal point-to-point signalling procedure is used for the establishment of a new point-to-multipoint call. The three-way handshake, however, is applicable only when the root initiates a point-to-multipoint call (see the next case). A two-way handshake is used for adding subsequent parties to an existing point-to-multipoint call.

- When a leaf uses LIJ capability to initiate a point-to-multipoint call, four-way handshake is used.
- If LIJ procedures are used, a three-way handshake is used (instead of two-way handshake) for adding subsequent parties to an existing point-to-multipoint call.

Henceforth, it is assumed that LIJ capability is not being used, unless otherwise mentioned. Discussion on LIJ is postponed for the present moment.

A point-to-multipoint call is established using normal point-to-point signalling procedures, as discussed in previous section. Once a point-to-multipoint call is established, subsequent parties are added by the root using the ADD PARTY message. Figure 11.8 illustrates the procedure for adding a leaf to an existing point-to-multipoint call. It may be noted that during the connection establishment phase, the connection parameters are fixed by the root and the first party. Hence, parameter negotiation for subsequent parties is not applicable.

FIGURE 11.8

Adding a Party to a Point-to-Multipoint Call

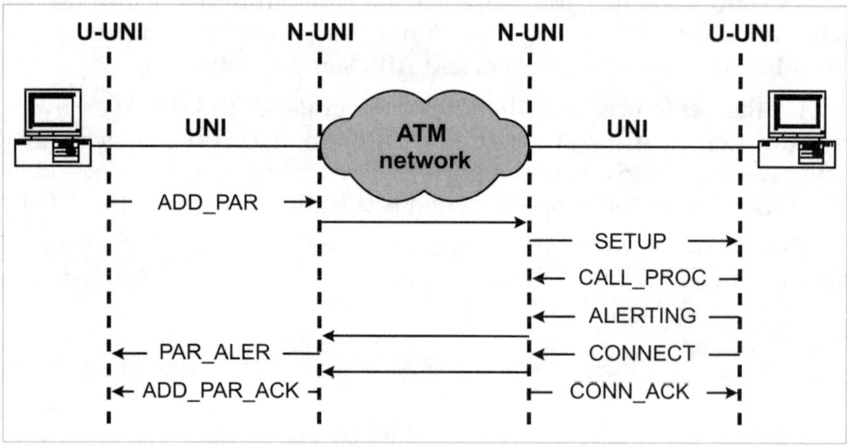

To add a leaf, the root sends an ADD PARTY message towards the N-UNI. The N-UNI subsequently forwards this ADD PARTY message to the leaf. Unlike the SETUP message, the ADD PARTY message is converted by the N-UNI at the leaf side into a SETUP message before forwarding it to the leaf. This conversion is done in order to allow the leaf to be a part of a point-to-multipoint call even if it cannot support Q.2971 procedures.

The leaf responds to this SETUP message by sending CALL PROCEEDING, ALERTING and CONNECT messages in that order, but not necessarily all. CALL PROCEEDING being a local message is not sent back towards the root by the N-UNI. However, both ALERTING and CONNECT message are sent back. In response to CONNECT message, the N-UNI sends a

CONNECT ACKNOWLEDGE message to acknowledge the successful addition of the leaf in the point-to-multipoint call.

The N-UNI at the root side converts ALERTING into PARTY ALERTING and CONNECT into ADD PARTY ACKNOWLEDGE before forwarding it to root. Here, the difference between ALERTING and PARTY ALERTING can be observed. Even though PARTY ALERTING message is the first message received by the root for a party, it does not contain the Connection Identifier IE. This is because between the root and its nearest network element (N-UNI), there is only one virtual connection. No new connection is made between the two once a point-to-multipoint call is established. Thus, Connection Identifier IE is not present in PARTY ALERTING MESSAGE. After the receipt of ADD PARTY ACKNOWLEDGE message by the root, procedure for adding the party is complete and the leaf becomes a part of the point-to-multipoint call.

Discussion until this point assumes that the party addition request is acceptable to all the intermediate network elements and even to the leaf. This, however, may not always be the case. If the request for adding a party is not acceptable to the N-UNI at the root side, it sends an ADD PARTY REJECT message towards the root. If the request is not acceptable to the leaf, it sends a RELEASE COMPLETE message towards the root. The N-UNI at the root side converts RELEASE COMPLETE message into an ADD PARTY REJECT and forwards it towards the root. On receiving this ADD PARTY REJECT message the root frees up any allocated resources for the party (not for the call).

The leaves can be dropped from a point-to-multipoint call either by the root or by the leaves themselves. If the root wants to drop a leaf, it sends a DROP PARTY message towards the leaf concerned (see Figure 11.9). The N-UNI, on receiving this DROP PARTY message, responds back with a DROP PARTY ACKNOWLEDGE message after freeing up its resources for that leaf. The N-UNI, also forwards the DROP PARTY message, towards the leaf. At the leaf side, the N-UNI converts the DROP PARTY message into a RELEASE message and forwards it to the leaf. Leaf responds with a RELEASE COMPLETE MESSAGE and frees its resources for the connection. By now, it must be clear that the message conversion is intended to make the leaf transparent of the Q.2971 procedures, in case it does not support it.

FIGURE 11.9

Root-Initiated Procedure for Dropping a Party

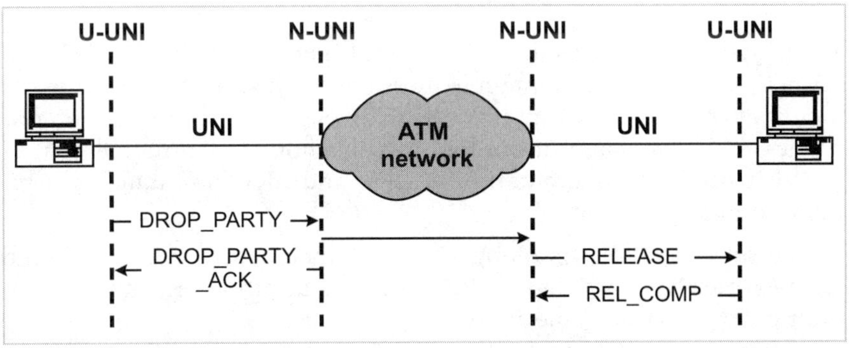

If the leaf wants to drop itself from the point-to-multipoint call, it sends a RELEASE message towards the root. The N-UNI responds with a RELEASE COMPLETE message and also forwards the message over the network. The N-UNI at the root side converts it into a DROP PARTY message before forwarding it towards the root. On receiving the DROP PARTY message, the root frees up the resources for the party and responds with a DROP PARTY ACKNOWLEDGE. On receiving the DROP PARTY ACKNOWLEDGE message, the N-UNI also frees up its resources. If this leaf happens to be the last party of the connection, the procedure at the root side is slightly different. The N-UNI at the root side sends a RELEASE message (instead of the DROP PARTY message) and the root responds to this by sending a RELAESE COMPLETE message (instead of DROP PARTY message). Here, the resources allocated for the whole call is freed both by the root and the N-UNI.

If the root wants to clear the whole point-to-multipoint call (including all the parties), it sends a RELEASE message toward the N-UNI. The N-UNI frees up its resources and forwards the message to all the parties. The network then sends the RELEASE message towards individual leafs and clears the whole call.

In order to maintain two notions of connection, one for the whole point-to-multipoint call and another for each party, two types of call states are maintained—link-state and party-state. Link-states define the overall state of a point-to-multipoint call and are similar to the call-states defined for point-to-point procedures. The names are different only to distinguish between point-to-point call-states and point-to-multipoint call-states. Since the link states do not specify the status of connections with the individual leaves, some other state variable is required to maintain this information. For this purpose, party states are used. However, unlike call-states, the definition for party-states is the same for both the user-side and the network-side.

When a new point-to-multipoint call is being established, both link-states and party-states change. However, once the connection is established, both the link-state and party-state become Active. For any party added subsequently to the call, the link-state remains Active. Again, while releasing the whole call, the link-state is changed only after the procedures for dropping of all the parties have been initiated.

11.6.1 Signalling Messages

The signalling messages for point-to-multipoint as defined in ITU-T Q.2971 have the same format as the Q.2931 signalling messages. Thus, all Q.2971 messages carry a CRV that identifies a point-to-multipoint call. However, in order to differentiate between different parties in the same point-to-multipoint call, another identifier, the End-point Reference Value (ERV) is used. The combination of CRV and ERV is unique for a party associated with a point-to-multipoint call.

The set of Q.2971 signalling messages includes all the Q.2931 messages. Additionally, six new messages are defined to allow the addition and dropping of parties in point-to-multipoint calls (see Table 11.5).

TABLE 11.5

Point-to-Multipoint Signalling Messages

Call Establishment Message	
ADD PARTY	Used to add parties to an existing point-to-multipoint call.
PARTY ALERTING	Used to indicate to the root that the called party has been alerted.
ADD PARTY ACKNOWLEDGE	Used to indicate to the root that the procedure for adding the party is complete.
ADD PARTY REJECT	Used to indicate to the root that the procedure for adding the party did not complete.
Call Release Message	
DROP PARTY	Used to drop a party from an existing point-to-multipoint call.
DROP PARTY ACKNOWLEDGE	Used to acknowledge the dropping of a party.

11.6.2 Information Elements

In addition to the information elements defined by Q.2931, two additional IEs are defined for the Q.2971 point-to-multipoint procedures. These two IEs are end-point state and end-point reference. The end-point state IE identifies the party-state of a particular party (leaf).

End-point reference IE is used to uniquely identify a party in a point-to-multipoint call. This IE is mandatory for all point-to-multipoint specific messages like ADD PARTY and ADD PARTY ACKNOWLEDGE. The IE is also present in SETUP, CALL PROCEEDING, ALERTING, and CONNECT messages during the addition of the first party of the point-to-multipoint call.

It is possible for a leaf to be unaware of the Q.2971 signalling procedures and still be a part of a point-to-multipoint call. For this, at the leaf side, only Q.2931 messages are exchanged;the only exception is the inclusion of IEs (like end-point reference) in some of the messages. The SETUP message sent from the N-UNI towards the leaf (U-UNI) carries, among other IEs, the end-point reference IE. If the leaf does not support point-to-multipoint procedures (i.e. it is unaware of Q.2971), it treats the end-point reference IE as an unrecognized IE and discards it. It treats the remaining SETUP message as a request for establishing a normal point-to-point connection and responds with a CALL PROCEEDING as per the point-to-point signalling procedures.

When the N-UNI side receives the CALL PROCEEDING message without an end-point reference IE, it assumes that the leaf does not support Q.2971 procedures. Hence, it does not include the end-point reference IE in all the subsequent messages that it sends to the leaf. The rest of the call establishment/clearing procedures are exactly identical to the Q.2931 signalling procedures. Thus, the leaf becomes a part of the point-to-multipoint call, without supporting the Q.2971 procedures.

11.6.3 Leaf Initiated Join (LIJ)

As mentioned earlier, the UNI 4.0 procedure allows a leaf to initiate the establishment of point-to-multipoint calls or even to add itself to an existing point-to-multipoint call. These twin capabilities are referred to as Leaf Initiated Join (LIJ) capability. LIJ procedures fall into the following two different categories:

- **Root LIJ:** The mode in which the leaf joins the connection by the intervention from the root, is termed as Root-prompted join or Root LIJ.
- **Network LIJ:** The mode in which the leaf joins without any intervention from the root, is termed as Leaf-prompted join (without root notification) or Network LIJ.

If a leaf wants to join a point-to-multipoint call, it uses the LIJ capability and sends a LEAF SETUP REQUEST message towards the root. If the root is unable to support this request, it sends a LEAF SETUP FAILURE to the leaf, indicating its inability to initiate connection establishment procedure. On the other hand, if the root is able to support the request, then it either initiates the point-to-multipoint procedures for adding the leaf to the point-to-multipoint call using ADD PARTY. Alternatively, if no point-to-multipoint connection exists for the requested connection, the leaf is added as per the point-to-point procedures.

If the point-to-multipoint call requires no root intervention (i.e. connection is Network LIJ connection), the LEAF SETUP REQUEST message sent by the leaf is not carried to the leaf, but is handled by the network. If the network is unable to handle the request, it sends a LEAF SETUP FAILURE to the leaf. On the other hand, if the request is acceptable to the network, then it adds the leaf as per the point-to multipoint procedures, with the exception that no messages are exchanged between the root and the network. All leaf-related information is maintained by the network and the root is unaware of the leaf being added to the point-to-multipoint call. A point-to-multipoint call is made a Network LIJ connection, when the first party is being added to the point-to-multipoint call. The SETUP message, traversing from the root towards the network, carries the information that the network should handle further requests from the leaves itself and not forward it to the root. Otherwise, the point-to-multipoint call is Root LIJ connection by default. Note that a network-initiated LIJ cannot initiate a point-to-multipoint call.

For supporting LIJ capability, the three additional IEs introduced are defined, which are Leaf Initiated Join Call Identifier IE, Leaf Initiated Join Parameters IE, and Leaf Sequence Number IE. Leaf Initiated Join Call Identifier IE is used to uniquely identify a point-to-multipoint call at a root's interface. It has a four-byte identifier value which may be obtained through the yellow pages or through some other directory service which contains identifiers for well-known point-to-multipoint services offered (like movie broadcast services).

The Leaf Initiated Join Parameters IE allow a leaf to create a network LIJ call by including this IE in SETUP message. The screening indication in this IE set to "Network Join without Root Notification".

The Leaf Sequence Number IE, is used by the joining leaf to associate a SETUP, ADD PARTY, or LEAF SETUP FAILURE response message with the corresponding LEAF SETUP REQUEST. It contains a four-byte leaf sequence number for maintaining the association.

11.6.4 Comparison between Q.2971 and UNI 4.0

ATM Forum's UNI 4.0 refers to ITU-T Q.2971 for the point-to-multipoint procedures, albeit with some changes. ATM Forum's UNI 4.0 adds an additional IE *Generic Identifier Transport* in the point-to-multipoint specific signalling messages. The most significant enhancement, however, is the introduction of the leaf-initiated join capability for the point-to-multipoint calls. Details of LIJ capability have already been discussed in the previous section.

CONCLUSION

ATM signalling procedures are a complex set of procedures. The complexity stems from the fact that in order to make ATM an integrated technology, capabilities must exist in the signalling protocol to carry appropriate information. Besides, standards permit interworking of ATM with other technologies. The interworking requirement introduces additional messages and IEs, which further increases the protocol complexity. Moreover, the regular modifications in ATM signalling standards do not allow the standards to stabilize, which affect the commercial deployments of ATM networks. In conclusion, the huge effort put in to provide a comprehensive set of services, has made the signalling protocol complex and implementation difficult. This has adversely affected the popularity of ATM signalling procedures. Thus, in most cases, Permanent Virtual Circuits (PVCs) are used instead of Switched Virtual Circuits (SVCs).

REVIEW QUESTIONS

Q 1. *Explain the basic design issues for any signalling protocol. With reference to ATM point-to-point signalling, discuss at least three important issues.*

Q 2. *Explain the ATM Signalling Reference Model and discuss the role of different protocols in this model.*

Q 3. *In ATM, why does call establishment require three-way handshaking while call release requires two-way handshaking? Explain the call establishment and call release flows with reference to the message used and the role they play.*

Q 4. *Explain the following procedures in brief. Also state the need for these procedures:*
- *Status Enquiry*
- *Restart*

Q 5. *List five important Information Elements (IEs) used in ATM signalling and explain the role they play.*

Q 6. *Differentiate between Call Reference Value (CRV) and End-point Reference Value (ERV).*

Q 7. *Explain the procedure of adding a party by the root in point-to-multipoint call.*

Q 8. *How can a leaf join a point-to-multipoint call?*

Q 9. *Differentiate between Root LIJ and Network LIJ model. Briefly compare the benefits and drawbacks of the two models.*

FURTHER READING

The contents of this chapter were based on plethora of standards, from both ITU-T and the ATM Forum. The important ITU-T specifications are [ITU-T Q.2931] for point-to-point signalling and [ITU-T Q.2971] for point-to-multipoint signalling. ATM Forum's [AMTF UNI 4.0] refers to these specifications and also provides additional features as discussed in this chapter. A recent update of UNI signalling published by ATM Forum is the [AMTF UNI 4.1]. For the lower layer SAAL protocol, one can refer to ITU-T specifications [ITU-T Q.2110], [ITU-T Q.2130] and [ITU-T Q.2140]. The book [Sig R. Onvural] is also a very comprehensive text on the subject.

Chapter 12

ATM Routing

12.1 Introduction

Routing is the act of forwarding packets from source to destination using intermediate network elements called *routers*. Routers are devices that use the destination address carried in the packet header to appropriately route packets to their respective destinations. For the purpose of routing, routers maintain routing tables that contain a list of possible destinations and the next-hop for each possible destination. Table 12.1 provides generic information on routing. For details on these topics, refer [Gen S. Kasera].

In the context of ATM, routing is seen in context with signalling. The previous chapter on signalling had explained how ATM end-systems, using signalling, establish switched virtual connections. The ATM network was treated as a cloud and no mention is made as to how signalling messages are internally exchanged (i.e. routed). The challenge of routing the signalling messages to the destination in ATM is comparable to routing IP datagrams. The only difference is that routing in ATM is restricted to signalling messages (i.e. restricted to call-setup period), while in IP, each datagram is routed. However, the core issues governing the design of routing protocol remain more or less the same for both IP and ATM. Nonetheless, routing in ATM is slightly more complicated than IP routing due to reasons cited below:

First of all, the scope of the problem in ATM is not just limited to finding a route to the destination, but also includes issues related to resource reservation and QoS-based routing. Every ATM switch in the path to the destination has to verify whether ample resources are available (i.e. has to perform Connection Admission Control). If available, the switch reserves adequate resources and forwards the call further. This is not the case in IP where the only point of concern is to find a route to the destination.

TABLE 12.1 Basics of Routing [Gen S. Kasera]

Definition	Routing is the act of forwarding packets from source to destination by using intermediate network elements
Requirements	• **Optimality:** Routes found must be optimal in nature (i.e. routes calculated must be best in terms of a given metric; for example, number of hops). • **Reducing exchange of routing information:** The flow of routing information (i.e. management overheads) must be minimal so that more bandwidth is available for carrying user traffic. • **Reducing processing load on routers:** The burden of processing routing information on the routers, so that more processing power is available to forward the packets. • **Faster convergence:** This refers to the ease with which routers achieve a consensus on optimal routes in the network. • **Scalability**: This refers to the ease with which a routing protocol can be extended to cover a larger network area. • **Robustness:** This refers to the ability of a routing protocol to tackle unforeseen or unusual circumstances.
Classification	• **Nature-wise:** Refers to the classification based on the extent to which dynamic changes in the network topology get reflected in the routing tables. This criteria leads to the two categories of routing—*static routing* and *dynamic routing*. • **Control-wise:** Refers to the classification based on the control (or power) that resides at a router to compute routes and leads to two major categories of routing—*distributed routing* and *centralized routing*. • **Scope-wise:** Refers to the classification based on the scope of the routing protocols, all routing protocols are classified as either *intra-domain protocol* or *inter-domain protocol*. • **Number-wise:** Refers to the classification based on the number of end-systems in the routing protocols and leads to two major categories of routing protocols—*unicast routing protocol* or *multicast routing protocol*.
Concepts	• **Summarization:** Refers to the process of using a single network prefix to represent a collection of nodes/subnets that begin with the same prefix. By summarizing routes, a single summarized route is advertised to the external world, which significantly reduces the routing complexity.

Contd.

Table 12.1

Contd.

	• **Hierarchy:** Refers to the hierarchical and summarized flow of routing information to allow routers to operate effectively without actually obtaining complete information of every end-system/link in the network. • **Route computation:** Refers to the two basic techniques for route computation, *link-state method* and *distance-vector method*. • **Path selection:** Refers to the two basic techniques for selecting a path till the destination, *hop-by-hop routing* and *source routing*. • **Parameterization:** Refers to the various parameters like hop count, bandwidth and delay among others used to define the optimality of a path.
Examples	• **Routing Information Protocol (RIP):** Based on distance-vector route computation technique. • **Open Shortest Path First (OSPF)**: Based on link-state route computation technique. • **Border Gateway Protocol (BGP):** Inter-domain routing protocol for routing between autonomous systems.

The second concern is the challenges associated with QoS-based routing. To recall from Chapter 7, ATM specifies a number of QoS parameters (CTD, CDV and CLR). The QoS parameters have to be considered while selecting a path to the destination. The greater the number of QoS parameters, the more complicated is the routing algorithm. Moreover, QoS-based routing is sensitive to load fluctuations and dynamic changes in the network. This demands that very accurate information be maintained about the state of the ATM network. QoS-based routing, when coupled with source routing, further complicates the matter. This is because in source routing, the ingress ATM switch calculates the path to the destination. For the ingress switch to calculate this path, it must have QoS information for all the switches in all the possible paths to the destination. This is required because of all the paths available to a given destination, the path that supports the requested QoS will be chosen. This warrants exchange of resource availability information between switches, which can be extremely complicated, given the granularity of QoS supported by ATM. In essence, QoS-based routing is a very challenging problem and still an active research area.

In order to provide the means to route signalling messages, the ATM Forum has come up with [ATMF PNNI 1.1] specification. The abbreviation PNNI, depending upon the usage, either refers to Private Network-Node Interface or to Private Network-to-Network Interface. The former term is used for the interface specification between two ATM switches/nodes, whereas the latter is used between two ATM networks. Both the scenarios are depicted in Figure 12.1. The protocol definition is, however, independent of the usage.

FIGURE 12.1

PNNI Interface

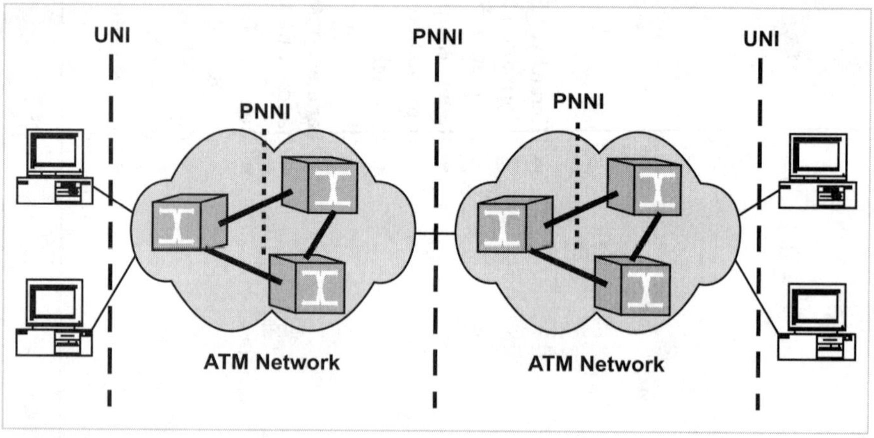

12.2 INTERIM INTER-SWITCH PROTOCOL (IISP)

The PNNI protocol, also called the Phase 1 protocol, is preceded by the Interim Inter Switch Protocol (IISP), a Phase 0 protocol. IISP is a rudimentary inter-switch communication protocol and does not support QoS routing, crankback and most other features of PNNI protocol. In essence, IISP was made to leverage the deployment of ATM networks and act as an interim protocol till the time PNNI was finalized.

12.3 PNNI Protocol

The PNNI protocol specification, [ATMF PNNI 1.1], was released by the ATM Forum in 1996. The protocol comprises of two parts: the first part (known as routing specification) defines the procedures for distributing topology information between switches and the mechanisms of using this information for computing routes. The routing specification also details guidelines for QoS-based routing and CAC. The second part of the specification (known as signalling specification) defines the message flows for point-to-point and point-to-multipoint signalling. Additionally, this part specifies the means for source routing and crankback, details of which are explained later in this chapter. PNNI signalling, however, does not support the Leaf Initiated Join (LIJ) capability, which is specified by UNI 4.0.

It may be reiterated that even though PNNI is explained in this chapter titled 'Routing', PNNI is not merely a routing protocol, but includes signalling specification as well.

The important features of PNNI routing protocol are as follows:

- **Hierarchical routing:** PNNI specifies a hierarchical structure to maintain topological information and to compute routes. The details of hierarchical routing are explained in the next sub-section.
- **Link-state routing:** PNNI follows the link-state method to exchange topological information and to compute optimal routes.
- **Source routing:** As against hop-by-hop routing, PNNI uses source routing.
- **QoS-based routing:** PNNI provides the means for QoS-based routing.
- **Scalability:** PNNI is designed to be a highly scalable protocol and all its aspects reflect this design philosophy. In fact, PNNI is intended to satisfy the goal of a global ATM network.
- **Flexibility:** PNNI is flexible, in the sense that it allows interworking with external routing domains, not necessarily using PNNI.

In the following sections, a bottoms-up approach is adopted to explain the PNNI routing specification. First, the elements of PNNI routing hierarchy are detailed. Also core routing concepts like routing information exchange, information summarization and path selection are elaborated. After explaining the essential parts of routing specification, the important aspects of signalling specification of PNNI are explained. Note that some definitions in this chapter are extracted from the [ATMF PNNI 1.1] which are copyrights, 1996, The ATM Forum.

12.4 PNNI Routing Hierarchy

PNNI specifies a hierarchical structure for maintaining topological information. A hierarchical structure reduces the overheads of a flat structure while providing an efficient mechanism for routing. In a flat structure, each node has to have knowledge of the complete topology of the network, which means maintaining reachability information for every other node in the network. This translates into huge routing tables. For a relatively small-sized network, routing tables can still be managed. But as the size of the network increases, managing the network becomes very difficult. Since scalability is a prime requirement for ATM networks, PNNI specifies an elaborate mechanism for defining routing hierarchies.

In essence, each node in the hierarchical PNNI network maintains a partial view of the network. (The *view* here refers to the extent of network topology visible to a node). This partial view is similar to the view of an internal router in Open Shortest Path First (OSPF) topology. The internal router has complete information about its own area, but only has a summarized view of other areas. This reduces the size of routing information database of an internal router. All inter-area communication takes place using Area Border Routers (ABRs). Thus, a reduction in the size of routing database is achieved at the cost of sub-optimal routing. The reduction occurs because a router does not maintain complete topological information, while the routing is sub-optimal because there could have been better paths between two areas that used routers other than the ABRs. The same concept is applied in PNNI routing, the only difference being that OSPF defines a two-level hierarchy, whereas in PNNI, there can be an N (< = 104) level hierarchy.

In PNNI routing hierarchy, sets of nodes are grouped into peer groups (akin to OSPF areas). Nodes in a peer group share information with each other and have complete information about the topology of the peer group. At the lowest level of the routing hierarchy, there are many peer groups. At the next level, each peer group is represented by a Logical Group Node (LGN). Collections of LGNs again form higher-level peer groups and this is a recursive procedure.

The following sub-sections elaborate upon the elements of PNNI routing hierarchy. Figure 12.2 is referred in the process.

12.4.1 Lowest-Level Node

At the bottom of the PNNI hierarchy is the lowest-level node, or simply, node. Formally, a node is defined as "a single instance of the PNNI routing protocol created in a switching system via configuration." In simple terms, a node can be viewed as an ATM switch that supports PNNI protocol. A collection of lowest-level nodes forms a peer group. In a peer group, each lowest-level node exchanges information with other members and has the same view of the group.

The lowest-level nodes are identified (i.e. addressed) using a 22-byte address structure. This address structure consists of three parts: The first byte represents the level indicator and specifies the level of the peer group to which the node belongs. The second byte is set to a value of 160. This value is chosen to distinguish between a node address and an LGN address. For an LGN, the second octet represents the most significant bytes of the peer group identifier, which cannot begin with this value. The remaining 20 bytes represent the ATM end-system addresses (AESA). These ATM end-system addresses are modelled after the

FIGURE 12.2

Partially Configured PNNI Hierarchy, Illustrating Various PNNI Elements

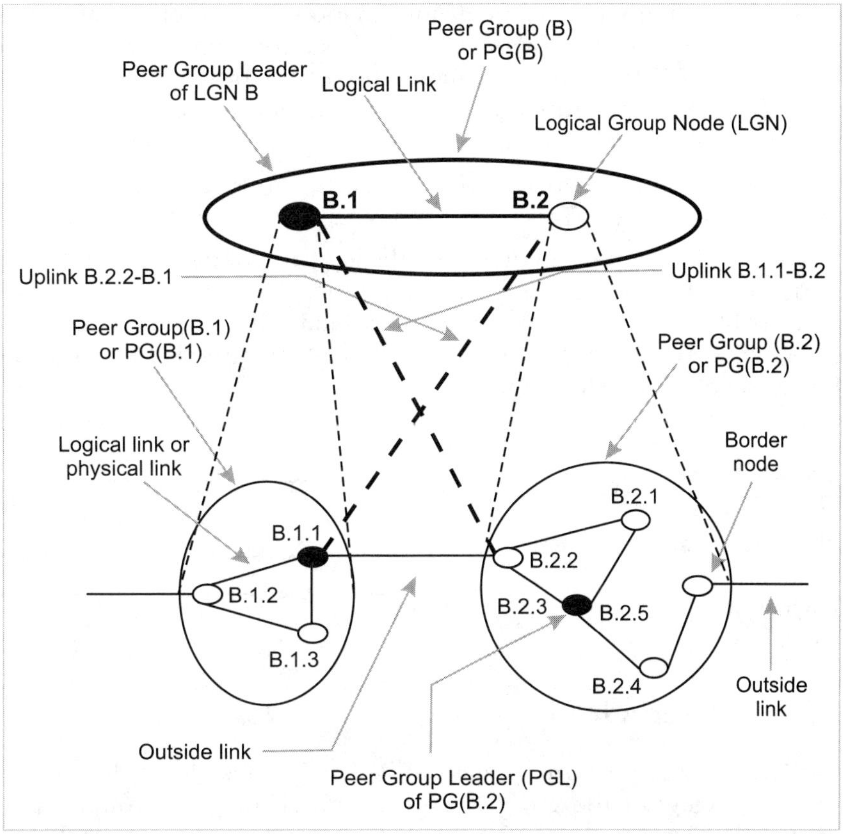

Network Service Access Point (NSAP) address format, as already explained. For the purpose of routing, only the first 19 bytes of the AESA are used. The twentieth byte (selector) has local significance only and is ignored by PNNI routing.

12.4.2 Logical Node

A logical node refers either to a lowest-level node or to a Logical Group Node (LGN). In the lowest hierarchical level, a lowest-level node is the logical node. In the given figure, B.1.1, B.1.2 and B.2.1 are all examples of lowest-level nodes. At all levels other than the lowest level, a logical node refers to an LGN (explanation of LGN follows).

12.4.3 Logical Group Node (LGN)

An LGN is defined as "an abstract representation of a peer group as a single point for the purpose of representing that peer group at a particular level (but not at the lowest) in the PNNI routing hierarchy." In other words, an LGN hides the internal details of its child peer group, but still provides adequate information to LGNs of the same peer group to reach every node that is a abstracted by the LGN. For example, in the figure, B.1 and B.2 are LGNs in the Peer Group PG (B). Now, B.1 hides the details of its child peer group PG (B.1), but still provides a summarized view to other LGNs of the peer group (i.e. to LGN B.2 of the peer group PG (B)). This summarized view is adequate for all nodes of B.1 to reach any node of B.2.

In essence, an LGN that resides at a particular level represents a group of interconnected LGNs of lower levels. This provides a recursive definition of LGNs.

An LGN is uniquely identified by the logical group node identifier. Figure 12.3 depicts the structure of the logical group node identifier. Note that both the lowest-level node address and the LGN address are referred to as Node Identifiers (or simply, node ID). Depending upon whether the second octet of the 22-octet structure is 160 or not, the node ID refers to the lowest-level node or to LGN.

FIGURE 12.3

PNNI Address Structure of Logical Group Node

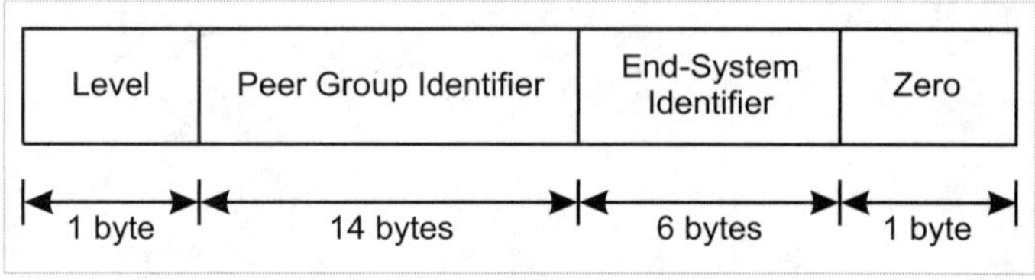

12.4.4 Peer Group

Earlier, the term peer group was defined as "a collection of lowest-level nodes, which have the same view of the group." This definition of peer group applies only to the lowest hierarchical level, whereas the concept of peer group is applicable to any level of the PNNI hierarchy. Generalizing the definition, a peer group is "a set of logical nodes which are grouped for the purpose of creating a routing hierarchy; the member nodes of the group exchange topology information between themselves, thereby maintaining a single view of the group." Now, since a logical node can either be a lowest-level node or an LGN, a peer group can either be a collection of lowest-level nodes or a collection of LGNs.

In the given figure, the lowest hierarchical level has two peer groups: PG (B.1) and PG (B.2). The PG (B.1) consists of B.1.1, B.1.2, and B.1.3 lowest-level nodes, while the PG (B.2) consists of B.2.1, B.2.2, B.2.3, B.2.4, and B.2.5 lowest-level nodes. At the next hierarchical level, PG (B.1) is represented by LGN B.1 and PG (B.2) by LGN B.2. The LGNs B.1 and B.2 together form the peer group PG (B). This example explains how a peer group can either be a collection of lowest-level nodes (e.g. PG (B.1)) or be a collection of LGNs (e.g. PG (B)).

Peer groups are identified using the address prefix field. An address prefix field can have a maximum length of 104 bits (13 bytes). This field, along with a one-byte level indicator field, forms what is known as a Peer-Group Identifier (PGID). The level indicator field identifies the number of bits that are significant within the address prefix of a PGID. For a peer group at level 'N', the identifier information field (i.e. the address prefix field) is encoded with

(104 – N) rightmost bits set to 0. Note that lower the value of the level indicator, higher is the level of the peer group. For example, a peer group with level 24 is at a higher level in the routing hierarchy, as compared to a peer group with level 56.

In the discussion so far, the difference between an LGN and a peer group is not very obvious. This is because they are merely two different ways of looking at the same thing. At a particular level, a peer group is a collection of lowest-level nodes or LGNs. At the next higher level, the same peer group is represented by an LGN in a higher level peer group. And this process continues till the top of the hierarchy is reached.

12.4.5 Peer Group Leader (PGL)

It is not enough to say that a peer group is represented by an LGN at a higher level. The question which then arises is: "What is an LGN besides being a logical abstraction of a peer group at a higher level?" If it is merely a theoretical concept, then why define it at the very first place? If not, then who performs the duties of an LGN? The answer to these questions lies in understanding the roles of a Peer Group Leader (PGL). A peer group leader is "A node of a peer group that performs the extra work of collecting, aggregating and building data that is suitable to represent the entire peer group as a single node in a higher-level peer group." Visualizing a PGL at the lowest level is simple enough. For example, B.1.1 is the PGL of PG (B.1), which means to say that B.1.1 collects data on behalf of its peer group members (B.1.1, B.1.2 and B.1.2.3) and makes it available in the PG (B). At a higher level, an LGN acts as a PGL for its peer group. For example, LGN B.1 acts as the PGL for PG (B), which means that LGN B.1 summarizes information in PG (B) and presents it to the higher layers. The point to note is that B.1 is itself represented by B.1.1 in PG (B). Thus, LGN B.1 is an abstraction of B.1.1 in PG (B).

In a peer group, there can be at most one PGL. The selection of a PGL in a peer group is done using a PGL election. The election is based on the node's leadership priority, with the node having the highest leadership priority being declared the winner. In case of ties, the node with a higher node identifier wins the election. The PGL election is a continuous process. When a node—which has a leadership priority higher than that of the existing PGL—becomes active, the PGL role is transferred to the newly active node. In case an active PGL fails or is removed, the node with the next highest leadership priority takes over the responsibility. In order to ensure stability, a successful candidate after becoming PGL, increases its leadership priority.

Note that a flat network, which has only one level, need not have a PGL at all, since there is no higher-level peer group where a PGL can represent its peer group.

12.4.6 Links

In the PNNI routing hierarchy, all links are referred to as logical links. A logical link represents the connection between two logical nodes and includes physical links and Virtual Path Connections (VPCs). A physical link refers to the duplex connection (symmetric or asymmetric) between two switching systems.

The logical link within a peer group is called a horizontal link, while the link connecting nodes of two different peer groups is called an outside link. Nodes connected via an outside link are called *border nodes*. In the given figure, B.1.2-B.1.1 is a horizontal link, whereas B.1.1-B.2.2 is an example of an outside link. B.1.1, B.2.2, B.2.5 and B.1.2 are examples of the border node.

The referred figure also illustrates the concept of uplinks. For example, B.2.2-B.1 and B.2-B.1.1 are examples of uplinks. An uplink represents connectivity from a border node to an upnode. An upnode is a neighbouring peer of one of the border node's ancestors. For the two uplinks mentioned above, B.1.1 and B.2.2 are border nodes, while B.1 and B.2 are upnodes. Moreover, B.1 (B.2) is a neighbouring peer of border node B.2.2's (B.1.1's) ancestor B.2 (B.1).

12.4.7 Complete PNNI Hierarchy

Since most of the elements of PNNI hierarchy have already been elucidated, the complete routing hierarchy (see Figure 12.4) is now explained, a part of which is depicted earlier. The configuration presented in Figure 12.4 will form the basis of subsequent discussions and it is recommended that the hierarchy be well understood.

The idea of creating PNNI routing hierarchies, based on peer groups and peer group identifiers, has its equivalent in telecommunication networks. In fact, ITU-T defines a five-tiered structure for telecommunication networks. Although citing details of this structure is beyond the scope of this book, the rationale behind creating hierarchies is worth exploring. A hierarchical network not only allows handling of heavy traffic, it also reduces database requirements.

To understand this, consider how telephone lines are laid out. Subscribers are connected to a local exchange. Each local exchange has a locally unique number and this, appended to the locally unique subscriber number, gives a telephone number. Here, the local exchange number is analogous to the peer group identifier. For example, in the telephone number 646-2279, the first part 646 is the local exchange number. This local exchange number is common to all subscribers belonging to that exchange. This concept is analogous to a group of end-systems sharing the same peer group identifier. Now, as an example, consider a call request generated from 646-2272 for 631-2279. To search for the destination number (631-2279), only the exchange number (631) is looked at. The 646 local exchange need not keep information of all subscribers attached to local exchange 631. In fact, 646 exchange need not even know whether 631-2279 number really exists or not. Only when the call request reaches exchange 631, the number is verified and attempts are made to connect to the destination end.

In other words, all a local exchange needs to know are details of its own subscribers and the means to reach other local exchanges. An exchange need not bother about the internals of another exchange. This is exactly how PNNI operates. In a peer group, each node has a complete view of its own peer group and an aggregated view of the rest of the network.

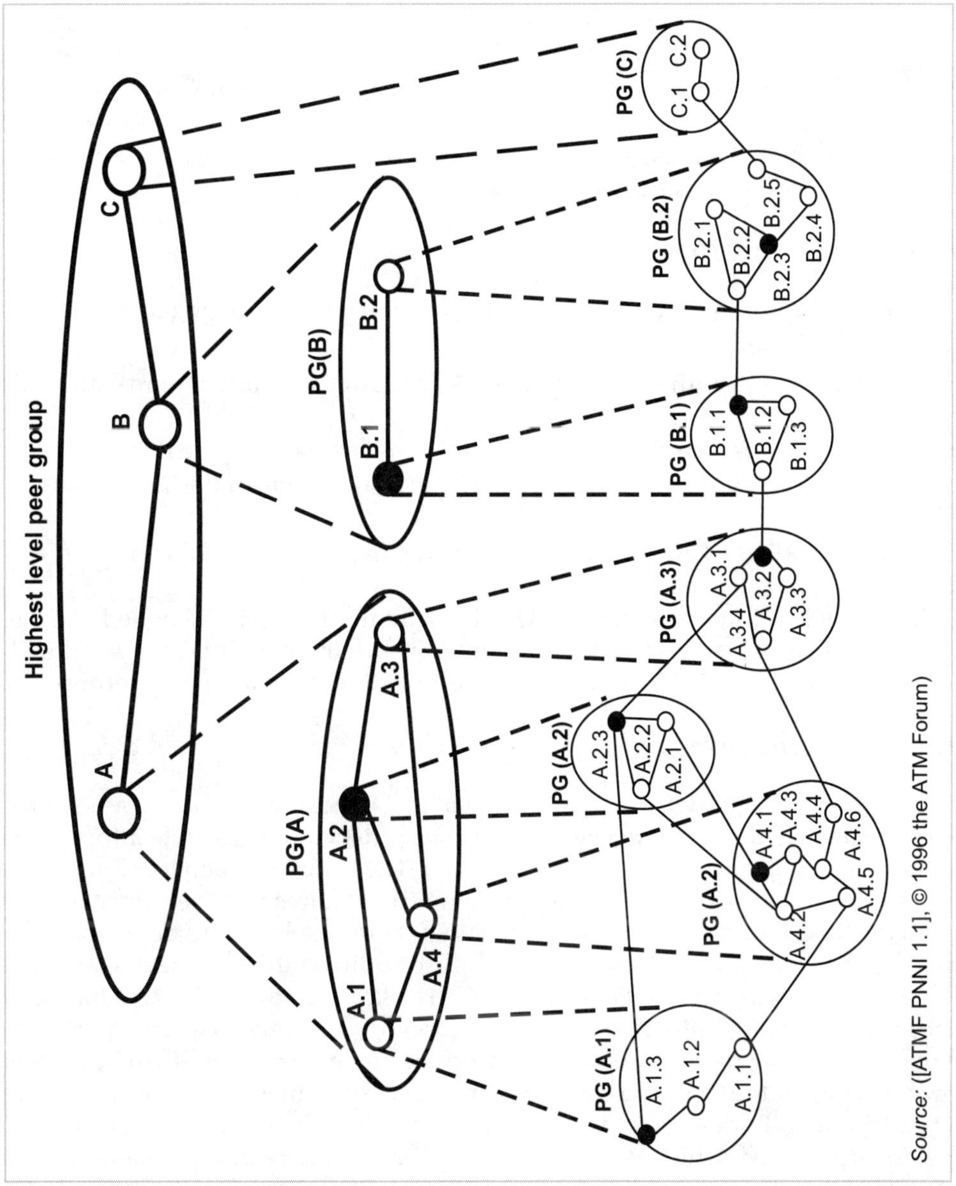

Source: ([ATMF PNNI 1.1], © 1996 the ATM Forum)

FIGURE 12.4 Complete View of a Hierarchically Configured PNNI Network

12.5 PNNI Topology Information

Each node in the PNNI hierarchy maintains topology information, which is used by nodes to perform routing functionality. The topology information is classified into three broad categories: nodal information, state information and reachability information. Each category of the topology information is explained below.

12.5.1 Nodal Information

As the name suggests, nodal information encompasses general information regarding a node. The elements of nodal information are as follows:

- **Node's ATM end-system address:** Refers to the ATM address required for reaching this node.
- **Leadership priority:** Refers to the desirability of the node to become the PGL.
- **Preferred PGL node id:** Refers to the node id of that node of which this node wants to be a PGL.
- **Nodal information flags:** This includes a set of five flags used for PGL election and various other purposes.
- **Next higher-level binding information:** This is maintained only if the node is a PGL. The information maintained includes the parent LGN id: the parent LGN's ATM end-system address, the parent PGID and the node id of PGL of the parent peer group.

12.5.2 State Information

State information is maintained to determine whether a new connection request can be accepted or not (i.e. to determine whether the requested QoS can be satisfied or not). State information parameters fall into two categories: metric and attribute. The difference between the two is that while each attribute is considered individually, a metric requires that the values of state parameters of all the links and nodes along a given path be combined before a decision on the combined value is made. Putting it differently, metrics are additive link parameters, whereas attributes are non-additive parameters. For example, Cell Delay Variation (CDV) is a topology metric, whereas the Available Cell Rate (AvCR) is a topology attribute. The first part is so because CDV is a cumulative measure, i.e. CDV introduced by intermediate switches, when added, determines whether the CDV of a connection request can be met or not. One particular switch cannot independently determine whether it can support the requested CDV or not. This is not the case for AvCR, which is used to check whether the requested bandwidth is available or not. This decision is independent of the state of other intermediate nodes/links. Table 12.2 provides the currently defined topology metrics and topology attributes.

12.5.3 Reachability Information

This identifies the nodes reachable through a PNNI routing domain. Formally, reachability information is defined as "topology information that binds reachable addresses to nodes

TABLE 12.2 Topology State Parameters [ATMF PNNI 1.1]

Topology Metrics

- **Cell Delay Variation (CDV):** Specified in microseconds, this metric is required for CBR and rt-VBR service categories.
- **Maximum Cell Transfer Delay (maxCTD):** Sum of the CDV and the fixed delay component across the link or the node. Specified in microseconds, this metric is required for CBR, rt-VBR and nrt-VBR service categories.
- **Administrative Weight (AW):** The value of this metric is set by the network administrator. AW is a dimensionless value required for all service categories and indicates the desirability (or the lack of it) of using a link or node. A higher value indicates a higher desirability and vice versa.

Topology Attributes

- **Cell Loss Ratio for CLP = 0 (CLR_0):** It is the fraction of cells belonging to CLP = 0 flow that are either not delivered to the destination or delivered after a pre-specified time.
- **Cell Loss Ratio for CLP = 0 + 1 (CLR_{0+1}):** It is the fraction of cells belonging to CLP = (0 + 1) flow that are either not delivered to the destination or delivered after a pre-specified time.
- **Maximum Cell Rate (maxCR):** Maximum capacity usable by connections belonging to the specified category. Specified in cells/second, this attribute is mandatorily required for ABR and UBR service categories and is optional for other categories.
- **Available Cell rate (AvCR):** This is a measure of the effective capacity available for CBR, rt-VBR and nrt-VBR service categories. For the ABR service category, this is a measure of the capacity available for Minimum Cell Rate (MCR) reservation. Specified in cells/second.
- **Cell Rate Margin (CRM):** This is a measure of the difference between the effective bandwidth allocation and the allocation for sustained cell rate. In other words, this indicates the safety margin allocated above the aggregate sustainable cell rate. Specified in cells/second, this attribute is required for the rt-VBR and nrt-VBR service categories.
- **Variance Factor (VF):** This is a relative measure of the square of the cell rate margin normalized by the variance of the sum of the cell rates of all existing connections. A dimensionless value, this is an optional attribute for rt-VBR and nrt-VBR service categories.

within the PNNI routing domain." Depending upon whether the PNNI protocol is used for routing exchange or not, reachability information is classified as either interior or exterior information. Interior information refers to local knowledge of reachability within the PNNI domain, whereas exterior information refers to the information derived from protocol exchanges outside the PNNI domain. Table 12.3 provides the elements associated with each internal and external reachable address.

TABLE 12.3

Reachability Information [ATMF PNNI 1.1]

Internal Reachable Addresses: Each internal reachable information group contains:

- Port ID
- Scope of Advertisement
- Address Information Length
- Address Information Count
- Pairs of Prefix Length and Prefix
- Optional Information Groups for Resource Availability

External Reachable Addresses: Each external reachable information group contains:

- Port ID
- Scope of Advertisement
- Address Information Length
- Address Information Count
- Pairs of Prefix Length and Prefix
- Optional Information Groups for Resource Availability
- Optional Transit Network Information Groups

12.6 ROUTING INFORMATION FLOWS AND ROUTE COMPUTATION

Earlier, the elements of the PNNI hierarchy were explained. This sub-section explains how various PNNI elements exchange routing information. The key procedures in PNNI network include Node Initialization, Database Synchronization, Flooding, Information Flow 'Up' the Hierarchy and Information Flow 'Down' the Hierarchy. These procedures are explained in the following sub-sections.

12.6.1 Node Initialization

As soon as a node comes up (i.e. becomes active), it exchanges state information with its immediate neighbours. The state information, which is carried in a 'Hello' packet, contains, among other things, the ATM end-system address, node ID and peer group ID (PGID). For the purpose of routing, the most important component of the message is the PGID. A node, upon receiving a Hello packet from its neighbour, compares the received PGID value with its own PGID. If the two values match, the node tries to synchronize its topology database with the neighbour who had originated the message.

12.6.2 Database Synchronization

Once neighbouring nodes realize that they belong to the same peer group, they proceed to synchronize their topology database. Through the synchronization process, neighbouring nodes belonging to the same peer group attain identical topology database. This topology database synchronization process between two nodes can be explained by the following steps:

1. The nodes involved in the synchronization process first exchange the PNNI Topology State Element (PTSE) header information. PTSE is a collection of PNNI information that is flooded among all logical nodes within a peer group.
2. When a node receives a PTSE header information, which it has not previously received, or which is more recent than the one it last received, the node requests for the advertised PTSE. Note that the PTSE header and PTSE are distinct entities. PTSE header is a 20-byte field and contains elements like the PTSE identifier, sequence number, checksum, remaining lifetime, etc. Using these elements, the PTSE header identifies a particular PTSE. PTSE, on the other hand, is a collection of topology information maintained by a node.
3. After receiving the requested PTSE, the node updates its topology database with the received PTSE.

12.6.3 Flooding

After the topology databases with neighbouring nodes are synchronized, a node in the peer group proceeds to propagate the topology information to all members of the peer group. This is achieved through a mechanism called flooding. Flooding is an integral part of all link-state routing protocols, because it enables a switch/router to have information of other parts of the network. In PNNI, flooding is the process through which nodes in a peer group exchange PNNI routing information and attain identical topology databases. The flooding process can be explained by the following steps:

1. The PTSEs (obtained after database synchronization) are bundled within a PNNI Topology State Packet (PTSP) and transmitted to all the neighbouring nodes. To improve transmission efficiency, as many PTSEs are bundled into a PTSP packet as much possible. However, care is taken to ensure that the PTSP size does not exceed the maximum permissible packet size of a link.
2. For each PTSE received in a PTSP packet, the receiving node acknowledges the receipt by sending an 'acknowledgement packet'.
3. If the received PTSE is more recent than the version previously-maintained, the node updates the topology database. Moreover, this PTSE is flooded to all neighbouring

nodes except the node from which the PTSE was received (this is required to ensure that the network is not flooded with duplicate packets).

4. The PTSEs are sent periodically, until an acknowledgement is received from the receiver.

Flooding is a continuous process because PTSEs age out and need to be retransmitted. Those PTSEs that age out, cannot be used for routing purposes. In such a scenario, the node that had originated the PTSE, updates the PTSE. No other node is permitted to update a PTSE that it had not originated.

12.6.4 Information Flow 'Up' the Hierarchy

The previous three sub-sections explained the procedures for information exchange at the lowest level. This sub-section and the next sub-section discuss information exchanges that take place across the PNNI hierarchy.

In the PNNI hierarchy, recall that a peer group, say PG (X.Y), is represented by its peer group leader (PGL) in the higher-level peer group PG (X). Putting it differently, the peer group PG (X.Y) is represented by its PGL as a Logical Group Node (LGN) in the peer group PG (X). Now, after the database synchronization and flooding is complete, the PGL (X.Y) has complete information of all the nodes of its peer group PG (X.Y). The PGL (X.Y) then summarizes the topology information of PG (X.Y) and propagates (floods) the information to its PG (X) peers. Conceptually, this is topology information flowing up the hierarchy. Note that the PTSEs received by PGL (X.Y) from the other nodes in PG (X.Y) do not directly flow up the hierarchy. Instead, PGL (X.Y) generates a summary of information received from all nodes of the PG (X.Y). This summarized information is then flooded through PTSEs to the LGNs contained in PG (X). Figure 12.5 depicts the conceptual flow of summarized information up the hierarchy.

12.6.5 Information Flow 'Down' the Hierarchy

Information does not only flow up the hierarchy, it also flows down. While upward information flow helps in creating the PNNI hierarchy, the downward movement is necessary for the lower-level nodes to know which addresses can be reached through the PNNI routing domain. An LGN receives downward information in the form of PTSEs flooded by other LGNs of the same higher-level peer group. The PGL of that LGN then floods the received PTSE to all members of the peer group. The concept of downward information flow is depicted in Figure 12.6. In the simplest of cases, downward information can be viewed as a set of address prefixes, which each lowest-level node must know in order to determine whether a given destination address (in a connection request) is reachable or not.

12.7 PNNI State Machines

In order to exchange routing information according to the procedures mentioned above, PNNI elements maintain a number of Finite State Machines (FSMs). This sub-section elaborates upon the important FSMs in PNNI protocol.

FIGURE 12.5

Information Flow Up the Hierarchy

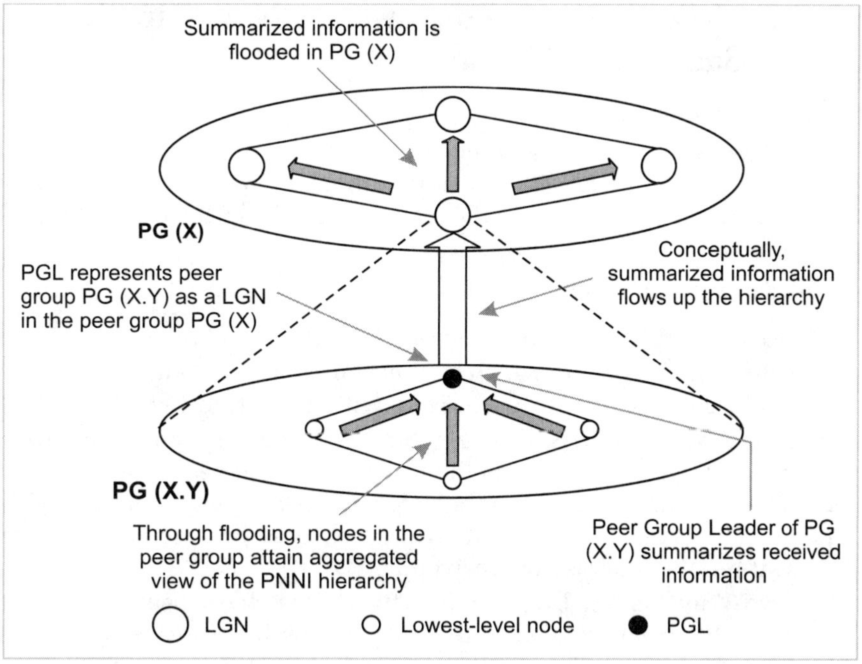

FIGURE 12.6

Information Flow Down the Hierarchy

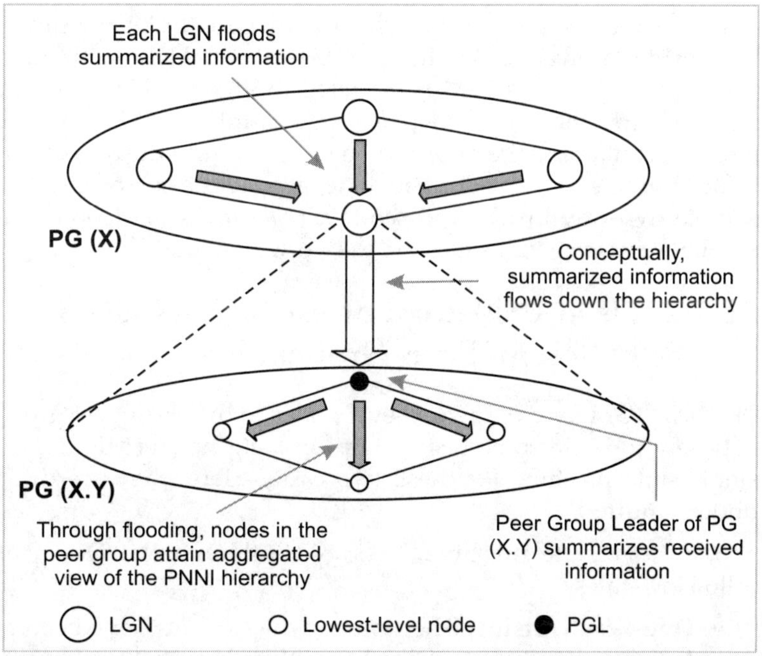

12.7.1 *Hello* FSM for Lowest-level Nodes in the Same Peer Group

When a node first comes up, the *Hello* protocol is used. At the lowest level, each physical link or VPC has its own instance of the *Hello* protocol, which is made of *Hello* data structure and *Hello* state machine. The *Hello* data structure contains elements including (but not limited to) the state, port ID, remote node ID, remote peer group ID, remote port ID, *Hello* Interval and *Hello* timer, among other things.

The *Hello* state machine contains the following states:

- **Down:** This is the initial state of FSM, when the link is unusable.
- **Attempt:** The state in which attempts are made to contact the neighbour by periodically sending the neighbour with *Hellos* with the period *Hello* Interval.
- **One-Way Inside:** The state in which *Hellos* have been received from the neighbour, and it has been established that they belong to same peer group. However, the remote node ID and remote port ID in the received *Hellos* were set to zero. This indicates that the neighbour does not know the values of these parameters and happens when a *Hello* is sent but the reply is not received.

- **Two-Way Inside:** The state in which *Hellos* have been received from the neighbour, and it has been established that they belong to same peer group. Also, the remote node ID and remote port ID in the received *Hellos* are correctly set. This state indicates that database synchronization can begin on this link.

Events that trigger the state changes include (but is not limited to) LinkUp, LinkDown, 1-WayInsideReceived, 2-WayInsideReceived, and HelloMismatchReceived. One possible set of state changes for this FSM is depicted in Figure 12.7(a). As shown, a node that comes up, sends a *Hello* message and reaches Attempt state. The neighbour responds with a *Hello* message containing the information received from the previous *Hello* message. It also changes its state to one-way Inside. The node, getting the correct neighbour information, changes its state to two-way Inside, and sends a *Hello* message. Upon receiving the *hello* message, the neighbour also changes its state to two-way Inside.

12.7.2 *Hello* FSM for Lowest-level Nodes in Different Peer Groups

The *Hello* FSM for two lowest-level nodes in different peer groups is similar to the *Hello* FSM between nodes in same peer group. In fact, the [ATMF PNNI 1.1] specification provides a single state machine for these two cases. Here, they are differentiated to facilitate easier understanding.

The *Hello* state machine, besides containing the Down and Attempt states, has the following states:

- **One-Way Outside:** The state in which *Hellos* have been received from the neighbour, and it has been established that they belong to different peer groups. However, the remote node ID and remote port ID in the received Hellos were set to zero. This indicates that the neighbour does not know the values of these parameters and happens when a *Hello* is sent, but a reply is not received.

- **Two-Way Outside:** The state in which *Hellos* have been received from the neighbour, and it has been established that they belong to different peer groups. Also, the remote node ID and remote port ID in the received Hellos are correctly set.
- **Common Outside:** This state is reached when a common level of routing hierarchy has been found. This state indicates that PTSEs can be advertised as uplinks to the upnode.

The events that trigger state changes include—but are not limited to—LinkUp, LinkDown, 1-WayOutsideReceived, 2 WayOutsideReceived, CommonHierarchyReceived, and Hello MismatchReceived. One possible set of state changes for this FSM is depicted in Figure 12.7(b). As shown, a node that comes up sends a *Hello* message and reaches the Attempt state. The neighbour responds with a *Hello* message containing the information received from the previous *Hello* message. It also changes its state to one-way Outside. The node, getting the correct neighbour information, changes its state to Two-way Outside, and sends a *Hello* message. Upon receiving the *Hello* message, the neighbour also changes its state to Two-way Outside.

12.7.3 Neighbouring Peer FSM

After two nodes in the peer group exchange *Hellos* and attain the Two-way Inside state, they proceed to synchronize their databases. Two neighbouring nodes may have multiple, parallel physical links and/or VPCs between them, each having a separate *Hello* data-structure/state-machine. However, for database synchronization and flooding, only one instance of data-structure/state-machine is maintained. This is referred to as neighbouring peer data structure and neighbouring peer state machine, respectively. A node creates an instance of these when the first link attains the Two-way Inside state, and clears when the last link leaves the Two-way Inside state.

The neighbouring peer data structure contains elements including (but not limited to) state, remote node ID, port ID list, master/slave, DS sequence number (used to identify individual database summary packets), and the peer retransmission list (list of PTSEs that have been flooded but not acknowledged by the neighbouring peer).

The neighbouring peer state machine contains the following states:

- **NPDown:** This is the initial state of FSM, when no link is active (i.e. in a two-way Inside state).
- **Negotiating:** The state in which neighbouring peers initiate the process of building adjacencies, and deciding who will be the master, and deciding upon the initial DS sequence number.
- **Exchanging:** The state in which a node describes its topology database by sending Database summary packets.
- **Loading:** The state in which Database summary packets have been exchanged and the required PTSEs are requested, but one or more requested PTSE is not yet received.

- **Full:** The state in which a node has received PTSEs available from the neighbour. The links can now be advertised in PTSEs.

The events that trigger state changes include, but is not limited to, AddPort (received when a link to peer reaches the two-way Inside state), DropPort (received when a link to peer leaves two-way Inside state), NegotiationDone, ExchangingDone, SynchDone and LoadingDone. One possible set of state changes for this FSM is depicted in Figure 12.7(c). As shown, the first database summary packet is exchanged followed by PTSE request and PTSPs.

FIGURE 12.7

PNNI Message Exchange and State Transitions

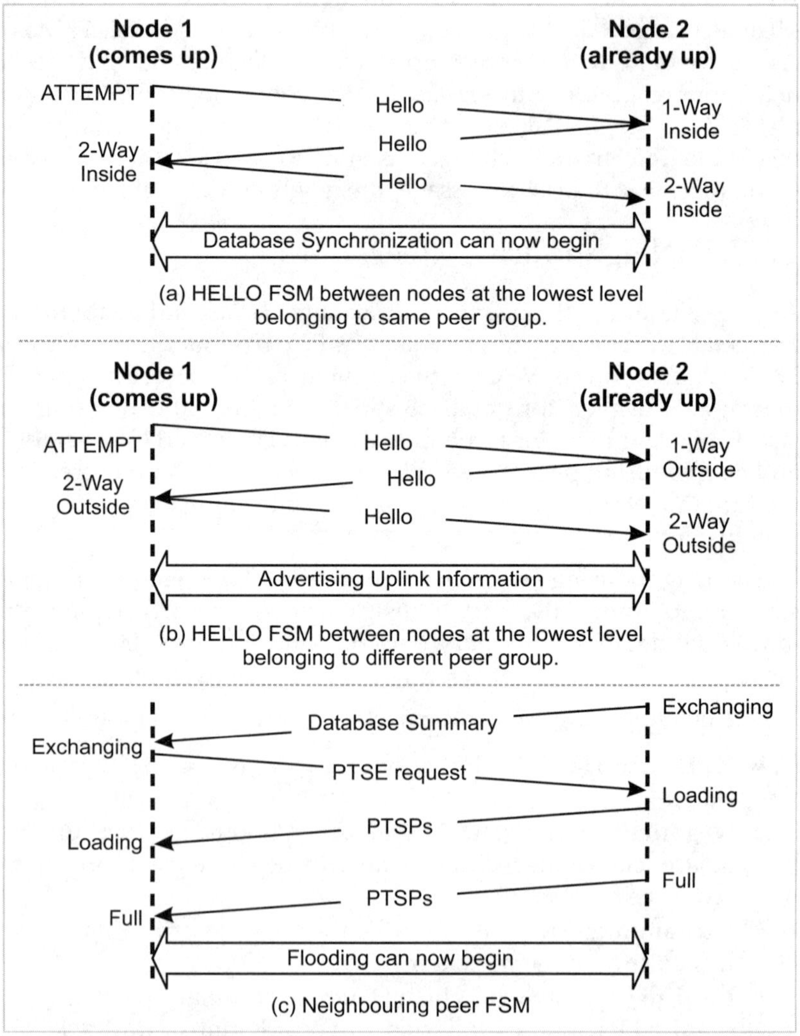

12.7.4 *Hello* FSM for LGNs

Like its two lowest-level node neighbours, two peer LGNs also exchange *Hellos* over the SVCC-based Routing Control Channel (RCC). The key difference between the two mechanisms is that LGN neighbours have only a single RCC between them. The RCC is established using the PNNI signalling protocol. After the SVCC is declared as 'up' by the signalling protocol, a *Hello* protocol instance is initiated. Note that the *Hello* protocol at the LGN level is only applicable within a peer group. Hence, states like One/Two-way Outside is not applicable to *Hello* protocol at this level.

The *Hello* FSM between LGNs is used for two purpose. First, it monitors the status of RCC between the two LGNs. Second, it is used to exchange information about the horizontal links that they will mutually advertise.

12.8 Summarization and Aggregation

It is generally observed that nearly 40 per cent of the network traffic is control information (like routing updates). This means that a significant portion of the network bandwidth is used just for managing the network. The PNNI specification aims to reduce the exchange of control information by summarizing information. Information summarization falls in following two categories:

- Address summarization and reachability
- Topology aggregation

12.8.1 Address Summarization and Reachability

Address summarization is "The process of using a single reachable address prefix to represent a collection of end systems and/or node addresses that begin with the same prefix." For example, if X.Y.1 and X.Y.2 are two address prefixes reachable via a given node, the summarized address is X.Y. It follows from this example that in order to achieve address summarization, the higher order bits of the summarized addresses must be the same. For ensuring this, there must exist an order in which the addresses are distributed (i.e. the distribution must not be random). The peer group and PGID concepts of PNNI hierarchy ensure that all nodes within a peer group have same higher-order bits (so that address summarization is possible).

In order to illustrate the concepts of address summarization and reachability, an example as provided by the [ATMF PNNI 1.1] specification is provided. The PNNI hierarchy for this example is depicted in Figure 12.8. In this example, A.2.1, A.2.2 and A.2.3 are three end-systems having a number of addresses associated with them. The first step towards summarization starts with each node flooding its reachable address prefixes. The advertised address prefixes are shown in Table 12.4.

FIGURE 12.8

PNNI Hierarchy for the Address Summarization Example

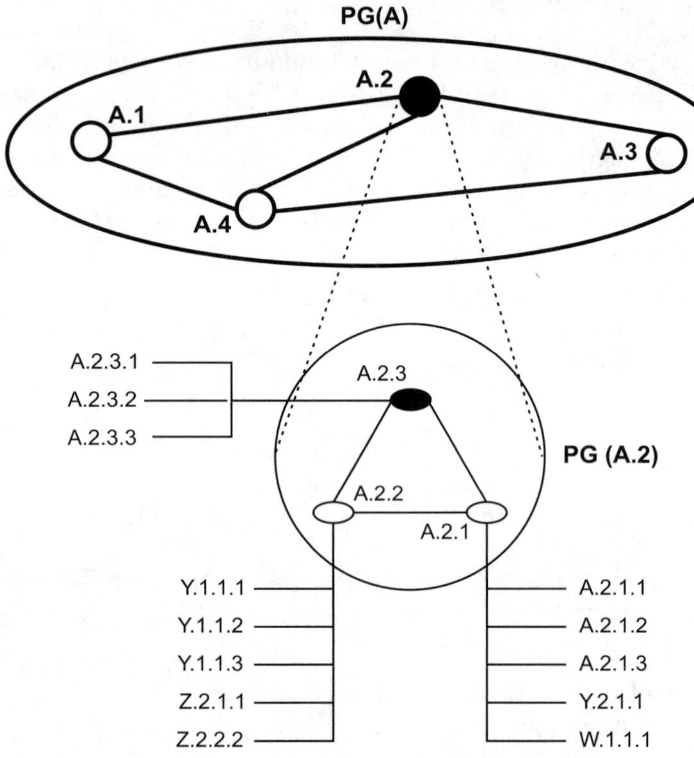

Source: ([ATMF PNNI 1.1], © 1996 the ATM Forum)

TABLE 12.4

Advertised Address Prefixes (Copyrights 1996 the ATM Forum)

Node	*Advertised Reachable Address Prefixes*
A.2.1	P<A.2.1>
	P<Y.2.1.1>
	P<W.1.1.1>
A 2.2	P<Y.1.1>
	P<Z.2.>
A.2.3	P<A.2.3>

Before flooding, one level of summarization has already taken place at each node. This fact is clear because for all the three nodes of the peer group, the number of address prefix advertised is less than the number of associated addresses. For example, though node A.2.3 has three associated addresses (A.2.3.1, A.2.3.2 and A.2.3.3), it advertises only one address prefix P<A.2.3>. Similar is the case with nodes A.2.1 and A.2.2.

Now, when PGL (A.2.3) receives the advertised reachable address prefixes from all the peer nodes, it produces the next level of summarized information. This summarized information is obtained by matching all the received address prefixes with a set of summary addresses using the longest prefix matching. Summary addresses are configured or default addresses maintained by the PGL for all nodes of the peer group. After the summarization is over, the information thereby obtained is flooded by the PGL to the higher-level peer group (summarized information flow up the hierarchy). This summarized reachability information is presented in Table 12.5.

TABLE 12.5

Reachability Information (Copyrights 1996 the ATM Forum)

Reachability Information Advertised by PGL A.2.3 in PG (A)
P<A.2>
P<Y>
P<Z.2>
P<W.1.1.1>

From Table 12.5, it follows that reachability information flooded by PGL A.2.3 in the PG (A.2) is different from what it floods in the PG (A). This happens because information flow up the hierarchy is always summarized. Thus, PGL A.2.3 presents a unified view of the whole peer group A.2 to the PG (A). The case within PG (A.2) is different, as PGL A.2.3 only presents the localized address summary to the rest of the members (not the complete summary of the whole peer group).

12.8.2 Topology Aggregation

Topology aggregation is the process of reducing topology information by summarizing nodal as well as link information. The motivation for aggregating topology is manifold. First, topology aggregation helps in hiding the internals of a peer group to other peer groups. This is necessary to maintain security of the network. Second, like address summarization, topology aggregation helps in reducing the flow of control information in the network. Moreover, aggregation makes maintenance of topology information easier.

Topology aggregation is classified into the following categories:

- **Nodal aggregation:** This is something that is already explained without making an explicit mention of aggregation. Nodal aggregation refers to the process of representing a child peer group by an LGN in its parent peer group.
- **Link aggregation:** To understand link aggregation, consider the logical link connecting LGNs A.2 and A.4 in Figure 12.4. That logical link is actually the aggregation of two links (A.2.1 - A.4.1) and (A.2.2 - A.4.2). Thus, link aggregation is defined as the process of representing a set of links between two peer groups as a single logical link.

Topology aggregation provides the means for a node to have a restricted, but adequate, view of the network. This view of a network from a node is called the single node perspective.

Figure 12.9 depicts the view of the network from nodes B.2.1, B.2.2, B.2.3, B.2.4, and B.2.5. Recall that each node in a peer group obtains an identical view of the network. Here also, nodes B.2.1, B.2.2, B.2.3, B.2.4, and B.2.5 attain the same view of the network.

FIGURE 12.9

Single Node Perspective of Nodes B.2.1, B.2.2, B.2.3, B.2.4, and B.2.5

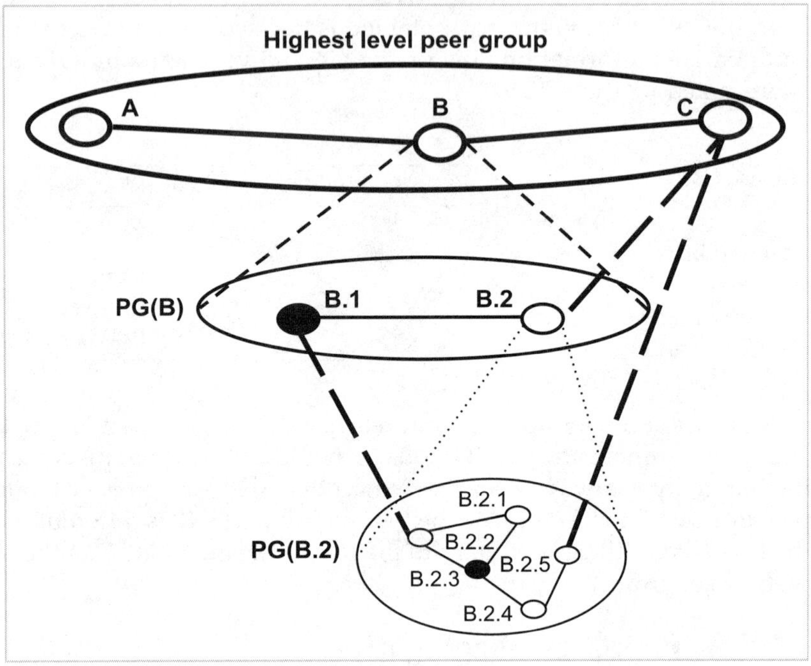

In order to understand the single node perspective, consider node B.2.1. At the time of initialization, this node comes to know about its peer group through flooding. After flooding is complete, B.2.1 has complete topological information of PG(B.2). In other words, B.2.1 knows exactly how to reach every node in its peer group and this is what is known as the complete view.

For other same-level and higher-level peer groups, B.2.1 has only a partial view. The first set of information received about other peer groups is that of PG(B.1). Connectivity to LGN B.1 is known through the uplink B.2.2-B.1, which was created and flooded by node B.2.2. This uplink provides all nodes in PG(B.2), including the PGL B.2.3, the information to reach LGN B.1. This uplink is used by LGN B.2 to establish a logical link with LGN B.1 and exchange summarized information. After the exchange, LGN B.2 floods the received information in peer group PG(B.2). After this is complete, all nodes in peer group PG(B.2) know that to reach peer group PG(B.1), node B.2.2 is to be used. Since every node in the peer group knows how to reach B.2.2, in effect, every node can also reach PG(B.1). The nodes in the peer group PG(B.2) need not know how to reach a particular node in peer group PG(B.1), because once the signalling message reaches PG(B.1), the nodes in that peer group have adequate information to forward the message to appropriate destination. Similarly, connectivity to LGN C is known through uplink B.2.5-C created by B.2.5.

Information about connectivity to other LGNs is obtained through information flows down the hierarchy. For example, LGN A can be reached through B.1 via uplink B.1-A (not shown in Figure 12.9). Hence, after flooding in peer group PG(B), B.2 knows that there is a uplink to A via LGN B.1. This information trickles down the hierarchy and node B.2.1 knows that to reach LGN A, it needs to route requests through B.1. This is how a node attains a partial view of the entire PNNI network.

12.9 Path Selection and Connection Admission Control

When a connection request arrives at the ingress of an ATM network, there exist two possibilities for routing the request, viz. hop-by-hop routing and source routing. In hop-by-hop routing, each node in the path to the destination inspects the 'called party address' and makes a forwarding decision using this address. In source routing, the ingress node specifies the complete path to the destination. Thus, the remaining nodes in the path make no routing decision, but only forward the request to the next hop. The merits and demerits of hop-by-hop routing and source routing have been explained earlier.

After appraising the advantages and disadvantages of both routing techniques, the PNNI working group of ATM forum preferred to adopt the source routing technique.

Here, two points related to source routing need to be clarified. The first point relates to the specification of the source route. Due to the hierarchical nature of the PNNI network, the ingress node does not have complete path information of other peer groups. Therefore, to specify a path (i.e. source route), the ingress node only details the hierarchical information known to it. This path is known as the hierarchically complete source route. This hierarchically complete source route is based on a node's view of the whole network. In the PNNI signalling discussed next, an example is presented to explain how the path information is specified by the ingress node.

The second point relates to the Connection Admission Control (CAC). Since the source route is determined by the ingress switch, it is mandatory that the ingress switch have the resource availability information of all other switches of the network. However, this makes scalability difficult and is not a feasible option. Now, as the CAC module is responsible for granting connection requests (based on resource availability information), to circumvent the above problem, CAC within an ATM network is divided into the following two categories:

- **Generic Connection Admission Control (GCAC):** The ingress node, while selecting a path for a new connection, performs GCAC. The GCAC algorithm is performed using the ingress node's topology state parameters maintained for the entire network, along with the service category, traffic characteristics and QoS requirements of the connection.

The goal of GCAC algorithm is to use the information mentioned above to predict the behaviour of switches along the transit.

- **Actual Connection Admission Control (ACAC):** The ingress node (through GCAC) can only predict the resource availability at other intermediate nodes. It is quite possible that while computing the source route, the ingress node does not possess latest information about other nodes. Therefore, in addition to the GCAC performed at the ingress node, each chosen node must also ensure that it has sufficient resources to support the new connection. This procedure, preformed at intermediate nodes to check resource availability, is referred to as ACAC.

CAC in ATM networks is not subject to standardization. Thus, each node is allowed to implement an independent CAC algorithm. However, each node must advertise the resource availability, using a set of topology state parameters (through PTSP flooding). These parameters are then used by the GCAC to make a rough guess about the resource availability for the entire path. Note that GCAC is just a prediction. It is through ACAC that a node confirms resource availability.

12.10 PNNI Signalling Protocol

PNNI signalling specification provides the means to exchange signalling messages between two ATM networks or between two ATM switches. The PNNI signalling specification is based on the ATM Forum's UNI specification (i.e. UNI 3.1 and UNI 4.0). In particular, PNNI supports all UNI 3.1 capabilities (e.g. signalling for point-to-multipoint connections) and some UNI 4.0 capabilities (e.g. supports Anycast, but does not support leaf-initiated join).

12.10.1 Signalling Messages and Information Elements

PNNI signalling specification does not define any new message. Instead, it uses the existing point-to-point messages (e.g. SETUP, CALL PROCEEDING, CONNECT, etc.) and point-to-multipoint message (e.g. ADD PARTY, ADD PARTY REJECT, etc.), as defined by the UNI specification.

All information elements defined by the UNI specification are applicable to PNNI as well. In addition, PNNI defines few more information elements which are necessary for carrying route information. Among the information elements defined, the two important ones are Designated Transit List (DTL) and crankback. While DTL IEs are used for source routing, Crankback IE provides the means for alternate routing. The next two sub-sections elaborate upon DTLs and Crankback.

12.10.2 Designated Transit List (DTL)

While discussing path selection techniques, it was stated that PNNI uses source routing for forwarding connection requests. Now, for source routing, the path information must be carried somewhere in the SETUP and ADD PARTY message. The Designated Transit List

(DTL) is the information element used to carry this path information. In essence, DTL is a list of logical nodes and logical links that a connection traverses while passing through a peer group. Each connection request message (i.e. SETUP or ADD PARTY) contains up to 10 instances of DTL information elements. These sequence of DTLs can be viewed as a last-in-first-out stack, representing the hierarchically complete source route.

To understand how DTLs are inserted in connection request messages, consider the following example in which end-system B.1.2.x wishes to establish a connection with end-system C.2.y (see Figure 12.10).

FIGURE 12.10

PNNI Hierarchy for Call routing Example

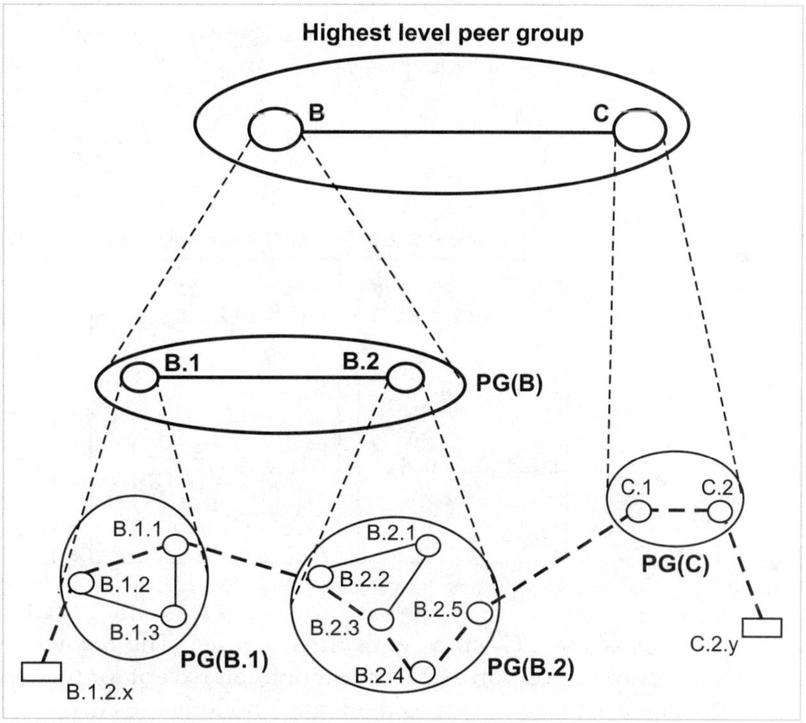

12.10.3 Call Routing Example

Step 1. End-system B.1.2.x sends a SETUP message with all the mandatory information elements to the nearest node B.1.2.

Step 2. Node B.1.2 builds the hierarchically-complete source route for the connection request, using its topology information. The hierarchically complete source route is represented by a DTL stack, as shown in Figure 12.11. For each level of hierarchy visible at B.1.2, a DTL entry exists in the DTL stack. Since there are three levels visible at B.1.2 (i.e., PG(B.1), PG(B), and the highest level PG), B.1.2

FIGURE 12.11

DTL Flow for Call-Routing Example

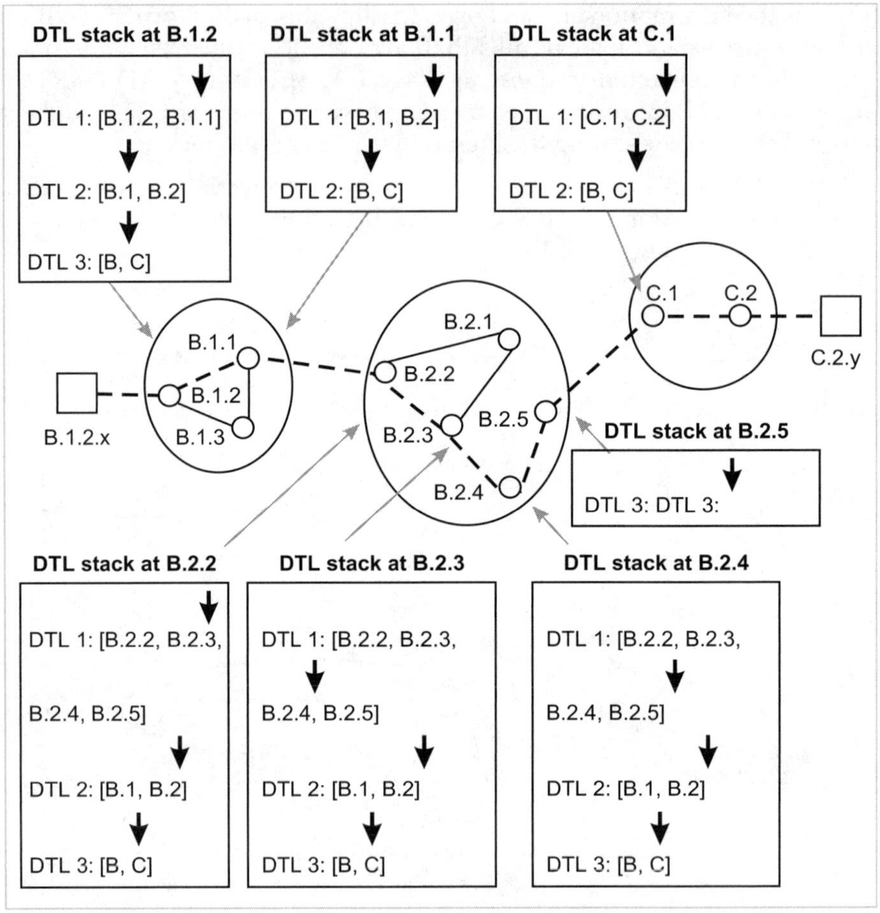

makes a DTL stack with three entries. The arrows in the figure indicate the current position of the transit pointer. Except for the top-level, the transit pointer at each level specifies the logical node being currently visited. At the top-level, the transit pointer indicates the node to be visited next. If one looks at DTL stack of B.1.2, then for DTL lists 2 and 3, the transit pointer points to logical nodes B.1 and B, respectively. This is because node B.1.2 is the child node of peer group B.1 and the grandchild of peer group B. In other words, since the request is at node B.1.2, the request is also at logical nodes B.1 and B. (Since B.1.2 originates DTL, it is called the DTL originator. In general, an ingress switch is the DTL originator and egress switch the DTL terminator).

Step 3. When node B.1.1 receives the connection request, it inspects the DTL stack and finds that the top-level DTL list is finished (i.e. the DTL pointer of list 1 has reached its end). It then proceeds to remove the top-level DTL entry. Node B.1.1. then looks at the DTL list 2 and finds that the next destination is B.2. Thus, it looks for paths to reach B.2. Since its neighbour B.2.2 belongs to B.2, node B.1.1 forwards the request to B.2.2. Note that no new entry is added because B.1.1. does not have internal information of the neighbouring peer group B.2. B.1.1 merely knows that by sending the request to B.2.2, the request reaches B.2.

Step 4. B.2.2 on receiving the connection request, checks DTL list 2 and finds that the connection request is to be routed to C. It then looks for the appropriate path in its topology information and finds a path via B.2.3-B.2.4-B.2.5. After finding the path and pushing a new DTL at the top of the stack, B.2.2 forwards the connection request to B.2.3.

Step 5. Node B.2.3 does not perform any search because the route for it has already been determined by B.2.2. This node merely right-shift the transit pointer of the DTL list 1 and forward the connection request to the next node.

Step 6. B.2.4 does not perform any search because the route for it has already been determined by B.2.2. This node merely right-shift the transit pointer of the DTL list 1 and forward the connection request to the next node.

Step 7. B.2.5, on receiving the connection request, finds that elements of DTL list 1 have exhausted, and so, it removes the first list. The node B.2.5 also realizes that DTL list 2 has reached its end and thus removes it also. Looking at DTL list 3, B.2.5 finds that the request has to be forwarded to LGN C. Through its topology database, it finds a link (B.2.5-C.1) to reach C.1. Right-shifting the pointer of DTL list 3, B.2.5 forwards the connection request to C.1.

Step 8. C.1, after receiving the connection request, inspects the called party address of the SETUP message. Note that the called party address is not looked at nodes other than the ingress/egress node. At C.1, this is necessary because the DTL stack, by itself, is insufficient to provide the subsequent route information and the only available information is the called party address. Using the summarized address information flooded by C.2, C.1 knows that the address is reachable via C.2. Hence, C.2 inserts another DTL list at the top of the stack and forwards the message to C.2.

Step 9. C.2 removes the DTL stack and forwards the request to end-system C.2.y. (Note that C.2 is called DTL terminator because it removes the DTL stack from the SETUP message).

This completes the call routing example.

12.10.4 Crankback

One of the drawbacks of source routing is that connection requests can get blocked anywhere in the computed path. Connections get blocked when GCAC at the ingress node suggests a particular path, but the ACAC at one of the intermediate node fails. Recall that GCAC is based on the information last obtained through PTSP flooding, and this information may not be precise enough. Since the ACAC algorithm is generally more accurate than the GCAC algorithm, it is used at intermediate switches to ensure that a switch has adequate resources to support the new connection. When the ACAC check fails, the connection gets blocked. Connections can also get blocked when there are transient link failures or when there is congestion. Whatever be the reason, the essence of the matter is that source routing becomes a potential bottleneck in call routing. In cases when this happens, to avoid clearing the call all the way back to the ingress node, crankback procedures are adopted.

Crankback is a mechanism for partial clearing of a connection. The idea of crankback is simple. When a node cannot process a connection request according to the connection's DTL list, the node cranks back the connection to the originator of that DTL, with appropriate error indication. The originator of the DTL may then look for alternate routes or may crankback the connection further. In this way, an alternate route is searched as close to the problem region as possible. Although crankback increases the connection-setup delay, it definitely improves the chances of a connection getting through.

Note that the crankback procedures are not invoked every time a connection request fails. For example, if the called end-system rejects the call, crankback procedures are not applied. The scenarios where crankback procedures are invoked, fall into two broad categories, viz. reachability errors and resource errors. Reachability errors occur when the path specified in the DTL list is unreachable. Resource errors occur when the selected path contains a bottleneck link/node, which cannot support the requested connection.

Figure 12.12 illustrates the use of crankback for solving the problem associated with source routing. After N3 determines that its link with N4 has failed, it crankbacks the connection request to N2. N2 then looks for alternate paths and finds a direct link with N4. This path is then used to route the connection request.

FIGURE 12.12

Scenario Depicting Crankback

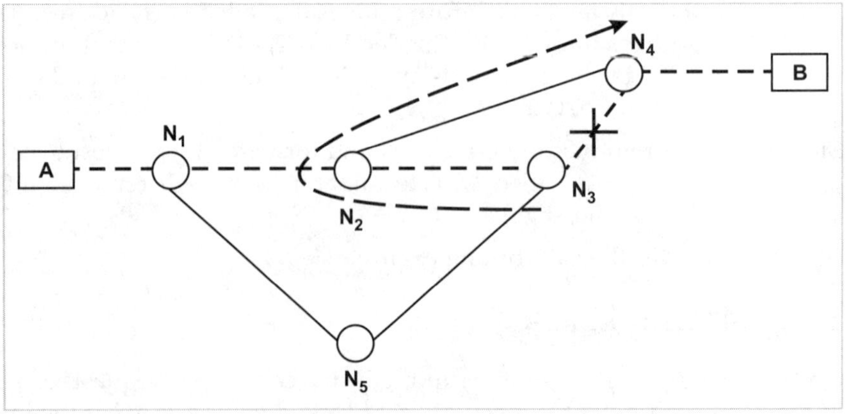

Conclusion

The PNNI routing protocol is one of the most complex routing protocols. Even if PNNI is compared to OSPF, which is one of the most comprehensive IP routing protocols, PNNI is much more elaborate and covers much more functionality. The complexity stems from two reasons: The first reason is that PNNI, besides distributing reachability information, also distributes the resource availability information for GCAC. Second, the routing hierarchy in PNNI is very deep and can extend up to many levels. Thus, conceptualizing and managing such complex hierarchies may be extremely difficult. This has implied that PNNI has not been as popular as the OSPF protocol. Nonetheless, PNNI is a routing protocol designed after a lot of effort.

Review Questions

Q 1. *What is IISP protocol? How is it different from PNNI?*
Q 2. *Describe the key attributes of PNNI protocol.*
Q 3. *Describe the hierarchical structure of PNNI. What is the rationale for having such a complicated hierarchy?*
Q 4. *What is the implication of the hierarchical structure on source routing?*
Q 5. *What are the different categories of information maintained by a PNNI node? What purpose does each category have?*
Q 6. *What is the difference between metrics and attributes? Describe the different metrics and attributes for PNNI.*
Q 7. *What are the various steps used by a PNNI node to obtain link-state information?*
Q 8. *Write shorts notes on the following:*
- *Path selection and Connection Admission Control*
- *Crankback*

Q 9. *Does PNNI support signalling? If yes, describe the signalling capabilities of PNNI protocol.*

Further Reading

This chapter is based on [ATMF PNNI 1.1] which provides a detailed specification of the PNNI protocol. As this protocol may be detailed and difficult to understand, readers may start with IP routing protocols. For this, the reader is referred to [RFC 2328], [RFC 2453] and [RFC 1771] which describe the OSPF, RIP and BGP protocols, respectively. [TCP/IP D. Comer] has dedicated chapters for the RIP and OSPF routing protocols.

Chapter 13

AAL2 Signalling

13.1 Introduction

The communication world is in a state of flux with rapid technological advancements, both in wireline as well as in wireless domain. In wireline domain, TCP/IP has become the de-facto networking protocol suite. ATM technology, which came into existence around a decade back, is still to attain the popularity in different networking solutions. However, certain applications, like Voice over ATM (refer to Chapter 20), Voice over DSL (refer to Chapter 21) and 3G Wireless (refer to Chapter 22), have brought ATM back into prominence.

In order to support these applications, there have been a lot of standardization efforts by ITU-T and ATM Forum. The AAL2 signalling standard is one of the steps in this direction. This standard provides the ability to dynamically establish and release AAL2 connections over existing ATM VCs. As there are VPI and VCI in ATM cell header to identify an ATM VC, similarly, an AAL2 channel is identified by 8-bit Channel Identifier field that is carried in the CPS-PDU header (refer to Chapter 6).

The standardization for AAL2 signalling is *graded* as three different levels of capabilities are defined, namely capability set 1, capability set 2 and capability set 3. Accordingly, there are three different ITU-T standards for these different capabilities, [ITU-T Q.2630.1], [ITU-T Q.2630.2] and [ITU-T Q.2630.3] respectively. The difference in capabilities relate to various aspects of AAL2 signalling protocol including the messages supported (for example, the Modify Req/Ack messages are not supported by capability set 1), information elements in a message (Path Type IE is not contained in any capability set 1 message) and the associated procedures. This chapter provides a simplified view of AAL2 signalling and thus uses [ITU-T Q.2630.1] as the reference standard. The readers may refer [ITU-T Q.2630.2] and [ITU-T Q.2630.3] for details on capability set 2 and capability set 3.

Irrespective of the capability set, the AAL2 signalling standard is quite often viewed as an extension of ATM signalling protocol, which is actually not the case. This chapter explains the difference between the two. Apart from this, this chapter provides an outline of AAL2 signalling. This chapter also addresses questions like why AAL2 signalling is required, what its architecture is, what its features are, and so on. Then, the differences between AAL2 signalling and ATM signalling is explained in Section 13.7. This section is quite comprehensive and explains the need for having a separate signalling protocol. This chapter is concluded by mention of the open issues of AAL2 signalling.

13.2 Relation between ATM Signalling and AAL2 Signalling

Generally, signalling is used between the user and the network, or between two network elements to establish, monitor, and release connections by exchanging control information. In Chapter 11, the ATM signalling procedures were elaborated. These procedures, based on [ITU-T Q.2931] or ATM Forum's UNI Signalling Specification [ATMF UNI 4.0], provide means to establish/release an ATM Switched Virtual Circuit (SVC). Formally, ATM signalling provides means to establish/release ATM Virtual Channel Connection (VCC) or Virtual Path Connection (VPC).

In the context of AAL2, the AAL2 signalling protocol encompasses the control plane functions to establish, clear and maintain AAL2 connections. An AAL2 connection can be viewed as the logical concatenation of one or more AAL2 links between two AAL2 service end-points. An AAL2 link is a communication path between two AAL2 nodes that is uniquely identified by a Channel Identifier (CID).

Contrary to popular perception, AAL2 signalling is totally independent of ATM signalling. It may be noted that out of all the AALs, only AAL2 has its separate signalling protocol. For other AALs (including AAL5), the paths are established either through configuration (i.e. Permanent Virtual Circuit or PVC) or through ATM signalling procedures (i.e. Switched Virtual Circuit or SVC). For AAL2, however, it was felt that extensions to ATM signalling would not be sufficient. This was primarily because of the following reasons.

- First, if ATM signalling protocol were to be extended, then AAL2 network would get coupled to the way the ATM connection is established (i.e. the network has to be SVC and not PVC). This was difficult to implement in scenarios where the ATM connections were static but the AAL2 connections had to be dynamic (e.g. Radio Access Network (RAN) in 3G networks).
- Further, since ATM network has many signalling protocols like ATM Forum's UNI 4.0 and PNNI and ITU-T's Q.2931/Q.2971, incorporating AAL2 into these specifications requires extensions in all these protocols. This is practically very difficult and time-consuming.
- Third, there are obvious consequences of the decoupling. The decoupling could allow existence of multiple AAL2 overlay networks operating over a single ATM network,

each having its own addressing and routing plan. The independence becomes more crucial when the operator of the ATM and AAL2 network is not the same organization, since it allows both parties to be in charge of the addressing within their own network.

- Finally, AAL2 defines multiplexing of many AAL2 channels within an ATM virtual connection. To establish and release these channels, a dynamic protocol is necessary. The requirement is fulfilled by AAL2 signalling. Similar requirement does not arise in other AALs because AAL1 and AAL5 do not have multiplexing support. The AAL3/4 has a multiplexing identifier but that can be ignored, as AAL3/4 is not very popular.

In summary, AAL2 signalling protocol is a protocol completely independent of ATM signalling. The former assumes the existence of ATM virtual channels on which AAL2 channels are then established. How these ATM virtual channels are established is not within the purview of AAL2 signalling.

13.3 AAL2 Signalling Features

The following points describe the features of AAL2 signalling protocol:

- **Scope and extent:** The scope of AAL2 signalling is restricted to control of AAL2 connections. It is assumed that the appropriate ATM resources must be present before an AAL2 connection can be established. Moreover, an AAL2 signalling endpoint is capable of controlling AAL2 links on more than one ATM VCC. These VCCs may be contained on different ATM VPCs, which, in turn, may be carried out on different ATM physical interfaces.
- **Addressing:** The AAL2 signalling supports addressing of AAL2 service end-points, connected via a network of AAL2 switches. The supported address formats are E.164 and NSAP.
- **Routing:** The AAL2 signalling supports hop-by-hop routing. The routing decision is based on addressing information, link characteristics reflecting the resources required, e.g. bandwidth and other information. The standards however do not define procedures whereby the AAL2 switches can exchange routing information. Thus, routing algorithms are implementation specific.
- **Symmetry of connection control:** The AAL type 2 signalling protocol and procedures reflect a peer-to-peer (rather than a master/slave) relationship among the AAL type 2 signalling endpoints. For example, both ends may request AAL type 2 connections with the same protocol procedures.
- **Asymmetric connection:** The AAL2 signalling support asymmetric AAL2 connections (i.e. the end-to-end connections are bi-directional where the information transfer capability in different direction might be different). Unidirectional AAL2 connections can, therefore, be also provided, i.e. connections that provide zero bandwidth in one direction.
- **Independence from underlying signalling transport:** The AAL2 signalling protocol is designed to be independent from the underlying signalling transport. This point is further elaborated in the next section. To achieve this independence, it is assumed

that the underlying signalling transport networks supports assured (error-free) data transfer, in-sequence delivery of PDUs and an indication of flow control.

Besides the features mentioned above, AAL2 signalling provides mechanisms for detecting and reporting signalling procedural errors or other failures detected by the AAL2 signalling endpoint to AAL2 management. There are also provisions for forward and backward compatibility. Moreover, there is a facility to reset all paths, a path and its channels or a particular channel in a path in an interface and to block and unblock AAL2 paths between adjacent nodes.

13.4 AAL2 Signalling Architecture

AAL2 signalling defines an overlay network over existing ATM network. The overlay network of AAL2 assumes an addressing, signalling and routing mechanism of its own. However, standards exist only for signalling. There are no standards for routing and addressing.

The high-level architecture for AAL2 signalling is shown in Figure 13.1. As shown in the figure, the AAL2 signalling entity either resides at AAL2 end-points or at intermediate AAL2 switches. At the end-point, the AAL2 signalling entity provides services (like connection establishment and release) to the served user. It is end-point in the sense that the AAL2 channels terminate here. In the intermediate nodes (i.e. AAL2 switches), the AAL2 stack provides bridging and routing support. In simple terms, the AAL2 switch switches AAL2 connections by swapping one CID to another.

Figure 13.1

High-level Architecture for AAL2 Signalling

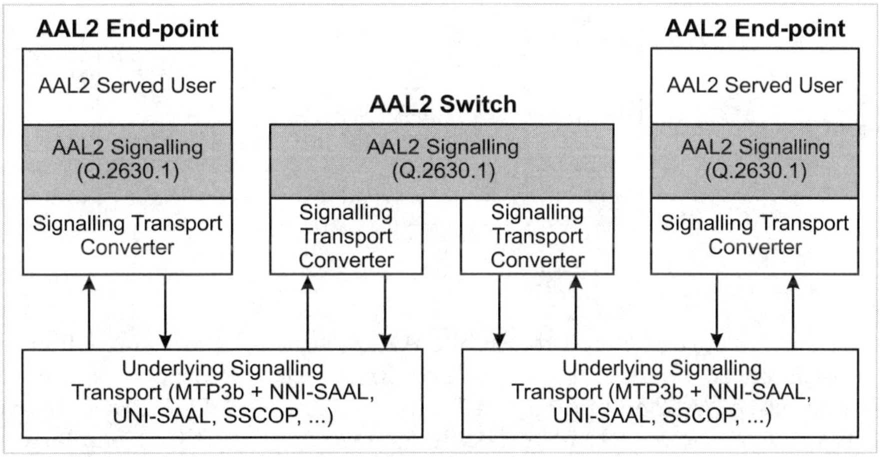

The AAL2 signalling messages can be carried over a variety of underlying transport options. Given this, one option is to adapt AAL2 signalling for every possible signalling transport. Alternatively, the services used by AAL2 signalling can be kept fixed; further, for each signalling transport, a converter function can be used that maps the services of the underlying transport to the generic services expected by AAL2 signalling. The generic expected by AAL2 signalling entity include indication for IN-SERVICE/OUT-OF-SERVICE, request for TRANSFER of signalling messages, indication of receipt of signalling messages and CONGESTION indication. It is the responsibility of Signalling Transport Converter (STC) to provide these generic services from the underlying transport services. For example, if AAL2 signalling stack resides over UNI-SAAL, the signalling transport converter so required is defined in [ITU-T Q.2150.2]. Similarly, for MTP3b/NNI-SAAL the signalling transport converter is defined in [ITU-T Q.2150.1]. For SCTP, the signalling transport converter is defined in [ITU-T Q.2150.3].

The protocol architecture for AAL2 signalling is shown in Figure 13.2. As shown in the figure, the signalling messages are reliably carried over SSCOP over AAL5. Note that even for AAL2 signalling, the underlying transport is based on AAL5 (and not AAL2). This can be attributed to the fact that AAL2 is for small low data-rate applications and not fit for carrying large signalling messages, something that is more suited to AAL5. Here, the ATM network based on AAL5 is merely an underlay network (i.e. a transport medium) for AAL2 signalling.

FIGURE 13.2

Protocol Architecture for AAL2 Signalling

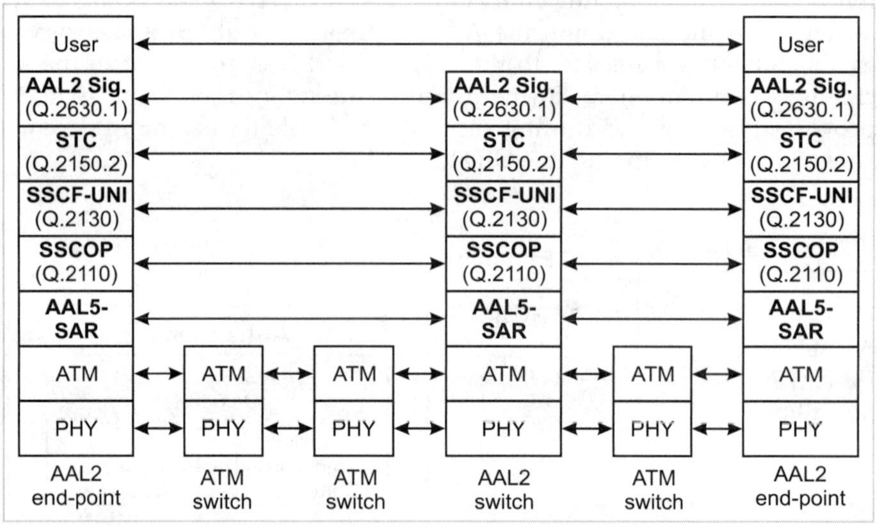

It may be further noted that while AAL5 is used to carry signalling messages, the user plane is based on AAL2. At this level, the AAL2 connection is used to carry CPS-PDU between two AAL2 end-points (see Figure 13.3). As the VPI/VCI are switched at intermediate ATM switches in an ATM Virtual Channel Connection (VCC), similarly the CID values are switched at intermediate AAL2 switches in an AAL2 connection. Considering the example in Figure 13.3, the intermediate ATM switch B swaps <VPI1, VCI1> to <VPI2, VCI2>. At C, the VCC1 terminates. Thus, C is the end-point for VCC1. However, for the AAL2 connection, it is an AAL2 switch and switches CID1 to CID2. The ATM switch D, like B, switches only the VPI/VCI. The AAL2 connection (CID1, CID2) terminates at E. Another AAL2 connection may

FIGURE 13.3

VPI/VCI and CID Translation

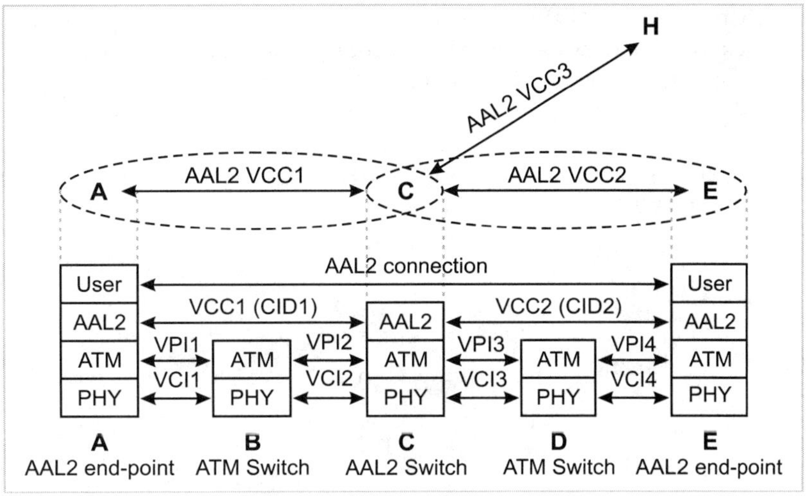

exist between A and H that uses the same ATM VCC (VCC1). Here, another CID (say CID3) may be used over VCC1 and CID4 over VCC3.

13.5 AAL2 Signalling Functions

AAL2 signalling is used to provide various functions. The important functions of AAL2 signalling are as following:

- **Connection management:** The primary function of AAL2 signalling entity is to provide means to establish and release AAL2 connections as identified by CID.
- **Reset:** In order to handle abnormal scenarios, AAL2 signalling provides a reset facility. Three types of reset are provided: reset all AAL2 paths associated with an end-point, reset a single AAL2 path and its channels, or reset a particular channel in an AAL2 path in an interface.
- **Block/unblock:** AAL2 signalling also provides the means to block and unblock AAL2 paths between adjacent nodes. A blocked path is not used to carry new connections other than test connections.
- **Error handling:** AAL2 signalling also provides mechanisms for detecting and reporting signalling procedural errors or other failures detected by the AAL2 signalling end-point to AAL2 management.

13.6 AAL2 Signalling Messages

Most message exchanges in AAL2 signalling involve a two-way handshake. The entity initiating the procedure sends a request message. The receiving entity responds with a confirm message. Table 13.1 shows important AAL2 signalling messages. The Establish Request (ERQ) message is used to initiate connection establishment. The peer entity responds by Establish Confirm (ECF). Thus, AAL2 signalling follows a simple two-way handshaking for call establishment and release (see Figure 13.4). The basic parameters of ERQ message are the connection element identifier that includes path identifier and channel identifier, service-specific information to define the type of SSCS residing over AAL2 and destination address in NSAP or E.164 format among others.

FIGURE 13.4

Call Establishment and Release Procedure in AAL2 Signalling

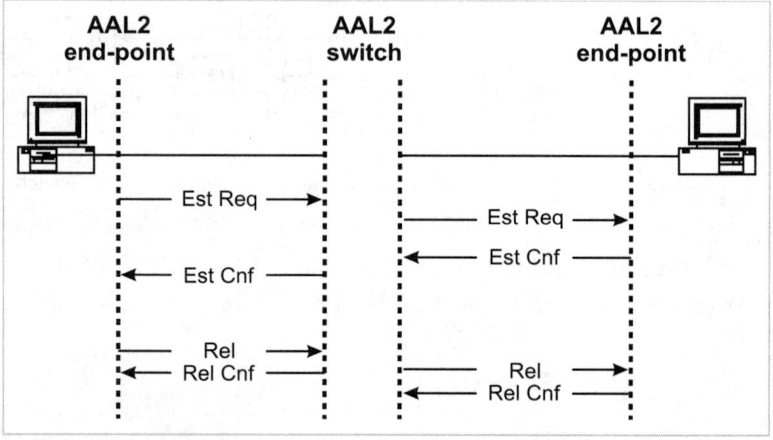

In order to release an established connection, the Release Request/Confirm messages are used. The release message has *cause* as the key element. Apart from connection establishment and release messages, other messages shown in Table 13.1 are used for layer management functions.

13.7 AAL2 Signalling versus ATM Signalling

The comparison between AAL2 signalling and ATM signalling offers interesting insight into the design of any signalling protocol. ATM Signalling protocol, documented in [ITU-T Q.2931], is derived from the ISDN signalling protocol [ITU-T Q.931]. Thus, the former has a similar look and feel as that of latter. On the other hand, AAL2 signalling protocol, as defined in [ITU-T Q.2630.1], is different. The following sub-section looks at some of the similarities and differences between the two signalling protocols.

13.7.1 Motivation

ATM virtual connections are established either statically (PVC) or dynamically (SVC). The ATM Signalling protocol provides for the establishment and release of a SVC. The signalling

TABLE 13.1

AAL2 Signalling Messages

Signalling Message	
Establish Request/Confirm	Used to establish an AAL2 connection between AAL2 end-points.
Release Request/Confirm	Used to release an AAL2 connection between AAL2 end-points.
Management Messages	
Block Request/Confirm	Used to block AAL2 paths from carrying new connections other than test connections.
Unblock Request/Confirm	Used to unblock a blocked AAL2 path so that it is available again for carrying new connections.
Reset Request/Confirm	Used to release all the affected channels and associated resources specified in the reset request.
Protocol Error Handling Messages	
Confusion	Used to indicate a protocol error to the sender of a message.

TABLE 13.2

A Snapshot of ATM Signalling Versus AAL2 Signalling

Topic	*ATM Signalling*	*AAL2 Signalling*
Motivation	Provides for the establishment and release of virtual connections.	AAL2 signalling is used to establish, release, and maintain AAL2 connections.
Architecture	Signalling and routing are done by separate protocols.	No separate routing protocol is defined.
Handshaking	A three-way handshake for call-establishment while a two-way handshake for call-release.	A two-way handshake for both call-establishment and call-release.
Call Identifier	Uses a single Identifier (i.e. CRV or Call Reference Value) and a flag (CRV flag) that is inefficient implementation.	More optimized solution by using two identifiers (originating SAID and destination SAID) where SAID is signalling Association Identifier.
Message Formatting	Formatting is done using type-length-value format.	Formatting is done using type-length-value format.
Parameter Negotiation	Limited parameter negotiation possible by inclusion of alter nate signalling parameters and minimum signalling parameters.	No possibility of negotiation because the handshake is two-way only. Modification of parameters is possible in newer capability sets of the protocol.

helps in obtaining VPI/VCI for the connection and defining the parameters (e.g. traffic descriptors) associated with the connection.

What ATM signalling is for an ATM virtual connection, the same is AAL2 signalling for an AAL2 connection. AAL2 signalling stack is used to establish, release, and maintain AAL2 connections. An AAL2 connection exists between two AAL2 end-points.

As mentioned earlier, AAL2 signalling protocol is a protocol completely independent of ATM signalling. The former assumes the existence of ATM virtual channels on which AAL2 channels are then established. How these ATM virtual channels are established is not within the purview of AAL2 signalling. The channels may be established statically or dynamically. The layer management ensures that the ATM virtual channel is available before a request for AAL2 connection is made.

13.7.2 Architecture

The architecture of ATM is somewhat different from AAL. ATM defines its own means for addressing based on AESA or ATM End-System Address (refer to Chapter 10), signalling based on Q.2931/Q.2971 (refer to Chapter 11), and routing based on PNNI or Private Network-Node Interface (refer to Chapter 12). Thus, ATM has well-defined components in its architecture.

AAL2 signalling does not have a routing protocol like PNNI. For addressing, it supports the E.164 and NSAP formats.

13.7.3 Handshaking

Handshaking refers to the basic exchange of signalling messages along with signalling parameters. The ATM signalling protocol follows a three-way handshake for call-establishment and a two-way handshake for call-release. The AAL2 signalling supports a two-way handshaking for connection establishment and release. Newer capability sets allow modification of connection parameters.

13.7.4 Call Identification

In ATM signalling, a 23-bit field called Call Reference Value (CRV) is used to identify a particular instance of a call. This is locally unique and is generated by the originating end. To ensure that the peer also does not generate the same value, a one-bit flag called CRV-flag is used. However, there is a problem associated with this strategy as described below.

In most implementations, the CRV chosen by the initiator of the call is done in such a way so as to simplify the search operations when a message is received back with this CRV. To take an example, information about the calls (active or inactive) within a system can be maintained as an array, and the index in the array can be used to generate the CRV Id. Now, since the CRV is a 23-bit identifier, it may not be feasible to keep an array of eight million entries (to have a one-to-one mapping between the CRV and the array index).

In order to keep the array size manageable, only some low-order bits of CRV (say 16 of them) can be formed using the index of the array. When a message is received back, a part of the CRV is masked out to get back the index in the array. Using this index, all information pertaining to the call can be easily retrieved. However, this flexibility of choosing a CRV is not available at the other end of the interface, since the CRV is not allocated there (i.e., CRV is allocated by remote end). Hence, the search strategy at the corresponding interface to determine the call related information associated with the CRV might not be very efficient. It is quite likely that a balanced tree is used with its associated searching overhead. The scheme used at the interface generating the CRV cannot be applied in this case, unless an array of eight million entries is maintained, or it is ensured that the remote user is not using the most significant bits (in which case a smaller array can be maintained). An array of eight million entries is expected to be sparse, and hence a one-to-one mapping at this interface may not be feasible.

In order to take care of this situation, a possible solution is the introduction of an optional 'Remote CRV' IE. Here, when the first signalling message is received at an interface, a local CRV is allocated (just as it was done at the sender-end of the interface) and sent back as part of the first response within the 'Remote CRV' IE. Note that the CRV that was allocated by the other end of the interface is still sent normally. Thus, each signalling message henceforth would have both the normal CRV IE used currently as well as the optional 'Remote CRV' IE. Now, even for remotely originated calls, the search is made using the CRV carried in 'Remote CRV' IE. Using this CRV, a direct search can be made thereby making search operations at the remote end also simple.

While the above proposal is not standardized in ATM signalling, in AAL2 signalling, the essence of this method has actually been adopted. There are two identifiers or Signalling Association Identifier (SAID), namely Originating SAID (OSAID) and Destination SAID (DSAID). At the time of call initiation, the originating end allocates an OSAID. On receiving the call initiation message, the receiving end binds the OSAID to a newly allocated DSAID. All further communication takes place using only that identifier which the peer had allocated. This solution is more optimized in the sense that no tree needs to be maintained, an array is sufficient.

13.7.5 Message Formatting

The message formatting for both ATM signalling and AAL2 signalling is by done using the standard type-length-value format using a two-level hierarchy. However, there are slight differences. In ATM signalling, the top-level hierarchy contains the CRV, message identifier, message length, and the contents. In AAL2 signalling, the length field is missing. It is assumed that the Assured Data Transfer (ADT) over which AAL2 signalling runs does not lose or gain octets, and that a primitive between AAL2 signalling and ADT provides length information. Apart from this difference, the basic formatting of messages at the top level is the same.

At the second level of hierarchy, the type-length-value is used. There are no major differences. The only difference is in the nomenclature. In ATM signalling, the groups of information are called IEs or information elements, while in AAL2 signalling these are called parameters.

13.7.6 Parameter Negotiation

Parameter negotiation refers to agreeing upon the parameters for communication. In ATM signalling, the key parameters are traffic descriptors, which define the nature of a connection. The ATM signalling protocol provides limited support for parameter negotiation. This is through the use of multiple instances of same IEs or through the use of alternate IEs. For example, the ATM Traffic Descriptor IE can have additionally an Alternate ATM Traffic Descriptor IE also present. The receiving entity can choose between the two in decreasing order of preference. Moreover, the sending entity can also send Minimum ATM Traffic Descriptor IE instead of Alternate ATM Traffic Descriptor IE. Inclusion of such IEs provides scope for parameter negotiation. There is no analogous support in AAL2 signalling. However, modification of parameters is possible in newer capability sets of the protocol.

13.8 Issues and Challenges in AAL2 Signalling

The AAL2 signalling protocol is a relatively new protocol. One of the important problems that remain unsolved in this protocol is the acquisition and distribution of routing information. As mentioned earlier, the solution to this problem is implementation-specific. This feature of has an important bearing on the scope of AAL2 signalling. This is because without the support of a routing protocol, static routing would be typically used. Thus, AAL2 signalling would be limited to small domains. Its primary function would then be to provide dynamic AAL2 connection setup and release on a virtual channel (i.e. basically, a mechanism to exchange channel identifiers).

Another aspect of AAL2 signalling that warrants notice is the use of generic signalling transport layers. While this provides the flexibility of using a number of underlying layers, it increases the complexity of protocol stack. And when only one of the many possible options is primarily used, such a generic mechanism is problematic rather than being beneficial.

Conclusion

In order to overcome the limitations of ATM signalling, AAL2 signalling was developed. The most important aspect of AAL2 signalling is that it decouples the way the ATM connection is established from the way an AAL2 channel is established i.e. Irrespective of whether an ATM connection is SVC or PVC, an AAL2 channel can be established over it. The advantages of this decoupling have already been explained. The other important feature of AAL2 signalling is allocation and de-allocation of channel identifiers (CID). This facility imparts greater dynamism to information transfer. Due to its features, AAL2 signalling has opened exciting possibilities. Voice over DSL and 3G Mobile Access Networks are two of many such application scenarios that are discussed in Chapters 21 and 22, respectively.

Review Questions

Q 1. *Why couldn't ATM signalling be extended to support AAL2 signalling? What are the benefits of decoupling the two?*

Q 2. *What is the relation between AAL2 connections and ATM connections? Explain with an example depicting ATM end-points and switches, as well as AAL2 end-points and switches.*

Q 3. *Can CID values change without terminating the AAL2 path? If yes, how? If no, why?*

Q 4. *What is the need for a Signalling Transport Converter (STC)?*

Q 5. *Compare the key aspects of ATM signalling with AAL2 signalling.*

Q 6. *How is the call identification procedure optimized in AAL2 signalling?*

Further Reading

For details of the topics covered in this chapter, the reader is referred to [ITU-T Q.2630.1]. The readers may refer to [ITU-T Q.2630.2] and [ITU-T Q.2630.3] for details on capability set 2 and capability set 3. The reader may also refer to [Wireless S. Kasera] to understand the applicability of AAL2 signalling in 3G Radio Access Network (RAN).

Chapter 14

ATM Network Management

14.1 Introduction

Network management is the process of managing a network. More formally, network management is the act of monitoring and modifying network operations with the purpose of detecting and solving problems, improving performance and collecting reports and statistics. There are five functional areas of network management, also called the FCAPS on the first alphabet of these functional areas. Table 14.1 provides generic information on network management. For details on these topics, refer [Gen S. Kasera].

This chapter discusses network management in ATM. The network management in ATM puts forth several new challenges not known to management systems of other networks. The traditional approach to network management is to first build the basic network architecture and then to add procedures for management. After all, only when the network is operational, can one have an accurate picture of the network management problems. However, this approach is not suitable for ATM networks because of their unique characteristics. First, the high speeds of operations require that the faults be detected and corrected immediately to limit loss of data. Moreover, not only the fault isolation/correction must occur in real-time, it must consume minimal network bandwidth. Achieving this goal requires careful design and planning, which is not possible if management systems are added as an afterthought.

Performance management also poses additional challenges in the form of service guarantees. The network management system not only needs to monitor the performance of physical links, but also of individual virtual connections. Monitoring various parameters like cell transfer delay and cell delay variation for individual virtual connections is a difficult task. Further, all networks cannot be migrated into an ATM network overnight. Therefore, integrating and interoperating with the existing networks, as the networks migrate

TABLE 14.1

Basics of Network Management [Gen S. Kasera]

Definition	Network management is the act of monitoring and modifying network operations with the purpose of detecting and solving problems, improving performance and collecting reports and statistics.
Goals	• To reduce the operational cost • To increase flexibility • To increase efficiency • To increase network availability • To provide security
Functional Areas (FCAPS)	• **Fault management:** Encompasses fault detection, isolation and the correction of abnormal operation. • **Configuration management:** Identifies, exercises control over, collects data from and provides data to open systems for initializing, starting, providing for the continuous operation of, and terminating interconnection services. • **Accounting management:** Enables charges to be established for the use of resources. • **Performance management:** Enables the behaviour of resources and effectiveness of communication activities to be evaluated. • **Security management:** Provides means to support security policies of various applications.
Models	• **Simple Network Management Protocol (SNMP):** It is an application layer protocol, designed for the exchange of management information between network elements. Model of SNMP consists of *agents, network management stations* and *MIBs*. • **Telecommunications Management Network (TMN):** It provides the means for transporting, storing, and processing the information necessary to support the management of telecommunication network and services. It architecture has different abstractions which include *functional architecture, physical architecture, information architecture* and *logical-layer architecture*.

systematically towards ATM, is also an issue that the ATM network management has to tackle.

14.2 Standardization of ATM Network Management Procedures

A lot of effort has been put in by various organizations to standardize network management procedures for ATM networks. Figure 14.1 depicts the important ATM network management standards. Among the multitude of standards defined, the following are discussed in this chapter:

- **Customer network management M3 specification defined by ATM Forum [ATMF CNM]:** This specification defines a reference model for ATM Network Management.
- **OAM standards defined by ITU-T in [ITU-T I.610]:** This standard defines use of special cells, termed as Operations and Management (OAM) cells, for fault management and performance management.
- **Integrated Layer Management Interface defined by ATM Forum in ILMI 4.0 [ATMF ILMI 4.0]:** Integrated Layer Management Interface or ILMI specifies how the Simple Network Management and ATM Interface MIB is used to provide any ATM device with status and configuration information.
- **AToMMIB defined by IETF in [RFC 2515]:** This RFC defines objects used to manage ATM-based interfaces, devices, networks and services.

Figure 14.1

ATM Network Management Standards

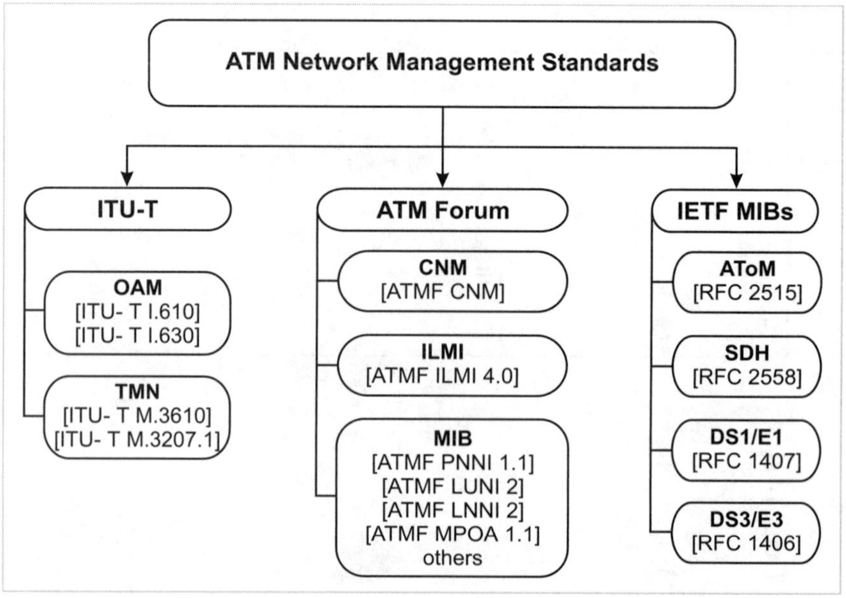

As far as the FCAPS functional areas of network management go, there is a lot to be done for ATM networks. Only fault management is the field where some standardization is achieved. In this regard, the OAM procedures provide extensive fault management procedures as well as some procedures for performance management. ILMI and AToMMIB also provide the means for fault management, performance management and configuration management. As far as accounting management and security management are concerned, very little standardization exists. Some procedures have been very recently defined for security management by ATM Forum. These are covered in Chapter 15.

14.3 ATM Network Management Reference Model

In order to standardize network management procedures, the ATM Forum standard [ATMF CNM] specifies a reference model as shown in Figure 14.2. This reference model developed by the ATM Forum provides a framework for end-to-end management of an ATM network. The various types of network management needed to support ATM devices, private networks, public networks, and the interaction between them is described as a part of this reference model. In this model, five areas are identified. Each area relates to a particular type of interface, dealing with the management of a different part and a different aspect of an ATM network. The five interfaces, termed M1 through M5, are as follows:

- **M1** is the interface between an ATM device and the corresponding private management system. The private management system lies at the customer site and is also known as an enterprise network management system.
- **M2** is the interface between enterprise network management system and a private ATM network. Both the interfaces, M1 and M2, are defined within an enterprise network management interface.
- **M3**, or the Customer Network Management (CNM) interface, defines the interaction between the customer's enterprise management system and the carrier management system (i.e. the public management system). M3 provides to the customer the means to monitor information related to configuration management, fault management and

Figure 14.2

ATM Network Reference Model

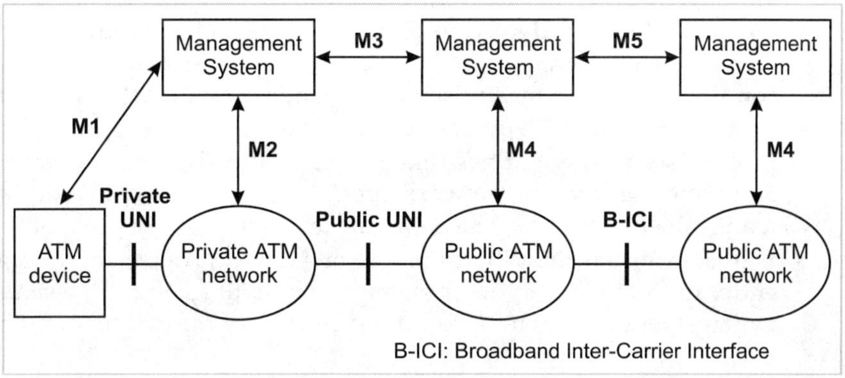

performance management. In other words, M3 provides to the customer a window into the carrier's network management system, through which the former can view the information relevant to it.

- **M4** defines an interface between a public ATM network and a public network management system. This interface is used to manage individual elements of a public ATM network, as well as to manage them in aggregation (e.g. sub-networks).
- **M5** is the management interface between two carrier network management systems (i.e. between two public management systems). Among the five interfaces, this is the most complex interface and no standards are available yet for this interface.

For example, MIBs applicable to the M1 and M2 interface include AToMMIB and LANE. AToMMIB is also applicable for M3 interface. For M4 (as well as M2) interface, ILMI MIB is applicable. SNMP and CMIP are the predominant protocols used to exchange management information over these interfaces.

14.4 Operations and Maintenance (OAM) Procedures

The virtual connections established in an ATM network are continuously monitored to identify and rectify any defects, and to ensure the continuity of the established connections. The monitoring of the connections is done at both the ATM layer as well as the physical layer. Procedures are defined which allow controlling of these connections remotely, determining the possible defects and informing the affected network entities about the defects. These functions clubbed in Operations and Maintenance (OAM) form the core of the connection management in ATM networks. These OAM procedures are detailed in [ITU-T I.610].

The OAM functions are allocated to the layer management of the ATM reference model. The aspects considered in the OAM procedures are as follows:

- **Performance management:** Performance management involves monitoring of the user data and calculating certain parameters like number of cells lost/mis-inserted, number of errored blocks, etc. In the OAM procedures, performance management is better known as performance monitoring because these procedures directly measure and control the parameters on which service guarantees are given to the users.
- **Defect and failure detection:** This entails periodic monitoring of network entities that are subject to failure. In case a defect or failure is detected in an equipment, an exception condition or an alarm is raised and the appropriate network entity is informed.
- **System protection:** To prevent a defected entity in the network from causing unnecessary problems for other entities (like sending false signals), the defective entity is isolated and removed from the network operations. If support for redundancy is provided, the management entity may also activate another entity in place of the defective entity.
- **Defect information:** When a failure or a defect is detected, the concerned management entity needs to inform other management entities about the same. This allows various management entities to have a uniform view of the network and helps in dealing with a

problem at a higher level. For example, a broken link may be taken care of by routing the data through an alternate path.

- **Fault localization:** The faults occurring in the network are better taken care of if the faults are localized to a region, and then isolated in the network. The information received from several management entities can be combined to localize the fault in the network.

The following sections describe how different OAM functions are realized in the ATM network.

14.4.1 OAM Hierarchy

OAM functions in the network are performed on five OAM hierarchical levels associated with the ATM layer and the Physical layer. For each of the hierarchical level, an OAM information flow is defined, which carries the necessary management information. The five flow levels defined are F1, F2, F3, F4, and F5 as shown in Figure 14.3. Flow levels F4 and F5 provide OAM functionality at the ATM layer. F1, F2 and F3 level flows provide OAM functionality at the physical layer at different levels of hierarchy within the physical networks. The cells associated with these OAM information flows are termed as OAM cells. The OAM levels defined are as follows:

- **Virtual channel level (F5):** OAM functions are performed over virtual channels. An end-to-end virtual channel connection consists of several virtual channel links. F5 level flows are defined on an end-to-end virtual channel connection (concatenation of virtual channel links) or on single/group of virtual channel links.
- **Virtual path level (F4):** A virtual channel link forms an end-to-end virtual path connection. The virtual path connection can itself have multiple virtual path levels. F4 level flows are defined on an end-to-end virtual path connection (concatenation of virtual path links) or on single/group of virtual path links.
- **Transmission path level (F3):** At the transmission path level, cell delineation and header error correction functions are performed. The OAM flow defined over transmission paths is F3 level flows.
- **Digital section level (F2):** A transmission path comprises of several digital sections. These digital sections are interconnected by line termination devices. Each digital section has a F2 level OAM flow.
- **Regenerator section level (F1):** A digital section has several regenerators in between to amplify the strength of the signal. The F1 OAM flow level is associated with a regenerator section level.

F5 flows are bi-directional in nature. The OAM cells corresponding to the F5 flow have the same VPI/VCI as the user cells on the connection. Two types of F5 flows are defined, namely end-to-end F5 flows and segment flows. End-to-end F5 flow is over the end-to-end virtual channel connections. The cells corresponding to these flows are inserted at the starting point of the connection and they travel till the destination of the connection. These cells cannot be removed at any intermediate network node. Segment F5 flow is defined over a single virtual channel link or multiple inter-connected virtual channel links.

FIGURE 14.3

OAM Information Flows

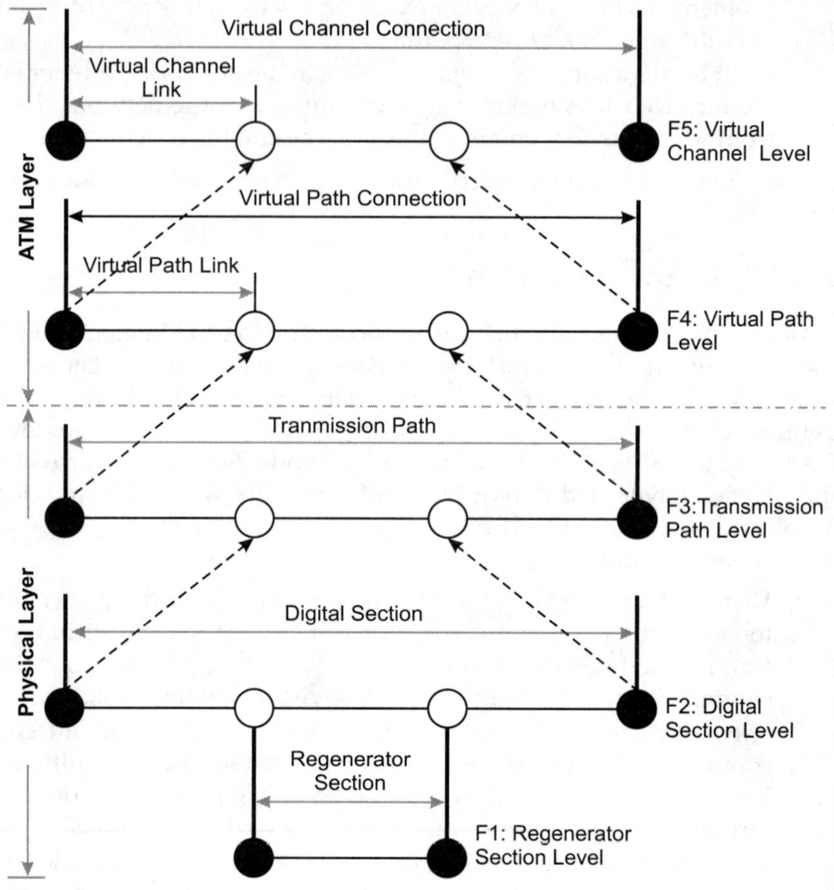

Source: [ITU-T I.160]

Figure 14.4 depicts the segment and end-to-end F5 flows. There exists an end-to-end F5 flow between the virtual channel connection end-points (A and D) and there is a segment F5 flow over a virtual channel link (A and C). The OAM channels of end-to-end F5 flow and segment F5 flow are differentiated on the basis of the value of Payload Type Indicator (PTI) field contained in the ATM cell header. The end-to-end and segment F5 OAM cells have their PTI fields set to 101 and 100, respectively.

F4 flows share quite a few characteristics of F5 flows. Like F5 flows, F4 flows are bi-directional in nature and can be segment flows or end-to-end flows. However, the use of VPI/VCI values in F4 flows is different from that of F5 flows. In F4 flows, the VPI value is the same as that of the user cells of the VPC, but a set of pre-assigned VCI values is used (unlike F5 flows). The pre-assigned values are VCI = 3 and VCI = 4 for segment F4 flows and end-to-end F4 flows, respectively.

FIGURE 14.4

Example of OAM Flows

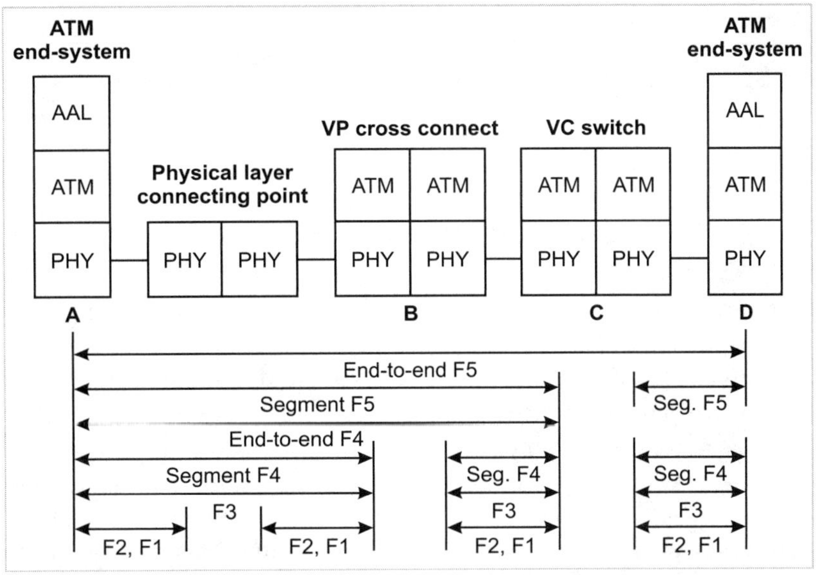

F1, F2 and F3 flows are defined at the *physical layer*. These three flows perform monitoring of various network entities at the physical layer. However, the hierarchical levels at which these flows operate are different. All three flows perform the functionality to detect and indicate the malfunctioning or unavailability of any physical layer entity. Other functions for the system management include performance monitoring to gather information about the network and network element behaviour in order to evaluate their performance, detection, isolation and correction of failed equipment, and verifying the connection continuity between the terminating equipments.

Since F5 and F4 flows operate at the ATM layer while F1 to F3 flows operate at the physical layer, the subsequent discussion is divided into two categories (i.e. the first category discusses OAM functions at the ATM layer, while the second category discusses OAM functions at physical layer).

14.4.2 OAM Cell Format

Before various OAM functions at the ATM layer are discussed, the OAM cell format is explained. Every OAM cell has a format as depicted in Figure 14.5. The following fields are common to all OAM cells:

- **Header:** This is the 5-btye header of an ATM cell. For F5 flows, two PTI values are reserved to identify end-to-end and segment OAM F5 flows. Similarly, two pre-assigned VCI values are used to distinguish between end-to-end and segment F4 flows.
- **OAM type:** This field indicates the type of management function performed.
- **OAM function type:** This indicates the actual management function performed. The possible values of OAM type and OAM function type are depicted in Table 14.2.

FIGURE 14.5

OAM Cell Format

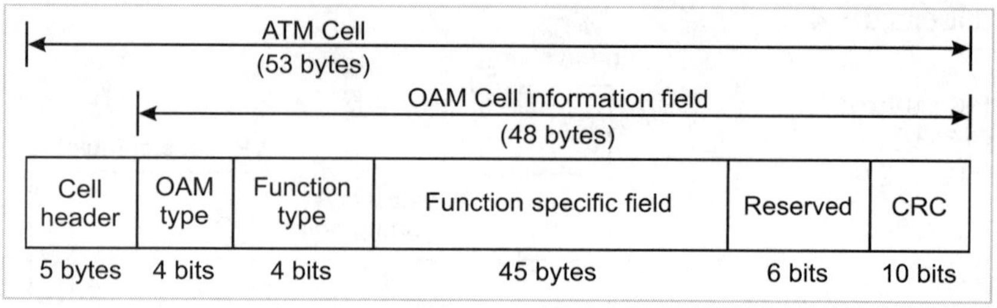

- **Function-specific field(s):** As will be discussed subsequently, this contains function specific field(s).
- **Reserved field:** This is coded to zero and reserved for future use.
- **Error detection code:** This field contains a CRC-10-error detection code.

From Table 14.2 it follows that the OAM functions at the ATM layer come primarily under the purview of fault management and performance management. The fault management functions include Alarm Indication Signal (AIS) and Remote Defect Indication (RDI), continuity checks, and loopback. All these functions are defined for both F5 level (virtual channel level) and F4 level (virtual path level). Performance management functions at the ATM layer include the use of monitoring cells to estimate performance in both forward and back directions. These functions are again applicable to both F5 and F4 flows. In the subsequent sections, the various fault management and performance functions are elaborated upon. Apart from this, the activation/deactivation of OAM cells is used to start or stop various performance management functions like performance monitoring or continuity check.

TABLE 14.2

OAM Type and Function Type Fields [ITU-T I.610]

OAM type	*Coding*	*Function type*	*Coding*
Fault management	0001	Alarm Indication Signal (AIS)	0000
	0001	Remote Defect Indication (RDI)	0001
	0001	Continuity check	0100
	0001	Loopback	1000
Performance management	0010	Forward Performance Monitoring (FPM)	0000
	0010	Backward Reporting (BR)	0001
APS Coordination protocol	0010	Group protection	0000
	0010	Individual protection	0001
Activation/deactivation	1000	FPM and associated BR	0000
	1000	Continuity check	0001
	1000	Forward Performance Monitoring (FPM)	0010

The APS coordination protocol is used for ATM Protection Switching (APS). Protection switching is a network survivability technique with dedicated protection resource to protect a part of VPs/VCs that need high reliability. Details of protection switching are available in [ITU-T I.630].

14.4.3 AIS and RDI Defect Indications

The occurrence of VPC/VCC failures is intimated to the concerned network management systems, using Alarm Indication Signal (AIS) and Remote Defect Indication (RDI) defect indications. VPC/VCC failures occur due to physical link failures or due to internal failures at the ATM layer (e.g. translation table problem). The Alarm Indication Signal is sent to a downstream node, when such a failure/defect is detected. Similarly, Remote Defect Indication is sent to an upstream node, indicating such a failure/defect. RDI was also referred to as Far End Reporting Failure (FERF). However, in current literature, the term RDI has replaced FERF.

Depending upon whether VPC has failed or VCC has failed, VP-AIS or VC-AIS, respectively, is sent in the downstream direction and VP-RDI or VC-RDI, respectively, is sent in the upstream direction. As long as the failure condition persists, the AIS and RDI cells are sent periodically to the downstream and upstream nodes, respectively. After recovery from the failure, the transmission of these cells is stopped. The AIS and RDI indications are sent using OAM cells. The AIS/RDI fault management cell carries along with it the defect type and defect location.

14.4.4 Continuity Check

In order to ensure that a fault in a link is identified quickly and the degradation in service is minimal, periodic check for continuity is done. Continuity check is used under two scenarios. In the first scenario, continuity check cells are inserted when there is no user cell to be transferred for a long duration on the connection. Typically, an inactivity period (i.e. when there is no user cell to send) of 1 second is considered nominal for inserting continuity check cells. In the second scenario, continuity check cells are inserted periodically at the intervals of 1 second, independent of user cells.

When continuity check is activated and no cell (user cell or continuity check cell) is received for a period of 3.5 s ± .5 s, the destination end-point assumes that connectivity is lost and sends an RDI indication to the source end-point. On receiving the user cell or the continuity check cells, the nodes along the path of the connection know about the recovery from fault and the return to normal state.

14.4.5 Loopback

Besides the two mechanisms mentioned above, loopback is another important mechanism for fault localization and connectivity verification. The loopback capability allows OAM-related information to be inserted at one location and looped back toward the source at another location. The loopback cell contains the following function specific fields:

- **Loopback indication:** This is a one-byte field, which has its least significant bit set to 1 by the source. The bit is reset to 0 at the loopback point where the cell is looped back. This helps to identify if the loopback cell that has arrived with a default location identifier has to be looped-back or the cell itself is one of the looped-back cells. This prevents infinite looping of the OAM cells in the network.
- **Correlation tag:** This field is used to correlate a transmitted loopback cell with the received loopback cell.
- **Loopback location ID:** This field specifies the location where the cell has to be looped back. The default value of all 1's for this field indicates an end-point of the connection or the end-point of the segment, depending upon whether the flow is end-to-end flow or segment flow.
- **Source ID:** An optional field, this identifies the source of the loopback cell.

14.4.6 Performance Monitoring

The OAM performance management function includes monitoring blocks of user cells by inserting monitoring cells, either at the end of the VPC/VCC or at the end of VPC/VCC segments. The performance management information flows upstream in the direction opposite to that of user cells. An upstream node requests a downstream node to initiate monitoring. The downstream node responds back with the results of performance management via the reporting cell. The performance management results include the number of cells of each type of user flow (CLP = 0 or CLP = 0 + 1 flow), the number of errored block in each flow and the number of lost/mis-inserted cells within a monitored block of cells. Use of monitoring cells for measuring parameters like CTD is under further study. The performance monitoring functionality directly measures the parameters on which the service guarantees is provided to the users. To ensure accurate performance monitoring, it is recommended that continuity check mechanisms be activated while monitoring. This ensures that results obtained are for normal operations of the network.

A performance monitoring cell insertion request is initiated by the upstream node after transmission of every N user cells. The nominal value of N is 128, 256, 512, 1024, 2048, 4096, 8192, 16,384 and 32,768, with a permitted variation of ± 50 per cent. The reporting cell is inserted by the downstream node at the first available cell location after receiving the request (see Figure 14.6). The fields of backward reporting cell include the following:

- Monitoring cell sequence number
- Total user cell number related to CLP_{0+1} user cell flow
- Total user cell number related to CLP_0 user cell flow
- Time stamp
- Reported Monitoring Cell Sequence Number (RMCSN)
- Severely Errored Cell Block Count (SECBC)
- Total received cell count related to CLP_{0+1} user cell flow
- BLock Error Result ($BLER_{0+1}$)
- Total received cell count related to CLP_0 user cell flow

Figure 14.6

Performance Monitoring Operations

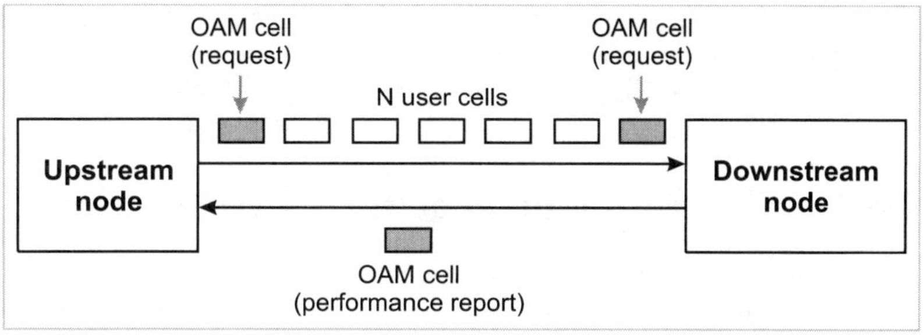

For the interpretation and use of individual fields, the reader is referred to recommendation [ITU-T I.610].

14.4.7 Activation/Deactivation Procedures

Provisions exist in ATM OAM standards to explicitly activate or deactivate performance monitoring and continuity checks. In order to activate either performance monitoring or continuity check, an ACTIVATE message is sent by the originating end. The receiving end-point responds by sending either ACTIVATION CONFIRMED or ACTIVATION REQUEST DENIED (see Figure 14.7(a)). This is a simple, two-way handshake protocol. The same is true for deactivation, which is initiated by sending a DEACTIVATE message. In this case, the receiving end-point responds by sending DEACTIVATION CONFIRMED (see Figure 14.7(b)).

The function-specific fields of activation/deactivation OAM cells are as follows:

- **Message ID:** Specifies the type of message (i.e. ACTIVATE, DEACTIVATE, etc.)
- **Correlation tag:** A tag used for correlating commands with responses.
- **Direction of actions:** Identifies the direction(s) of transmission to activate/deactivate OAM functions.

Figure 14.7

Activation/ Deactivation Handshakes

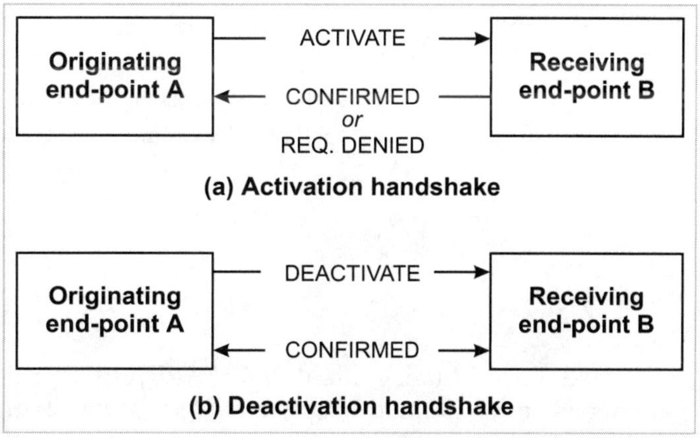

- **PM block sizes A-B:** Specifies the performance-monitoring block size (i.e. the value of N which can be 128, ... or 32,768) in A-B direction.
- **PM block sizes B-A:** Specifies the performance-monitoring block size (i.e. value of N which can be 128, ... or 32,768) in B-A direction.

14.4.8 OAM Functions at the Physical Layer

OAM functions at the physical layer are performed using F1, F2 and F3 level flows. As already mentioned, F1 flows operate at the regenerator section level, F2 at the digital section level, and F3 at the transmission path level. The mechanisms to provide OAM functions and to generate cells of the three flows depend upon the underlying transport mechanism. Depending upon whether cell-based or SONET/SDH-based transmission system is used, the OAM functions and OAM flows differ.

For cell-based transmission systems, the OAM information at the physical layer is carried in special cells called the PL-OAM cells, which are distinguished by a special pattern in cell header. In these systems, only F3 and F1 flows are defined. The F3 flows are used to indicate failures/defects like loss of cell delineation, loss of F3 PL-OAM recognition and cell rate decoupling defect. F1 flows indicate error conditions like loss of signal, loss of F1 PL-OAM cell recognition and unacceptable error performance.

For SONET/SDH-based transmission systems, all three flows are applicable. Flows F1 and F2 are carried in the transport overhead bytes of the SONET/SDH overhead. F1 flow is carried in the section overhead and F2 is carried in line overhead bytes of the transport overhead. The F3 flow is carried in the path overhead bytes in the SONET/SDH payload envelope. The mapping of physical layer flow information in the SONET overhead is shown in Table 14.3.

TABLE 14.3

Mapping Physical Layer Flow Information in SONET Overhead

Functions	*SONET overhead bytes*
Line Error Monitoring	B2, Z2(18-24)
Path Error Monitoring	B3, G1(1-4)
Section Error Monitoring	B1
STS Path AIS (Alarm Indication Signal)	H1, H2, H3
STS Path RDI (Remote Defect Indicator)	K2(6-8)
Line AIS and RDI	K2(6-8)
Connectivity Verification	J1

14.5 Integrated Layer Management Interface (ILMI)

Integrated Layer Management Interface (ILMI) provides the status and configuration parameters to an ATM device (e.g. ATM end-system, or ATM switch) necessary for its functioning. These status and configuration parameters primarily relate to Virtual Path

Connections, Virtual Channel Connections, ATM Network Prefixes and ATM Addresses associated with an ATM interface. ATM interface in the current context either refers to a physical interface (i.e. physical link) or to a virtual link (VPC/VCC).

Associated with each ATM device connected via an ATM interface, is an Interface Management Entity known as IME. The IME of an ATM device can access, via the ILMI protocol, the ATM Interface MIB associated with its adjacent IME. For example, the IME of an ATM end-system can know the UNI signalling version, running at its adjacent ATM switch through the peer IME via ILMI (see Figure 14.8).

FIGURE 14.8

ILMI Reference Configurations

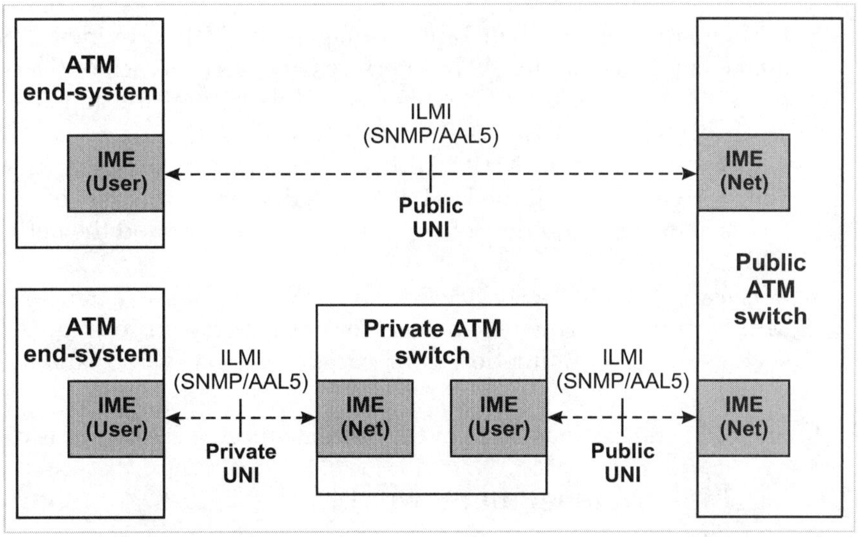

In earlier specifications, ILMI was referred to as Interim Layer Management Interface because it was intended to be an ad-hoc specification, to be used in the short-run only. However, the ILMI specification is now stable and the term 'interim' in ILMI is replaced with 'integrated' to form the Integrated Layer Management Interface. The specification of ILMI is given in ATM Forum's specification [ATMF ILMI 4.0]. Initially ILMI was defined for public UNI and private UNI. ILMI 4.0 extends the capability of ILMI to include public NNI.

The ILMI communication protocol is an open management protocol. It uses SNMP and AAL5 to exchange management information. The message format used is that of SNMPv1 [RFC 1157]. The use of SNMPv2 is for further study. Under default configuration, SNMP messages of size up to and including 484 octets, are acceptable to any ILMI implementation. Higher values of the messages can be negotiated between the sender and the receiver. For exchanging management information through ILMI, a dedicated VCC must be always available. While the VPI/VCI of this VCC is configurable, the default value is VPI = 0 and VCI = 16.

The ILMI attributes, organized as a standard MIB (called ATM interface MIB), are maintained separately for each interface. The rationale behind keeping a separate set of information for each interface is to provide flexibility of configuring parameters of each interface individually. In association with ATM Interface MIB, ILMI provides an ATM device the status and configuration information about physical and ATM layer parameters. ILMI also enables two adjacent ATM devices to automatically configure the operating parameters of the common ATM link, between them. Further, it provides mechanisms for bi-directional communication across the ATM link, enabling both sides to verify subscription parameters and to configure new traffic services such as ABR.

The ILMI MIBs fall into the following categories:*

- **Link management MIB:** Link Management MIB provides a general-purpose link management facility for ATM interfaces. It describes each of the objects in the MIB, the procedures for the use of the objects and then gives the actual syntax of the MIBs. In essence, this is the actual ATM interface MIB.
- **Address registration MIB:** Address Registration MIB specifies the procedures for address registration at the UNI. It allows dynamic exchange of addressing information between the user and the network, allowing the user and the network to agree on ATM address in effect.
- **Service registry MIB:** Service Registry MIB provides an extension to the ATM Interface MIB to provide a general-purpose service registry for location ATM network services such as the LAN Emulation Configuration Server (LECS) and the ATM Name Server (ANS).

In the following sections, each of the aforementioned categories is discussed.

14.5.1 Link Management MIB

As mentioned earlier, Link Management MIB provides a general-purpose, link management facility for ATM Interfaces. The link management information is clubbed into several groups, which are as follows:

- **Per-system attributes:** These are defined in [RFC 1213]. The 'system' group is defined as being mandatory for all systems such that each managed entity contains one instance of each object in the 'system' group. Thus, those objects apply to the entity even if the entity's sole functionality is the forwarding of ATM cells. The MIB attributes include sysDescr, sysObjectID, sysUpTime, sysContact, sysName, sysLocation and sysServices.
- **Per-physical interface attributes:** This group consists of various attributes associated with the individual physical interfaces. The physical interface is identified by the interface index (atmfPortIndex). The MIB attributes include interface index and adjacency information.
- **Per-ATM layer interface attributes:** This group (known as atmfAtmLayerGroup) contains various attributes associated with individual interfaces of the ATM layer. The

*Portions extracted from [ATMF ILMI 4.0] (Copyrights, 1996, The ATM Forum).

MIB attributes include maximum number of active VPI/VCI bits, maximum number of VPC/VCC, maximum SVPC VPI, maximum SVCC VPI/VCI, ATM Interface type (public/private). ATM device type (User/Node), ILMI version, UNI signalling version and NNI signalling version. On initialization, an IME retrieves the value of this object from its peer. The IME then compares the retrieved value from the locally maintained value. The actual value is set to the lower of the two values.

- **Per-virtual path attributes:** This group (atmfVpcGroup) consists of various attributes associated with individual virtual paths on an interface. Note that only permanent virtual path connections are represented in this group (i.e. information on switched VPCs are not maintained). The MIB attributes include VPI Value, Operational status, Transmit/Receive Traffic Descriptor, Best Effort Indicator, Transmit/Receive QoS class and Service Category.
- **Per-virtual channel attributes:** This group (atmfVccGroup) consists of various attributes associated with individual virtual channels on an interface. Only permanent virtual channel connections are represented in this group, including all standard permanent VCCs like signalling, ILMI and LECS VCCs, and other non-standard permanent VCCs that are configured for use. The MIB attributes include VPI/VCI Value, Operational status, Transmit/Receive Traffic Descriptor, Best Effort Indicator, Transmit/Receive QoS class, Service Category and Transmit/Receive Frame Discard Indication.
- **Per-virtual path ABR attributes:** ABR virtual path connections require additional attributes to be defined because traffic descriptors like PCR and MCR are not sufficient to manage an ABR connection. These additional attributes are defined in the Virtual Path ABR Group (atmfVpcAbrVpi). The MIB attributes include VPI Value and ABR Operational parameters (see Table 14.4).
- **Per-Virtual channel ABR attributes:** Additional attributes have been defined (in atmfVccAbrGroup) for the ABR virtual channel connections as well. MIB objects includes, interface index, VPI, and ABR operational parameters. The ABR operational parameters for VCCs are the same as those used with VPCs.
- **Link management traps:** Two traps (atmfVpcChange and atmfVccChange) have been defined for ILMI. (Traps are asynchronous messages sent by agents to report the occurrence of special events.) One of the traps indicates that a permanent VPC is configured, modified or deleted. The trap message generated for this event carries the VPI value of the affected VPC. Other trap indicates that a permanent VCC is configured, modified or deleted. The trap message generated for this event carries the VPI and the VCI values of the affected VCC. This completes our discussion on Link Management MIBs.

14.5.2 Address Registration MIB

In order to establish an ATM connection at the UNI, both the user and the network must know the ATM address(es) which are in effect at that UNI. These addresses are used in the signalling messages exchanged on that interface. Address registration allows the user and network to exchange addressing information between them. Address registration occurs

TABLE 14.4

Operational ABR Parameters (Copyrights 1996 the ATM Forum)

ABR Parameter	*Description*
Initial Cell Rate (ICR)	Upper-bound on the source's transmission rate, imposed at initial start-up and after an idle period. The unit is an integer number of cells/second.
Number of data cells per forward RM-cell (Nrm)	The maximum number of data cells a source may send between forward RM-cells.
Maximum time between forward RM-cells (Trm)	Upper-bound on the time between forward RM-cells for an active source (in milliseconds).
Cut-off Decrease Factor (CDF)	Controls the required decrease in source transmission rate associated with lost or delayed backward RM-cells.
Rate Increment Factor (RID)	Controls the allowed increase in source transmission rate associated with the receipt of a backward RM-cell which indicates no congestion in the network.
Rate Decrease Factor (RDF)	Controls the required decrease in source transmission rate associated with the receipt of a backward RM-cell indicating congestion in the network.
ACR Decrease Time Factor (ADTR)	Allowed time between the transmission of forward RM-cells, before the source is required to decrease its transmission rate to ICR.
RM-Cells before Cut-off (CRM)	Limits the number of forward RM-cells which may be sent in the absence of received backward RM-cells.

when the network is setup, when a new end-system is added, or at other times as required. The End System Identifier (ESI) and the Selector (SEL) fields are supplied by the user to the network. The network, in turn, supplies the user the network prefix. The ESI and SEL field, appended to the network prefix, forms an ATM address.

The capabilities that fall under the purview of address registration are as follows:

- Initialization-time exchange of addressing information,
- Restrictions on network-prefix/user-part combinations,
- Acceptance of Unassigned Network-Prefixes,
- Rejection of unacceptable values, either rejection of a specific network prefix by the user, or of a specific user part by the network,
- Dynamic addition/deletion of additional network prefixes and user parts,
- De-registration of addresses when ILMI connectivity is lost,
- Indication of support or non-support for address registration at an interface.

For the purpose of address registration, three additional MIB groups have been defined—NetPrefix Group, Address Group and Address Registration Admin Group. Each of the above groups is discussed below.

- **NetPrefix group:** Net Prefix Group (atmfNetPrefixGroup) is mandatorily implemented by the IME on the user side of the private UNI and optionally by the user side of public UNI. The MIB attributes include Interface Index, Network Prefix and Network Prefix Status. The individual entries in the MIB, contained in the Network prefix table, are indexed using the Interface Index and the Network Prefix. The Network Prefix Status object provides an indication of the validity of network prefix at the ATM interface. To configure a new network prefix, the network-side IME uses a SetRequest to set the Network Prefix Status object for that prefix to be valid. To delete an existing network prefix, the network-side IME uses a SetRequest to set the Network Prefix Status object for that prefix to be invalid.
- **Address group:** Address Group (atmfAddressGroup) is mandatorily implemented by the IME on the network side of the private UNI and optionally by the network side of public UNI. The MIB attributes include Interface Index, ATM Address, ATM Address Status and ATM Address Organizational Scope. The individual entries in the MIB, contained in the Address table, are indexed using the Interface Index and the ATM Address. The ATM Address Status object provides an indication of the validity of ATM address at the ATM interface. Procedures for configuring or deleting ATM addresses is similar to that defined for configuring/deleting network prefixes. The ATM Address Organizational Scope object indicates the organizational scope for the associated address. An organizational scope may apply to individual addresses as well as group addresses. The associated address is not distributed outside the indicated scope. This is particularly useful when using anycast capability while establishing connections. The calling user indicates to the network the organizational scope of the connection request. This connection request, containing a group address, reaches a group member within this organizational scope, but not to any group member outside the organizational scope.
- **Address Registration Admin Group:** All IME's are required to implement the Address Registration Admin Group (atmfAddressRegistrationAdminGroup). The MIB attributes include Interface Index and Address Registration Admin Status. The individual entries in the MIB, contained in the Address Registration Admin table, are indexed using the Interface Index. The Address Registration Admin Status object provides an indication of support for the Prefix and Address Groups. The Prefix and Address Groups are supported only if the Address Registration Admin Status object on both sides of the UNI indicate such support. This completes our discussion on Address Registration MIBs.

14.5.3 Service Registry MIB

Through the use of Service Registry MIB, ILMI provides a general-purpose service registry for locating ATM network services such as the LAN Emulation Configuration Server (LECS) and the ATM Name Server (ANS). The MIB attributes include Interface Index, Service Identifier, ATM Address, Address Index and Parameter.

The service identifier uniquely identifies the type of service at the address provided. ATM Address is the full address of the service. IME may use this address to establish a connection with the service. Address Index is used to differentiate between different ATM addresses providing the same service on the same port. The parameter represents an octet string whose interpretation depends on the type of service used.

14.6 AToMMIB

AToMMIB, as specified by [RFC 2515], defines objects used to manage ATM-based interfaces, devices, networks and services. Specifically, AToMMIB defines managed objects for the management of PVCs. Some aspects of managing SVCs are also considered in this standard. There are quite a few objects common to AToMMIB and the MIB defined by ILMI. However, while ILMI is limited to use at UNI, AToMMIB is used between ATM switches.

The AToMMIB defines the following MIB tables:

- **ATM interface configuration table:** This table contains information on ATM cell layer configuration of local ATM interfaces. The elements in this table include Port identifier, Physical transmission types, Administrative status, Active VPI/VCI fields, Maximum number of VPI/VCI bits, Maximum number of VPCs/VCCs, Configured VPCs/VCCs, Operational status, Number of received/transmitted cells, Number of cell header validation errors, and Number of cells with HEC error.
- **ATM interface DS3 PLCP and TC sub-layer tables:** This table provide performance statistics of the DS3 PLCP and TC sub-layer of local ATM interfaces on a managed ATM device. (Note that DS3 PLCP and TC sub-layer are currently used to carry ATM cells respectively over DS3 and SONET transmission paths.) The elements in these tables include DS3 PLCP severely errored framing seconds, DS3 PLCP unavailable seconds, DS3 PLCP alarm state, DS3 PLCP out of cell delineation events and TC alarm state.
- **ATM traffic parameter table:** This table contains elements required to define the traffic and service requirements of an ATM connection. The elements in these tables include Transmit/Receive traffic/service category parameters, AAL type, Transmit/Receive AAL5 SDU size, Receive AAL5 SDU size, AAL5 encapsulation type, Connection topology type and Use of call control.
- **ATM interface virtual link (VPL/VCL) configuration tables:** This table provides the means for managing ATM VP links (VPL) and VC links (VCL). The elements in these tables include VPI or VPI/VCI value, VCL or VPL operational status, VCL/VPL administrative status and VCL/VPL last change status.
- **ATM VP/VC cross-connect tables:** This table provides the means for managing VP/VC cross-connects. The elements in these tables include Cross-connect identifier, Port identifier of one end, Port identifier of the other end, VPI or VPI/VCI value of one end, VPI or VPI/VCI value of other end, VP/VC cross-connect operational status, VP/VC cross-connect administrative status and VP/VC last change status.

- **AAL5 connection performance statistics table:** This group is defined for an end-system, switch, or network that terminates AAL5 protocol. The elements in this table include PDUs discarded for CRC errors, PDUs discarded due to reassembly time out, PDUs discarded due to large SDUs, Number of transmitted PDUs, Number of transmitted PDUs, PDUs discarded due to protocol errors and Set of configuration/state parameters.

Conclusion

Network management (NM) procedures in ATM are still in their preliminary stages. Even though a number of standards are defined by ITUT, ATM Forum and IETF, some of which were explained in this chapter, a lot still needs to be done. In the field of fault management, due to the high-speed nature of ATM, it is critical that NM systems have the adequate processing power to analyze received management data. Besides, the NM console must provide filtering capabilities so that only critical information is presented to network managers.

Multiple service classes make performance management difficult. Means must exist so that it can be verified whether the QoS parameters guaranteed by the network are being provided or not. Numerous QoS parameters complicate the verification of Service-Level Guarantees (SLAs). All this, coupled with a large number of virtual connections, recessitates elaborate performance management procedures.

The factors mentioned above also complicate accounting management. In the ideal case, a user must be charged for the amount of data transferred, and not upon a flat rate. This requires that the ingress switch must monitor the usage of each virtual connection.

Very little has been done with regards to security management. One critical issue is that the security-related management tools are designed for slower speed links and may not support the high-speed links of ATM. Some of the security aspects are discussed in Chapter 15.

Like security management, very little standards exist for configuration management. The functions within the purview of configuration management include configuring switched, routers, physical circuits and setting up and tearing down PVCs/SVCs. The key issue is to have a uniform platform which can used by switches from different vendors.

Review Questions

Q 1. What is the current status of standardization for network management in ATM?

Q 2. What is Customer Network Management (CNM) model for network management in ATM? What are the five different interfaces defined by this model? What is the relevance of these interfaces?

Q 3. What are the important fault management procedures in ATM as defined by [ITU-T I.610]? What is the difference between them?

Q 4. Write notes on

- *Alarm Indication Signal (AIS)*

- *Remote Defect Indication (RDI)*
- *Continuity check*
- *Loopback*
- *Activation/Deactivation procedures*

Q 5. *How does the performance-monitoring mechanism work in ATM by using OAM cells?*
Q 6. *What is the importance of ILMI? What are the important categories of MIB defined by ILMI?*

Further Reading

There is a dearth of good literature on network management in ATM. Thus, one has to depend upon the ITU-T and ATM Forum specifications for further reading. The OAM procedures are nicely described in [ITU-T I.610] and provide detailed description of F1 to F5 flows, along with the fault and performance management procedures. The [ATMF CNM] specification provides a model for network management. There are other specifications from ATM Forum that focus on the M4 interface, for which the reader is referred to http://www.atmforum.com. For MIBs of different protocols, one needs to refer to the base specification of that protocol. For example, PNNI MIBS are described in [ATMF PNNI 1.1]. Then there is ILMI protocol described in [ATMF ILMI 4.0].

Chapter 15

ATM Security

15.1 Introduction

As organizations grow and unscrupulous elements try to disrupt the proceedings, there is clear need to protect networks from various kinds of attack. There can be various types of attacks on a network. As an example, consider eavesdropping: eavesdropping involves monitoring network traffic to gain illegal access, or to view sensitive information. One of the examples of eavesdropping is when an end-system operates in promiscuous mode and listens to even those frames that are not meant for it. While this chapter does not attempt to discuss the different types of attack, it is clear that there is need to protect networks and the information contained therein from unauthorized access (read/write/modify). The means to protect networks to provide secure transfer of information from one node in the network to another comes under the purview of network security. Table 15.1 provides the basics of network security. For details on these topics, refer [Gen S. Kasera].

Security is a relatively new concept in ATM networks. Till some time back, typical threats for a networking technology was not found to be important in the context of ATM networks. However, in recent times, there has been increasing security violations and there are growing concerns for secured corporate environment. Given this, ATM Forum has worked towards incorporating security features in the ATM networks as well. This chapter discusses the objectives and requirements for security in ATM networks and some of the solutions to tackle these challenges.

15.2 Security Objectives for ATM Networks

Before a framework or solution is made for providing security features in ATM networks, the requirements that govern such effort must be specified. Note that the requirements are

TABLE 15.1 Basics of Network Security [Gen S. Kasera]

Definition	*The means to provide secure transfer of information from one node in the network to another*
Threats	• **Masquerade (spoofing):** Refers to the pretense by an entity to be a different entity. • **Eavesdropping:** A breach of confidentiality by monitoring network traffic to gain illegal access, or to view sensitive information. • **Unauthorized access:** An attempt to access data in violation to the security policy in force. • **Loss or corruption of information:** Refers to compromising with the integrity of data by unauthorized deletion, insertion, modification, reordering, replay or delay. • **Repudiation:** Refers to the denial of an event by an entity that was involved in some communication. • **Forgery:** Refers to fabricating information and claiming that such information was received from another entity or sent to another entity. • **Denial of service:** This occurs when an entity prevents other entities from performing their functions (e.g. by flooding the server with fake requests thereby preventing genuine users to get service).
Elements	• **Data integrity:** Means to ensure that the contents of data packets are not altered in an unauthorized manner. • **Origin authentication:** Refers to the corroboration that the source of data is as claimed. • **Authentication:** Data integrity and origin authentication together is referred to as authentication. • **Confidentiality:** Refers to the property that information in a data packet is not made available or disclosed to unauthorized entities. This is achieved by encrypting the data using ciphering techniques. • **Replay-protection:** Refers to the techniques that ensure that a captured packet loses its relevance when stored over a pre-defined period of time.
Encryption techniques	• **Symmetric (Secret Key) technique:** Works on the principle that a secret key is used for encryption. The encryption process consists of an algorithm (each for encryption and decryption) and the secret key. For e.g. Data Encryption Standard (DES) and Advanced Encryption Standard (AES) algorithms. • **Asymmetric (Public Key) techniques:** Each user who wishes to establish a secure communication channel

Contd.

Table 15.1

Contd.

	needs to have a pair of keys. One of these keys, known as the *public* key is made public and is known to all. The other key, known as the *private* key, is known only to the user, and is kept secret from the other users. For example, RSA algorithm and Elliptic Curve Cryptography.
Key Management	• **Physical delivery:** One of the communicating parties chooses a secret key, and physically delivers the key to the other party. • **Secure Channel:** The communicating parties can share the keys to be used for a communication session over another secure channel. For example, Kerberos system. • **Exchange of public keys prior to communication start:** Using this scheme, two parties wishing to engage in a secure communication can exchange their public keys with each other prior to start of secure communication. For example, Diffie-Hellman algorithm. • **Public key certificates:** Certified authorities issue these certificates, which contain the user identity and public key of a user.
Data integrity	• **Hashing:** Technique commonly used for providing data integrity (or message authentication), which implies that the message is authentic and is not tampered with. For example, Message Digest 5 (MD5) algorithm and Secure Hash Algorithm 1 (SHA1).
Others	• **Firewalls:** A network entity that resides between a private network (for example, the LAN) and a public network (for example, the Internet).

independent of implementation and thus do not dictate how a requirement is achieved; it merely states what is to be achieved.

ATM Forum has prepared a framework specification [ATMF SEC FW 1.0] that provides objectives and requirements for providing security in ATM. The following points summarizes the security objectives in ATM networks:

- **Functional objectives:** The functional objectives refer to the capabilities in terms of security mechanisms. The capabilities include the two key elements, namely confidentiality and data integrity. Apart from this, the ATM network should have accountability (whereby any entity should be responsible for any actions initiated, and availability (whereby legitimate entities should have correct access to ATM facilities).
- **Customer objectives:** Depending upon the needs of the customer (personal or corporate, or any other classification), each customer has its set of objectives. However, certain common objectives could include data integrity, data confidentiality, privacy protection, service subscription management (including activation/deactivation), availability of

functional ATM network services, correct and verifiable billing and capability to use a service anonymously.

- **Operator objectives:** The operations seek to maximize their revenues and reduce cost. Thus, the key security objectives for an operator could include a availability of network services and network management system, correct and verifiable billing (with little possibility of fraud), non-repudiation for all ATM network services used, preservation of reputation (above all preservation of customers' and investors' trust), accountability for all activities, data integrity, data confidentiality and privacy protection.

15.3 Security Requirements for ATM Networks

While security objectives listed above are slightly abstract, the ATM security framework specification [ATMF SEC FW 1.0] provides numbered security requirements for ATM networks. These security requirements are listed in Table 15.2. Each requirement has a title, a description and the means to achieve these requirements. For example, the identities of the user can be verified using authentication mechanisms. Similarly, confidentiality for stored data can be ensured by access control and for communicated data using confidentiality/ encryption mechanisms. The exact means depend upon user plane and control plane, details of which is available in [ATMF SEC FW 1.0].

15.4 ATM Security Model

In ATM, the security services are provided by a Security Agent (SA). The agents may reside at end systems or at intermediate switches. Figure 15.1 shows the simple case where the agents reside at the end systems. The security agents may also reside at ATM switch where the switch may act as a firewall. In such a scenario, the agent on the switch provides security service to one or more users of the switch.

Figure 15.1

A Model for ATM Security

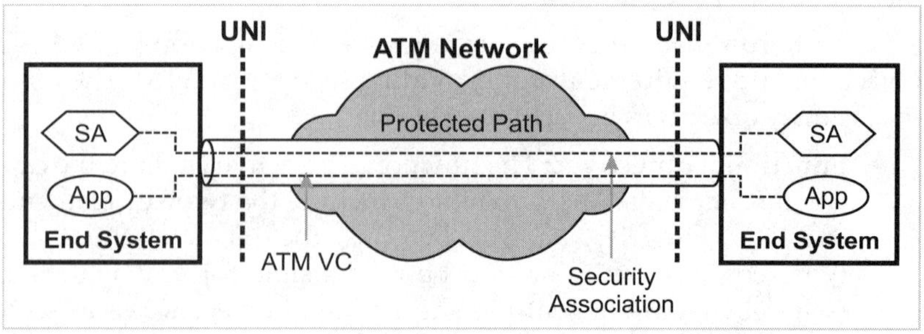

In order to provide secured communication between any two entities, a security association is necessary. In the context of ATM, "a security association is the distributed contextual information (that includes the algorithm type, security keys, mode of operation, etc.), which

TABLE 15.2 Security Requirements for ATM Networks [ATMF SEC FW 1.0]

Title	*Requirement*	*Security service*
Verification of Identities	The ATM network shall support capabilities to establish and verify the claimed identity of any actor in an ATM network.	Authentication
Controlled Access and Authorization	The ATM network shall support capabilities to ensure that actors are prevented from gaining access to information or resources they are not authorized to access.	Access Control
Protection of Confidentiality	The ATM network shall support the capability to keep stored and communicated data confidential.	Access Control/ Confidentiality
Protection of Data Integrity	The ATM network shall support granting the integrity of stored and communicated data.	Access Control/ Integrity
Strong Accountability	The ATM network shall support the capability that an entity cannot deny the responsibility for any of its performed actions as well as their effects.	Non-repudiation
Activity Logging	The ATM network shall support the capability to retrieve information about security activities stored in the Network Elements with the possibility of tracing this information to individuals or entities.	Security Alarm, Audit Trail and Recovery
Alarm Reporting	The ATM network shall support the capability to generate alarm notifications about certain adjustable and selective security related events.	Security Alarm, Audit Trail and Recovery
Audit	The ATM network shall support the capability to analyze and exploit logged data on security relevant events in order to check them on violations of system and network security.	Security Alarm, Audit Trail and Recovery
Security Recovery	The ATM network shall support recovery from successful and attempted breaches on security.	—
Management of Security	The ATM network shall support capabilities to manage the security services derived from the security requirements listed above.	—

controls the nature of the security service to be provided for a given VC". A security association is required for every service security service. Authentication, confidentiality, and data integrity are examples of security service.

The negotiation of services with a peer security agent takes place via *Security Information Exchange*. This exchange could be realized through a number of means. One of the options is the Security Message Exchange (SME) protocol. The protocol could be *signalling-based* and realized through the extension of UNI [ATMF UNI 4.0 SEC] and PNNI protocol [ATMF PNNI 1.0 SEC] for transfer of security information (see Figure 15.2). It may seem that there have been significant enhancements done in these two protocols to support security. However, all they do is to add the Security Services Information Element (SSIE) information element in all messages. This IE is passed transparently to the receiving SA without any modification. Another option for SME is Inband where the information is carried over the data channel that is pre-established or in the process of being established. After the security association is established, the OAM security cells could be used to maintain the state of the security associations. The role of these special cells is to exchange additional connection session keys for confidentiality and data integrity as needed (note that the initial session keys are established during SME and not through the OAM cells) and also to maintain synchronization of encryption engines relying on the counter mode of operation.

Figure 15.2

Extensions to UNI and PNNI for Security

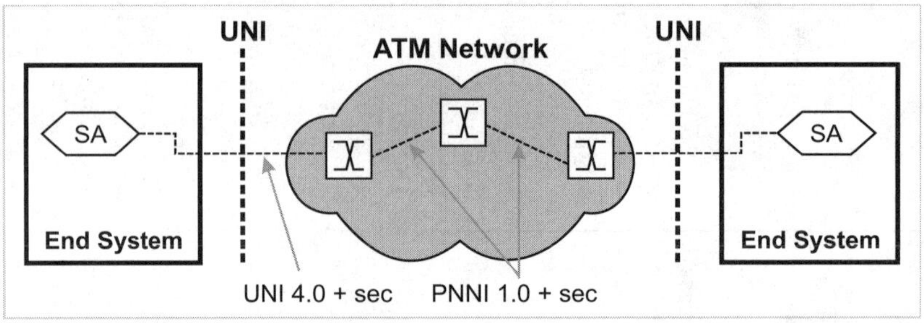

It follows from above that the key functions of the Security Agent is to interact with a peer security agent to perform the Security Message Exchange, provide the actual authentication, integrity check, and confidentiality services, determine when and how security must be applied, and other functions such as OAM cell processing.

15.5 User Plane Security

In the user plane, the ATM Forum specification [ATMF SEC SPEC 1.1] specifies support for identity authentication, confidentiality, data origin authentication and data integrity and access control. Each of the service is realized through a specific mechanism. As an example, Figure 15.3 shows the reference model for authentication support in the user plane. A number of symmetric and asymmetric techniques could be used for the actual authentication, some of which are RSA algorithm, DES algorithm, Triple DES algorithm and digital signature.

FIGURE 15.3

Reference Model for Authentication Support in the User Plane

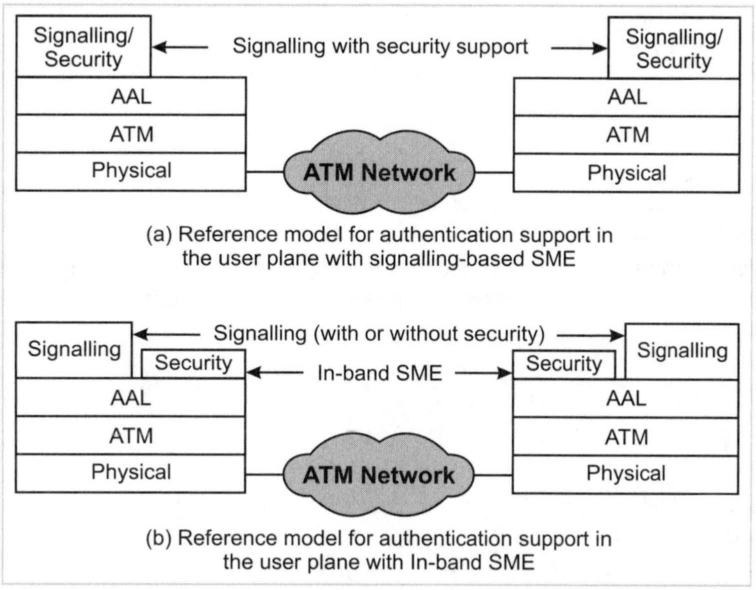

Similarly, for each of the other service including confidentiality, data origin authentication and data integrity and access control a reference model along with the supported algorithms is defined in [ATMF SEC SPEC 1.1].

15.6 Control Plane Security

In the control plane, the ATM Forum specification [ATMF SEC SPEC 1.1] specifies support for data origin authentication and data integrity (no confidentiality). This service is realized through the use Message Authentication Code (MAC) and is same as that used in the control plane. The MAC is appended to the SAAL messages exchanged between two signalling entities (Figure 15.4). The two ends of the signaling channel have a pre-configured shared secret master key and initial session keys that is used for computing MAC. The session keys can be updated by using the OAM cells.

FIGURE 15.4

Reference Model for Data Integrity Support in the Control Plane

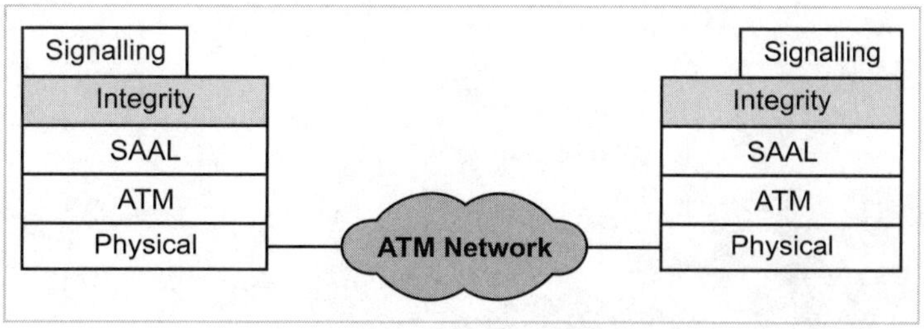

Additional mechanisms for providing security in control plane are provided in [ATMF SEC C-PLANE].

Conclusion

This chapter provided a brief overview of security management in ATM. The concept is relatively new to ATM and it is to be ascertained whether the techniques so defined are being used or not. ATM Forum has concentrated on using the existing algorithms rather than developing new algorithms. The focus is thus on providing a framework where existing algorithms can be reused.

Review Questions

Q 1. What are the threats to network security? What are the means to tackle these threats?
Q 2. What are the objectives of ATM network security?
Q 3. What are the requirements for ATM network security? How are they addressed?
Q 4. What are the functions of a security agent? Describe the techniques used to establish security association between two security agents.
Q 5. What are the security features supported in the user plane? How are they realized?
Q 6. What are the security features supported in the control plane? How are they realized?

Further Reading

[Secu W. Stallings] is an excellent reference for learning more about cryptography and related topics. [Secu Kahate], [Secu W. Stallings 2] and [Secu R. Bragg] also provide an overview on the subject. For ATM, since security is a relatively new concept, there is inadequate literature on the subject. However, some specifications from the ATM Forum do exist. The basic document is the ATM security framework specification [ATMF SEC FW 1.0]. This only provides the objectives and requirements for ATM security. The exact means are provided in [ATMF SEC SPEC 1.1]. Additional mechanisms for providing security in the control plane are provided in [ATMF SEC C-PLANE].

PART 4

Interworking with ATM

The previous nine chapters elaborated upon various facets of the ATM technology, including the ATM traffic/service parameterization, traffic management features, switching architectures, addressing, signalling capabilities, routing techniques, AAL2 signalling, network management standards and security aspects. This part of the book uses the underlying concepts, as elucidated in the previous chapters, to explain the interworking of ATM in various environments including LAN, MAN and WAN.

Talking about interworking, an important fact that must be kept in mind is that migration towards ATM is a long and steady process. Thus, in the early phases of deployment of ATM technology, ATM must interoperate with existing link-layer protocols (e.g. the Ethernet) and network-layer protocols (e.g. IP). For successful interoperability, it is mandatory to have well-defined standards that govern such interoperability.

In an attempt to provide well-defined standards, ATM Forum and IETF have approved a number of interworking standards like LAN emulation or LANE [ATMF LUNI 2], Classical IP over ATM [RFC 2225] and Multiprotocol over ATM or MPOA [ATMF MPOA 1.1]. Not only are these standards explained in this part, the motivation behind the approval of the standards is also explained.

LAN emulation provides a means for traditional LAN technologies like Ethernet and Token ring to co-exist with ATM LANs in a heterogeneous LAN environment. Using LAN emulation, two non-ATM end-systems can communicate not only with each other, but also with end-systems directly attached to an ATM network. ATM in LAN environment and LAN Emulation (LANE) are explained in Chapter 17.

Classical IP over ATM provides mechanisms to transfer IP datagrams over ATM networks. This standard also defines the means for address resolution. This technique is explained in Chapter 18.

Another technique for IP over ATM is Multiprotocol over ATM (MPoA). MPoA uses LAN emulation to techniques and the Next Hop Resolution Protocol (NHRP) to provide the means to establish a direct VC connection between end-systems belonging to different sub-nets. The MPOA and NHRP techniques are explained in Chapter 18.

Before explaining key protocols like LANE, Classical IP over ATM, NHRP and MPOA, Chapter 16 first describes the ATM network architecture and important interfaces defined for interworking. Some of the important interfaces discussed in Chapter 16 include ATM Inter-Network Interface (AINI), B-ISDN Inter-Carrier Interface (B-ICI), Data Exchange Interface (DXI), Frame-based UNI (FUNI), Frame-based ATM over SONET/SDH Transport (FAST) and Frame-based ATM Transport over Ethernet (FATE).

Chapter

16

ATM Network Interfaces and Architecture*

16.1 Introduction

This chapter presents a view of ATM network components and interfaces. The ATM network interfaces are of different types (see Figure 16.1). Prominent among these is the User-Network Interface (UNI). Two variations of UNI exists, namely private UNI and public UNI. The private UNI is the interface between an ATM end-system and an ATM switch—both of which lie completely within a private ATM network. A public UNI provides the means for a private ATM network to interface with a public network. Public UNI can also be an interface between an ATM end-system and a switch belonging to a public ATM network. The prominent difference between public and private UNI pertain to the nature of physical interfaces supported, administrative control operations (like traffic policing) and management complexity. The UNI signalling procedures were discussed in Chapter 11.

As UNI provides an interface between an ATM end-system and an ATM-switch, Network-Node Interface (NNI), also known as Network-Network Interface, provides an interface between two ATM switches. NNI can again be private NNI (i.e. the switches lie within a private ATM network) or public NNI (i.e. the switches lie within a public ATM network). The differences between public and private NNI are similar in flavour to those between public and private UNI. The private NNI interface is also referred to as PNNI interface. The PNNI signalling and routing procedures were discussed in Chapter 12. The UNI and PNNI interface is not detailed any further in this chapter.

There is another specification called the ATM Inter-Network Interface (AINI) that is based on PNNI. The goal of AINI protocol is to facilitate interworking of two networks running

* This chapter is contributed by Nishit Narang.

FIGURE 16.1

ATM Interfaces

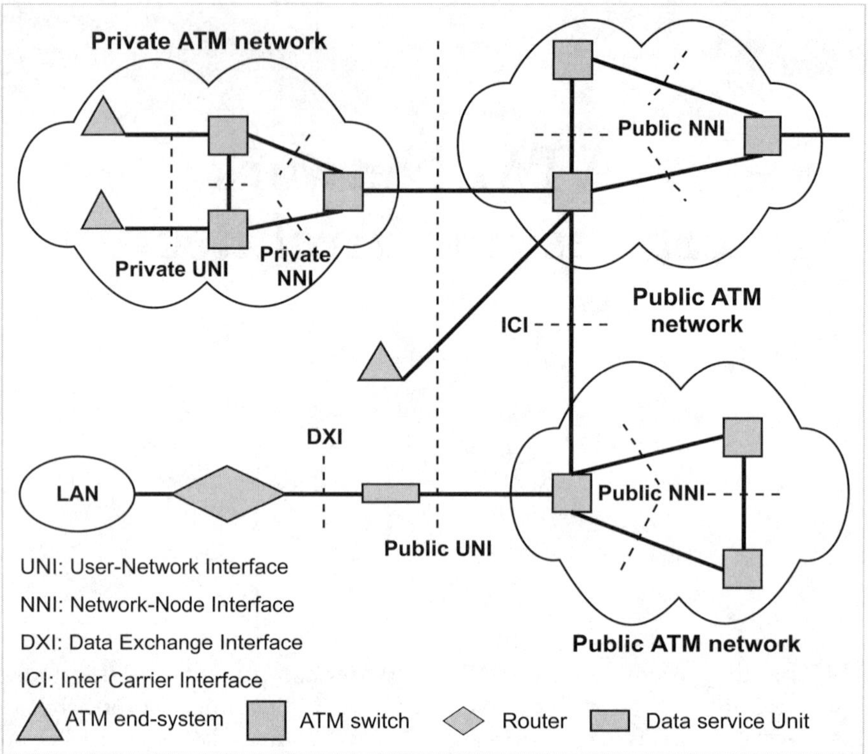

PNNI internally in disjoint PNNI routing domains. Thus, AINI can be viewed as inter-domain routing protocol akin to Border Gateway Protocol (BGP) used in Internet. AINI is very briefly explained in this chapter.

Besides UNI, PNNI and AINI, the B-ISDN Inter-Carrier Interface (B-ICI) is the interface between two public ATM networks. Standardization of B-ICI helps in achieving interoperability between two public ATM networks belonging to different carriers. B-ICI and its multi-service aspects are explained in this chapter.

While the UNI, NNI and B-ICI define a cell-based interface between the network entities involved, ATM 'cell tax' issues lead to the specification of frame-based ATM interfaces for providing data services. Amongst the initial specifications in this regard was the Data Exchange Interface (DXI). DXI defines a standardized mechanism to interface between frame-based devices (e.g. IP-based routers) and ATM network elements. The DXI specifications were later evolved to a more elaborate specification, called Frame-based UNI (FUNI). FUNI can be considered as an enhancement on the DXI, and in fact, the two share a lot of commonalities. Both DXI and FUNI are elaborated in this chapter.

Besides the DXI and the FUNI specifications, two other specifications were aimed at standardizing frame-based interface between network elements supporting conventional transmission techniques like Ethernet and SONET. These specifications were called Frame-based ATM over SONET/SDH Transport (FAST) and Frame-based ATM Transport over Ethernet (FATE). As the names suggest, FATE defines a mechanism to carry frame-based ATM traffic over Ethernet, while FAST defines the same for SONET/SDH. FAST and FATE are explained towards the end of this chapter.

16.2 ATM Inter-Network Interface (AINI)

The AINI specification [ATMF AINI 1.1] defines the signalling procedures for dynamically establishing and clearing the ATM connections between ATM networks. The networks on either side of AINI are assumed to be running PNNI internally, and assumed to fall in disjoint PNNI routing domains. In some sense, the AINI specification can be considered as a very thin addendum to the PNNI specification, which also defines the protocol interworking between AINI and PNNI. The AINI interworking scenario with PNNI is depicted in Figure 16.2. For details of AINI, refer to [ATMF AINI 1.1].

Figure 16.2

AINI Interworking Scenario

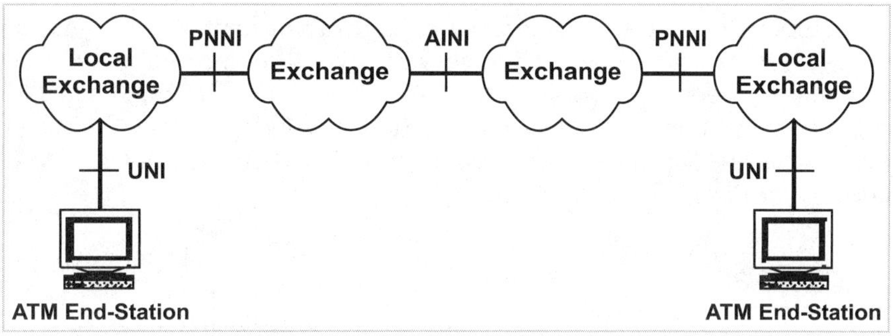

16.3 B-ISDN Inter-Carrier Interface (B-ICI)

To provide end-to-end ATM services spanning regional and national boundaries, it is required that ATM-based public networks belonging to different carriers be interconnected with each other. To accomplish this need, the B-ISDN Inter-Carrier Interface (B-ICI) specification is defined in [ATMF BICI 2.0]. The B-ICI specification is intended as an implementation agreement between ATM public network providers, which when followed, will lead to interoperability between the carriers. The agreement spans the following aspects of the ATM network:

- ATM physical layer characteristics (includes the supported physical layer options like 155.520 Mbit/s STS-3c, 622.080 Mbit/s STS-12c and 44.736 Mbit/s DS3, their frame structure, HEC handling rules, synchronization, timing and jitter aspects, and other physical layer issues),

- ATM layer characteristics (includes cell structure, encoding/decoding rules, and ATM layer functions like cell rate decoupling, cell discarding and loss priority indication among others),
- Traffic management and performance aspects (includes performance considerations, traffic contract management, connection admission control, network parameter control, priority control and explicit forward congestion indication),
- Operations and maintenance aspects (includes physical layer network management, ATM layer management and OAM functions),
- Signalling aspects (includes the signalling network configuration, signalling protocol architecture, point-to-point signalling and point-to-multipoint signalling), and
- Various PVC-based services (cell-relay services, circuit emulation service, frame-relay services, and switched multi-megabit data service).

Strictly speaking, [ATMF BICI 2.0] is a compliance specification that mentions how various ATM specifications are applicable for B-ICI interfaces. By itself, it does not introduce new procedures. Thus, for example, [ITU-T I.150] defines the ATM layer functions. The [ATMF BICI 2.0] only defines which ATM layer functions are applicable over B-ICI interface and in what manner.

Figure 16.3 depicts the reference configuration for the B-ICI. As depicted in the figure, the B-ICI interface can be used to connect two different public ATM networks belonging to different carriers via a public transit ATM network. The B-ICI can also be used to directly connect the public ATM networks, without involving a transit network in between. This is depicted in the figure as a direct connection between the public ATM networks of carrier A and carrier B.

FIGURE 16.3

B-ICI Reference Configuration

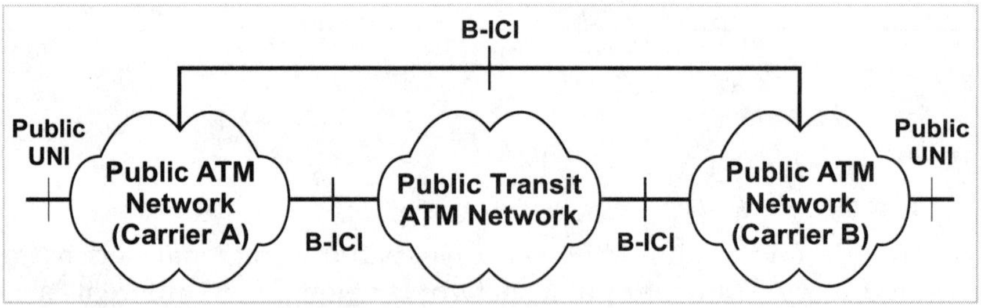

16.3.1 B-ICI and NNI

While the public NNI defines the interface between two ATM switches within the same public ATM network, the B-ICI defines the interface between two ATM switches in two different public ATM networks, belonging to different carriers. However, there are a few similarities between the NNI and the B-ICI specifications. Figure 16.4 depicts the relation between the NNI and the B-ICI interfaces.

FIGURE 16.4

Relation Between NNI and B-ICI

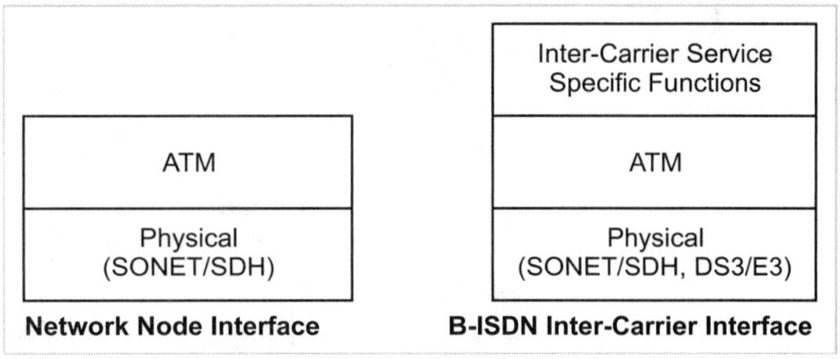

The physical layer of the B-ICI protocol stack is based on the NNI specifications defined by the ITU-T. However, in addition to the SONET/SDH-based physical layer defined in the NNI specifications, B-ICI specifications allow the usage of the DS3/E3-based physical layer. The decision to allow both electrical and optical media was taken in order to allow carriers to meet their near-term and long-term business requirements.

While the NNI specifications do not define anything above the ATM layer, the B-ICI specifications specify the need to have inter-carrier service specific functions above the ATM layer. These functions are required to transport, operate and manage a variety of inter-carrier services across the inter-carrier interface. Examples of such services include Circuit Emulation Service (CES), Frame Relay Service (FRS) and Switched Multi-megabit Data Service (SMDS). A description of the multi-service aspects of the B-ICI is provided in the following section.

16.3.2 Multi-service B-ICI

As mentioned in the previous section, multi-service B-ICI supports inter-carrier services such as Cell Relay Service (CRS), Circuit Emulation Service (CES), Frame Relay Service (FRS) and Switched Multi-megabit Data Service (SMDS). This is depicted in Figure 16.5. Inter-carrier service is a sequence of the following three steps:

1. Accepting service-specific traffic at the ATM switch. The service-specific traffic consists of ATM cells (for CRS), DSn Frames (for CES), HDLC Frames (for FRS) and L3_PDUs (for SMDS).
2. Converting the service-specific traffic into standard 53-octet ATM cell. The conversion is not required for CRS traffic, since the service-specific traffic is already in the form of ATM cells.
3. Transport of the ATM cells carrying service-specific information across the inter-carrier interface over SONET/SDH or DS3/E3.

Table 16.1 describes the conversion function required to convert service specific traffic into standard 53-octet ATM cells, before transmission over the inter-carrier interface. As described in Table 16.1, AAL forms one of the components of the inter-carrier service-specific functions depicted in Figure 16.4.

FIGURE 16.5

Multi-service B-ICI

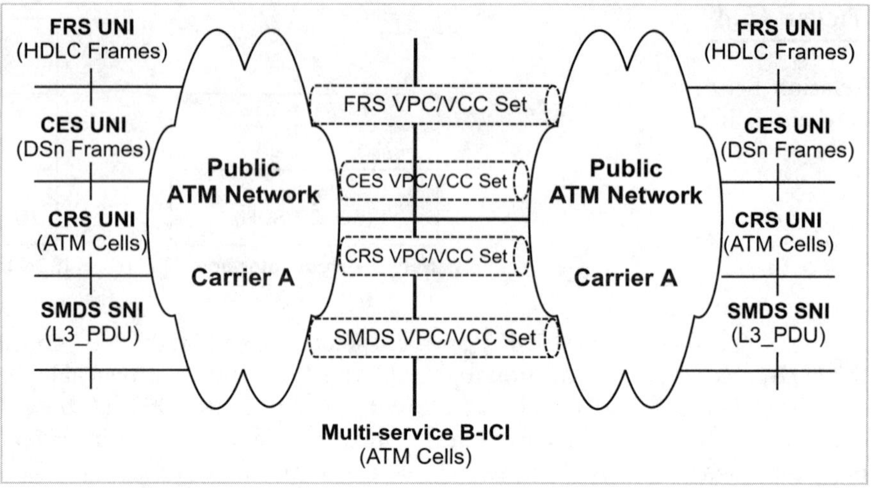

TABLE 16.1

Conversion of Service Specific Traffic

Service	*Traffic format*	*Description of the conversion function*
CRS	ATM cells	No conversion required.
CES	DSn frames	CES DSn frames are encapsulated in AAL type 1 PDUs, and then mapped to ATM cells.
FRS	HDLC frames	FRS HDLC Frames are encapsulated in AAL type 5 PDUs and then mapped to ATM cells.
SMDS	L3_PDUs	L3_PDUs encapsulated into Inter-Carrier Interface Protocol (ICIP) Connectionless Service (ICIP_CLS) PDUs, and then in AAL type 3/4 PDUs, and then mapped to ATM cells.

Figure 16.5 depicts a multi-service B-ICI connecting two public ATM networks, belonging to Carrier A and Carrier B. On the B-ICI, traffic from different services, after conversion, is mapped to different ATM VPC/VCCs. Thus, the B-ICI in the figure is shown to be transporting four different sets of ATM VPC/VCCs, one set for each category of service.

Figure 16.6 depicts an example ATM network supporting multiple services. As shown in the figure, an interworking function is required to connect different networks (FR, SMDS, CES) to the ATM network. The interworking function includes network interworking (i.e. interworking between different types of network) and service interworking (i.e. interworking between different types of services). However, the service interworking functionality is not considered within the scope of the B-ICI specifications. As an example, for FRS-ATM interworking, the Data Link Connection Identifier (DLCI) are mapped to ATM VCI, the frame are segmented into ATM cells and reassembled at the receiver, there are mapping for congestion indication bits, and other mapping for header contents.

Service-specific functions exist on ATM switches, which may not necessarily be the switch providing the B-ICI interface. In other words, service-specific functions may exist on switches

FIGURE 16.6

Example Network Depicting B-ICI and IWF

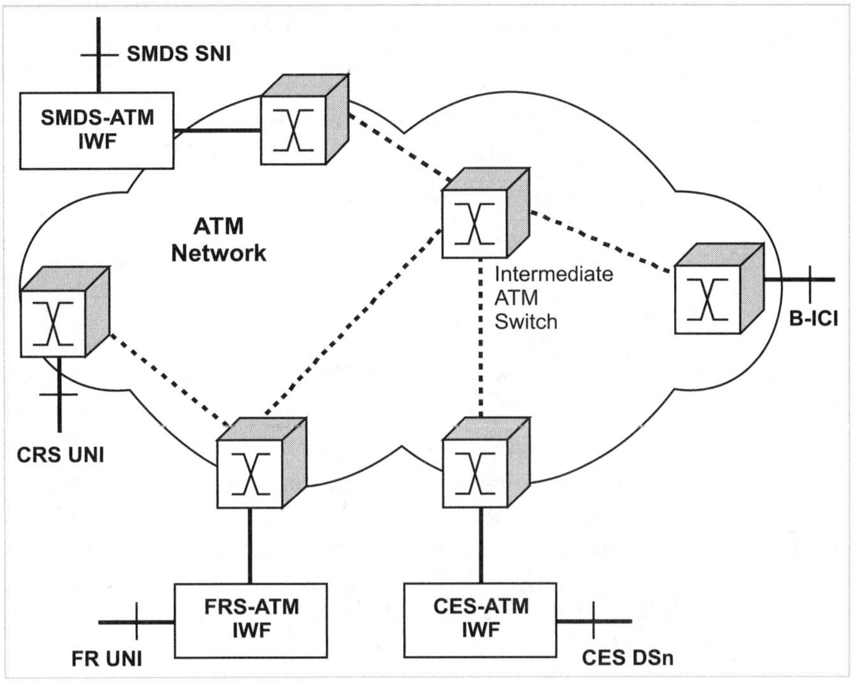

that are physically distant from the switch actually providing the inter-carrier connection (B-ICI). Intermediate ATM switches may exist between the switches providing the service specific functions, and the switch providing the inter-carrier connection (B-ICI interface).

For details on the functionality of the multi-service B-ICI, the reader is referred to the B-ICI specification [ATMF BICI 2.0].

16.4 DATA EXCHANGE INTERFACE (DXI)

The Data Exchange Interface (DXI) was introduced to define a standardized interface between frame-based devices (e.g. IP-based routers) and ATM network elements. Using DXI, it is possible to connect frame-based devices to each other via a high-speed ATM backbone, with minimal upgrade to the software on the frame-based devices.

Figure 16.7 depicts how an IP-based workstation/router can connect to an ATM network via DXI. DXI uses a frame-based communication mechanism over serial interfaces and transfers the SAR functionality to another network element, outside of the frame-based device. A High Speed Serial Interface (HSSI) link connects the IP workstation/router to an external ATM equipment, which is called the Data Service Unit (DSU) or the Channel Service Unit (CSU). The DSU/CSU and the IP workstation/router exchange DXI frames with each other. The DSU/CSU performs the task of segmentation of the DXI frames to ATM cells, before transmission on the UNI, to the ATM network. In the reverse direction, the DSU/CSU re-

FIGURE 16.7

Data Exchange Interface (DXI)

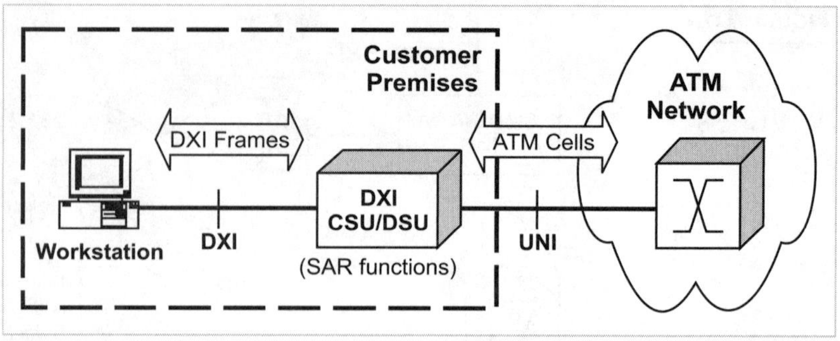

assembles the ATM cells received over the UNI to form DXI frames, before transmission to the IP workstation/router.

Since most IP-based routers support the HSSI interface, and since conversion to/from DXI frames does not require much upgrade in the software, DXI is one of the most suitable mechanism to connect existing IP routers to each other via an ATM-based backbone. In the specifications, the term Data Terminal Equipment (DTE) is used to denote frame-based devices, such as routers, and the term Data Circuit-Terminating Equipment (DCE) is used instead of DSU/CSU. The next few sections also use this terminology.

The DXI is defined in the [ATMF DXI] standard.

16.4.1 Modes of Operation

The DXI specifications define three different operating modes: Mode 1a, Mode 1b and Mode 2. A comparison between the three modes of operation is provided in Table 16.2. Each of these is further elaborated in the following subsections.

TABLE 16.2

Modes of Operation of DXI

Mode	*# of virtual connections*	*AAL supported*	*Service data unit size*	*FCS mechanism*
Mode 1a	Up to 1023	AAL5 only	Up to 9232 octets	16-bit FCS
Mode 1b	Up to 1023	AAL3/4 on at least one VC	Up to 9224 octets (for AAL3/4)	
		AAL5 for other VC	Up to 9232 octets (for AAL5)	16-bit FCS
Mode 2	Up to16,777,215	AAL3/4 and AAL5	Up to 65,535 octets	32-bit FCS

16.4.1.1 DXI Mode 1a

Mode 1a of operation of DXI supports up to 1023 virtual connections between the DTE and the DCEs. Since the DTE are frame-based devices, only AAL5 is supported for packet data transfer. The DTE transforms a Service Data Unit (SDU) to be transported across the ATM network, into an ATM DXI frame, and sends it to the DCE. The DCE then performs the functionalities of the AAL5 CPCS, AAL5 SAR and ATM layer. The processing at the DTE and the DCE is as depicted in Figure 16.8.

FIGURE 16.8

Mode 1a Processing for AAL5

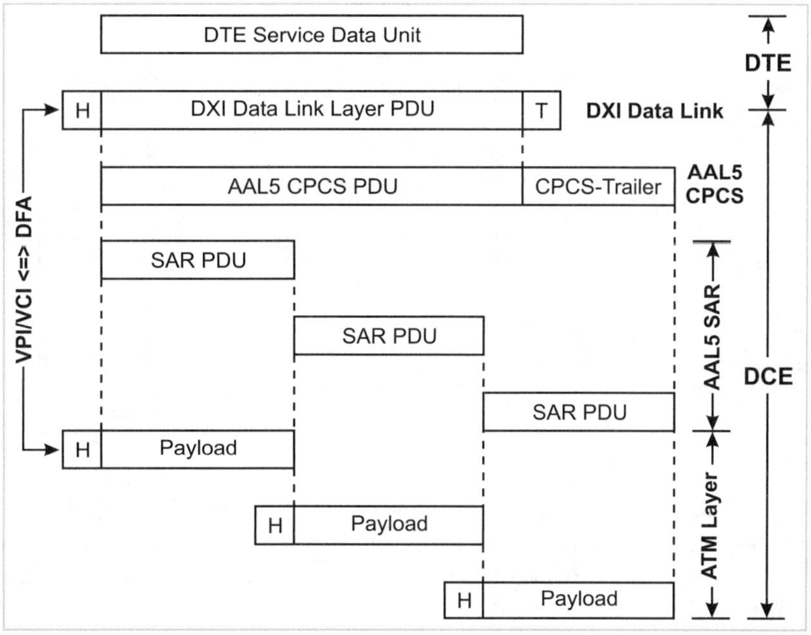

The DXI data link layer PDU consists of a Header (H) and Trailer (T). The header of the DXI data link layer PDU contains a DXI Frame Address (DFA), which maps to the VPI/VCI value to be filled in the ATM cell header. In the reverse direction, the VPI/VCI value received in the ATM cell header is used to fill the DFA in the DXI data link layer PDU. Since the DFA is only 10 bits in Mode 1a of operation, only a maximum of 1023 virtual connections can be supported. The rest of the functions performed in the DCE are those of the AAL5 and the ATM layer.

16.4.1.2 DXI Mode 1b

The Mode 1b of operation of DXI is similar to Mode 1a, with the addition of AAL3/4 as the alternate AAL that can be used in Mode 1b. When using AAL5, Mode 1b of operation is same as that of Mode 1a (see Figure 16.8). However, Mode 1b defines that at least one virtual circuit must exist between the DTE and the DCE that supports AAL3/4. For AAL3/4, the DCE includes the additional functionality of AAL3/4 CPCS, while the DTE performs the AAL3/4 SAR and ATM layer functionality. Note that in mode 1a, there is no AAL related functionality at the DCE. The processing at the DTE and the DCE for AAL3/4 is as depicted in Figure 16.9.

FIGURE 16.9

Mode 1b Processing for AAL3/4

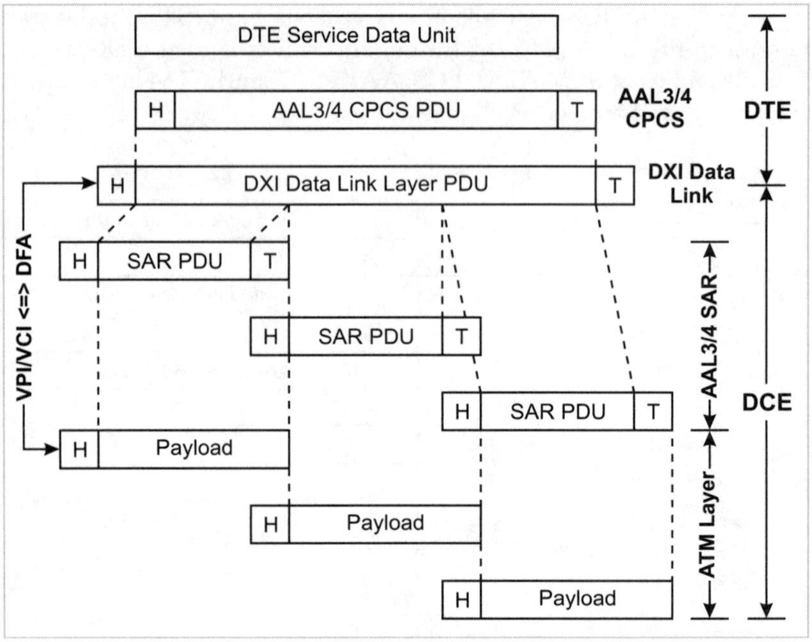

Just like Mode 1a, Mode 1b uses the DFA in the DXI data link layer PDU header to map to the VPI/VCI value in the ATM cell header, and vice versa. Mode 1b also defines a 10-bit DFA value, thus limiting the maximum number of virtual circuits that can be supported to 1023.

16.4.1.3 DXI Mode 2

Just like the Mode 1b of operation, Mode 2 also supports both AAL5 and AAL3/4. However, the processing for AAL5 at the DTE and the DCE in case of Mode 2 of operation is different from the processing done for Mode 1b. In Mode 2 of operation, the DTE always performs the AAL3/4 CPCS encapsulation, irrespective of whether AAL5 or AAL3/4 is to be used on the ATM virtual circuit. The DCE then verifies whether the virtual circuit supports AAL3/4 or AAL5. In case AAL3/4 is used, the processing at the DCE is similar to the processing done for Mode 1b. However, if AAL5 is used on the ATM virtual circuit, the DCE strips off the AAL3/4 CPCS header and trailer, and then performs the AAL5 CPCS, AAL5 SAR and ATM layer functions. The DXI Mode 2 processing for AAL3/4 is same as that for Mode 1b (see Figure 16.9).

Like the Mode 1a and Mode 1b of operation, Mode 2 also uses the DFA in the DXI data link layer PDU header to map to the VPI/VCI value in the ATM cell header, and vice versa. However, Mode 2 defines a 24-bit DFA value, thus increasing the limit on the maximum number of virtual circuits that can be supported to 16,777,215.

16.4.2 DXI Protocol Architecture

As discussed above, there are three modes of operation. For modes 1a and 1b, where AAL5 is used, the protocol stack used is depicted in Figure 16.10. On the DXI interface, there is a physical layer and a data link layer. For the physical layer between the DTE and the DCE, [ATMF DXI] define different options including V.35, EIA/TIA 449/530 and EIA/TIA 612/613 (HSSI). The HSSI specification (EIA/TIA 612/613) defines the concept of clock 'gapping', which can be used for flow control of data to the DCE. Clock rates between the DTE and the DCE are left open as an implementation option. At the DXI data link layer, a frame-based structure is defined, that has start and end flags, a header field (containing the DFA) and the frame check sequence.

FIGURE 16.10

Modes 1a and 1b Protocol Architecture for AAL5

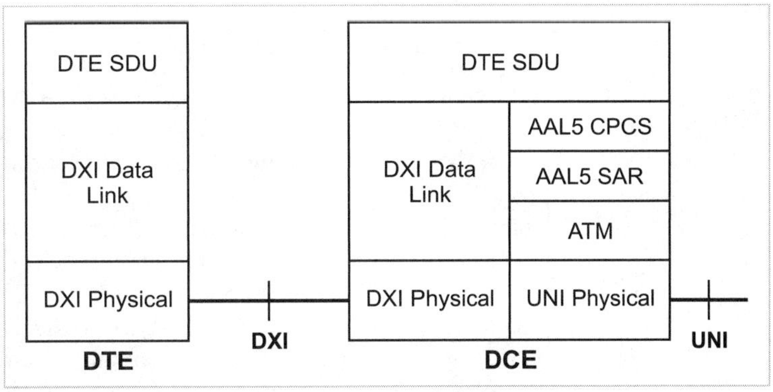

When the AAL3/4 is used, the AAL3/4 CPCS layer also resides at the DCE above the DXI data link layer.

16.4.3 Local Management Interface

The ATM DXI Local Management Interface (LMI) defines the protocol for exchange of management information across the DXI. Management information exchanged using LMI includes DXI-specific information, AAL-specific information and ATM UNI-specific information.

Figure 16.11 depicts the DXI Local Management Interface (LMI) configuration. The DXI LMI is designed to support both a network management station running SNMP, and a switch running ILMI. It is assumed that the DTE will include an SNMP proxy agent and an ILMI proxy agent, which determine when to query the DCE in regards to a SNMP/ILMI request received from the network management station/ATM switch.

The DXI LMI defines five Protocol Data Unit (PDU) types. These PDU types are described in Table 16.3. These PDU types are quite similar to the messages used for SNMP. The DXI LMI MIB defines objects that are required to support global management mechanisms, such as

FIGURE 16.11

DXI Local Management Interface (LMI)

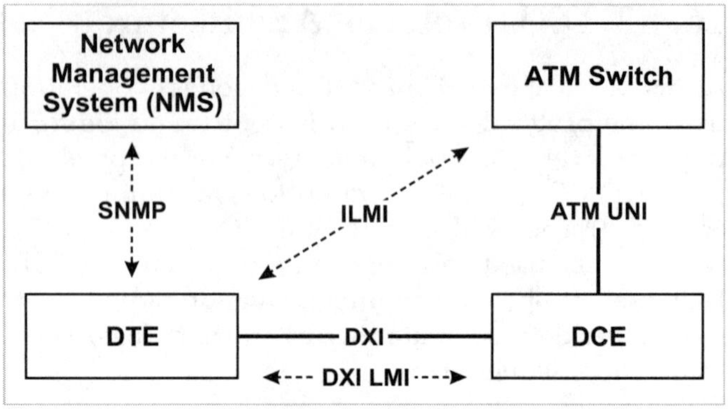

TABLE 16.3

LMI Protocol Data Unit (PDU) Types

PDU Type	*Description*
GetRequest	Originated from the DTE, this is used to fetch value(s) of management object(s) from the DCE. The management objects are defined by the DTE using the object identifiers.
GetNextRequest	Originated from the DTE, this is used to fetch value(s) of management object(s) that occur after a particular object in the MIB.
SetRequest	Originated from the DTE, this is used to modify the value of a management object in the DCE.
GetResponse	Originated from the DCE, this is used to respond to a GetRequest, GetNextRequest or SetRequest message from the DTE.
Trap	Originated from the DCE, this is sent autonomously to the DTE in alert conditions.

SNMP and ILMI. Physical layer MIBs is also supported. Further, vendors of DXI can extend the DXI LMI MIB as more specifications become standardized.

16.5 FRAME-BASED UNI (FUNI)

The Frame-based UNI (FUNI) specification can, in some sense, be considered as an evolution of DXI. Both FUNI and DXI use a frame-based communication mechanism over serial interfaces to connect with existing frame-based devices. However, while DXI is merely a protocol for the local interface within the customer premises, FUNI is a true user-network interface protocol. FUNI can easily be used across wide-area links, and includes management capabilities and functionality for SVC signalling. Another difference between FUNI and DXI arises in the location of the SAR functionality. While DXI places the SAR functionality in a DSU/CSU located within the customer premises, FUNI pushes this function to within the ATM network, as part of the ATM switch functionality (see Figure 16.12). This helps FUNI to solve an additional problem, as discussed in the following paragraph.

FIGURE 16.12

Frame-based UNI (FUNI)

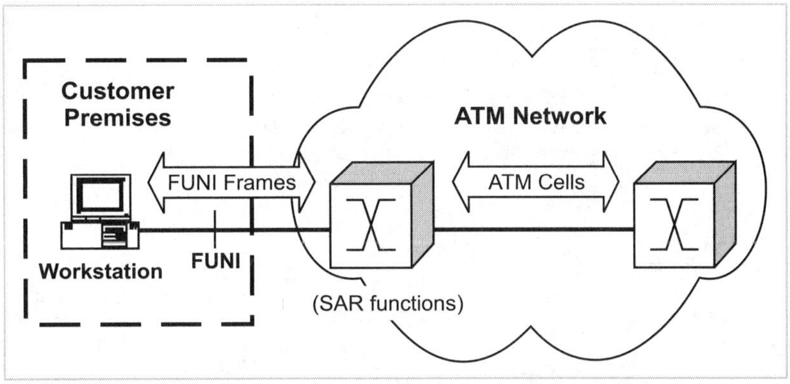

The fixed length small size (53 bytes) ATM cells help in reducing queuing delays at switches and improving the overall processing efficiency. This makes ATM extremely suitable for carrying voice traffic. However, when carrying large variable-sized data payloads, the small size ATM cells do not offer the most efficient solution. In fact, ATM has quite often been criticized for its 'cell tax', which becomes noticeable when carrying data.

ATM cells include a 5-byte header to a 48-byte payload. This means that if 4800 bytes of higher layer packet data were to be transported using ATM, the upper layer packet data would have to be segmented into 100 smaller ATM payloads of 48 bytes each. Each of these payloads would require a separate 5-byte header. Thus, transporting 4800 bytes of higher layer packet data would require 500 bytes of header overhead. This is indeed quite a high overhead when compared to some of the other virtual-circuit based technologies, such as Frame Relay.

While both DXI and FUNI require that variable-sized frames be segmented into 48-byte ATM payloads before transmission over the ATM network, the placement of the SAR functionality provides the difference between the two. Since DXI requires that the SAR functionality be executed within the customer premises, the lower access speeds coupled with the ATM header overheads leads to lower effective user data rates. By pushing the SAR functionality to the edge ATM switch within the ATM network, FUNI ensures that the header overhead does not come into the picture when transmitting over the lower bandwidth (narrowband) access links.

The FUNI is defined in the [ATMF FUNI 2.0] standard. Apart from this, there is the [ATMF FUNI MSE] defined for carrying real-time and multimedia applications (e.g., voice) over a FUNI interface. However, the latter has not been detailed in this chapter.

16.5.1 FUNI Reference Model

The reference model for FUNI is depicted in Figure 16.13. On the FUNI interface, the FUNI specification defines a new data link layer protocol, which is similar to the DXI data link layer protocol. Just like in DXI, FUNI defines four different modes of operation: Mode 1a, Mode 1b, Mode 3 and Mode 4. The details of each of these modes of operation are omitted, and the reader is referred to [ATMF FUNI 2.0] for details. The encapsulation/decapsulation for the AAL5 FUNI is similar to the AAL5 processing for Mode 1a in DXI (see Figure 16.8). Similarly, the encapsulation/decapsulation for the AAL3/4 FUNI is similar to the AAL3/4 processing for Mode 1b in DXI (see Figure 16.9). The only difference is that the DXI data link layer frame is replaced with a FUNI data link layer frame.

FIGURE 16.13

FUNI Reference Model

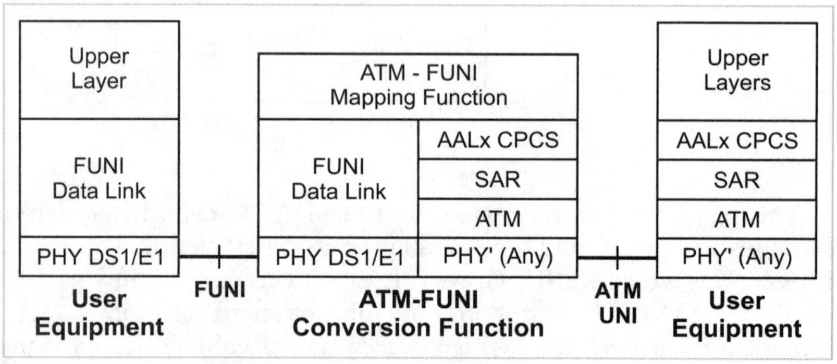

The physical layer supported over the FUNI interface is DS1 or E1. The user's FUNI payload may be carried by using full DS1/E1 line bandwidth, or a fraction thereof. The specifications treat any fractional DS1/E1 bandwidths as a single FUNI interface. However, the multiplexing/demultiplexing of multiple such FUNI interfaces within the same DS1/E1 physical layer is left outside the scope of the FUNI specifications. AAL5 is the adaptation layer used over the FUNI interface. However, the specifications allow implementations to optionally support AAL3/4 over the FUNI interface.

16.5.2 Signalling and ILMI Over FUNI

FUNI specifications define a mechanism for UNI-based SVC signalling over the FUNI interface. The protocol stack for SVC signalling is depicted in Figure 16.14. The signalling stack on the FUNI interface consists of the UNI Q.2931, SSCF and SSCOP stack elements over the FUNI data link layer using AAL5. Since the SSCOP stack entity assumes AAL5 as its lower layer, [ITU-T Q.2119] specification defines a convergence function for mapping the SSCOP messages to the FUNI data link layer. The VPI/VCI values for the signalling channel are kept consistent with the UNI 4.0 specifications (default to VPI = 0 and VCI = 5). The ATM signalling terminating end is connected to the ATM-FUNI conversion function via the ATM–

FIGURE 16.14

Signalling Over FUNI

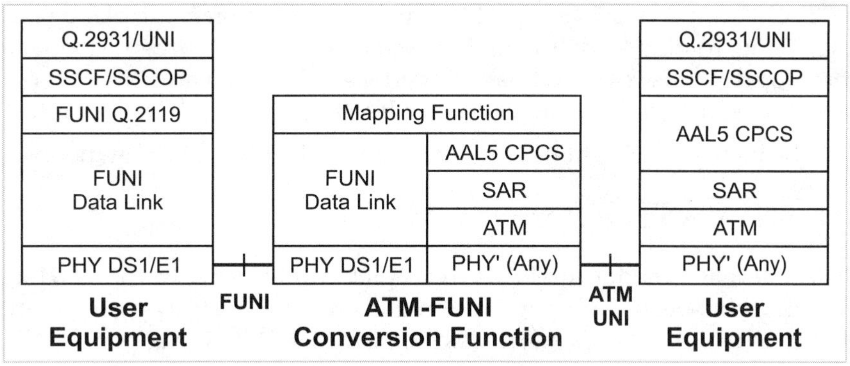

UNI interface. It is transparent of the fact that the user equipment used a FUNI interface for its SVC signalling.

Similar to the signalling over FUNI shown in Figure 16.14, there is a protocol stack for ILMI over the FUNI interface. The UNI/Q.2931 and SSCF/SSCOP stack entities are replaced with the stacks for SNMP/ILMI. [ITU-T Q.2119] defines the conversion function for SNMP/ILMI over the FUNI Data Link Layer using AAL5.

16.6 FRAME-BASED ATM OVER SONET/SDH TRANSPORT (FAST)

In order to overcome the inefficiencies within ATM due to 'cell-tax', the DXI and FUNI specifications define a frame-based access interface to an ATM network. The approach used in DXI and FUNI is to address the issue of 'cell-tax' when operating over narrowband access speeds. When operating at broadband speeds, however, the 'cell-tax' of ATM can be tolerated, since there is surplus bandwidth. Hence, this was not one of the areas of concern when DXI and FUNI specifications were defined. However, later it was realized that with higher broadband speeds (speeds greater than 622 Mbps/OC-12 speeds), cells were in fact no longer required to meet the latency requirements of applications, and frame-based communications would work well even for real-time applications. This led to the definition of an approach to carry frame-based ATM over broadband speeds, as provided by SONET/SDH.

Frame-based ATM over SONET/SDH Transport (FAST) defines the mechanism to transport variable length datagrams, called ATM frames, over high-speed SONET/SDH transmission facilities. Using FAST, up to 65535 bytes of data can be transmitted using only four bytes of header information, thus significantly reducing the header overhead.

FAST is very similar to the DXI and FUNI specifications, except for two differences. Firstly, DXI and FUNI interfaces are designed to operate over low-speed plesiochronous transmission facilities. However, FAST is designed to operate over high-speed SONET/SDH facilities. Secondly, DXI and FUNI specifications provide access to an ATM network. However, FAST may be used to provide inter-switch trunking, in addition to providing network access.

FAST supports existing ATM control and management plane functions, which include signalling, routing, addressing, ILMI and OAM. This ensures inter-operability with capabilities developed for use over cell-based interfaces. Both UNI and NNI interfaces are defined by the FAST specification, and are called the FAST-UNI and FAST-NNI interfaces.

The FAST specifications are defined in the [ATMF FAST] standard.

16.6.1 FAST Reference Model

The reference model for FAST is depicted in Figure 16.15. FAST-UNI is defined as the interface between the User Equipment and the ATM network node, while the interface between two ATM network nodes supporting FAST is called FAST-NNI.

FIGURE 16.15

FAST Reference Model

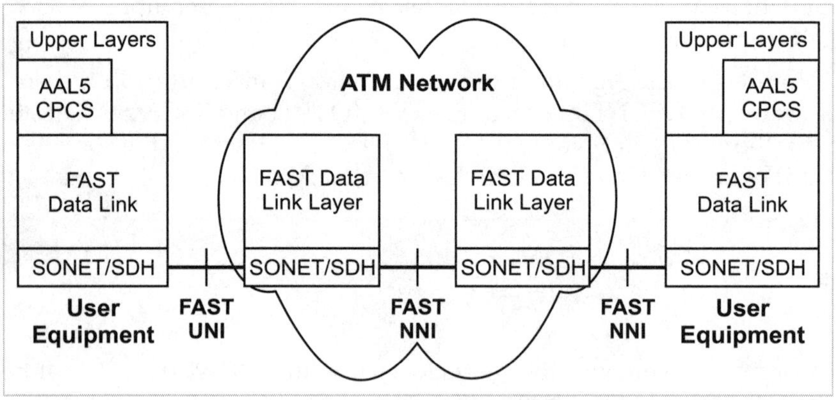

The physical layer on the FAST-UNI and FAST-NNI interfaces is SONET/SDH. Above this layer, FAST defines a complete data link layer protocol, which provides services to the higher layers. The FAST data link layer frame format is a simplified version of the frame structure defined in 'PPP in HDLC-like Framing' [RFC 1662]. FAST omits the address, control, protocol and padding fields defined in [RFC 1662], and mandates the use of a 32-bit Frame Check Sequence (FCS).

The Upper Layers at the user equipment can access the FAST Data Link Layer directly, or via the modified AAL5 CPCS layer. The interface provided by the AAL5-CPCS to the upper layers is not modified, and is as specified by specification I.363.5. However, the AAL5-CPCS lower layer interface is modified to accommodate the FAST Data Link Layer, rather than the AAL5 SAR layer.

FAST supports two modes of operation: Mode 0 and Mode 1. At a very high level, the difference between the two modes is that Mode 1 allows for fragmentation and reassembly of large frames, while Mode 0 does not support fragmentation and reassembly. Communication between the two AAL5 CPCS entities on the user equipments is supported only for mode 1 of operation, and not for mode 0.

16.6.2 FAST Interworking Function

In order to support the establishment of ATM connections across a mixture of FAST, cell-based and FUNI-based interfaces, a FAST Interworking Function (IWF) is required. The FAST IWF provides the mapping between the data link layers in a manner that preserves the ATM layer semantics.

Figure 16.16 depicts three locations where a FAST IWF may be deployed. At the edge of the ATM network, the FAST IWF connects a cell-based UNI to a FAST-NNI. Within the ATM network, the FAST IWF connects a FAST-NNI to a cell-based NNI. At the other edge of the ATM network, the FAST IWF connects a cell-based NNI to a FAST-UNI. Further, a FAST IWF may also be required at the edge of an ATM network to connect a FUNI-based interface with a FAST-NNI interface (not shown in figure). However, this is not considered a new configuration, but considered as a special case of the configuration connecting a cell-based UNI to FAST-NNI interface.

16.6.3 FAST Encapsulation Formats

FAST defines a mechanism to transport variable length ATM frames over high-speed SONET/SDH transmission facilities. Using FAST, up to 65535 bytes of data can be transmitted using only four bytes of header information. Two types of encapsulation formats are defined for use over FAST-based interface. These are frame encapsulation and cell encapsulation. Frame encapsulation supports the transmission of variable length ATM frames within a FAST frame. It supports Virtual Channel Connection (VCC) and all ATM layer service categories using AAL5 as the adaptation layer. The maximum SDU size that can be supported by using frame encapsulation differs for various implementations, but is in the range of 9216 octets to 65535 octets.

Cell encapsulation encapsulates the payload of a single ATM cell within the standard FAST frame. It supports both Virtual Path Connection (VPC) and Virtual Channel Connection (VCC). All ATM layer service categories are supported using cell encapsulation, using any of the AALs.

Specifications mandate support for both frame and cell encapsulation at the FAST-NNI. For FAST-UNI, frame encapsulation is required to be supported for transport of user data. However, a FAST-UNI interface may support the transfer of user data using cell encapsulation. OAM-related information in FAST is always transported using Cell Encapsulation. For more details on the encapsulation formats in FAST, the reader is referred to [ATMF FAST].

16.7 Frame-based ATM Transport over Ethernet (FATE)

Like FAST, Frame-based ATM Transport over Ethernet (FATE) defines a mechanism to provide ATM-based services over an Ethernet medium by transporting ATM data within

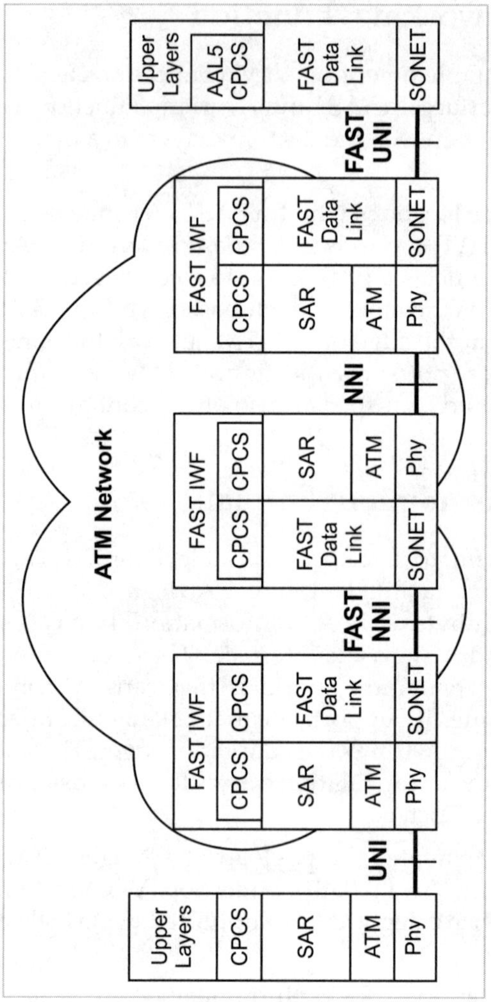

FIGURE 16.16

FAST Interworking Deployment Scenarios

Ethernet Frames. FATE preserves support for ATM control mechanisms such as QoS, signalling and management, and is therefore able to provide the benefit of ATM services. The FATE specification is currently defined only for the UNI, but is expected to be extended to support the NNI as well. The FATE UNI interface is defined in the ATM Forum specification [ATMF FATE].

16.7.1 FATE Reference Model

The FATE UNI reference model is depicted in Figure 16.17. As depicted in the figure, multiple frame-based end-stations connected to an Ethernet LAN communicate with a FATE UNI converter via the FATE-UNI. The FATE UNI converter provides connectivity to an ATM network by implementing the function for conversion and relay of the traffic received over the FATE-UNI to the UNI and vice versa.

FIGURE 16.17

FATE Reference Model

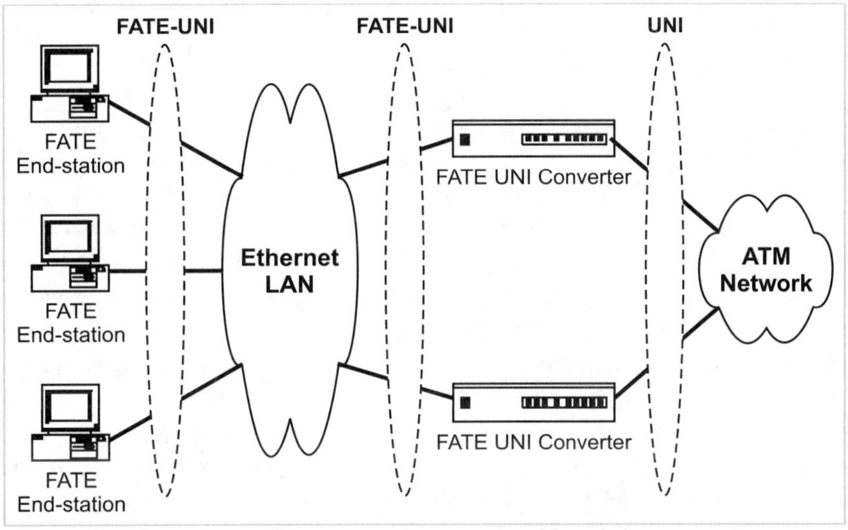

The FATE specifications discuss two different environments where FATE can be used. The first environment is where the ATM network consists of an ADSL-based access network that supports the 'ATM-mode' of ADSL and provides connectivity to a public ATM network. In this environment, the FATE UNI converter can be located at the customer premises within the ATU-R external modem. The FATE end-stations can be typical TCP/IP based personal computers that connect to the ATU-R via an Ethernet LAN. Figure 16.18 depicts the protocol stack model at the PC and the ATU-R resident FATE UNI converter. Note that the figure assumes that only one end-station is connected to the FATE UNI converter. If more than one end-station required connectivity to the ATM network, then, the FATE UNI converter will be required to support additional functions such as PVC/SVC-Multiplexing, ILMI and ATM SVC signalling.

FIGURE 16.18

FATE-UNI Over Ethernet

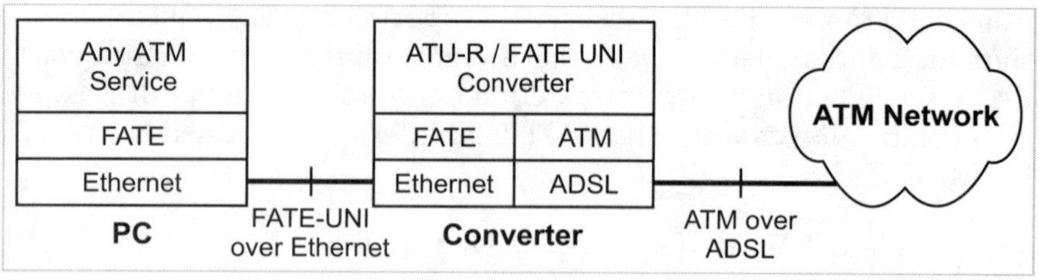

The second environment where FATE can be supported is a corporate LAN, which uses an ATM-based backbone to connect Ethernet-based LANs together and to provide access to a public ATM network. In such an environment, the ATM network consists of the ATM-based backbone and the ATM public network to which the backbone connects. The FATE UNI converter can be resident in an Ethernet switch/hub that has an additional ATM interface towards the backbone. As in the first environment, FATE end-stations can be typical TCP/IP based personal computers connected to an Ethernet-based LAN.

16.7.2 FATE Procedures

FATE specifications define the mechanism used for the transmission of FATE UNI frames over Ethernet. Besides the 6-octet source and destination MAC addresses, Ethernet frame header consists of a two octet 'Ethertype' field that indicates the higher layer protocol whose payload the Ethernet frame is carrying. FATE specifications define a new 'Ethertype' value (0x8884), which is used in any Ethernet frame carrying a FATE UNI payload.

FATE specifications also define a 'Discovery Procedure' for FATE end-stations to discover dynamically the FATE-UNI converters providing connectivity to the ATM network. A different 'Ethertype' value (0x884C) is used for the FATE discovery messages. The FATE discovery procedure consists of the following messages:

- **DISCOVER-REQ message:** The DISCOVER-REQ message is used by the FATE end-station to discover a FATE UNI converter by multicasting the discovery request message. More than one FATE-UNI converters may be providing service to end-stations in a particular environment. Therefore, the destination MAC address used in the Ethernet frame carrying the discovery request message is a multicast address on which the FATE UNI converters will be listening to. Within the discovery request message, the FATE end-station includes the Maximum Segment Size (MSS) value that is wishes to support when communication with the FATE UNI converter.
- **DISCOVER-ACK message:** On receipt of a FATE DISCOVER-REQ message, a FATE-UNI converter responds to the FATE end-station using a DISCOVER-ACK message. If the MSS indicated by the FATE end-station is not acceptable to the FATE UNI converter, it also includes the MSS value that it is wishing to support. Since multiple FATE-UNI converters may respond to the multicast discovery request message, the FATE end-station may receive multiple DISCOVER-ACK messages. In such a case, the FATE end-station chooses the most suitable FATE UNI converter based on its internal logic.

For more details on the FATE procedures, the reader is referred to the FATE specification [ATMF FATE].

Conclusion

This chapter discussed the ATM network components and interfaces. The ATM network interfaces fall into multiple distinct categories, prominent amongst which are the UNI and the NNI. While UNI defines the interface between an ATM end-system and an ATM switch, NNI defines an interface between two ATM switches belonging to the same ATM network. Besides UNI and NNI, the B-ICI defines the interface between two public ATM networks and its standardization helps in achieving interoperability between two public ATM networks belonging to different carriers.

While 'cell-based' ATM transport is ideal for carrying voice traffic, providing data services over ATM leads to transmission inefficiencies due to the ATM cell header overheads. This lead to the specification of frame-based ATM interfaces for providing data services. Prominent amongst these are the DXI and the FUNI interfaces, which define a frame-based access interface between a frame-based system and an ATM switch. Besides DXI and FUNI, two other specifications, FAST and FATE were defined to standardize transport of frame-based ATM traffic over conventional interfaces. While FATE defines a mechanism to carry frame-based ATM traffic over Ethernet, FAST defines the same for SONET/SDH.

Review Questions

Q 1. ATM network interfaces fall into multiple distinct categories. Draw a network diagram depicting different ATM network interfaces and explain their purpose.

Q 2. While the public NNI defines the interface between two ATM switches within the same public ATM network, the B-ICI defines the interface between two ATM switches in two different public ATM networks. Compare and contrast the NNI and the B-ICI protocol reference models.

Q 3. Describe what conversion functions will be required to translate service-specific traffic into cell-based traffic before transmission over the B-ICI.

Q 5. Describe the differences in the mechanism used to transport AAL5 frames over DXI using Mode 1b and Mode 2 of operation.

Q 6. Compare the DXI and the FUNI approaches towards defining an frame-based ATM access interface.

Q 7. With the help of a diagram, describe how you would connect an FUNI-based interface with a FAST-NNI interface.

Q 8. Describe two types of environments where FATE can be used to connect frame-based end-stations to an ATM network.

Further Reading

This chapter explains different interworking techniques and is based on ATM Forum specifications. The ATM Inter-Network Interface (AINI) is specified in [ATMF AINI 1.1], B-ISDN Inter-Carrier Interface (B-ICI) in [ATMF BICI 2.0], Data Exchange Interface (DXI) in [ATMF DXI], Frame-based UNI (FUNI) in [ATMF FUNI 2.0], Frame-based ATM over SONET/SDH Transport (FAST) in [ATMF FAST] and Frame-based ATM Transport over Ethernet (FATE) in [ATMF FATE] ATM Forum specifications.

Chapter 17

ATM in LAN

17.1 Introduction

ATM was designed to be the desktop-to-the-core networking solution. In the Wide Area Network (WAN) where there was a steady rise in demand for bandwidth, ATM required the means to support fast switching. Being a virtual circuit switching technology, fast cell switching was a simple proposition. However, getting into the LAN environment, where a well-entrenched market existed for Ethernet was not an easy job. The slow progress of ATM in LAN made people realize that migration towards ATM is a long and steady process. Thus, during the early phases of deployment of ATM technology, it was necessary for ATM to interoperate with the existing link-layer protocols (e.g. the Ethernet). The idea was not to provide an alternative that required a complete overhaul but a solution that could co-exist in parts. Towards this end, the LAN emulation or LANE was standardized by the ATM Forum in [ATMF LUNI 2]. LAN emulation provides a means for traditional LAN technologies like Ethernet and Token ring to co-exist with ATM LANs in a heterogeneous LAN environment. Using LAN emulation, two non-ATM end-systems can communicate not only with each other, but also with end-systems directly attached to an ATM network. This chapter elaborates upon the LAN emulation technique and other aspects related to the deployment of ATM in a LAN environment.

17.2 Traditional LAN Technologies

Till recent times, LAN technologies like Ethernet and Token ring were considered adequate to satisfy the needs of most LAN applications. Even today, these technologies are, to some extent, effective in supporting LAN applications. Moreover, as long as the predominant use of LANs is for bursty, time-insensitive data applications, Ethernet will continue to dominate the LAN market.

However, with the growth in the size of networks and user requirements, there is a serious need to augment the existing network capacity and provide more bandwidth to the users. This increased demand for bandwidth is tackled by migrating from shared LANs to switched LANs. In a shared LAN, multiple hosts share a common communication channel, as shown in Figure 17.1(a). Thus, for a 10 Mbps Ethernet LAN with 10 hosts, each host, on an average, gets 1 Mbps of bandwidth. However, this is true only if the LAN operates at 100 per cent capacity. This, unfortunately, is not the case because the Carrier-Sense Multi-Access Collision Detection (CSMA/CD) technique used in Ethernet reduces the effective throughput due to collisions. During heavy load conditions, the numerous collisions further reduce the throughput. Moreover, in a shared LAN, a frame is broadcast to all the hosts, which requires that hosts filter unwanted frames. Besides, broadcasting frames to all hosts also makes the network traffic susceptible to security violations.

A switched LAN (e.g. switched Ethernet) seeks to alleviate some of these problems by providing a dedicated bandwidth to each host. Each host in a switched LAN is directly connected to a switch (or an intelligent hub) and thus gets the entire channel bandwidth for itself, as shown in Figure 17.1(b). Since the switch forwards frames according to the destination address, the filtering overheads and security concerns as in shared LAN, is eliminated. Moreover, fault diagnosis becomes easier because the switch can monitor each link individually. In essence, a switched LAN provides higher throughput and better manageability. However, this comes at a cost, because each host in the LAN has to get direct connectivity with the switch.

FIGURE 17.1

Shared and Switched LAN

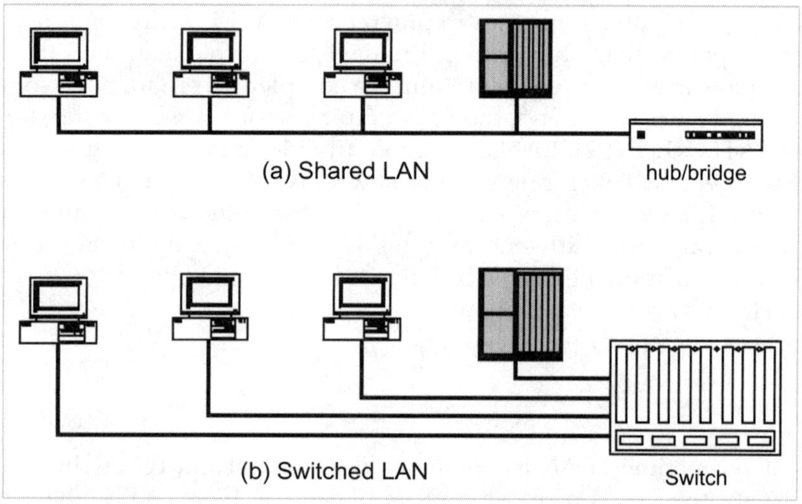

17.3 ATM IN LAN

An ATM LAN consists of ATM capable end-systems attached to ATM switches (i.e. *switched* LAN). ATM end-systems communicate with each other by using Permanent Virtual Connection (PVC) or Switched Virtual Connection (SVC). ATM switches are connected by using a complete or a partial mesh, and communicate with each other using by PNNI or other routing protocols. A typical ATM LAN is depicted in Figure 17.2.

FIGURE 17.2

A Typical ATM LAN

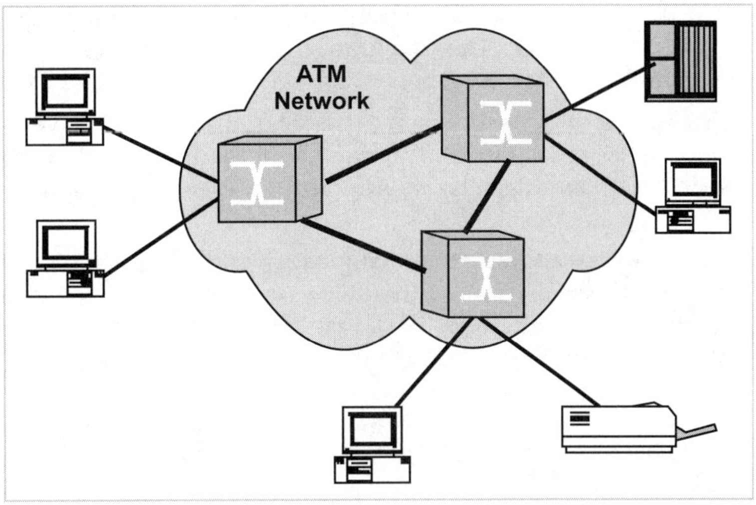

The previous section mentioned some of the benefits of a switched medium over the shared medium. Since ATM is inherently a switched transfer medium, it offers similar benefits. Apart from these benefits, there are several other benefits of using ATM in the LAN environment, some of which are as follows:

- **Greater bandwidth:** Technologies like Ethernet and Token ring offer LAN connectivity at 4, 10, 16, 100 and, of late, 1000 Mbps. In contrast, ATM (depending upon the requirements) offers a wide range of bandwidth, ranging from as low as 25 Mbps to a few Gbps per second.
- **QoS guarantees:** Ethernet and Token ring provide connectionless mode of data transfer and one of the fundamental drawbacks of any connectionless service is the lack of QoS guarantees. In contrast, ATM offers a connection-oriented service. Such a service provides users with the capability to reserve resources, according to the needs of their applications by choosing one of the many service classes (CBR, VBR, ABR, GFR and UBR). Moreover, as the transfer is connection-oriented in nature, two end-systems may have more than one active virtual connection active between them. These virtual connections can belong to different service classes as well.

- **Security aspects:** Again, because of the connection-oriented nature of ATM, a frame is sent to an end-system only if the frame is destined for it. This is in sharp contrast to the broadcast mechanism of shared Ethernet and Token ring LANs wherein broadcast and frame filtering makes the network susceptible to security breaches like eavesdropping.
- **Seamless networking:** The traditional LAN technologies are inapplicable in a WAN environment, wherein the bandwidth is precious and distance an important factor. The LAN protocols are developed with the assumption that bandwidth is free and hence conserving it is not a prime concern. This is not true in a WAN environment wherein bandwidth is precious and wherein network managers try to utilize the bandwidth to any extent possible. ATM, in this regard, offers a seamless way to integrate a LAN and WAN network. This integration offers numerous benefits like reduced costs due to single infrastructure, better manageability and higher scalability.

In essence, with the pervasive use of bandwidth-extensive multimedia applications, ATM is finding its way in the LAN market, hitherto dominated primarily by Ethernet, and to some extent, by Token ring networks, and provides an excellent communication medium for most LAN applications.

Nonetheless, a few factors that go against the large-scale deployment of ATM LANs are a very strong market base of Ethernet, economies of scale (i.e. high costs) and the development of high-speed alternatives like Gigabit Ethernet (1 Gbps). These factors have a detrimental effect on the popularity of ATM in a LAN environment.

17.4 LAN Emulation (LANE)

Technologically, ATM is capable of supporting a wide base of current and emerging applications. However, the commercial success of ATM depends upon many factors other than its technological strengths. Notable amongst them is the ease with which applications in existing LANs are migrated to ATM. Significant investments have been made to build the current LAN infrastructure. Therefore, reuse of the existing LAN infrastructure also becomes quite crucial. Moreover, in a heterogeneous environment where traditional LANs and ATM co-exist, interoperability becomes another prominent issue. In order to address these issues, the ATM Forum has introduced a standard called LAN emulation, or simply LANE in [ATMF LUNI 2].

In essence, LANE provides a means for traditional LAN technologies like Ethernet and Token ring to co-exist with ATM LANs in a heterogeneous LAN environment. Using LAN emulation, non-ATM end-systems can communicate with each other and also with end-systems directly attached to an ATM network. The underlying idea behind defining LANE is to leverage the initial deployment of ATM networks in a traditional LAN environment. Adequate care has been taken to propose minimal changes so that the existing infrastructure can be reused with relative ease. The following sections elaborate upon the requirements of LANE, components of LANE, the LANE reference model and procedures associated with LANE.

17.5 Issues and Requirements for LAN Emulation

As briefly mentioned, LAN emulation proposes minimal changes in the existing LAN infrastructure so that ATM can co-exist with other LAN technologies. This co-existence demands certain requirements that must be fulfilled by ATM. This section mentions such requirements for LAN emulation. Specifically, the means to provide traditional LAN features (e.g. connectionless service, broadcast/multicast support and address resolution) in an ATM LAN environment are elaborated.

17.5.1 Programming Interface

One of the most important challenges that must be addressed by LAN emulation is to define a place for ATM in the protocol stack. Putting it differently, LAN emulation must define the type of Application Programming Interfaces (APIs) that ATM will provide to its users. In this regard, ATM can provide the following alternatives with respect to APIs (see Figure 17.3):

- **Low-level interface:** Under this model, ATM provides users with services similar to those provided by the IEEE MAC sub-layer. Network Driver Interface Specification (NDIS) and Open Data-link Interface (ODI) are examples of standardized MAC layer interfaces. Providing low-level interface is attractive because no changes are required in the existing software applications and thus, the deployment of ATM becomes simple. However, a low-level interface has a major drawback, that is, instead of providing native capabilities (i.e. capabilities that provide a direct interface between ATM and the user-level applications), ATM provides a connectionless service, thereby making the QoS capabilities of ATM invisible to user applications. Despite the drawback of this model, the ATM Forum's LAN emulation standard has chosen this model to facilitate early deployment of ATM technology.
- **High-level interface:** In order to effectively utilize the native capabilities of ATM, a high-level interface is a much better choice. A high-level interface is also performance-efficient because the overhead involved in emulating MAC sub-layer characteristics is

Figure 17.3 Programming Interfaces

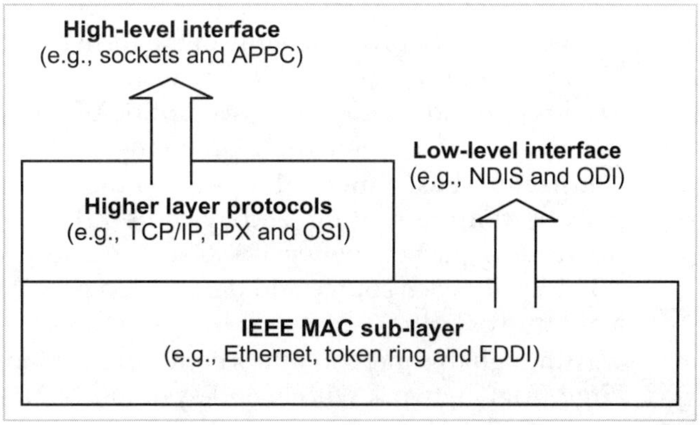

done away with. Several alternatives exist for ATM to provide a high-level interface. The simplest option is to consider ATM as a link layer protocol (like frame relay) and adapt it to the existing network layer protocols. Classical IP over ATM [RFC 2225] is an example of this technique. However, Classical IP over ATM defines the means to carry only IP traffic over ATM. Apart from IP, there are a multitude of other network layer protocols (e.g. IPX, NetBioS and Appletalk), the porting all of which would involve a considerable standardization effort. Another option is to treat ATM as either a network layer or transport layer protocol. Here again, a significant change is required in existing software along with a spate of standardization activities to make this idea technologically feasible. Some of these options are elaborated in Chapter 3.

Considering the pros and cons of both alternatives, it seems that in the short term, a low-level interface is attractive. In the long run, however, the use of high-level interfaces is expected to become more widespread. Moreover, with the spread of ATM technology, native mode applications, which directly utilize ATM programming interfaces, will become common.

17.5.2 Connectionless Services

Today's LAN technologies like Ethernet and Token ring provide connectionless services, that is, no connection is established prior to data transfer. This makes data transfer in traditional LANs extremely simple. The same, however, cannot be said about data transfer in ATM, wherein a connection has to be explicitly set up (in SVCs) or configured (in PVCs).

For LAN emulation, ATM must provide connectionless service over a connection-oriented framework. There exist two basic techniques to achieve this [LAN H. Troung]. These are:

- **Server-based method:** In this method, additional network entities in the form of connectionless servers are used to provide connectionless services. These entities either reside in the switch themselves (in which case the switch design needs to be modified) or are physically distinct entities. Each ATM end-system has at least one connection (SVC or PVC) with the connectionless server. The servers are also connected amongst themselves by using ATM VCs. Packets received from higher layers are transmitted by the end-systems to the connectionless server, after appending the destination address. The servers then forward the packets to the appropriate end-system(s).

  This approach offers benefits in the form of reduced requirements of VC connections. However, the server is liable to cause bottlenecks in terms of performance. [LAN H. Troung] provides a detailed description of the problems associated with this approach and the means to tackle these problems.
- **Connection-based method:** In this method, a connection is established between every pair of communicating end-systems. The connectionless service is emulated by forwarding packets between the source and destination end-systems over a separate ATM VC for each source and destination pair. The VC connections can be SVC, PVC or a mixture of both.

  Although this method offers better performance, the number of VCs required is quite large. For a network with N end-systems, $N(N-1)/2$ duplex connections are required.

However, this number can be reduced because an end-system does not communicate with every other end-system simultaneously. This reduction can be done by having active connections with only those end-systems for which packets are being transmitted. If a packet is received for a given destination and no active connection for that destination exists, a connection is established. This process of connection establishment is called rapid connection setup. If a connection is inactive for sufficient interval, it is terminated.

ATM Forum's LAN emulation standard uses both the methods to provide connectionless service. Initially, when a connection with the destination end-system is not established, all packets destined for that end-system are sent to a server (called Broadcast and Unknown Server or BUS). The BUS then forwards the packet to the appropriate end-system. Once a direct connection with the end-system is established, packets are forwarded by using the direct connection.

17.5.3 Broadcast/Multicast Services

The current LAN technologies (Ethernet/802.3 and Token ring/802.5) support broadcast and multicast services. The widespread use of these services is because these LANs usually have a shared medium wherein broadcasting and multicasting are inherently simple. Even while using a switched medium, broadcasting/multicasting is not difficult because of the connectionless nature of the frame transfer. Broadcasting/multicasting is done by using the 6-byte address format as defined by IEEE. The Individual/Group (I/G) bit set to '1' indicates a *group* address. An address with all bits set to one indicates a broadcast address (0×FFFFFFFFFFFF). Multicast addresses also have their I/G bit set to 1. For multicast addresses, the remaining bits indicate address of the multicast group.

In ATM, emulating broadcast/multicast is quite challenging because packets are sent only to those end-systems for which they are due. Still, the following, two techniques exist to provide this feature in ATM [LAN H. Troung]:

- **Point-to-multipoint connections:** A point-to-multipoint connection utilizes the capability of an ATM switch, whereby a cell arriving on an input port is replicated on one or more output ports. A point-to-multipoint connection is established by the root and leaves are added either by the root or by the network. Details of signalling procedures for point-to-multipoint connections are provided in Chapter 11.

  Two techniques exist for providing broadcast/multicast service using point-to-multipoint connections. In the first technique, each end-system establishes a point-to-multipoint connection with every other end-system in the network (i.e. there are multiple point-to-multipoint connections in the network). The biggest limitation of this scheme is number of virtual connections required, which is N*(N–1). Moreover, for every addition of a node, the network is flooded by signalling messages. For a network with N end-systems, the addition of a node requires the establishment of N point-to-point connections and one point-to-multipoint connection.

  To reduce virtual connection requirement, and to simplify procedures for add/drop operations, another technique exists where a multicast server is used to broadcast/

multicast packets (see Figure 17.4). The leaves establish a point-to-point connection with multicast server, which in turn establishes a point-to-multipoint connection with all the leaves. As compared to the previous technique, which requires N point-to-multipoint connections, this technique requires only N point-to-point connections and one point-to-multipoint connections per multicast group. The add/drop procedure is also simple in this case.

Figure 17.4

Multicast Server Method for Broadcast/ Multicast

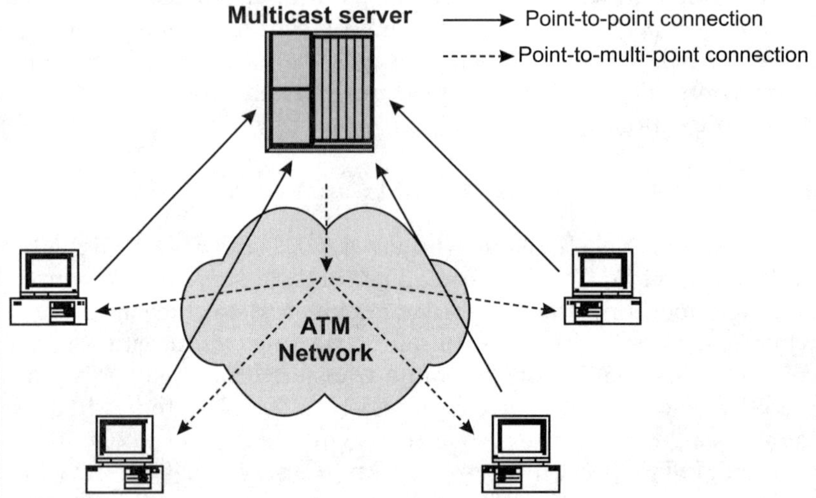

Source: ([LAN H. Troung] © 1995 IEEE)

- **Multipoint-to-multipoint connection:** Instead of establishing multiple connections, it is definitely better to have one multipoint-to-multipoint connection wherein any end-system can send or receive packets from any other end-system. The key challenge for this type of connection is the identification problem that arises due to inter-leaving of cells from multiple sources. Putting it differently, an end-system must be able to recognize the identity of the sender for each cell received. One of the options is to use VCI as the multiplexing identifier and use VP connections for data transfer. Another option is to use the Multiplexing Identifier (MID) field of AAL3/4 cell to demultiplex cells from different sources. However, no standards on multipoint-to-multipoint connection currently exist.

ATM Forum's LAN emulation standard uses the multicast server method to provide broadcast/multicast services. In LANE terminology, the multicast server is called Broadcast and Unknown Server (BUS).

17.5.4 MAC Driver Interface in ATM Stations

As mentioned earlier, the main objective of the LAN emulation service is to enable existing applications to access an ATM network via existing network layer protocols like IP, IPX, etc.

as if they were running over traditional LANs. Since the network layer protocol stacks communicate with a MAC driver, the LAN emulation service needs to offer the same MAC driver service primitives, thus keeping the upper protocol layers unchanged.

The three key industry standards for MAC driver interfaces are Network Driver Interface Specification (NDIS), Open Data-Link Interface (ODI), and Data Link Provider Interface (DLPI). Each of the standard interfaces specifies its own primitives and parameter sets, but the essential services remain the same. LAN Emulation provides these interface and services to the upper layers.

17.5.5 Address Resolution

Address resolution is a mechanism to bind a high-level protocol address (like IP) to a low-level hardware address (like Ethernet address). In TCP/IP networks, Address Resolution Protocol (ARP) is used to achieve this binding. The mechanism of ARP protocol is depicted in Figure 17.5. In the figure, node S knows the IP address of destination node D. In order to find the hardware address of node D, node S broadcasts the ARP request to all nodes in the sub-net. In case node D resides on the same sub-net (which is the case in the figure), it responds back with the address binding information containing its physical and IP addresses. Otherwise, the default router for the sub-net responds back with its own address.

FIGURE 17.5

ARP Protocol

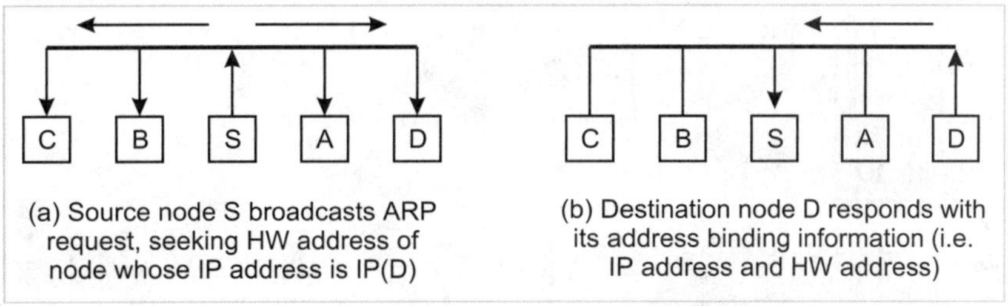

As can be observed from the above explanation, address resolution in traditional LANs is simple because of their inherent broadcast nature. In ATM, a similar broadcast-based method can be used. An end-system broadcasts an address resolution request to all other end-systems in the ATM sub-net. All end-systems check the destination MAC address in the received frame and respond if their address matches with that contained in the frame. The ease with which this method can be adopted is directly related to the ease with which the broadcast is supported.

Another alternative is to use an address resolution server, which maintains a centralized database. The database contains a mapping between ATM addresses and MAC or higher layer address for each ATM end-system in the network. This mapping is obtained at the time an end-system registers itself with a central server that maintains such a database. ATM end-systems send their address resolution requests to the address resolution server, which responds back with appropriate mapping.

ATM Forum's LAN emulation standard uses the address resolution server method to service address resolution requests. The task of address resolution is performed by LAN Emulation Server (LES).

17.5.6 Emulated LANs

In some environments, there is a need to configure multiple, separate domains within a single network. This requirement leads to the definition of an 'emulated LAN', which consists of a group of ATM-attached devices. This group of devices is logically analogous to a group of LAN stations attached to an Ethernet/IEEE 802.3 or Token Ring/IEEE 802.5 LAN segment.

Several Emulated LAN (ELAN) can be configured within an ATM network. Membership in an emulated LAN is independent of where an end system is physically connected. The multiple emulated LANs over a single ATM network are logically independent. Therefore, a broadcast frame originating from a member of a particular emulated LAN is distributed only to the members of that emulated LAN. Figure 17.6 depicts an ATM network having two emulated LANs. As can be seen from the figure, an ATM switch can provide connectivity to ATM end-systems belonging to different ELAN. This decouples the physical location of an end-system from its logical location. For example, if an end-system is to be moved from ELAN1 to ELAN2, this can be achieved by using software reconfiguration, without any physical reconfiguration of the network. An end-system could also belong to multiple emulated LANs. For example, as depicted in the figure, a high-end server is attached to both ELAN1 and ELAN2, providing logical connectivity at the cost of a single physical connection.

FIGURE 17.6

Emulated LANs

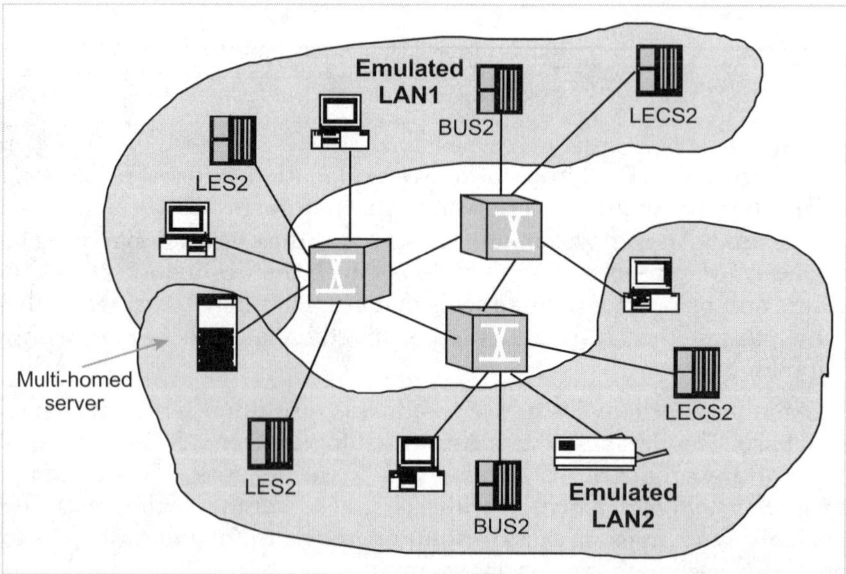

17.5.7 Interconnection with Existing LANs

In existing LANs, the interconnection between different LAN segments is achieved by using bridges. A bridge is a device that operates at the data link layer and provides the means to connect different LAN segments. There are two basic techniques of bridging, namely transparent bridging and source-route bridging. Following are brief descriptions of these bridging techniques:

- **Transparent bridging:** In this technique, the operation of a bridge is transparent to the end-systems. All the capability to perform bridging resides in the bridge itself and the end-systems remain unaware of the fact that a bridge is being used. In order to perform transparent bridging, techniques like flooding and adaptive learning are used. When a frame are received, a bridge looks at the source address of the frame and searches a hash table. If an entry does not exist, the frame is flooded to all connected LANs and an entry in the hash table corresponding to that address is created. This entry identifies the LAN segment to which frames containing that address as the destination address are to be forwarded. Thus, when frames bearing the same address in their destination address field are received, the entry in the hash table is used for forwarding. Old entries are removed periodically, so that only the latest information is contained in the hash table. In order to avoid forwarding loops, a spanning tree is maintained.
- **Source-route bridging:** In contrast to transparent bridging, source-route bridging assumes the involvement of end-systems in the bridging process. In source-route bridging, an end-system wishing to communicate with another end-system must know the complete route (called the source-route) to the destination. This source-route is a sequence of LANs and bridges that a frame must traverse to reach the destination. In order to find the source route, discovery frames are used.

LAN emulation is defined in such a way that both transparent bridging and source-route bridging are supported. There are various scenarios where bridges can be used to provide interconnectivity between ATM and non-ATM LANs. Figure 17.7 depicts two scenarios where bridges interconnect ATM and non-ATM LANs. In Figure 17.7(a), a bridge which has both ATM and non-ATM interfaces, provides connectivity between an end-system connected to ATM LAN and an end-system connected to Ethernet LAN. Figure 17.7(b) depicts the case wherein multiple bridges are used to provide connectivity between two Ethernet LANs over an ATM network.

17.6 LANE Components

The components of LANE are defined for a single emulated LAN. Each emulated LAN has at least one LE Client (LEC), and the components of the LE Service. LE Service consists of the mandatory components that includes LE Server (LES) and Broadcast and Unknown Servers (BUS) and the optional component LE Configuration Server (LECS). Figure 17.8 depicts two emulated LANs having their respective LANE components.

FIGURE 17.7

Scenarios of Interworking between ATM LANs and non-ATM LANs

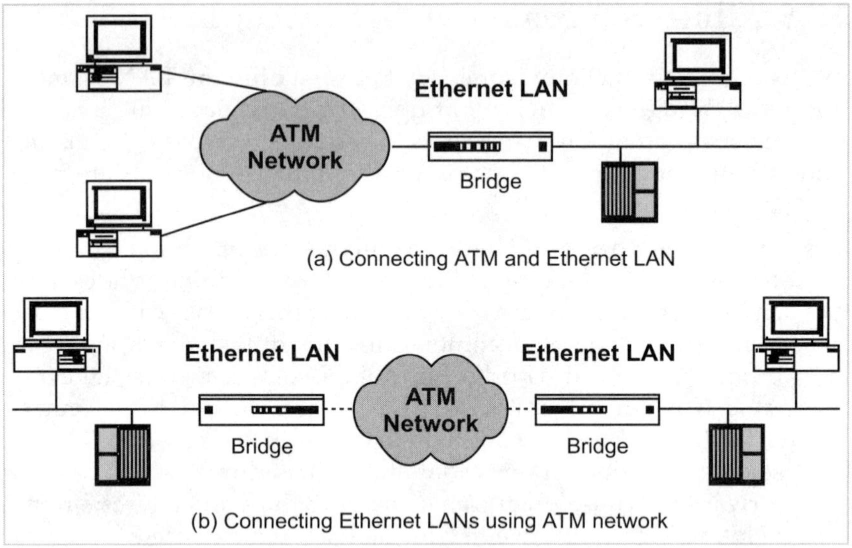

FIGURE 17.8

LAN Emulation User Network Interface (LUNI)

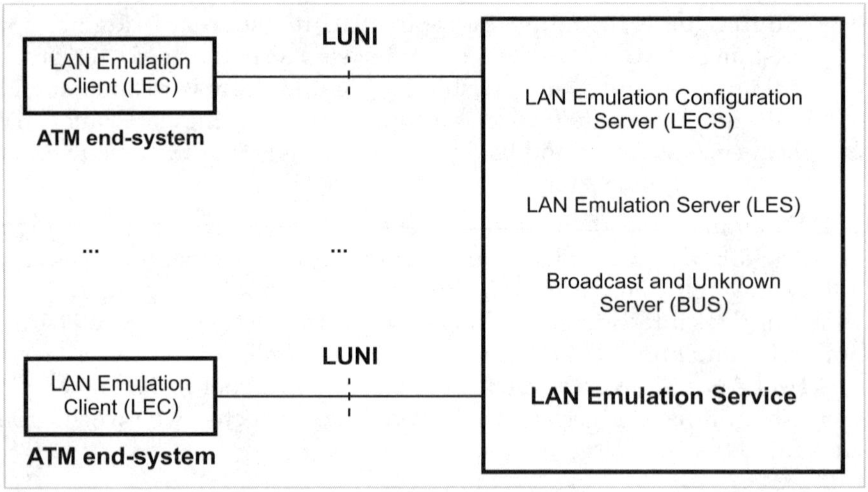

The interface between LEC and LE service falls under the purview of LAN Emulation User Network Interface (LUNI) protocol specified in [ATMF LUNI 2]. The LUNI interface is depicted in Figure 17.8. The interface amongst LECS, BUS and LES comes under the umbrella of LAN Emulation Network Network Interface (LNNI) protocol specified in [ATMF LNNI 2]. LNNI defines procedures necessary for distributed and reliable LAN emulation service. It should be noted that LES and BUS service exactly one emulated LAN. One LECS, however, can service more than one emulated LAN. The LANE service components are interconnected by VCCs for configuration, database synchronization, control and data forwarding. The following sections elaborate upon the various LANE components.

17.6.1 LAN Emulation Client (LEC)

Any network element (e.g. end-systems, bridges and routers) attached to an ATM network and participating in LAN emulation, contains a LAN Emulation Client (LEC). The primary function of an LEC is to emulate MAC level Ethernet/IEEE 802.3 or Token Ring/IEEE 802.5 service interface to higher-layer software. In order to provide this emulation, an LEC uses data forwarding capability of the AAL5 layer and signalling capability of UNI signalling protocol [ITU-T Q.2931]. Besides MAC-level emulation, an LEC also provides address resolution capability whereby the ATM address of a node is determined by using its MAC level address.

An LEC is associated with a single ELAN only. Since a node can be a part of multiple ELANs, a separate LEC is configured for every ELAN of which the node is a part of. An LEC is identified by a unique ATM address and is associated with one or more MAC addresses reachable through that ATM address. For an ATM end-system, the LEC is typically associated with a single MAC address only. However, for a LAN bridge that connects a number of non-ATM hosts to the ATM network, the LEC residing in the bridge is typically associated with more than one MAC address. Specifically, the LEC is associated with all MAC addresses reachable through the ports of the bridge and which are a part of a particular ELAN.

Figure 17.9 depicts the position of LEC in the protocol reference model. The LEC entity, shown as LANE in the figure, resides over AAL5 and provides to the higher layer standard interfaces like NDIS and ODI. The figure also depicts how an ATM end-system communicates with a non-ATM end-system, using an 802.1D bridge.

17.6.2 LAN Emulation Server (LES)

The LE Server (LES) implements the control co-ordination function for the emulated LAN. It provides a facility to register and resolve unicast and multicast MAC addresses to ATM addresses. In order to achieve this, the LES maintains a database of registered LAN destinations and their associated ATM addresses. The entries of the database are obtained when an ATM end-system (LE Client) registers its address(s) with the LES. The LES then uses the database to resolve MAC addresses to ATM addresses.

17.6.3 Broadcast and Unknown Server (BUS)

The Broadcast and Unknown Server (BUS) provides a facility for centralized forwarding of broadcast frames, and of those frames whose destination address is not registered with the LES. It is because a BUS provides the means to forward frames to unknown destinations that it is called an unknown server. Inclusion of the word 'broadcast' is obvious because BUS

FIGURE 17.9

Protocol Reference Model for LANE

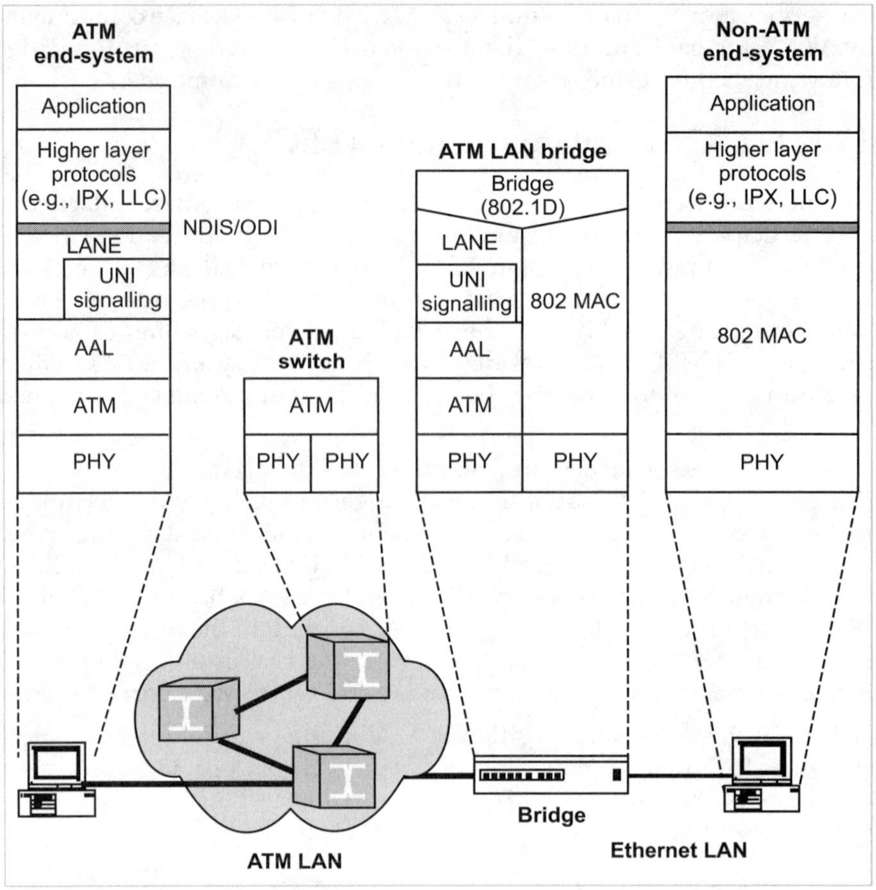

forwards broadcast frames. A BUS also shares in responsibility for the forwarding of sustained streams of multicast data.

Since no protocol is defined for LES/BUS interactions, each BUS is assumed to be logically paired with an LES. Within the LES associated with the BUS, the broadcast MAC address (0×FFFFFFFFFFFF) is mapped to the ATM address of the associated BUS.

The data frames from the LE Clients are received completely by the BUS before being forwarded to the destinations. This is done to prevent interleaving of AAL5 cells of outbound frames from different sources.

17.6.4 LAN Emulation Configuration Server (LECS)

The LE Configuration Server (LECS) acts as a central repository of configuration information and assigns individual LE Clients to an Emulated LAN. LECS also provides the LE Clients, ATM address of the LES corresponding to that ELAN. The assignment of an LE Client to an LES is based upon LECS' policies or the configuration information provided by LE Clients.

LECS is also used by the components like LES for configuration. LECS returns to these entities the ATM addresses of the neighbouring service entities. A BUS does not configure with the LECS, rather a BUS is assumed to be logically associated with an LECS, and may communicate with LECS to obtain its configuration.

17.7 LANE Connections

Various types of connections exist between LANE components for exchanging configuration information, forwarding data, and synchronizing various databases. While some of these connections are point-to-point, a few others are point-to-multipoint in nature. On the basis of the type of traffic carried, two distinct categories of VCCs exist: *configuration and control connections*, and *data connections*. The various VCCs belonging to these two categories are explained below (see Figure 17.10).

Figure 17.10

Various Types of Connections in LANE Environment

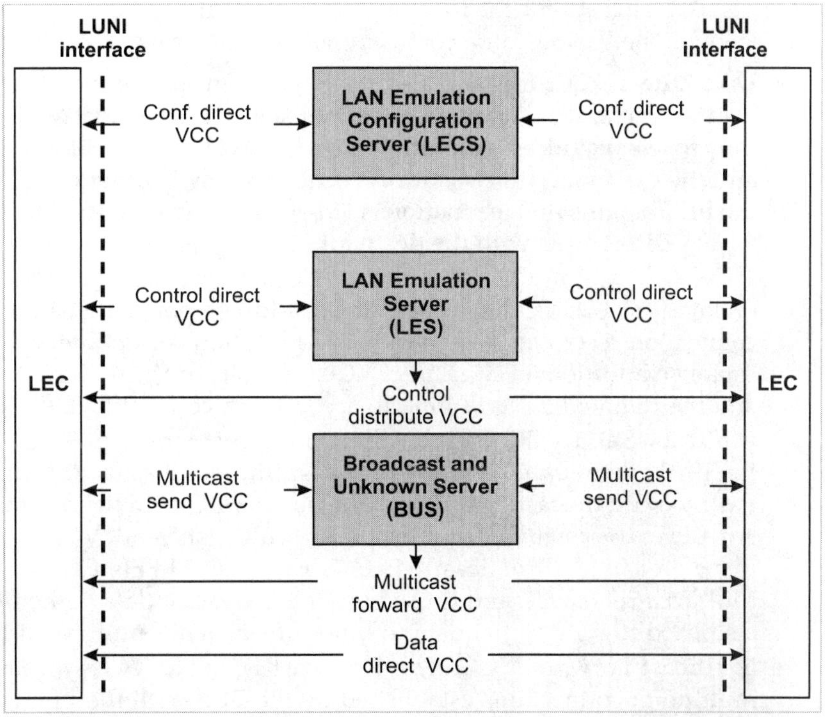

The control and configurations connections are used by LANE components to exchange configuration and control information. For example, an LEC may obtain the ATM address of its LES using one such connection. The various control and configuration connections are as follows:

- **Configuration Direct VCC:** This is a bi-directional, point-to-point VCC established by LEC with the LECS, to obtain configuration information (including the address of the LE Server) from LECS. After obtaining the necessary configuration information from the LECS, an LEC may terminate this VCC connection.
- **Control Direct VCC:** An LEC establishes this bi-directional point-to-point with the LES to send and receive control information. The control information here refers to address resolution requests generated by LEC and responses sent by LES. This VCC is established at the time of initialization after the LEC has obtained the ATM address of the LES from the LECS. The Control Direct VCC remains active as long as LEC and LES are a part of the Emulated LAN.
- **Control Distribute VCC:** This unidirectional, point-to-multipoint VCC is established by the LES with the LECs to distribute control information. The establishment of this VCC by LES is optional. However, if established, this VCC must not be terminated as long as LECs and LES are a part of the Emulated LAN.

The data connections are used by LEC and BUS to exchange unicast and broadcast/multicast data. Except for the flush message, no control information is exchanged over these connections. The various data connections are as follows:

- **Data Direct VCC:** This is a bi-directional, point-to-point VCC set up between LECs that want to exchange data. An LEC establishes a Data Direct VCC with another LEC after the process of address resolution is over. To elaborate, when an LEC has a frame to send and the destination ATM address is not known, it generates an LE_ARP request. Once the LEC obtains the destination ATM address through the LE_ARP reply, it establishes a Data Direct VCC with the destination.

  Two LECs typically use a single VCC to carry all packets between them, rather than opening a new VCC for each MAC address pair between them, so as to conserve connection resources and to reduce connection set-up latency. However, an LEC may establish additional Data Direct VCCs if it has traffic with particular QoS requirements and the remote LEC is willing to accept such connections.
- **Multicast Send VCC:** This bi-directional, point-to-point VCC is set-up by an LEC with the BUS and is used to send broadcast/multicast data. A Multicast Send VCC is also used to send unicast data till a Data Direct VCC with the destination is not established. An LEC can establish one or more Multicast Send VCC with the BUS. The VCC associated with broadcast MAC address (0×FFFFFFFFFFFF) is called the Default Multicast Send VCC and is used to send broadcast data to the BUS. This default VCC is also used to send initial data to other unicast and multicast destinations.
- **Multicast Forward VCC:** The Multicast Forward VCCs are uni-directional, point-to-multipoint connections established by the BUS with the LECs, to distribute data from the BUS. At least one Multicast Forward VCC is required by an LEC to participate in the emulated LAN.

Apart from control connections and data connections, there are synchronization connections that LE service components use to synchronize their databases. For details, the reader is referred to [ATMF LNNI 2].

17.8 LANE Operations

For an ATM LAN to emulate a traditional LAN, the LANE components go through several phases. These phases are as follows:

1. Initialization,
2. Registration,
3. Address Resolution, and
4. Data Transfer.

Each of the phase is elaborated in the following sections.

17.8.1 Initialization

During initialization, an LE client first discovers its ATM address either through configuration or through some other means. After this, the LE Client determines the ATM address of LECS. The address of LECS may be obtained using ILMI procedures or LECS may even be assigned a well-known address known to all LE Clients. Once the LE Client obtains the address of LECS, the Configuration Direct VCC is established with the LECS. If the LE Client cannot establish a VCC to the ATM address of the LECS, then well-known permanent connection of VPI = 0, and VCI = 0, is also used for Configuration Direct VCC.

After establishing a connection with LECS, the LE Client requests the LECS to provide the necessary configuration parameters. The LECS responds with the parameters necessary for the LE Client to connect to the ELAN. The information provided by the LECS includes LAN type being emulated (Ethernet or Token ring), the maximum frame size of ELAN, emulated LAN name and the address of the LES. The LES assigned to the LE Client may depend upon LECS policies, configuration database or upon information provided by LE Clients in its request.

17.8.2 Joining

After obtaining the address of LES, the LE Client establishes Control Direct VCC with LES. Once this VCC is established, LE Client sends a request to the LE Server to add it to the ELAN. This requests contains, among other information, the client's ATM address and the name of the emulated LAN obtained from LECS. The request also carries the information whether the LE Client is acting as a proxy for other unicast MAC addresses (as in case of a LAN switch). If the request of LE Client is successful, LES assigns a unique LEC Identifier (LECID) to the LE Client, establishes Control Distribute VCC with the LE Client, and sends back a suitable response to the LE Client containing the LECID. LE Server also registers the mapping between MAC address and ATM address for that LEC. After the LES has confirmed the successful joining of the LE Client to the ELAN, LE Client may then register a number of LAN destinations with the LES, for which it is acting as a proxy. This is in addition to the single unicast LAN Destination that it initially registered with the LES.

After Control Direct VCC and (optionally) Control Distribute VCC are established, the LE Client can use LE ARP (LE_ARP) procedures for finding the ATM address corresponding to a MAC address. When an LES receives a resolution request from a LEC, it may formulate a response and send it immediately if the mapping for the MAC address mentioned in the resolution request is registered with LES. However, if no such mapping is registered with LES, it forwards the request to all LE Clients on the Control Distribute VCC. The request is forwarded to all LE Clients, because there are LE Clients who act as proxis for several hosts. Since it is not mandatory for an LE Client to register all the addresses for which it is acting as a proxy, it is possible that an MAC address may not have a mapping in the LES table and still be there in the network.

An LE Client, on receiving the forwarded resolution request from the LES, checks if the MAC address received is the one of the addresses for which it is acting as a proxy. If this is so, the LE Client responds to the LES on the Control Direct VCC. The LES then forwards the reply to the requesting LE Client.

As an implementation-dependent option, an LES, on receiving a resolution request for which it does not have a mapping, may just forward the request to only those clients which are acting as a proxy to one or more LAN destinations. In such a case, LES needs to establish additional Control Distribute VCC connecting it to all the LE Clients acting as a proxy for LAN destinations.

After the initial registration is over, the LE client also establishes a connection with the BUS. For finding the address of BUS, LE Client sends an address resolution request (LE_ARP request) to LES with broadcast MAC address (all ones). The response to this request provides the LE Client with the ATM address of the BUS. The LE Client then establishes the Default Multicast Send VCC with the BUS. When the Multicast Send VCC is established, BUS automatically establishes the Multicast Forward VCC with the requesting LE Client for forwarding any multicast frames to the LE Client. The LEC is now ready for data transfer to the various LAN destinations present in the ELAN.

17.8.3 Data Transfer

During the data transfer phase, the LE Client receives frames from the higher layer for transfer to remote LAN destinations. If the LE Client does not know the ATM address of the LAN destination, it finds this out using LE_ARP procedures. Once the ATM address of the destination is obtained, it establishes a Data Direct VCC, using the obtained ATM address. This VCC is now used to transfer any frame between the LEC peers. The LEC may also locally cache the mapping between the MAC address and the ATM address. If another request to send a frame to the same destination is received and there is no active Data Direct VCC corresponding to that destination, then instead of sending out an LE_ARP request to the LES, the address resolution may be done by referring to the local table.

When the LE Client has sent the LE_ARP request and is waiting for a response to the request, it may also choose to transmit frames over the Default Multicast Send VCC to the BUS. This is necessary to prevent the LE Client's buffer from overflowing and to prevent inordinate delays resulting from buffering. The frames forwarded to the BUS are sent to all the LE Clients in the ELAN. This is necessary because there can be a passive device that can be reached through a transparent bridge and the bridge does not know the whereabouts of that destination. The location of this device can be known only when this device responds to the flooded packet. Therefore, the frame needs to be transferred to all such bridges so that they can flood that frame further.

After the LE_ARP response is received, the LE Client establishes a Data Direct VCC with the destination. After establishing the Data Direct VCC, the LE Client discontinues sending frames to the BUS and uses the Data Direct VCC instead. During the migration period, it is possible that some frames forwarded through BUS reach the destination after the frames which were sent through the Data Direct VCC. This loss of sequence results due to the different paths taken. To ensure that the destination receives the frames in sequence, a flush protocol is defined.

Flush protocol ensures an in-sequence delivery for unicast frames, when the transmission path taken changes. In this mechanism, a control packet is sent on the first path after the last packet is transmitted. The peer, on receiving a control packet, sends back a response on the same path. On receiving the response control packet, the sender is ensured that all the frames sent are delivered to the destination and the source may now switch the flow to the second path. However, the same does not hold in the sequence delivery of multicast frames forwarded to BUS. In order to minimize out-of-order delivery of multicast frames, the LE Client can delay when switching between paths.

17.8.4 Address Registration

The address registration mechanism provides LE Clients with the means to register additional LAN destination addresses that were not registered during the time of initial registration. It also provides LE Clients with the mechanism to deregister any LAN destinations registered earlier. The address registration procedures can be invoked only in the data transfer phase. The procedures for adding a client are the same as those for adding multiple LAN destinations during the initialization phase. Deregistering requests, just like registration requests, are sent over Control Direct VCC. The response of this request is sent at either the Control Direct VCC or the Control Distribute VCC.

17.8.5 Effect of Topology Changes

As mentioned, LE Clients maintain a cache in which they locally store the mappings between MAC addresses and ATM address. However, topology changes may result in erroneous forwarding. An external MAC address that was once reachable through an LE Client (LAN switch) may, at a later point of time, be unreachable due to reconfigurations in the network. Therefore, there is a possibility that LECs use obsolete mappings.

In order to prevent LE Clients from using obsolete mappings, LANE protocol provides an LE-Topology-Request message. On receiving the LE-Topology-Request, an LEC either discards the entries in its cache or reduces the ageing period so that the entries in the cache are discarded quickly. This ensures that the LEC obtains the mappings through LE_ARP request frequently and has the latest information.

Another mechanism to remove obsolete entries from the local caches of the LE Clients and LES is provided. When an LE Client detects that a particular mapping is no longer applicable, it sends an LE-NARP message to LES. The LES removes the mapping from its own table and forwards this message to all the LE Clients. This message expedites the process of updating the mappings throughout the LANE System.

Conclusion

This chapter provided a brief overview of the applicability of ATM in LAN environment. The focus of the chapter was on LAN Emulation (LANE) and how it could be achieved. The existing LAN techniques are connectionless and can easily support broadcast/multicast capabilities. This chapter discussed how these could be provided in an ATM environment, which is by no means a simple proposition. Migration towards ATM is a difficult and tedious process. The key issue is the hurdle posed by existing infrastructure. The vastly different nature of ATM from existing LAN technologies makes it very difficult to make an incremental transition towards ATM. Rather, the transition must be steep. This makes it necessary for ATM offer a distinct advantage, in the absence of which the transition stands unjustified. Despite the inherent strengths of ATM, the cost factor seems so predominant that the transition is financially challenging. Stop-gap measures like LANE hide the strengths of ATM, thereby providing no benefits to the ATM user. In essence, the effort put into publishing ATM internetworking standards is not commensurate with the popularity the standards occupy.

Review Questions

Q 1. What are the benefits of using ATM in LAN environment?

Q 2. Discuss the requirements and proposed solution for the following aspects in LAN Emulation:
- *Address Resolution*
- *Broadcast/Multicast Services*
- *Address Resolution*

Q 3. What are the components of LANE? What role does each component play?

Q 4. Explain the need for having different types of connections in LANE.

Q 5. Describe the different phases of LANE. What is the impact of topology changes in LANE environment?

Further Reading

Information on LANE is available in two documents published by the ATM Forum, one for LUNI [ATMF LUNI 2] and other for LNNI [ATMF LNNI 2]. Apart from this, [LAN H. Troung] provides a good description for emulating LAN using ATM. Other references for ATM in LAN are [LAN I. A. Akylidiz], [LAN B. Vickers], [LAN E. Biagoini], [LANG. Parolkar] and [LAN N. Finn].

Chapter 18

IP Over ATM

18.1 Introduction

Out of the many possible models to deploy ATM in the existing LAN infrastructure, LANE provides one of the means. LAN emulation provides a means for traditional LAN technologies like Ethernet and Token ring to co-exist with ATM LANs in a heterogeneous LAN environment. Using LAN emulation, two non-ATM end-systems can communicate not only with each other, but also with end-systems directly attached to an ATM network. However, one of the serious shortcomings of LANE is that it fails to exploit the QoS capabilities of ATM. The applications have no mechanism to specify their QoS requirements because the programming interface an application gets is left untouched. In fact, existing applications are transparent to the fact that ATM, and not Ethernet or token ring, is being used at the MAC layer.

However, in the emerging communication paradigm, some of the new applications (like video-conferencing) require real-time transport and tremendous amount of bandwidths. Certain new network layer protocols, like IPv6 and RTP, provide service guarantees required by such applications. For these protocols to use QoS capabilities of ATM networks, a change in the model is required and new programming interfaces for them must be defined.

This chapter looks at new models for use of ATM. In particular, the Classical IP over ATM and Multi-Protocol Over ATM (MPOA) techniques are discussed. Classical IP over ATM, specified in [RFC 2225], provides mechanisms to transfer IP datagrams over ATM networks. This standard also defines the means for address resolution. However, one major drawback of this scheme is that when nodes belong to different sub-net, packets traverse through one or more router, despite the fact that a direct connection could be possible.

Multi-Protocol Over ATM (MPOA) uses LAN emulation techniques and the Next Hop Resolution Protocol (NHRP) to provide the means to establish a direct VC connection between end-systems belonging to different sub-nets. In other words, MPOA integrates LANE and NHRP to preserve the benefits of LAN emulation, while allowing inter-sub-net, inter-network layer protocol communication over ATM VCCs without requiring routers in the data path. MPOA is defined in [ATMF MPOA 1.1].

Another industry development for the interworking between IP and ATM is Multi-Protocol Label Switching (MPLS). The MPLS is explained in Chapter 19.

18.2 Classical IP Over ATM

Classical IP over ATM, defined by IETF's IP over ATM working group, is one of the means whereby network layer protocols can map the service requirements of higher-layer applications to the QoS provided by ATM. Besides exploiting QoS capabilities, Classical IP over ATM also does away with another overhead of LANE, that of dual address resolution. In the LANE environment, IP addresses are first resolved to MAC addresses, which, in turn, are resolved to ATM addresses. Classical IP over ATM gets rid of the intermediate mapping to MAC addresses and instead, maps IP addresses directly to ATM addresses. Classical IP defines a method to carry IP datagrams over ATM networks. In essence, classical IP is an address resolution mechanism whereby ATM address resolution requests and responses are carried over AAL5. Classical IP over ATM is defined in [RFC 2225]. This RFC obsoletes the earlier version defined in [RFC 1577].

The primary motivation behind defining the classical model is to leverage the initial deployment of ATM in a classical IP network. Here, a classical model refers to the treatment of the ATM host adapter as a networking interface to the IP protocol stack operating in a LAN-based paradigm. In particular, the classical model, as defined by [RFC 2225], is characterized by the following factors:

- One IP sub-net is used for many hosts and routers. Each VC directly connects two IP members within the same Logical IP Sub-network (LIS). LIS is formally defined in Section 18.3.
- IP addresses are resolved to ATM addresses by the use of an ATMARP service within the LIS (i.e. ATMARP stays within the LIS). ATMARP services are explained in Section 18.4.
- Default Logical Link Control (LLC)/Sub-network Attachment Point (SNAP) encapsulation of IP packets. The LLC/SNAP encapsulation mechanism is explained in Section 18.5.
- The same Maximum Transmission Unit (MTU) size is used for all VCs in an LIS. Issues related to MTU size are detailed in Section 18.6.
- End-to-end routing architecture stays the same.

The classical model suggests two possible application areas for the deployment of ATM. The primary use of ATM will be to replace traditional LAN technologies (like Ethernet). As another use, ATM will be used to interconnect routers.

18.3 Logical IP Sub-network (LIS)

Classical IP over ATM defines a logical entity called Logical IP Sub-network (LIS). An LIS consists of ATM attached hosts and routers and is configured in a manner similar to the way any IP sub-net is configured. The secific requirements for hosts/routers in an LIS include the following:

- All members of the LIS have the same IP network/sub-net number and address mask.
- All members of the LIS are directly connected to the ATM network.
- All members of an LIS have a mechanism for resolving IP addresses to ATM addresses via ATMARP services (based on [RFC 826]) and vice versa via In ATMARP (based on [RFC 2390]), when using an ATM Switched Virtual Connection (SVC).
- All members of an LIS have a mechanism for resolving VCs to IP addresses via In ATMARP (based on [RFC 2390]) when using an ATM Permanent Virtual Connection (PVC).
- All members of an LIS are able to communicate via the ATM with all other members in the same LIS; i.e. the VC topology underlying intercommunication among the members is fully meshed.

Each LIS in an ATM network operates and communicates independently of the other LIS within the network. The scope of ATMARP services is restricted to an LIS. Nodes belonging to different LIS communicate using a router.

18.4 Address Resolution in Classical IP Environment

A node wishing to communicate with another node in an LIS needs to know the ATM address corresponding to the destination node's IP address. To achieve this mapping, [RFC 2225] defines an ATMARP service. The ATMARP service is quite similar to the Address Resolution Protocol (ARP) defined for broadcast-based mediums like Ethernet and token ring. The exact method depends upon whether SVCs or PVCs are used. In the following sections, address resolution for both SVCs and PVCs are explained.

18.4.1 Address Resolution with SVC

In an SVC environment, an ATMARP server is used for address resolution. The ATMARP server maintains a centralized database of IP and ATM addresses of all ATM nodes contained in its LIS. This database is used to resolve the address resolution queries of ATM nodes.

At the time of initialization, a node in the LIS establishes a connection with the ATMARP server. The address of the ATMARP server is pre-configured. The node then sends normal ATMARP request packets to the server. The ATMARP server examines each ATMARP-request packet for the source protocol and source hardware address information to build its ATMARP table cache. This information is used to generate replies to any ATMARP requests it receives.

Subsequently, when the node wants to send an IP packet to another node on the same LIS, it sends an ATMARP request to the ATMARP server, specifying the destination IP address. The ATMARP server looks for an entry corresponding to the destination IP address. If the destination node lies in the same LIS, and if the destination node has already registered, the ATMARP server sends an ATMARP response, which includes IP and ATM addresses of the destination. The source node then establishes a direct connection using the ATM address received in the ATMARP response. The complete scenario is depicted in Figure 18.1. In the figure, note that the registration procedures of node B has not been depicted. If the destination node does not lie within the LIS, then all frames to it are forwarded to a default router, which then forwards the frame appropriately.

FIGURE 18.1

ATM Address Resolution

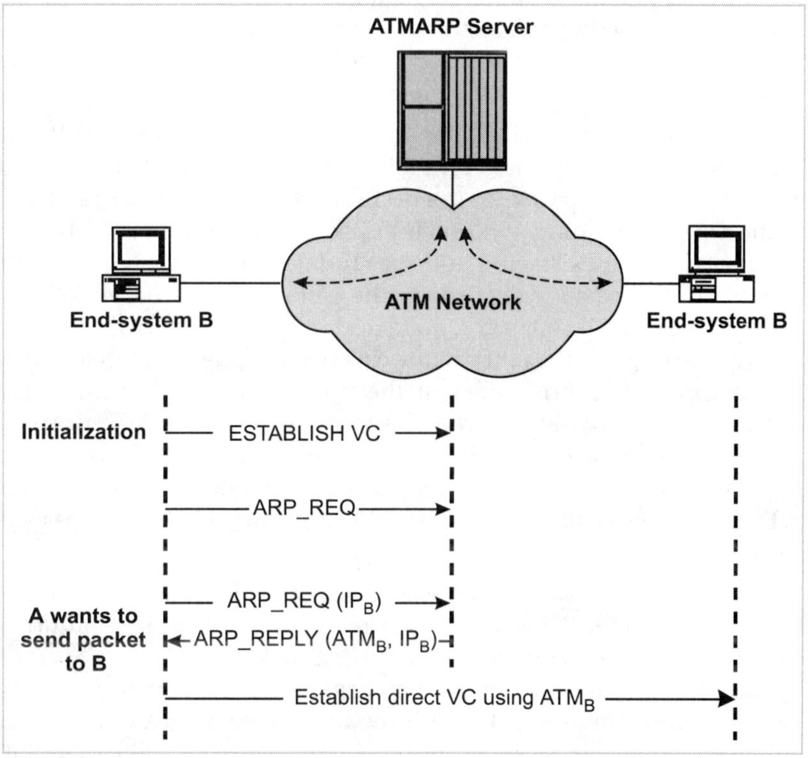

18.4.2 Address Resolution with PVC

In a PVC environment, a node does not have the capability to establish a VC with ATMARP server and resolve address-mapping requests. Hence, under such a scenario, Inverse ATMARP is used. Inverse ARP protocol, defined in [RFC 2390], provides a mechanism for a node to know the IP address of the destination node, corresponding to an open PVC connection. Using the Inverse ARP mechanism, a node can know all the IP addresses reachable through it. The inverse mechanism is a slightly better option than the option of statically configuring reachable IP addresses.

The inverse ATMARP is similar to ARP except that instead of broadcasting the request, the request is sent on each PVC connection having the source IP and hardware address along with the destination hardware address. The destination responds with its IP address.

18.4.3 Caching

Like in all address resolution protocols, significant performance benefits can be achieved by locally storing IP-ATM address mappings for a given period of time. This process of locally storing entries for subsequent use is termed as caching. Caching prevents the unnecessary overhead of generating address resolution requests for every transmitted packet. Caching is also used for incomplete entries, i.e. for which an address resolution request is sent, but a response is not received. This is useful in preventing multiple address resolution requests from being generated for the same node.

In Classical IP over ATM, a node caches mappings between IP and ATM addresses locally. The entries in the cache are valid for a period of 15 minutes after which the entries age out (i.e. expire). After an entry expires, a node checks for the presence of any open VC corresponding to the expired mapping. If no such VC exists, the entry in the table is deleted. However, if a VC for that mapping exists, the node revalidates that entry before sending out any more packet on such a VC. The node revalidates the table entry by querying the ATMARP server. If a valid reply is received, the querying node updates its entry. However, if the ATMARP server cannot resolve the request, the entry in the cache of the querying node is deleted and all open VCs for that mapping are terminated. If the node is unable to contact the server, and if the SVC corresponding to the expired mapping is LLC/SNAP encapsulated, the node tries to revalidate the mapping by sending an Inverse ATMARP request. Upon receiving a valid Inverse ATMARP reply, the querying node revalidates the entry. However, if a valid Inverse ATMARP reply is not received, the SVC is terminated and the associated mapping is removed from the cache.

Like a node, an ATMARP server also maintains a cache and ages out entries every 15 minutes. When the ATMARP server receives an address resolution request, it revalidates the mapping corresponding to the source IP and ATM address. This is done by resetting the timer associated with the corresponding cache entry. Also, prior to aging an entry, the ATMARP server sends an inverse ATMARP request on the open VC associated with that mapping. If a suitable response is received, the ATMARP server revalidates that table entry. However, if there is no open VCC for that entry, the entry from the cache is deleted.

18.5 Multi-Protocol Encapsulation Over AAL5

In a layered architecture model, generally, each PDU at a particular layer carries the protocol identifier of the higher-layer PDU encapsulated in it. For example, an IP datagram header contains the protocol field that mentions whether the payload contains a TCP segment, a

UDP segment or contains PDUs of other protocol. However, in an AAL5 frame, no such field exists which could be used to identify the higher layer PDU contained in the AAL5 frame. In order to tackle this problem, multi-protocol encapsulation over AAL5 is defined. This technique is defined in [RFC 2684], which obsoletes the earlier version defined in [RFC 1483].

In multi-protocol encapsulation over AAL5, [RFC 2684] defines two methods for carrying multi-protocol traffic over AAL5. The first method, called LLC encapsulation, allows packets from different protocol to be multiplexed over a single VC connection. The second method, called VC-based multiplexing, does higher-layer multiplexing implicitly, using ATM VCs. Both these encapsulation methods are explained below.

18.5.1 Logical Link Control (LLC) Encapsulation

In Logical Link Control (LLC) encapsulation, packets belonging to multiple protocols are multiplexed over a single VC. The need for multiplexing arises when a source wants to use a single VC connection to carry packets from several applications bound to the same destination. A header placed in front of each PDU identifies the protocol being carried in the AAL5 frame. The format of the header depends upon the type of protocol carried.

[RFC 2684] classifies all PDUs into routed PDUs (i.e. PDUs of layer 3, for example IP) and bridged PDUs (i.e. PDUs of layer 2, for example 802.3/Ethernet). Frame formats for both types of PDUs are explained as follows:

- **Routed PDUs:** These are again classified as routed ISO PDUs and routed non-ISO PDUs. Routed ISO PDUs correspond to protocols that are under the purview of ISO. Examples of such protocols include ISO CLNP and ISO ISIS. The routed ISO PDUs are identified by a 3-byte LLC header 0×FE-FE-03 (which indicates that a routed ISO PDU follows) followed by a one-byte NLPID field. The NLPID or Network Layer Protocol ID is a one-byte field administered by the ISO and ITU-T, which tells the receiver the type of protocol that follows. For example, NLPID value of 0×81 indicates that the frame contains ISO CLNP PDU.

  Routed non-ISO PDUs are identified using an extension mechanism via the Sub-Network Attachment Point (SNAP) header. LLC value of 0×AA-AA-03 indicates that a 5-byte SNAP header follows. The SNAP header comprises of a 3-byte Organizationally Unique Identifier (OUI) and a 2-byte Protocol Identifier (PID) field. OUI identifies an organization that administers the meaning of the PID field. Protocol Identifier (PID) identifies the actual protocol. OUI and PID collectively identify a routed or a bridged protocol. For example, an OUI value of 0×00-00-00 and PID value of 0×08-00 indicates the IP protocol. The encapsulation of IP PDU in an AAL5 frame for this example is depicted in Figure 18.2.
- **Bridged PDUs:** The bridged PDUs are identified using the SNAP header. Here again, the presence of SNAP header is indicated using LLC value of 0×AA-AA-03. In the SNAP header, the OUI field corresponds to 802.1 organization code, which is 0×00-80-C2. The PID field specifies the type of bridged PDU carried. For example, Figure 18.3 depicts the AAL5 frame format for bridged Ethernet/802.3 PDUs.

Figure 18.2

Encapsulation of IP PDU in an AAL5 Frame Using LLC encapsulation

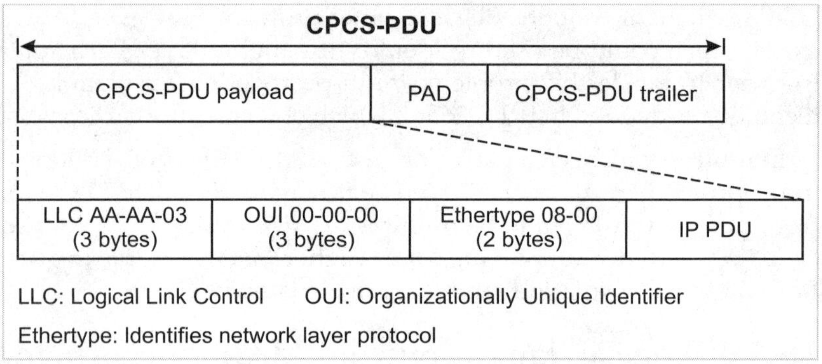

Figure 18.3

Encapsulation of Bridged Ethernet/ 802.3 PDU in an AAL5 Frame Using LLC Encapsulation

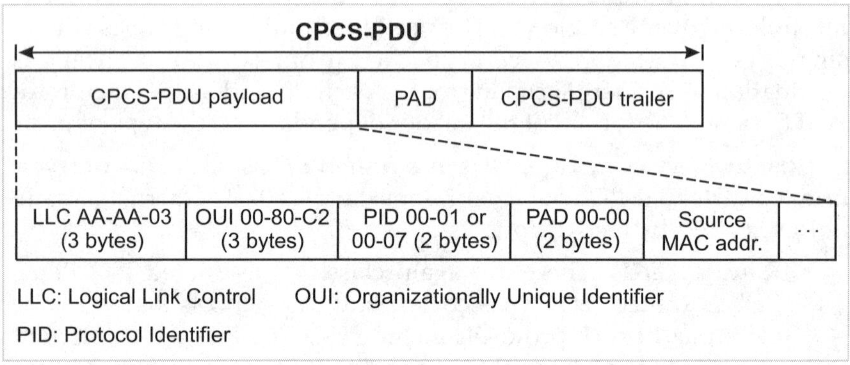

18.5.2 VC-based Multiplexing

In VC-based multiplexing, the protocol of the packets is implicitly identified by the VC connecting two ATM end-systems. That is, each protocol is carried over a separate VC and the VC implicitly identifies the protocol being carried. The protocol carried is either manually configured or negotiated dynamically during call establishment. There is, therefore, no need to include multiplexing information in every transferred packet. This results in minimal overheads in terms of bandwidth and processing requirements. However, this also imposes additional burden of creating and terminating large number of VCs. Thus, VC-based multiplexing method is useful in scenarios where the creation of a large number of VCs is fast and economical. On the other hand, for a network supporting only PVCs, VC-based multiplexing is not preferable. In such networks, the LLC encapsulation scheme is used.

VC-based multiplexing also distinguishes between routed and bridged PDUs. Routed PDUs are carried as it is in the payload part of the CPCS-PDU. For bridged PDU, frame format is similar to that used in LLC encapsulation, except that LLC, OUI and PID fields are absent.

18.6 Maximum Transmission Unit (MTU) Size

The TCP/IP protocol suite operates over a multitude of link-layer technologies, each differing in the maximum size of frames it can carry. For example, Ethernet allows a maximum frame size of 1500 bytes, while ATM allows a much larger frame size. This limit is referred to as Maximum Transmission Unit (MTU). If a datagram is transmitted by using the smallest frame size possible, then this scheme leads to inefficiency, because there will be several sub-nets that can carry frames of size more than the minimum size. On the other hand, if the datagram size is greater than the maximum size the underlying link layer can carry, then the link layer may not be able to forward the datagram.

In order to handle such a scenario, IP specifies a fragmentation and reassembly mechanism whereby a larger datagram is fragmented into several smaller datagrams. Although the fragmentation/reassembly mechanism enables IP to operate over different link-layer technologies, it still leads to certain inefficiencies. Consider a case where a sub-net Z with MTU equal to 128 bytes lies between two sub-nets X and Y, each having an MTU of 1500 bytes. Once the datagram is fragmented by Z, even if Y can carry larger datagrams, Z is forced to carry smaller datagrams. Fragmentation also increases the burden on intermediate routers because the size of a datagram hardly makes any difference to the time required to forward it. In order to avoid cases of datagram fragmentation, [RFC 1191] specifies a path *MTU discovery* mechanism. Here, path MTU refers to the minimum MTU of any sub-net traversed by a datagram between two hosts. The path discovery mechanism utilizes the 'don't fragment' bit of the IP header. If a datagram has this bit set and if the IP layer of a particular sub-net cannot forward this datagram, then an error indication is sent to the source. By choosing a large MTU and then decrementing it on receiving an error message, the source can find the path MTU.

Classical IP over ATM defines the default IP to be 9180 bytes. This is kept same as the default MTU for IP over SMDS to allow interoperability and reduce the incidence of IP fragmentation. The use of path MTU discovery mechanism is also mandatory for the hosts for reducing IP fragmentation and ensuring efficient performance. It may be noted that Classical IP over ATM does not solve the inefficiencies cropping up, arising in the example cited earlier.

18.7 Next Hop Resolution Protocol (NHRP)

Classical IP over ATM has limitations in terms of data forwarding when nodes belong to different LIS. Even if a direct connection between nodes belonging to different LIS is possible, all packets traverse through one or more router. This is because ATMARP services (as defined by Classical IP over ATM) works only for nodes within an LIS; i.e. for intra-sub-net communication, not for inter-sub-net communication. For a node to establish a direct connection with another node belonging to a different LIS, the source node must have the mapping between the ATM address and the IP address of the destination node. To provide this mapping, IETF's Network Working Group has come up with Next Hop Resolution Protocol (NHRP). NHRP protocol is detailed in [RFC 2332].

NHRP essentially provides a mapping between IP and the ATM address of a destination node. This applies for the case when the destination node is directly attached to an ATM network. However, if the destination node is not directly connected to an ATM network, NHRP provides the means to determine the egress point from the ATM network to reach the destination. Note that [RFC 2332] neither mentions the term IP address and nor the term ATM address. Rather, IP address is referred to as protocol address in the said RFC and ATM address as NBMA address (discussed below). This is correct in the sense that NHRP is a generic protocol and can be used for protocols other than IP and ATM. However, in the following sub-sections, the terms IP address and ATM address are used to facilitate easier understanding.

NHRP is useful in host–host, host–router and router–host communications. However, when used in router–router communication, NHRP can produce routing loops. This occurs when routed paths are unstable. A loop may also exist when the there exists two paths to reach a particular node, one through an NHRP network and other through an alternate network. [RFC 2332] mentions a few more instances when routing loops are created and also recommends that NHRP be not used under such circumstances.

18.7.1 NBMA Networks and their Modelling

Next Hop Resolution Protocol (NHRP) introduces a concept of logical Non-Broadcast, Multi-Access (NBMA) networks. An NBMA network, unlike the traditional Ethernet and Token ring networks, does not support broadcasting. An NBMA network consists of a set of nodes, interconnected by using point-to-point or point-to-multipoint links. Examples of NBMA network include X.25 and ATM networks.

The NBMA networks are modelled using the concept of logically independent IP Sub-nets (LIS). The concept of LIS has been detailed earlier. Another mechanism of modeling NBMA networks is to use the notion of Local Address Group (LAG). The LIS and LAG models differ in the way local/remote forwarding decision is made. In the LIS model, local/remote decision refers to the process whereby a node determines whether the destination address contained in a packet identifies an interface that the node is currently attached to, or not. In case the packet's destination interface has the same network address and the prefix as that of the node's interface, the outcome is local else it is remote. In case the outcome is local, the node resolves the IP address to an ATM address (using ATMARP) and sends the packet directly to the destination using an ATM VC. Else, the packet is forwarded to the node's first-hop router.

The local/remote forwarding decision in the LAG model is slightly different. In the LAG model, the local/remote decision is decoupled from the addressing information and is based solely on QoS and/or traffic characteristics. For example, even if a destination node is on the same IP sub-net, a source node may choose to forward its packet to a default router. This may be due to the fact that the packet is for a low-priority application (requiring best effort transfer) and it is not desirable that a VC be explicitly setup. In essence, besides the QoS and/or traffic requirements, LAG model also takes into account the cost of establishing and releasing VCs, thereby providing a better and efficient SVC management mechanism. For details on issues related to local/remote forwarding, the reader is referred to [RFC 1937].

18.7.2 NHRP Components

NHRP defines two entities in the network—the Next Hop Server (NHS) and the NHRP Client (NHC). A NHS is an entity within the logical sub-network that receives NHRP address resolution requests from NHC and responds by sending the mapping between ATM and IP addresses. This ATM address may correspond to the destination node (when the destination node is directly connected to the NBMA network) or may correspond to the egress router that is nearest to the destination node. An NHS also interacts with NHS entities of other sub-nets to service the resolution requests it receives. Typically, an NHS resides in a router or an edge device.

An NHRP Client (NHC) is an entity that initiates NHRP requests in order to obtain access to various NHRP services. All requests initiated by an NHRP Client are sent to the NHS. The address mappings received in reply are used to establish direct connections with the destination node (or the egress router, whichever is the case).

Each NHS has associated with it a set of NHCs for which it provides address resolution services. In order to avail the services of an NHS, an NHC first registers itself with an NHS. At the time of registration, the NHC provides its IP and ATM address to the NHS. This information is then used by the NHS to service requests of other NHCs. Figure 18.4 shows the NHRP components. Each sub-net has an NHS associated with it, to which all resolution

FIGURE 18.4

NHRP Components

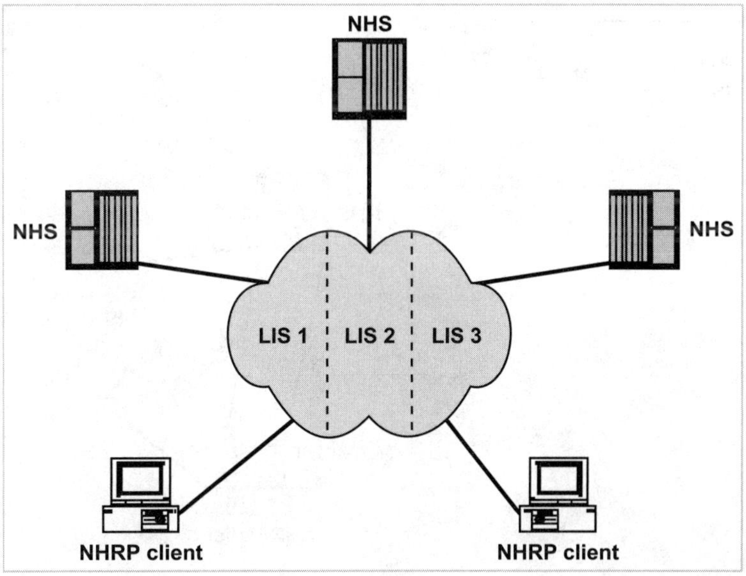

requests are forwarded. The NHS, in different sub-nets co-ordinate with each other to service resolution requests of various NHRP clients.

18.7.3 NHRP Address Resolution

The NHRP address resolution procedure is normally initiated by a node, say S, when it needs to transmit a packet to another node, say D. In order to do so, node S must have the ATM address of node D (S already has the IP address of node D, provided by the IP layer). Hence, node S first checks its local cache for the existence of a mapping between the IP address of D and the corresponding ATM address. If such a mapping exists, the obtained ATM address is used to establish a VC connection (if one does not already exist) and the packet is forwarded.

Otherwise, an NHRP resolution request is sent to the NHS. NHS searches its own cache for the mapping. If a mapping is available, an NHRP resolution reply is sent to the NHRP client, otherwise, NHS forwards the request to another NHS. The manner in which the next-hop NHS is chosen, is similar to the way an IP datagram is forwarded (using routing protocols like OSPF). The IP address, for which the mapping is requested, is used to determine the next-hop NHS. The next-hop NHS, to which the request is forwarded, examines the request message and checks if it serves the destination address present in the request. If the NHS does not, and if a mapping for that IP address does not exist in the local cache, the request is forwarded further towards the destination. When the request reaches the NHS serving the destination, an NHRP Resolution Reply is formed and sent towards the originating NHS along the path from which the request was received. The originating NHS then forwards the reply to the requesting NHRP Client. The whole mechanism is depicted in Figure 18.5.

FIGURE 18.5

NHRP Address Resolution

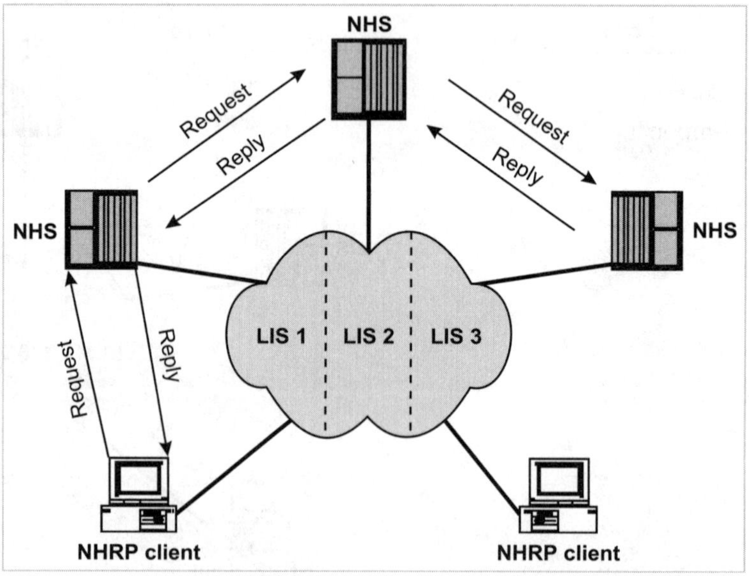

While the NHRP Client is waiting for the address to be resolved, it may still be receiving the packets that need to be transmitted. The NHS may perform one of the following actions on the received packets:

- Drop the packet.
- Keep the data in the buffers until the NHRP Resolution Reply arrives.
- Forward the packet along the routed path to the destination.

An NHRP Client may perform any of the actions mentioned above. However, choices 2 and 3 are preferable. If the time required for the address resolution is large, buffering scheme (choice 2) may also prove to be unfeasible, as large buffers will be required to store packets received during the interim period. Hence, the third option is most commonly used. This option ensures that even when the address resolution time is large, a packet reaches its destination without requiring buffering capabilities.

18.7.4 Caching and Purging

Like in other address resolution protocols, significant performance benefits can be achieved in NHRP by maintaining an address cache. A cache is maintained by both NHC and NHS. The NHC, upon receiving an address resolution response from its NHS, caches the received mapping. If any packet is to be sent to a node whose mapping already exists in the NHC's cache, then an address resolution request is not initiated. Rather, the cached mapping is used to establish a connection with the destination node. NHC also caches an address resolution request so that subsequent requests for the same destination node is not generated.

The NHS also maintains its cache to store mappings between IP and ATM addresses. The NHS cache contains entries from two sources. The first set of entries corresponds to the mapping of the hosts, which the NHS serves and which is received at the time of host registration. The second set of entries corresponds to mappings obtained while interacting with other NHS entities.

An intermediate NHS, while forwarding an NHRP resolution reply, also caches the mapping contained in the reply. If an address resolution request for that address is subsequently received, the NHS may return the NHRP resolution reply itself, rather than forwarding the request to the NHS serving that address. This significantly reduces the time required to resolve an address. However, it is possible for an NHRP client to change its NHS rendering the entries in other NHSs caches incorrect. This may result in inaccurate reporting of address mapping and a failure to connect to the actual destination. In order to tackle these situations, an NHRP client may demand 'authoritative' replies. A client specifies in the resolution request that it requires resolution from 'authoritative' NHS. Only the NHS that serves the destination for which the resolution is required, generates the response to this request. A NHS with cached information cannot reply to authoritative requests.

It is required that all entities in a NHRP network contain correct IP and ATM address mappings. In order to ensure that an incorrect entry does not create connection problems, a mechanism called purging is defined. The purging mechanism uses the NHRP purge request message. An NHRP purge request is sent by an NHS to invalidate a cached entry in the network. This message may be triggered by a change in network topology or when the NHS finds that some of its entries are no more valid. The purge request may also be sent by a NHC to the NHS to invalidate its registration (this typically happens when the NHC wants to change its NHS). When a node receives a purge request, it sends an NHRP Purge Reply to acknowledge that all cached entries corresponding to the address specified in the request are deleted. If a node wishes to reestablish communication with the destination shortly after receiving an NHRP purge request, it makes an authoritative NHRP Resolution Request in order to avoid any stale cache entries that might be present in intermediate NHSs.

18.8 Multi-Protocol Over ATM (MPOA)

LANE provides the mechanism for traditional networks to co-exist with ATM. Since LANE provides to the network layer an interface similar to the one in the traditional LANs, the applications in the host are totally unaware of the underlying ATM network. ATM is used as just another transfer mode without regards to its capabilities to support a variety of traffic having varying characteristics and different QoS requirements. In order to solve this problem for IP networks, Classical IP over ATM is recommended. To enable the ATM to be used efficiently with other network layer protocols including IP, Multi-Protocol Over ATM (MPOA) is defined in [ATMF MPOA 1.1].

In a LANE environment, if two entities connected to two different emulated LANs want to interact, the data is routed through the routers. This model of data transfer is called the sub-net model and is depicted in Figure 18.6. In this model, the routers connecting different sub-nets are subjected to heavy loads and thus, can cause congestion. Moreover, because routing is a hop-by-hop forwarding technique, there is an inherent latency involved in the forwarding process, which disrupts the service guarantees of the connection. In contrast to this, the cut-through model offers a more efficient and predictable forwarding mechanism. The cut-through model is depicted in Figure 18.7. This model involves establishing direct virtual connections with the destination as soon as a significant flow is detected. A

Figure 18.6

Sub-net Model

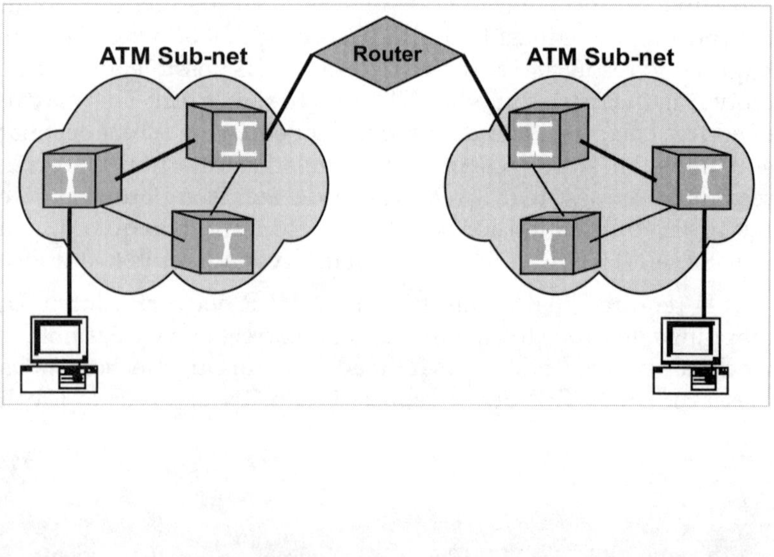

FIGURE 18.7

Cut-through Model

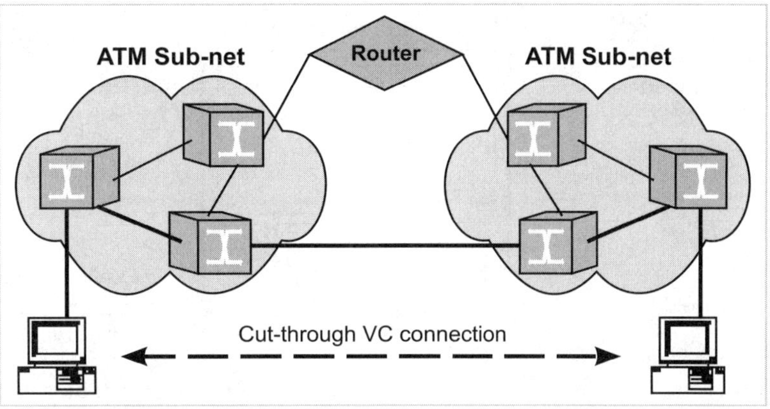

significant flow implies that at least a given number of packets are sent to a particular end-system within a specified time interval.

Multi-Protocol Over ATM (MPOA) uses the NHRP protocol as described earlier to establish cut-through connections. A cut-through connection allows bypassing the intermediate routers in the data path. MPOA also employs the LANE standard for intra-sub-net communication. Apart from LANE and NHRP, another concept called virtual routing is central to understanding the MPOA protocol. In a typical routing environment, each router in the path has the complete route computation and forwarding capability. This not only increases the cost of a router, it also makes network management difficult, because each router must be individually configured. In contrast, the virtual routing technique allows splitting of the route computation and forwarding capability. Virtual routing offers several benefits like easier manageability, reduced costs and greater scalability. In essence, a virtual router provides all the functionality of a typical router, but does away with the processing and management of overheads of the hop-by-hop routing.

18.9 MPOA Components

The primary goal of MPOA is to enable efficient transfer of data by establishing 'cut-through' routes across the routers on the path. To provide this functionality, two entities are defined: MPOA Client and MPOA Server. These two entities operate in a client-server framework and are connected in the network via LANE. Thus, in the protocol reference model, the MPOA protocol sits over LANE protocol in both MPOA Client (MPC) and MPOA Server (MPS). Components of MPOA systems are shown in Figure 18.8.

18.9.1 MPOA Client (MPC)

An MPOA Client (MPC) resides in a host attached to the ATM network or in edge devices that forward data from non-ATM networks to ATM networks. In essence, MPOA clients act as

FIGURE 18.8

Components of an MPOA System

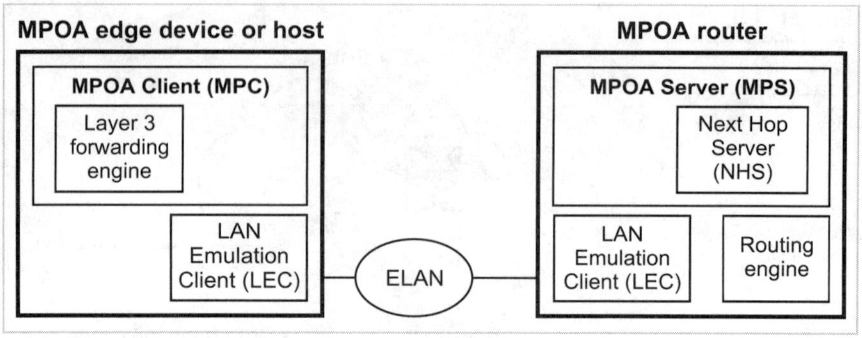

entry and exit points in the MPOA network i.e. the shortcut paths begin and end at an MPOA client. The MPOA client at entry point is called ingress MPOA client and the MPOA client at exit point is called egress MPOA client.

At the entry point, the role of the MPOA client is to detect a significant flow between two end-systems and initiate procedures to establish shortcut routes. These shortcuts are established by NHRP-based query-response protocol between the MPOA Client and MPOA Server. These short-cut connections remain active as long as there is data flowing over the connection. As soon as the inactivity period of the connection exceeds certain threshold value, the connection is terminated.

At the exit point, the role of the MPOA client is to receive the data and forward it to its local interfaces/users. The egress MPOA client also adds an encapsulation header (called the Data Link Layer (DLL) encapsulation) to the packets received from the shortcuts and gives them to the higher layer. The DLL encapsulation is stored in the cache of the egress MPOA Client. The egress cache comes to know about the DLL encapsulation when the shortcut is established.

18.9.2 MPOA Server (MPS)

An MPOA Server (MPS) is a component of the router that provides necessary forwarding information to MPOA Clients. The MPOA Server also includes a Next Hop Server (NHS). The MPS interacts with the NHS and the underlying routing functions to respond to the queries of the ingress MPOA Client. The MPOA Server also interacts with the egress MPOA Client and provides it the DLL encapsulation information.

18.10 MPOA CONNECTIONS

Various types of information flows exist between the MPOA entities. These information flows are shown in Figure 18.9. The MPOA information flows are divided into two categories: MPOA control flows and MPOA data flows. The MPOA control flows include the following:

- **Configuration flows:** MPOA server and MPOA clients establish a connection with the LECS to obtain their configuration information.

FIGURE 18.9

Information Flows in an MPOA System

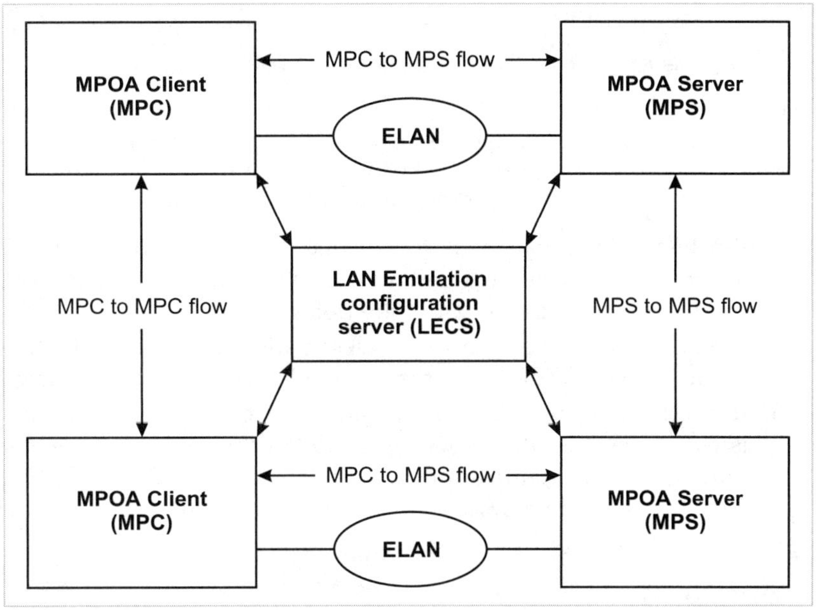

- **MPC–MPS control flows:** This bi-directional connection is established between MPOA Clients and MPOA Server to exchange information using the request–response protocol. This connection is also used by the MPS to provide the egress MPC with necessary information (e.g. encapsulation information).
- **MPS–MPS control flow:** The bi-directional connection is used between two MPOA Servers to exchange routing information using standards routing protocols like OSPF. This connection is also used between two MPOA Servers to exchange NHRP messages.
- **MPC–MPC control flows:** This connection is used by the MPOA Clients to handle misdirected packets. The egress MPC, on receiving a misdirected packet, sends a purge request to ingress MPC. On receiving this purge request, ingress MPC invalidates its erroneous cache information.

The information flows mentioned above are used by MPOA component to exchange control information and are hence called control flows. Besides control flows, MPOA components establish data flows to forward data. The MPOA data flows include the following:

- **MPC–MPC data flow:** This connection is used to transfer data between MPOA clients over shortcut connections.
- **MPC–NHC data flows:** This connection is used by the MPOA client to forward unicast data to Next Hop Client (NHC). The NHC then routes this data to the destination MPOA Client. This flow is used when the short-cut connection has not yet been established.

18.11 Establishing Shortcut Connection and Forwarding Data Packets

In this section, the mechanism to forward data packets and to establish shortcut connection is explained. A packet enters the MPOA system at the ingress MPOA Client (see Figure 18.10). To forward a packet, the MPOA Client has the following options:

1. By default, the packet is forwarded to the default router. The default router then forwards the packet further and the packet ultimately reaches the destination. The path used in this process is called the default path.
2. If a shortcut connection to the destination already exists, the Data Link Layer (DLL) encapsulation of the packet is stripped and the packet is forwarded using the shortcut.
3. If no shortcut exist, but the requisite number of packets for a particular destination is exceeded, (i.e. there is a significant flow), short-cut establishment procedures are initiated. For a significant flow to be detected, [ATMF MPOA 1.1] specifies that at least 10 packets must be transmitted within a time interval of one second.

Figure 18.10

Establishment of Shortcut

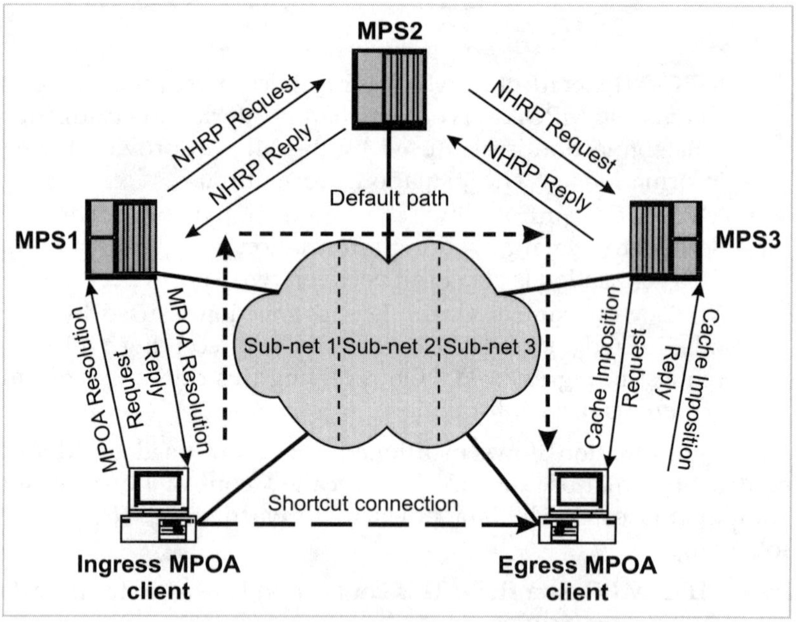

Assuming that case 3 applies, the MPOA Client sends an MPOA Resolution Request to the MPOA Server. The MPOA server responds back with the ATM address of the destination (if the destination is directly connected to an ATM network) or the address of the edge device which connects to the destination. The MPS uses its own database if the destination is locally connected. Otherwise, the MPS queries other MPS using NHRP request message. Each MPS on the path of the destination repeats this step until the egress MPS server is reached. The egress MPS, on receiving an NHRP request, sends an MPOA cache imposition request message to the egress MPC, which responds back with MPOA cache imposition reply. This message exchange provides MPS the necessary information to respond to the NHRP request, while it provides the MPC necessary encapsulation information. Once the NHRP reply propagating through various MPS reaches the ingress MPC, the MPC establishes a cut-through connection.

Conclusion

This chapter looked at options for carrying IP over ATM, including Classical IP over ATM, NHRP and MPOA. Among these, techniques like MPOA are not very popular and remain more of a standardization initiative. Classical IP is simpler and more popular. Chapter 19 looks at newer schemes like Multi-protocol Label Switching that could provide the co-existence of IP and ATM.

Review Questions

Q 1. Why is Classical IP named so?

Q 2. What are the key aspects of Classical IP over ATM? Explain with reference to the following:
- *Logical IP sub-net*
- *Address resolution*
- *Caching*
- *MTU size*

Q 3. What is Multi-Protocol Encapsulation over AAL5? What are the various options for carrying different PDU over AAL5?

Q 4. What is the drawback of Classical IP? How is this solved by using NHRP?

Q 5. What is NBMA network?

Q 6. Describe the different types of connections in an MPOA network.

Q 7. Explain how a cut-through connection is established by using MPOA.

Further Reading

The concepts presented in this chapter are derived from standardization from the ATM Forum and IETF. The Classical IP over ATM is described in [RFC 2225]. The Multi-Protocol Encapsulation over AAL5 is described in [RFC 2684]. There are associated RFC for address resolution. The NHRP is described in [RFC 2332]. The MPOA is described in [ATMF MPOA 1.1].

PART 5

Application of ATM Network

Any discussion of ATM is incomplete without a mention of the applications where ATM technology can be applied. This part of the book looks at the application areas of ATM networks. Owing to its unique features, ATM has a potential for widespread use, out of which four key application areas are considered in this book. These include application of ATM in Multi-Protocol Label Switching (MPLS) networks, Voice over ATM solutions, application of ATM in Digital Subscriber Line (DSL) networks and application of ATM protocols in Third Generation (3G) networks.

Chapter 19 looks at the role of ATM in MPLS networks. Multi-Protocol Label Switching (MPLS) is a recent industry development standardized by the IETF for the interworking between IP and ATM. MPLS can be best described as a synergistic approach between IP and ATM. MPLS merges the flexibility of the IP routing protocols with the speed that ATM switches provide to introduce fast packet switching in frame-based IP networks. IETF started the MPLS standardization process with the intention of introducing label switching within IP networks. While the initial focus was on IP networks alone, the MPLS framework is flexible enough to work with multiple protocols. Due to its close resemblance to the ATM protocol, MPLS has many possible applications when used in conjunction with ATM. Amongst the two main areas of research involving ATM and MPLS is the support of IP over ATM using MPLS, and the concept of ATM–MPLS network interworking. Both these techniques are covered in Chapter 19.

Chapter 20 looks at Voice over ATM solutions. The ATM Forum has defined a number of techniques for Voice and Telephony over ATM (VTOA). The initial standards by the ATM Forum were based on AAL1 while the newer ones are based on AAL2. AAL1 provides Circuit Emulation Service (CES) including its low-speed variant Low Speed Circuit Emulation Service (LSCES) and more efficient Dynamic Bandwidth Circuit Emulation Service (DBCES). Apart from circuit emulation, ATM also provides trunking solutions based on AAL1 and AAL2. All these techniques are described in Chapter 20.

Chapter 21 looks at the developments in the last mile where Digital Subscriber Line (DSL) is increasingly becoming popular with users of Public Switched Telephone Network (PSTN). For decades, residential users have been using the dial-up modem to access the Internet via the PSTN. However, with the theoretical upper bound of 56 kbps on the bandwidth available using dial-up modems, other broadband access technologies are being explored. Amongst the proposed solutions for broadband access, Digital Subscriber Line (DSL) is being seen as a technology that will overcome the above limitations of cost to offer a cost-effective broadband access solution. DSL proposes to use the existing infrastructure of the Plain Old Telephone Systems (POTS) to provide a high-rate access solution. DSL is designed to provide typical data rates of up to 2 Mbps over the existing copper cables installed for the POTS network. An overview of DSL along with the role of ATM in DSL networks is provided in detail in Chapter 21.

Another development of recent times is the gradual migration of voice users from fixed wireline networks to wireless mobile cellular networks. The wireless networks too have evolved from Second Generation (2G) networks like Global System for Mobile communications (GSM) and General Packet Radio Service (GPRS) to newer Third Generation (3G) networks like

Universal Mobile Telecommunications System (UMTS). As the networks evolve, there is associated requirement for higher bandwidth and better service capabilities. Towards this end, the ability of ATM to provide speed transfer with service guarantees emerges as a suitable choice for the transfer of voice and data in wireless networks. An overview of 3G along with the role of ATM in 3G networks is provided in detail in Chapter 22.

Chapter 19

ATM and MPLS Networks*

19.1 Introduction

In Chapter 18, two models for carrying IP over ATM, namely Classical IP over ATM and Multi-Protocol Over ATM (MPOA) techniques, were discussed. Classical IP over ATM, specified in [RFC 2225], provides mechanisms to transfer IP datagrams over ATM networks. One major drawback of this scheme is that when nodes belong to different subnet, packets traverse through one or more router, despite the fact that a direct connection could be possible. Multi-Protocol Over ATM (MPOA) uses LAN emulation techniques and the Next Hop Resolution Protocol (NHRP) to provide the means to establish a direct VC connection between end-systems belonging to different subnets.

A recent industry development for the interworking between IP and ATM is Multi-Protocol Label Switching (MPLS). This chapter looks at some of the concepts related to MPLS and also the interworking aspects between ATM and MPLS networks.

19.2 Overview of Multi-Protocol Label Switching (MPLS)

Multi-Protocol Label Switching is a new industry development standardized by the IETF. From the phrase 'Multi-Protocol' one might imply that MPLS provides support for multiple different protocols. However, the reality is that the emphasis of MPLS has till date been only on supporting the Internet Protocol. The emphasis on IP has primarily been because of the

* This chapter is contributed by Nishit Narang.

huge success of the Internet. While IP is a connectionless best effort protocol that works effectively in data networks with no QoS requirements, ATM is popular for providing network services where QoS is of prime concern. MPLS is being standardized as a technology that can be best described as a synergistic approach between IP and ATM. MPLS merges the flexibility of the IP routing protocols with the speed that ATM switches provide to introduce fast packet switching in frame-based IP networks. The next sub-section describes the reasons for the interest in MPLS. This is followed by a discussion on MPLS concepts in section 19.2.2.

19.2.1 Motivation for MPLS

The requirements for MPLS have been largely related to the observed deficiencies in the Internet Protocol. Some of the main motivations for the development of MPLS are as follows:

- **Efficient forwarding mechanism:** The traffic on the Internet has been increasing rapidly. To support this increase in traffic, the bandwidth of the links carrying this traffic is increased alongside. However, the speed of the IP routers to forward this increased traffic is fast becoming a bottleneck. The prime reason for this situation is the complexity in the algorithm that IP routers implement to forward packets. IP routers implement an algorithm that is based on the *longest-match prefix* approach to determine the next-hop for a packet. This is an extremely time consuming algorithm, which is repeated at each node for each packet along the path to the destination. A need is hence being felt to overcome this bottleneck by developing a more efficient forwarding mechanism.
- **Support for traffic engineering and explicit routing:** Traffic Engineering is the process of utilizing the links in the network optimally. This requires load balancing on links so as to ensure that a situation does not arise in which some links are congested and some under-utilized. A related concept is the concept of explicit routing, whereby a node can explicitly specify the path to be followed to reach the destination, instead of making a hop-by-hop decision at each node.

  Unfortunately, IP does not have an elegant mechanism for supporting either of these concepts. In IP, forwarding of packets is still being done mainly on the basis of the number of hops to reach the destination, without much concern about the bandwidth of the link or the congestion conditions on the link. Typically, load balancing is not done in IP networks, even if multiple paths exist towards the destination and the packets are routed to the best route. Further, even though the IP options field in the IP packet header can be used to specify the explicit path, routers generally do not implement this feature. This is because implementing this option would require extra processing for each packet at each node, and would thus add to the already high processing done for route table look-up.
- **Support for quality of service:** IP traffic has traditionally been designed to provide *best-effort* service. However, lately, Internet users are becoming more and more demanding, especially in terms of the quality of service they receive from the service provider. Traditional IP networks had no means to provide users with any QoS guarantees.

However, IETF has now standardized the Differentiated Services (Diffserv) standards to define and achieve QoS in IP networks. MPLS is also being developed to use a concept similar to Diffserv to provide QoS in IP networks.

19.2.2 How MPLS Works

In order to understand how MPLS works, one first needs to understand how a traditional IP router works. When an IP router receives a packet, it first maps the packet into a Forwarding Equivalence Class (FEC). Normally, the mapping into an FEC is done solely on the basis of the destination IP address of the packet. Once the classification is done, the route table is looked-up to determine the next hop for the FEC. This process of classification into an FEC and the subsequent route table look-up to forward the packet is done at each router node in the path from the source to the destination.

MPLS is designed to overcome this hop-by-hop process of classification and forwarding. In MPLS, the classification of a packet into an FEC is done only at the ingress point in the MPLS network (at the *ingress MPLS router*). In other words, the packet classification is done only at the time when the packet is entering the MPLS domain. Based on this one-time classification, a label is attached to the packet. Subsequent nodes within the MPLS domain only perform a label-table look-up to switch the packet towards the destination.

To a large extent, the concept of label and label tables in MPLS is similar to the concept of VPI/VCI and VPI/VCI translation tables in ATM. Just like ATM signalling protocols work to populate the VPI/VCI switching tables across ATM switches, a protocol is required in MPLS to populate the label tables across MPLS routers within an MPLS domain. This functionality is provided by the Label Distribution Protocol (LDP). Besides LDP, IETF has also extended the Resource Reservation Protocol (RSVP) protocol to achieve the same purpose of label distribution. The MPLS network operator, based on multiple technical and non-technical factors, chooses either RSVP or LDP for label distribution. Paths set up using the LDP or RSVP protocol is called Label Switched Path (LSP) in MPLS. LSP in MPLS is similar to the concept of SVCs in ATM.

Figure 19.1 depicts an example MPLS network, with two core MPLS routers, and two boundary routers, also called the *edge routers*. The edge router handling the flow of traffic coming into the MPLS domain is called the *Ingress MPLS router*, while the edge router handling the flow of traffic going out of the MPLS domain is called the *Egress MPLS router*. Due to the nature of the function they perform, MPLS routers (Edge and Core) are also called Label Switching Routers (LSR).

As depicted in the figure, the Ingress LSR maintains a Classification Table, using which incoming packets are classified into an FEC and assigned a label. Once the classification is done, core LSRs within the MPLS domain only switch the packets based on a look-up into the label table. It is evident that the process of label table look-up is much simpler than IP routing table look-up, performed by an IP router. Unlike the longest-prefix-match process followed in IP routing table look-up, MPLS label table look-up involves finding the first exact match of the label in the table. Thus, this is much faster than the route table look-up performed by IP

FIGURE 19.1

MPLS Label Switching

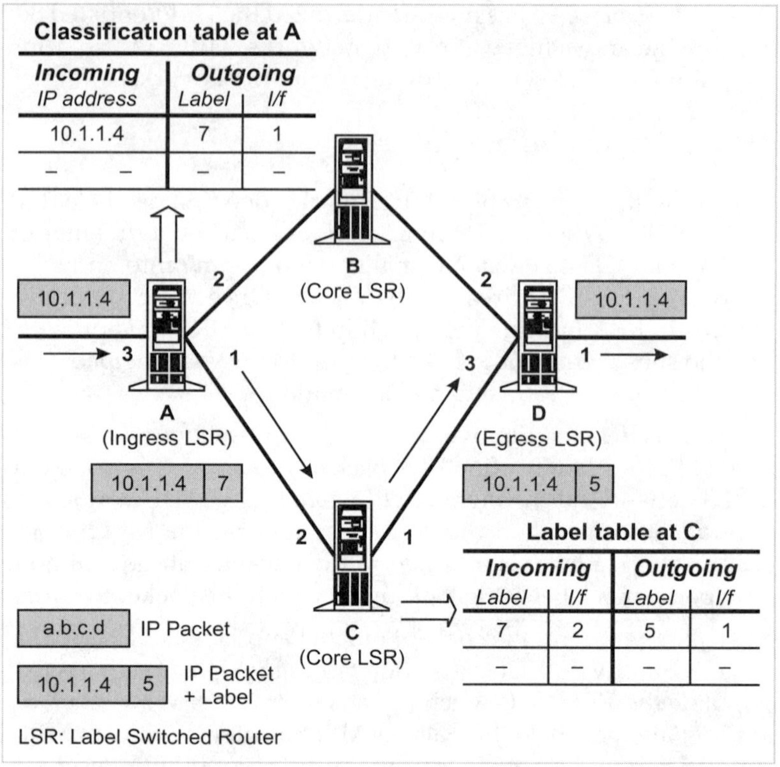

routers. The classification/label tables maintained by MPLS LSRs are themselves filled by using signalling protocols like LDP/extended-RSVP.

19.2.3 Quality of Service (QoS) in MPLS

Multiple features help MPLS to support and provide QOS. The classification of a flow of packets into a FEC and the corresponding setup of the Label Switched Path can be done by taking into account the QoS requirements of the packet flow. Two flows between the same source and destination nodes can be classified into different FECs depending upon the QoS required by them, and they can be then switched along different paths. The reduced processing required for packet forwarding also aids in reducing the per-hop processing time, and thereby making it feasible to meet the timing requirements of traffic flows. Further, the use of traffic engineering and explicit forwarding feature of MPLS, described in the next section, also helps MPLS in achieving the guaranteed QoS.

19.2.4 Traffic Engineering in MPLS

The use of LSPs in MPLS can help administer the traffic on the network links evenly. Not only can the LSPs be configured to take paths other than the ones dictated by the Interior Routing

Protocols (for example, the shortest paths according to the Interior Gateway Routing Protocol (IGRP)), but by moving a part of the traffic on to other available LSPs reduces the traffic load from the IGRP paths. Thus, it helps in uniformly distributing the traffic on the other links in the network, which are not parts of the IGRP shortest path.

LSPs in MPLS can also be established using the explicit path specified by the source. RSVP, for instance supports an explicit route option, in which the source can specify the path that it wants the LSP to take. Once this path has been established, the traffic would flow along this explicit routed LSP.

19.3 ATM and MPLS

IETF started the MPLS standardization process with the intention of introducing label switching within IP networks. While the initial focus was on IP networks alone, the MPLS framework is flexible enough to work with multiple protocols. Due to its close resemblance to the ATM protocol, MPLS has many possible applications when used in conjunction with ATM. Amongst the two main areas of research involving ATM and MPLS is the support of IP over ATM using MPLS, and the concept of ATM-MPLS network interworking. Each of these concepts is discussed in detail in the following sections.

19.4 MPLS for IP Over ATM

The flexibility of IP and the advantages of ATM cannot be over-emphasized. Both IP and ATM have their own advantages, and it is certain that both IP and ATM are bound to stay. Lately, a lot of work has been going on to merge the advantages of both IP and ATM, by providing IP network services over an ATM network. Chapter 18 described the Classical IP over ATM model, as well as the ATM Forum-proposed Multi-Protocol Over ATM (MPOA). However, many argue that the MPOA specification proposed by the ATM Forum might not be able to find much acceptance. Amongst the major hindrances felt for the deployment of MPOA is the feeling that the MPOA protocol is too complex to deliver effective interoperable solutions.

Given this, MPLS is projected as a synergistic approach between IP and ATM. Since MPLS uses the flexibility of the IP routing protocols, and a label switching paradigm similar to ATM, it can play a significant role in supporting IP over ATM. Realizing this, the IETF is working on specifying procedures for supporting MPLS in ATM environments.

Figure 19.2 depicts the IP over ATM model using MPLS. Multiple MPLS-based IP networks can be connected to each other via an ATM backbone. Within the ATM backbone, MPLS packets (IP packets with MPLS labels) are switched towards the destination MPLS domain using ATM switching. At the interface between the MPLS and the ATM domain there are interworking nodes that can work as an MPLS edge LSR as well as an ATM switch. These interworking nodes are called ATM LSRs.

FIGURE 19.2

IP Over ATM Using MPLS

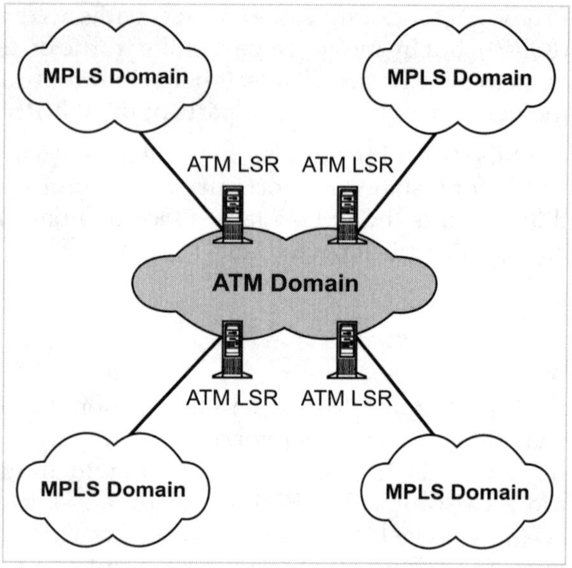

ATM LSRs perform the following broad functions (see Figure 19.3):

- **Frame to cell translation:** The ingress ATM LSR encapsulates the incoming MPLS packet as an AAL5 PDU. This packet is then segmented into one or more cells, depending upon the size of the packet. The ingress ATM LSR then uses the value of the label in the incoming MPLS packet to encode the ATM VPI/VCI field of each cell formed using the encapsulated MPLS packet. The egress ATM LSR performs the reverse process of re-assembly of the AAL5 PDU, and the de-encapsulation of the MPLS packet from within the AAL5 PDU.
- **ATM cell switching:** After cell formation, the ingress ATM LSR switches the cells to within the ATM network. The ATM switches within the ATM network switch the ATM cells, as they would do for any other ATM traffic. ATM switches within the ATM domain are not aware of the contents of the ATM cells. Hence, this model is also sometimes called the *Ships-in-the-Night* model, to signify how cells with MPLS-based payload pass through transparently, just like ships sail through in the night.

FIGURE 19.3

MPLS LSPs and ATM VCCs

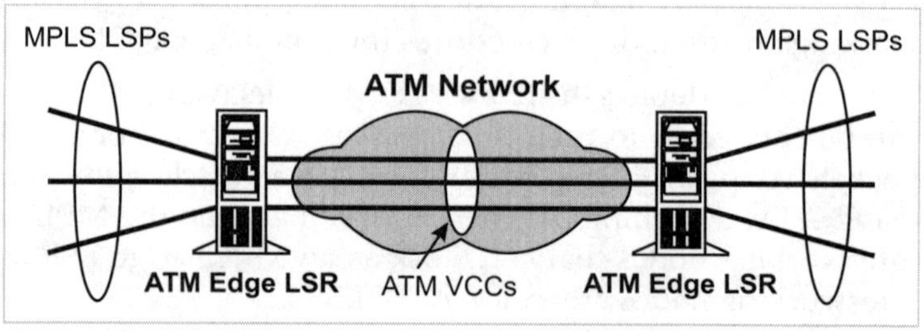

Using the aforementioned approach, it is possible to have an IP over ATM solution, whereby MPLS can be used to carry IP packets via an ATM-based backbone. However, it is too early to predict whether this approach of carrying IP traffic over ATM will at all become successful. Meanwhile, research is proceeding on another aspect of ATM-MPLS interworking, and this is the topic of discussion of the next section.

19.5 ATM-MPLS Network Interworking

As discussed in the previous section, ATM networks can be used as backbone networks to connect multiple MPLS networks and to provide IP over ATM services. Just like ATM networks can be used as backbone networks to connect multiple MPLS networks, the reverse scenario is also possible. This leads us to the concept of ATM-MPLS network interworking, which is standardized by the ATM Forum and is documented in the ATM Forum specification [ATMF MPLS 2.0].

As depicted in Figure 19.4, the ATM–MPLS network interworking is aimed to provide a transparent transport of ATM connections across an MPLS network. In other words, the MPLS network acts as a backbone to connect multiple ATM networks. To the ATM networks, the MPLS network appears to be a set of one or more ATM links within another ATM network. These ATM links are, however, in reality, MPLS Label Switched Paths or LSPs, which carry the ATM traffic towards the destination ATM network. Just like ATM LSRs are required to interwork between MPLS and ATM networks in an IP over ATM solution, special Internetworking Network Elements (INEs) are required to perform the interworking between the ATM and MPLS networks to achieve ATM–MPLS network interworking. An INE includes interworking functions for the user plane, control plane and the management plane of ATM and MPLS networks. The INE itself can be a stand-alone network element, or be a part of an Edge ATM switch or an Edge MPLS LSR.

Figure 19.4

ATM–MPLS Network Interworking

19.5.1 ATM-MPLS Encapsulation Format

At the entry point to the MPLS network, ATM cell-payload is encapsulated in a frame format as depicted in Figure 19.5. The ATM-MPLS encapsulation frame format contains the following fields in the frame header:

- **Transport label:** The four-octet transport label identifies a transport LSP in the MPLS network that is used to transport the traffic between two ATM-MPLS INEs. This label is

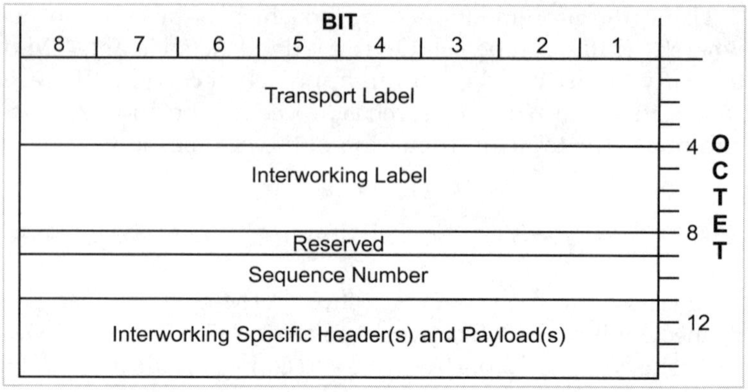

FIGURE 19.5

ATM–MPLS Encapsulation Format

used by the MPLS core LSRs to switch the packets carrying ATM payload towards the destination ATM–MPLS INE (and hence, towards the destination ATM network).

- **Interworking label:** The four-octet interworking label identifies an interworking LSP nested within the transport LSP (see Figure 19.6). One interworking LSP carries ATM traffic for one ATM virtual connection. Thus, if traffic from multiple ATM connections is to be carried between the same pair of ATM–MPLS INEs, then this is done by having multiple nested interworking LSPs within one transport LSP. Within the MPLS domain, the MPLS core LSRs will only look at the transport label to forward the packets towards the destination ATM–MPLS INE. The destination MPLS-INE will then use the interworking label to segregate the traffic for different ATM connections. In other words, two ATM–MPLS INEs connected via a transport LSP use the interworking label as context information to identify the ATM connection type (VPC or VCC) and the corresponding VPI/VCI values for individual ATM connections within the same transport LSP.

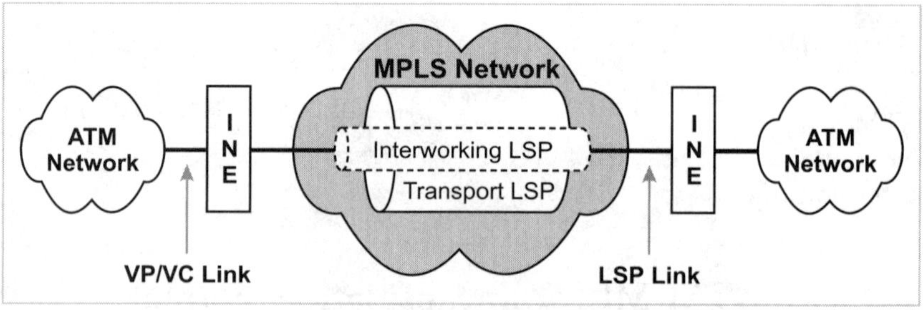

FIGURE 19.6

Transport LSP and Interworking LSP

- **Sequence number:** The sequence number is a two-octet field that is optionally used to verify in-order delivery of packets across the intermediate MPLS network. It can also be used to identify lost packets.
- **Interworking specific header and payload:** The ATM payload of the ATM cells is carried as part of this field. An Interworking Specific Header (ISH) is attached on top of the ATM payload, to include certain protocol specific information. As part of this

protocol specific information in the ISH is the information about the encapsulation mode that is used to carry the ATM payloads. Four different encapsulation modes are defined for carrying ATM traffic within MPLS packets, and these are discussed in the next section. Depending upon the encapsulation mode used, more than one occurrence of the ISH and ATM payload might exist in the ATM–MPLS encapsulation frame, as depicted in Figure 19.5.

19.5.2 ATM–MPLS Encapsulation Modes

Four different modes of encapsulation are defined in the ATM Forum specification [ATMF MPLS 2.0]. These are categorized as either *cell encapsulation modes* (in which ATM cells are exchanged) or *frame encapsulation modes* (in which AAL5 PDU is exchanged).

The cell encapsulation mode can be of three types. The first type is the single cell encapsulation mode that is also known as the one-to-one mode. In single cell encapsulation mode, only one ATM cell is encapsulated in an ATM-MPLS encapsulation frame. This is as depicted in Figure 19.7.

FIGURE 19.7

Interworking Specific Header (ISH) and Payload for Single Cell Encapsulation

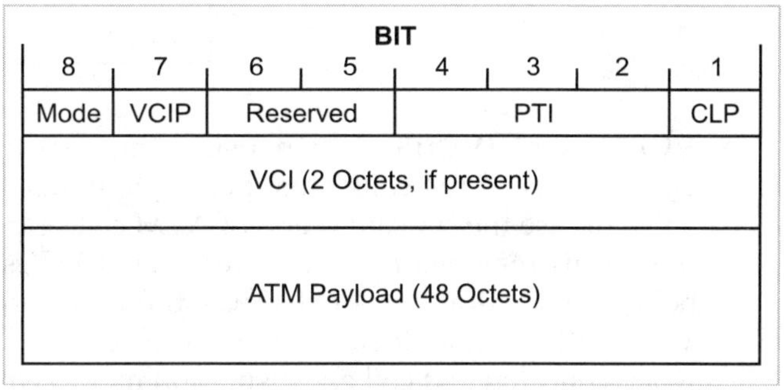

The second mode is the concatenated cell encapsulation mode (with no VCIP optimization). In this mode, more than one ATM cell is encapsulated in a single ATM–MPLS encapsulation frame. Thus, in this case, more than one occurrence of the ISH and the ATM payload will exist in the ATM–MPLS encapsulation frame, as depicted in Figure 19.8.

The third mode, which is a variant of second mode, is the concatenated cell encapsulation (with VCIP optimization). This mode of encapsulation is similar to the concatenated cell encapsulation mode (with no VCIP optimization), except that in this mode, some optimization is performed when encapsulating cells of different VCI but same VPI. This is discussed a little later, when discussing the VCIP field. Encapsulation scheme for this mode is same as depicted in Figure 19.8.

The fields of the Interworking Specific Header (ISH) for cell encapsulation mode is described below:

- **Mode:** This one bit field is used to identify if cell or frame encapsulation mode is used. For all cell encapsulation modes, this field is set to 0.

FIGURE 19.8

ISHs and Payloads for Concatenated Cell Encapsulation

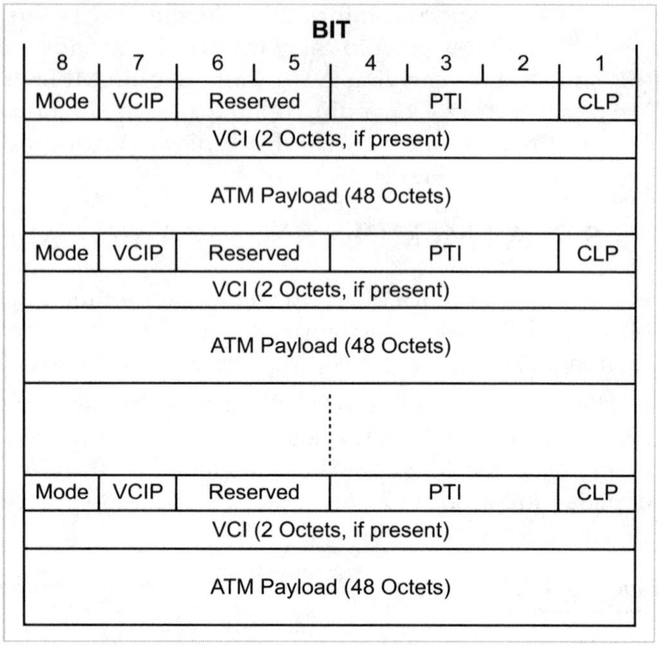

- **VCI Present (VCIP):** When using the concatenated cell encapsulation mode, the encapsulated ATM cells must belong to either the same ATM VCC or the same ATM VPC. In case the concatenation of ATM cells is performed for an ATM VPC connection, these cells may belong to different ATM VCCs within the VPC. To identify the cells belonging to different ATM VCCs, the concatenated cell encapsulation mode (with no VCIP optimization) includes the two-octet VCI field with each ATM payload. However, when concatenated cell encapsulation mode (with VCIP optimization) is used, the two-octet VCI field is only included in two conditions:
  - In case the cell is the first cell in the MPLS frame
  - In case the previous cell in the MPLS frame belongs to a different VCC.

- For the other cells, the VCI field is not included. This helps in saving some bytes in the MPLS frame, when contiguous cells of the same ATM VCC are encapsulated. The VCIP field is thus used to indicate if the two-octet VCI field is included in the ISH or not.
- **Payload Type Identifier (PTI):** This field is used to carry the PTI value of each encapsulated ATM cell.
- **Cell Loss Priority (CLP):** This field is used to carry the CLP value of each encapsulated ATM cell.
- **VCI:** This two-octet field is optionally included, if the VCIP field indicates so. If included, this field identifies the VCC of the encapsulated ATM cell, and of all following ATM cells (in case of concatenated cell encapsulation mode with VCI optimization) till a new VCI value is encountered.

The fourth encapsulation mode, that is different from cell encapsulation mode, is the frame encapsulation mode. This encapsulation mode is also known as the *AAL5-PDU-mode* encapsulation. In this mode, an ATM-MPLS INE at the entry point of the MPLS network first reassembles ATM cells into an AAL5 PDU, before transportation via the MPLS network. This reassembled AAL5 PDU (or a fragment of this AAL5 PDU) is then carried in one ATM-MPLS encapsulation frame (see Figure 19.9).

FIGURE 19.9

ISH and Payload for Frame Mode Encapsulation

BIT							
8	7	6	5	4	3	2	1
Mode	VCIP	Reserved			UU	EFCI	CLP
Payload (multiple of 48 Octets)							

For the frame encapsulation mode, the Interworking Specific Header (ISH) contains the following fields:

- **Mode:** This one bit field is used to identify if cell or frame encapsulation mode is used. For frame encapsulation modes, this field is set to 1.
- **VCI Present (VCIP):** When using the frame encapsulation mode, this field is set to 0 to indicate that the two-octet VCI field is not present in the ISH.
- **User-to-User (UU):** This bit contains the least significant bit of the PTI field of the last ATM cell encapsulated within the MPLS frame.
- **Explicit Forward Congestion Indicator (EFCI):** This field carries the middle bit of the PTI field of the last ATM cell encapsulated within the MPLS frame.
- **Cell Loss Priority (CLP):** This field carries an aggregated CLP value of the encapsulated ATM cells within the MPLS frame.

In case of the frame encapsulation mode, the payload following the ISH (see Figure 19.9) carries either the complete reassembled AAL5 PDU, or a fragment of the reassembled AAL5 PDU. The payload length is always kept as a multiple of 48 octets.

Conclusion

This chapter discussed the interworking aspects of ATM and MPLS technology. MPLS is a new paradigm introduced by the IETF to allow fast packet switching in IP-like packet switched networks. Due to its close resemblance to the ATM technology, MPLS has may applications when used in conjunction with ATM. Two such applications, namely, IP over ATM using MPLS and ATM–MPLS network interworking were discussed in this chapter. In the first case, ATM provides means for realizing MPLS networks. In the second scenario, MPLS networks provide interworking between two ATM networks.

Review Questions

Q 1. Compare and contrast the MPLS framework with the ATM networking principles.

Q 2. Describe some of the major aspects of the MPLS technology that allow it to provide QoS guarantees when used in a best-effort network like IP.

Q 3. Describe the Ships-in-the-Night model for carrying IP traffic over ATM. What components of an ATM switch can be retained for use in the ships-in-the-night model? What new components will have to be added to an ATM switch for this model?

Q 4. To some extent, the concept of Transport Label and Interworking Label used in the ATM–MPLS encapsulation frame is similar to that of VPI and VCI in ATM frames. Discuss.

Further Reading

The concepts presented in this chapter are derived from standards produced by the ATM Forum and the IETF. The MPLS-related concepts are described in [RFC 3031] and [RFC 3032]. Concepts related to use of MPLS for IP over ATM are described in [RFC 3035] and [RFC 3038]. The ATM–MPLS network interworking concepts are described in [ATMF MPLS 2.0].

Chapter

20

Voice Over ATM

20.1 Introduction

The growth of ATM is leveraged by the requirement to have a universal platform that can support different types of applications. Primarily, the networking requirements are classified into two broad categories, viz. voice communication and data communication. Voice communication requires a fixed bandwidth and low delay and is tolerant to occasional information loss. On the other hand, data communication is bursty in nature, and although data is tolerant to higher transit delays, it is susceptible to information loss. This basic difference in the requirements for voice and data communication makes the design of a universal platform difficult. Thus, prior to the deployment of ATM, the networking fraternity was divided into two groups—the telecommunication world and the datacommunication world. The telecommunication world caters to services required for voice conversation. The services include basic telephony and facsimile (fax). The datacommunication world caters to services required for data transfer. The services include data transfer in LAN and WAN environment. However, during the last decade or so, the distinction between a telecommunication network and datacommunication network has become very hazy. This is due to packetization of voice and introduction of integrated service networks.

Given this, two worlds will ultimately converge into one, a phenomenon termed as *convergence*. ATM is developed to expedite the process of convergence and provide a synergy between the telecommunication and datacommunication world. Support of voice over ATM is a step in this direction. This chapter looks at the challenges in carrying voice over ATM and the different solutions available for the same.

20.2 Technical Challenges

Among different traffic types, voice is among the most difficult to carry owing to the real-time considerations and other related issues like echo handling/cancellation, maintaining synchronization, and ensuring a good Mean Opinion Score (MOS), a subjective test to measure the quality of voice transmission. In circuit-switched networks wherein there is a dedicated connection, it is easier to provide a good transmission medium for voice transfer. However, for a packet-switched network (including IP in which packets are routed and ATM in which cells are switched using virtual circuits), there are many technical challenges in carrying voice over it. This section looks at the technical issues encountered in carrying voice over ATM.

20.2.1 Delay and Echo

The first problem in carrying voice is the delay encountered in carrying voice. For circuit-switched network where there is bit-oriented transmission, the issues related to delay is considerably less significant (unless satellite links are used where there is considerable pauses between a statement and its response). However, in packet-based networks, there are a number of sources that contribute to the delay, which are as below:

- **Encoding delay:** This is the time taken by an algorithm to encode the analog voice signal to digital form. Due to progress in codecs, there are algorithms currently available like the Low Delay Code Excited Linear Prediction (LD-CELP) that allow toll-quality encoding to be achieved at bit rates as low as 16 Kbps. However, these codec require less than 10 milliseconds for the encoding process. Typically, the better is the voice compression (implying lesser bandwidth needed for voice), higher is the encoding delay. In other words, a better encoding algorithm would require more time to encode the speech more efficiently.
- **Packetization delay:** While voice samples in circuit-switched samples are transmitted as soon as they are available, the same is not the case in packet-switched networks where the packets need to be filled before they can be transmitted. This was one of the reasons to keep the cell size small such that the packetization delay could be controlled. However, for low rates, even 48 bytes could take sufficient time to be filled. For a 16 Kbps channel, the packetization delay would be 24 ms. In order to solve this problem, partially filled cells can be used but this would mean less efficient utilization of bandwidth. Another option that is used in AAL2 is to allow a payload to be used for several voice channels.
- **Transmission delay:** Another source of delay is the time taken to transmit signal from source to destination. This parameter is dependent on the distance between the two.
- **Switching and queueing delay:** This includes the delays encountered during the cell relaying process and also includes the time taken inside the buffers/queues of the switches.
- **Buffering delay:** In order to play the packets at the destination smoothly and to remove the jitter at the destination, some delay may be introduced to buffer the received contents before sending to the actual recipient. This causes additional delay.

The delay caused by the factors above lead to various problems. First of all, at 30 ms of delay, the echo becomes significant and echo cancellers are required. Even with echo cancellers, after 150 ms, the voice conversation is hindered, as people do not get the response in the expected time.

20.2.2 Signalling Support

Voice communication requires not only voice samples to be delivered in bounded delay and jitter, but also considerable support for signalling. The signalling includes the called party number, the on-hook/off-hook status of the call, and possibly a variety of other information. As discussed in Chapter 11, the signalling can be Common Channel Signalling (CCS), where a separate channel is used solely to carry signalling information for a number of connections or can be Channel Associated Signalling (CAS) where signalling information is embedded within each discrete voice channel. Packetization of voice requires that at times the network needs to understand all applicable signalling protocols (that include standardized and proprietary implementation). At times it is observed that some signals are not available in one type of network due to interoperability issues (e.g. the Interactive Voice Recognition (IVR) may not decode the inputs fed from a packet phone). While standardization bodies attempt to remove such issues, the possibilities of such challenges continue to exist.

20.2.3 Other Considerations

Another challenge in carrying voice in asynchronous networking solutions like ATM is maintaining synchronization (i.e. timing relation between sender and receiver). In Chapter 6, the Synchronous Residual Time Stamp (SRTS) Method and Adaptive Clock Method were discussed for clock recovery in AAL1. While these techniques have their own limitations, they fail when multipoint services are used. Under such circumstances, the sender and receiver must synchronize with an external clock source.

Lastly, since the voice conversation has a characteristic that speakers speak for not more than 40 per cent of the time, transfer can be optimized by using silence suppression techniques.

20.3 Carrying Voice Over ATM

Even with the challenges mentioned above, carrying voice over ATM can be quite an attractive proposition. First of all, unlike IP or frame relay that have large packet/frame sizes, ATM has a small and fixed size (53 bytes) that results in lower packetization delay and a predictable switching/buffering delay. Further, unlike IP that provides best-effort service, ATM provides reliable service with QoS guarantees making it easier for carrying voice. The different traffic management facilities including CAC, policing and prioritization facilitates guarantee of QoS.

ATM has are different types of AAL for carrying different services. For real-time services including voice and video, there is AAL1 and AAL2. While AAL1 provides efficient circuit emulation services, AAL2 provides excellent means for low rate voice transfer. Even with extra headers, AAL2 can be more efficient than AAL1 (see [AAL2 GDC]).

Another advantage of carrying voice over ATM is that it allows the convergence of different networks for voice and data thereby reducing infrastructure costs and management costs.

It follows that carrying voice over ATM can be beneficial. The main target segments include the enterprise segment as well as the public service provider.

20.4 VTOA Standards

The ATM Forum has standardized a number of solutions for VTOA, which are listed in Table 20.1. The initial standards by ATM Forum were based on AAL1 while the newer ones are based on AAL2. AAL1 provides Circuit Emulation Service (CES) including its low-speed variant Low Speed Circuit Emulation Service (LSCES) and more efficient Dynamic Bandwidth Circuit Emulation Service (DBCES). The CES scheme is discussed in Section 20.5 and DBCES scheme is discussed in section 20.6.

Apart from circuit emulation, ATM also provides very good trunking solutions. The AAL1-based trunking solution is discussed in Section 20.7 while the AAL2 based trunking solution is elaborated in Section 20.8. Other specifications are not discussed in this chapter any further.

Table 20.1

ATM VTOA Standards

Standard	*Description*
[ATMF CES]	In this scheme, ATM emulates the behaviour of a circuit by providing a transparent transport pipe for carrying voice traffic.
[ATMF LSCES]	This standard provides CES for low-speed applications (less than 64 Kbps).
[ATMF DBCES]	This standard removes the drawback of CES such that idle slots are not carried.
[ATMF VTOA Desktop]	This describes the functions to be performed by a native ATM terminal and an IWF to provide service between B-ISDN terminals and/or N-ISDN terminals.
[ATMF AAL1 Trunk 1.1]	This standard, based on AAL1, provides the means to dynamically carry voice between narrowband networks.
[ATMF AAL2 Trunk]	This standard, based on AAL2, provides support to a broad range of applications involving interconnection of an IWF with narrowband and broadband facilities.
[ATMF AAL2 LES]	This standard provides efficient transport mechanism to carry voice, voice-band data, fax traffic, ISDN B/D-channels over a broadband subscriber line connection such as xDSL.

20.5 Circuit Emulation Service (CES)

One of the earliest techniques to carry voice over ATM is the Circuit Emulation Service (CES) standardized in [ATMF CES]. In this scheme, as the name suggests, ATM emulates the behaviour of a circuit by providing a transparent transport pipe for carrying user traffic. The circuit here could be *structured* or *unstructured*. In structured service, one or more 64 Kbps channels (referred as Nx64 service) are provided to the user. This structured service thus provides a fraction of DS1/E1 capacity and is thus modeled after a Fractional DS1/E1 circuit. The structure provides means to use as much bandwidth as necessary. In contrast, the unstructured service provides a data pipe of DS1 (1.544 Mbps), E1 (2.048 Mbps), DS3 (44.736 Mbps) or E3 (34.368 Mbps) capacity.

Figure 20.1 shows how two CBR equipments can connect each other using the CES service. The CES service is provided by an Interworking Function (IWF). The IWF has a CBR interface (e.g. DS1 or E1) on one side and ATM interface on the access network side. The CBR equipment is not aware of the fact that such an interworking is taking place. In this sense, the CES provides transparent service to CBR end-points.

Figure 20.1

ATM CES Interworking Function (IWF) Between Two CBR Equipments

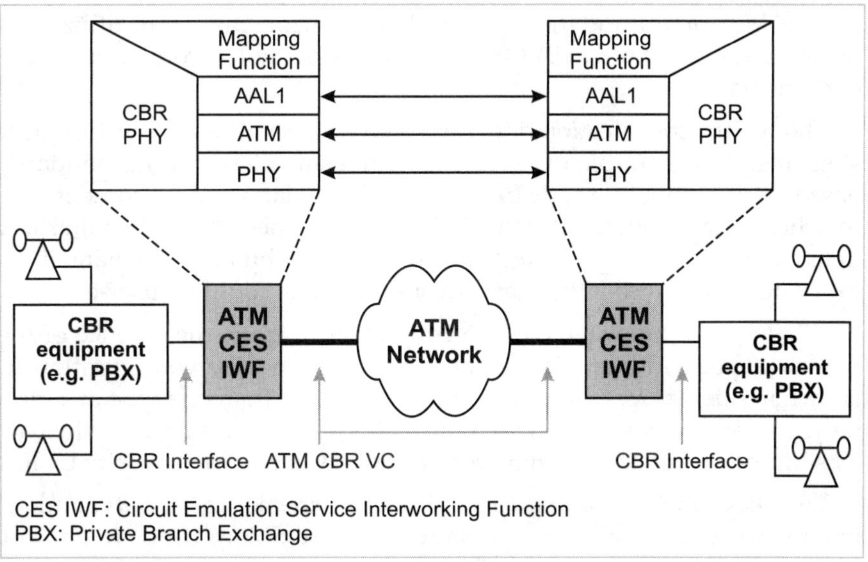

The CES uses Constant Bit Rate (CBR) service and operates over AAL1. Typically, the CBR service gets the first priority at the ATM switches. Thus, one can assume that the service requires fixed allocation at the switches. The [ATMF CES] specification essentially specifies the service in terms of framing, timeslot assignment in ATM cells, clock recovery, jitter and wander control, carriage of signalling bits, cell coding and bit ordering, the applicable traffic parameters (e.g. PCR), QoS parameters (e.g. CDV) and the support required for signalling (e.g. the coding of various information elements in the SETUP message as an example is depicted in Table 20.2).

CES is a simple solution and its transparency makes its implementation easier. Further, one physical link with the service provider can provide requisite capacity without really any changes in the hardware. However, the scheme has a couple of serious drawbacks. First of all, as the cell header is part of each payload, there is cell tax of about 10 per cent. Another drawback of this scheme is that it nullifies the use of statistical multiplexing because it does not differentiate between idle and active time slots and treats both of them as user data.

20.6 Dynamic Bandwidth Circuit Emulation Services (DBCES)

As mentioned in the previous section, the CES service does not distinguish between idle and active time slots. Thus, the CES does not use the bandwidth of ATM network efficiently. To remove this drawback, the Dynamic Bandwidth Circuit Emulation Services (DBCES) was standardized by the ATM Forum in [ATMF DBCES]. The basic principle behind the scheme is that when an inactive state is detected in a specific time slot, the time slot is dropped from the next ATM structure and the bandwidth it was using may be re-utilized for other services. The technique applies both to SVC and PVC and does not introduce any new signalling messages or elements.

Though the specification does not standardize a particular technique for detecting inactive slots, there are a couple of recommendations available in the standard. The first scheme is based on detecting idle code patterns. In particular, the data content of this time slot must be matched for a specified interval against a predefined idle code pattern. Another technique to detect idle slots is to use the A and B signalling bits of the Channel Associated Signalling. The values of A and B bits that indicate idle slots could be user-configurable.

Whatever be the means of detecting inactive state, the essence of the matter is to utilize network bandwidth efficiently. Figure 20.2 shows the configuration for using DBCES. As shown in this figure, DBCES is useful when more than one type of service is provided by the network element such that when the CBR service is not needed, the bandwidth can be used by some other service (e.g. data service over AAL5 or low priority UBR service).

Two key concepts associated with this scheme are the dynamic sizing and dynamic bandwidth utilization. The Dynamic Structure Sizing (DSS) refers to the ability of a CES interworking function to dynamically adjust the size of the AAL1 structure up or down based on the number of active time slots contained in the DS1/E1 trunk undergoing circuit emulation. However, a maximum size is defined which is used when all slots are active. Another important concept is the Dynamic Bandwidth Utilization (DBU). This is useful when some of the slots are inactive. Note that the CES is allocated some bandwidth which is used when all slots are active. In case some of the slots are inactive, then the DSS sends a dynamic structure that requires lower rate. For this, the DBU functionality is used such that the newly created bandwidth is used by some other service (e.g. UBR).

TABLE 20.2

SETUP Message Contents in for N × 64 CES Service

Information Element	*Fields*	*Specified Value*
ATM Traffic Descriptor	Forward peak cell rate CLP = 0 + 1	If partial cell fill is not used $\lceil(8000 \times N)/46.875\rceil$ If partial cell fill is used $\lceil(8000 \times N)/K\rceil$ where K is number of user octets
	Backward peak cell rate CLP= 0 + 1	- do -
Broadband Bearer Capability	Bearer class	'1000 0' BCOB-X
	Traffic type	'001' Constant bit rate
	Timing requirements	'01' End-to-end timing required
	Susceptibility to clipping	'00' Not susceptible to clipping
	User plane connection Configuration	'00' Point-to-point
Quality of Service Parameter	QoS class forward	'0000 0001' QoS Class 1
	QoS class backward	'0000 0001' QoS Class 1
ATM Adaptation Layer Parameters	AAL Type	'0000 0001' AAL Type 1
	Sub-type	'0000 0010' Circuit Transport
	CBR rate	'0000 0001' 64 kbit/s '0100 0000' Nx64 kbit/s, N > 1
	Multiplier	The value 'N' for Nx64 kbit/s. Omit field for 64 kbit/s case.
	Structured data transfer block size	Size in octets for structured data transfer
	Partially filled cells method	K, the number of AAL-user octets filled per cell. Omit if partial cell is not used.
Broadband Low Layer Information	User-information layer 3 Protocol	'01011' ISO/IEC TR 9577
	ISO/IEC TR 9577 Initial protocol Identifier (IPI) (octet 7a, 7b)	IPI is coded '1000 0000' to indicate IEEE 802. 1 SNAP identifier. Hence, octets 7a and 7b are coded as '0100 0000' and '0000 0000', respectively.
	Organizational Unit Identifier (OUI)	'x'00 A0 3E' ATM Forum OUI
	Protocol Identifier (PID)	'x'00 06' DS1/E1/J2 Nx64 Basic Service

FIGURE 20.2

Configuration for DBCES Service

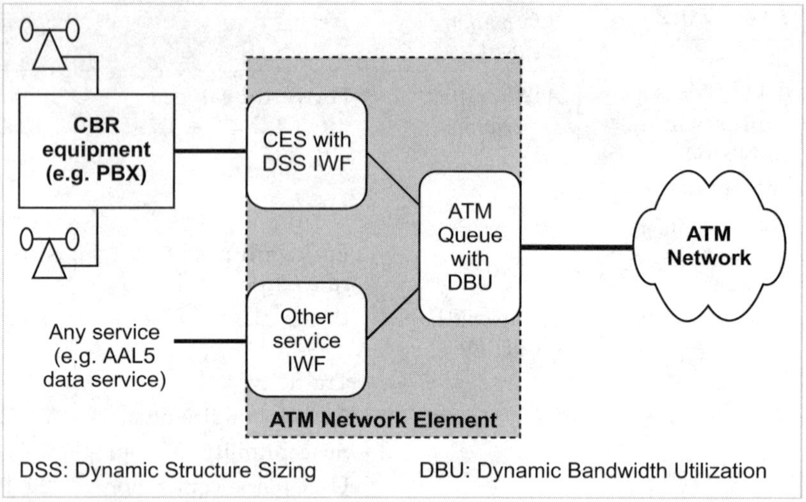

20.7 ATM Trunking Using AAL1 for Narrowband Services

While circuit emulation services discussed above provides data pipes for circuit emulation, there is a need for greater efficiency in the ATM backbone networks and in cases where there is interworking required between broadband (e.g. ATM) and narrowband (e.g. ISDN) networks. For this purpose, ATM Forum has standardized a solution for ATM Trunking using AAL1 for Narrowband Services in [ATMF AAL1 Trunk 1.1]. A trunk line generally refers to a pipe with high bandwidth. In this context, ATM trunk applies to the interconnection of two narrowband networks via an ATM network. Formally, "An ATM trunk is defined as one or more ATM virtual connections that carry a number of 64 Kbps narrowband channels and associated signalling between a pair of IWFs."

The key features of this AAL1 trunking solution are dynamic allocation of resources and call-by-call routing (as against static allocation of resources). In simple terms, this implies that there is no circuit emulation or static resource reservation. The VCC are established either when a new call set-up is required or is based on predetermined traffic patterns. Further, the call can be routed on a per-call basis (through the use of dynamic signalling procedures). This allows the bandwidth to be shared by multiple services.

The basic network configuration for AAL1 trunking solution using a trunking IWF is depicted in Figure 20.3. The three vertical stacks on ATM interface side is explained as follows. The *mapping function* provides means to extract voice samples from time slots and transmit them as ATM cells. However, to set-up the AAL1 connection for this transfer, the *Q.2931* signalling protocol is required (third vertical column). Another connection is set-up between the two IWFs such that the IWF can exchange signalling messages. This IWF-IWF signalling essentially provides means for exchange of narrowband messages between the

FIGURE 20.3

ATM Trunking Interworking Function (IWF) Between Two Narrowband Equipments

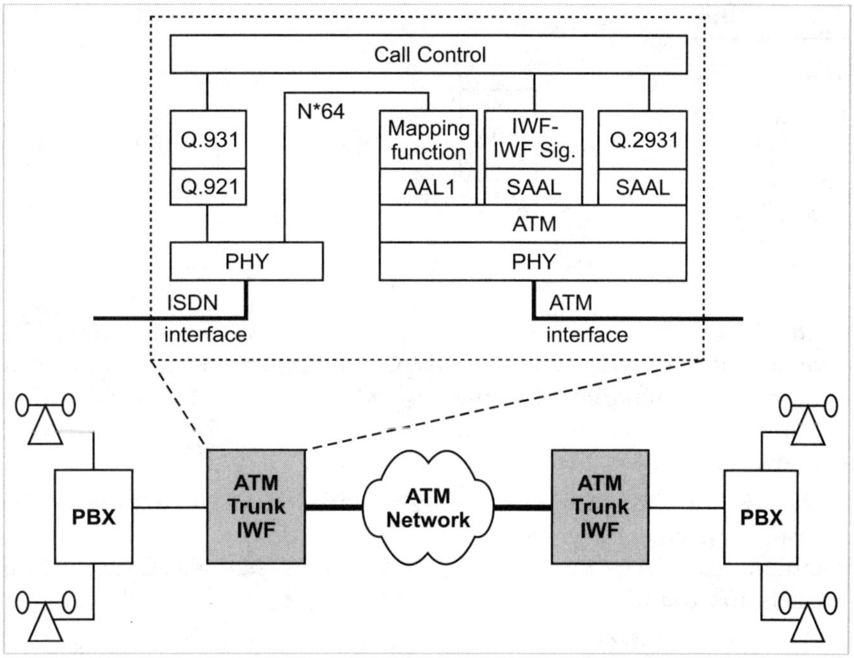

IWFs. In this sense, the narrowband signalling is maintained end-to-end. The destination IWF is determined through the called-party address in the narrowband signalling message.

20.8 ATM Trunking Using AAL2 for Narrowband Services

The ATM trunking solution specified in [ATMF AAL2 Trunk] provides support for a broad range of applications involving interconnection of an IWF with narrowband and broadband facilities and interworking with various other telecommunications devices including PBXs and ATM network switches. Figure 20.4 shows the trunking Interworking Function (IWF) between PBX and PSTN based on AAL2.

In this solution, two modes of trunking are supported, namely, *switched trunking* and *non-switched trunking*. In switched trunking, there is an analysis of the signalling that accompanies an incoming narrowband call. Based on this analysis, the bearer information is routed to an AAL2 channel within a VCC between IWFs. Similar analysis and routing is required for incoming calls from the ATM network as well. After the narrowband call has ended, subsequent calls occupying the same narrowband channel (TDM timeslot) may be switched to different AAL2 channels and VCCs. In summary, there is no permanent relationship between a narrowband channel and an AAL2 channel.

FIGURE 20.4

ATM Trunking Interworking Function (IWF) between PBX and PSTN

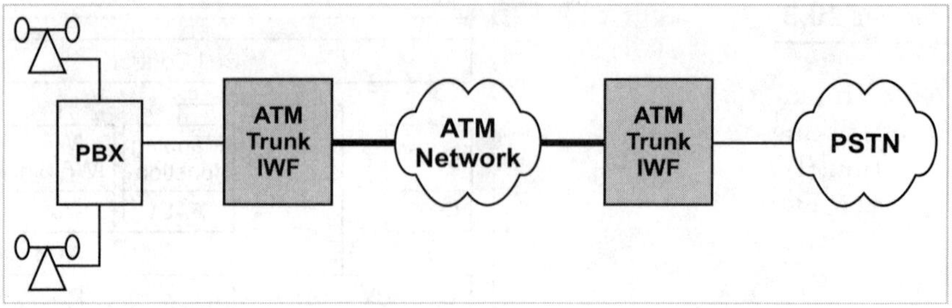

In non-switched trunking, there is a permanent correspondence between a narrowband channel and the AAL2 channel and VCC designated for its support. Non-switched trunking involves no termination of signalling and no routing of narrowband calls in the IWFs. Thus, the information stream of a narrowband channel is always carried on the same AAL2 channel within the same VCC and vice versa.

The IWF in this solution not only provides voice but also provide special capabilities to transport the voice band data through modem detection, fax data through demodulation and remodulation, circuit mode data for Nx64 kbit/s channels, DTMF information through DTMF packets and frame mode data through SAR functionality.

The protocol stack to provide these features is shown in Figure 20.5. As shown in the figure, the services are provided through the ATM layer. The AAL above could be AAL2 to provide bulk of the service or could be AAL5 for carrying CCS signalling information. Above AAL2, Common Part Sub-layer (CPS), there are different Service Specific Convergence Sub-

FIGURE 20.5

ATM Protocol Architecture for Providing Different Services

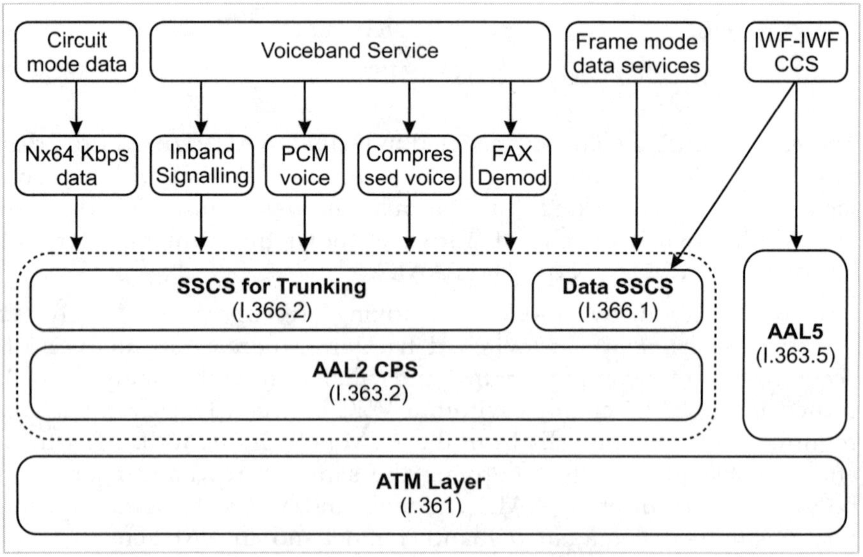

layer (SSCS) for trunking [ITU-T I.366.1] and data [ITU-T I.366.2]. These SSCS layers provide enhanced features (e.g. compression) over AAL2.

Conclusion

This chapter discussed the solutions for VTOA. Different standards were discussed, each having its purpose and associated advantages and disadvantages. While CES provides leased link services, there are modifications for dynamic bandwidth utilization in DBCES. The trunking solutions could be provided through AAL1 and AAL2, but the latter is preferred due to its higher efficiency.

Review Questions

Q 1. *What are the major challenges in carrying voice over any packet network? How are these issues addressed in ATM networks?*

Q 2. *Which of the AALs, AAL1 or AAL2, is better for carrying voice, and under what circumstances?*

Q 3. *What is Circuit Emulation Service (CES)? What are its drawbacks? How are these removed in Dynamic Bandwidth Circuit Emulation Service (DBCES)?*

Q 4. *What is trunking? How is voice trunking done in ATM? How is trunking done using AAL1 and AAL2?*

Q 5. *What is the role of the Service-specific Convergence Sub-layer (SSCS) in voice trunking using AAL2?*

Further Reading

The concepts presented in this chapter are derived from standards produced by the ATM Forum as listed in Table 20.1. Apart from this, [VTOA D. Minoli] is a very good book dedicated to the topic. [VTOA R. Jagannathan], [VTOA S. Kasera] and [VTOA ATM Forum] are also good papers/presentations on the subject.

Chapter 21

ATM and DSL Networks*

21.1 INTRODUCTION

For decades, residential users have been using the dial-up modem to access the Internet via the Public Switched Telephone Network (PSTN). However, with the theoretical upper bound of 56 Kbps on the bandwidth available using dial-up modems, other broadband access technologies are being explored. Not all solutions proposed for broadband access provide an easily deployable solution. Most solutions for broadband access involve huge infrastructure costs in installation of new network elements and networking of these elements by cables. This means that the solutions are not cost-effective, and beyond the reach of the common users.

Amongst the proposed solutions for broadband access, Digital Subscriber Line (DSL) is being seen as a technology that will overcome the above limitations of cost to offer a cost-effective broadband access solution. DSL proposes to use the existing infrastructure of the Plain Old Telephone Systems (POTS) to provide a high-rate access solution. DSL is designed to provide typical data rates of up to 2 Mbps over the existing copper cables installed for the POTS network. This chapter provides a description of the DSL technology, and goes on to discuss the importance of ATM as an enabler technology for DSL.

21.2 OVERVIEW OF DIGITAL SUBSCRIBER LINE (DSL)

Digital Subscriber Line (DSL) is a technology that is used to provide an access solution for the home and business environments. It has the potential to become the most promising

* This chapter is contributed by Nishit Narang.

broadband access technology. For its operation, DSL uses the existing copper infrastructure that is installed for the Plain Old Telephone System (POTS) network. For decades, POTS networks have been utilizing only the lower 4 KHz spectrum out of the approximately 1 MHz spectrum available on the copper cable. Usage of frequencies above this 4 KHz spectrum causes a lot of near-end cross talk, and hence, these frequencies have been left unutilized for decades (see Figure 21.1).

FIGURE 21.1

POTS Bandwidth Within the Copper Cable

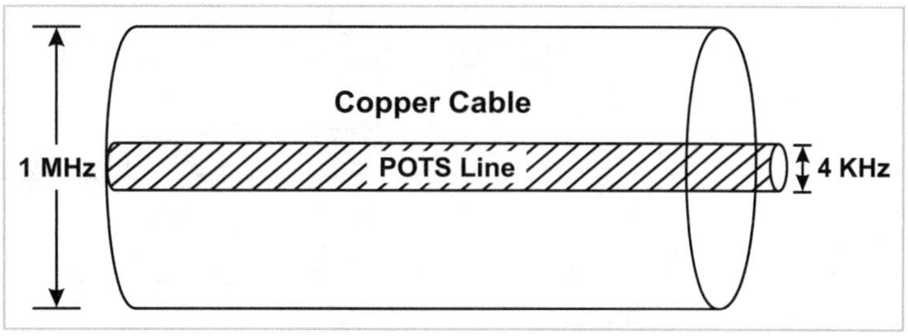

DSL technology uses highly sophisticated techniques to limit the near-end cross-talk when using higher spectrum frequencies. The higher spectrum frequencies are used in DSL to carry data in digital form. Conventionally, dial-up modems have been used from residences and small offices for Internet access. Dial-up modems make use of the POTS spectrum for carrying the data in analog form, and thus the telephone line is engaged during periods of Internet access. On the other hand, DSL leaves the POTS spectrum free for use for voice communication, and uses only the unutilized higher frequencies available on the copper infrastructure for data transfer. In other words, DSL technology allows the simultaneous access of the Internet and usage of the telephone/POTS line.

21.2.1 DSL Network Architecture

The DSL network augments the existing POTS network by introducing two new network elements (see Figure 21.2). These network elements are DSL modem and DSL Access Multiplexer (DSLAM).

The DSL modem is installed within the customer premises and is used to originate and terminate the data traffic over the local loop. The DSL modem uses the available frequency spectrum on the copper cable, after servicing the POTS voice traffic, to send and receive data in digital form. The DSL modem connects to the Network Interface Device (NID), which is a simple hardware unit that terminates the POTS local loop.

Within the POTS network, a *splitter* is used to segregate the audible spectrum from the spectrum used by the data traffic. While the POTS voice traffic is forwarded by the splitter towards the telephone switch, the data traffic is forwarded towards the DSLAM. The DSLAM serves as a packet concentrator, which receives data traffic from multiple customers over DSL links, and sends this aggregate traffic towards the data network. Since the principle service

FIGURE 21.2

DSL Network Architecture

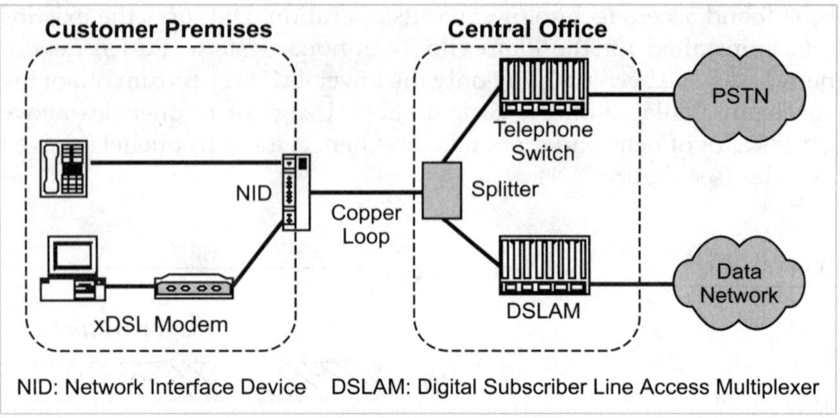

that is expected to be offered to DSL subscribers is Internet access, the data network connected to the DSLAM is most often the Internet.

In this regard, the DSL network architecture is a very simple architecture in the sense that it simply augments the existing POTS network architecture. DSL subscribers are only required to purchase a DSL modem to begin using the DSL service, while service providers are required to install DSLAMs at the many telephone exchanges within a POTS network.

21.2.2 DSL Flavours

The previous sections defined the high-level concepts for DSL. Multiple flavours of the DSL technology are defined. Some of these flavours are as follows:

- **Asymmetric DSL (ADSL):** ADSL is the most popular flavour of the DSL technology. As the name suggests, ADSL divides the available DSL bandwidth asymmetrically, such that more bandwidth is available downstream (from network to customer premises) than in upstream (from customer premises to network). Because of this asymmetric partitioning of bandwidth, ADSL is considered most suitable for providing Internet access to Small Office/Home Office (SOHO). ADSL deployments can support up to 12 Mbps traffic in the downstream and up to 2 Mbps traffic in the upstream direction.
- **Symmetric DSL (SDSL):** SDSL, as the name suggests, apportions the available DSL bandwidth symmetrically between downstream and upstream traffic. Hence, SDSL is not considered suitable for providing Internet-access type of applications. However, SDSL is more suitable for supporting applications that require equal bandwidths in either direction. Examples of such applications include video telephony and multimedia conferencing among others. SDSL deployments are available that support up to 2 Mbps traffic in both upstream and downstream directions.
- **Very-High-Rate DSL (VDSL):** VDSL technology is similar to the ADSL technology with respect to the partitioning of the DSL bandwidth. VDSL supports much higher data rates than the ADSL technology (up to 50 Mbps in the downstream). However, VDSL

technology has a severe limitation on the maximum possible distance between the DSL modem and the DSLAM. VDSL is designed to operate over extremely small distances (approximately 1000 feet). Hence, it is expected that the VDSL deployments will not occur independently, but in conjunction with optical fiber deployments, such that VDSL will be used along with Fiber-To-The-Curb (FTTC) technology.

- **High-Bit-Rate DSL (HDSL):** Unlike the rest of the DSL flavours (ADSL, SDSL and VDSL), HDSL does not support the transfer of traditional POTS traffic. HDSL uses two pairs of copper cable, instead of one, and provides T1 (1.5 Mbps) or E1 (2 Mbps) bandwidth in both upstream and downstream directions. Thus, HDSL is similar to SDSL with respect to providing symmetrical bandwidth in both directions.

Because of its multiple variants, DSL technology is also sometimes referred to as xDSL, to jointly refer to its flavours ADSL, SDSL, VDSL and HDSL.

21.3 ATM AND DSL

ATM technology plays an important role in providing broadband access to DSL subscribers. This section explores the use of ATM in a DSL network. Until about a couple of years back, DSL technology was considered synonymous with Internet access. However, recently, a variant of conventional DSL technology, called Voice over DSL (VoDSL) has evolved, which provides multiple voice communication channels along-with data access over the same DSL link. ATM plays an equally important role in VoDSL networks as well, and this is explored in further detail in Section 21.4. But first, this section discusses the role of ATM in the conventional DSL network.

Conventional DSL networks, as introduced in Section 22.2, are based on two broad models. While one model uses a pure ATM-based approach for broadband access, the other model uses the IP over ATM approach also called IPoA. These models are explored in the following two sections. Note that only the models discussed in the following sections explore how broadband access is provided over DSL using ATM. Access to conventional POTS-based voice services remain unaffected (as discussed in Section 22.2), and is not discussed in the following sections.

21.3.1 ATM-based Model for DSL

In the ATM-based model for DSL, ATM technology is used as the layer 2 and layer 3 (i.e. data link layer and network layer) technology for accessing the services provided by an ATM service network. The DSL link is treated as a physical link/media, over which ATM protocols can operate. The network architecture for an ATM-based model is depicted in Figure 21.3.

The network architecture for an ATM-based model as depicted in Figure 21.3 is similar to the generic DSL network architecture depicted in Figure 21.2. However, a few points are worth mentioning. First, in this model, the xDSL modem is also sometimes referred to as the ADSL Terminal Unit for the Remote site (ATU-R). The ATU-R may be used to connect a single

FIGURE 21.3

End-to-End ATM-Based Model for xDSL

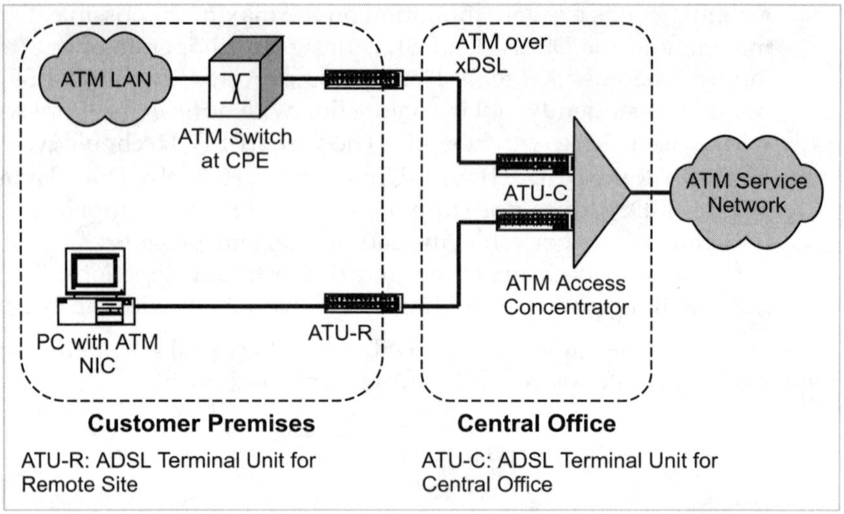

PC via a network interface card on the PC, or an entire ATM LAN via an ATM switch at the customer premises.

In case of an ATM-based model, the DSLAM can be seen as a combination of the ADSL Terminal Unit for the Central Office (ATU-C) and the ATM Access Concentrator. The ATU-C terminates the ADSL line originating from the customer premises at the central office, while the ATM Access Concentrator receives data traffic from multiple customers over DSL links, and switches this traffic towards the ATM service network.

The protocol stack architecture at the Customer Premises Equipment (CPE)/ATU-R and at the ATU-C/ATM Access Concentrator is depicted in Figure 21.4. The stack architecture at the CPE is similar to the ATM stacks existing on any ATM end-node. At the ATU-C/ATM Access Concentrator, the protocol stacks are similar to those that exist at any intermediate ATM switch. The DSL link between the ATU-R and the ATU-C thus serves as the physical layer carrying the ATM traffic.

FIGURE 21.4

Protocol Stack Architecture for ATM-Based Model

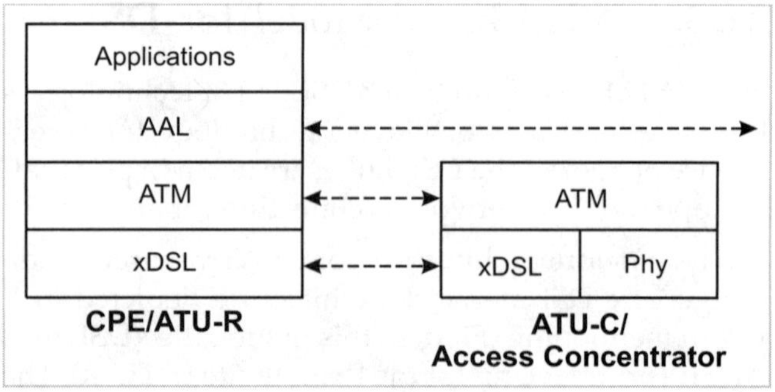

21.3.2 IPoA-Based Model for DSL

In the IP over ATM (IPoA) based model for DSL, the ATM technology is used as the layer 2 technology (i.e. data link layer) and IP is used as the layer 3 protocol (i.e. network layer). An IP service network provides the packet-data services, and an intermediate ATM network provides means for accessing the services provided by the IP service network. As in the case of an ATM-based model, the DSL link is treated as a physical link/media, over which ATM and IP protocols can operate. The network architecture for an IPoA-based model is depicted in Figure 21.5.

FIGURE 21.5

IP over ATM-Based Model for xDSL

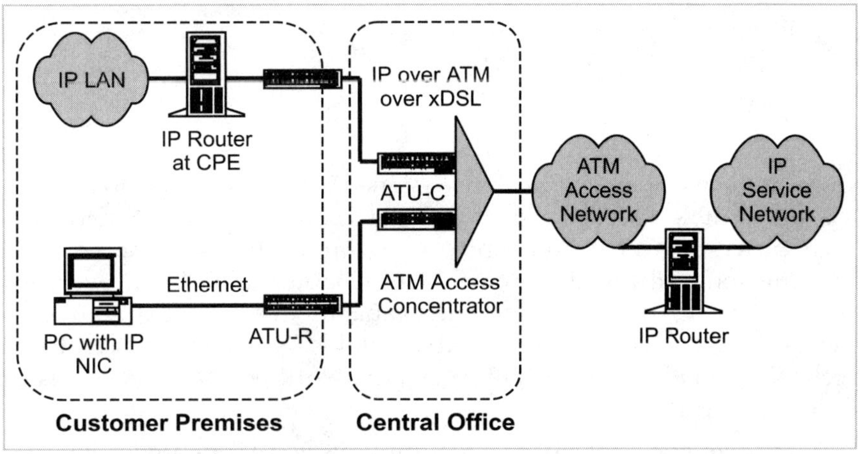

The network architecture for an IPoA-based model for DSL is very similar to the ATM-based model described in Section 21.3.1. However, there are some important differences between the two models. In the IPoA-based model, the IP protocol is used at the customer premises over any data link layer technology (e.g. Ethernet, which is most popular), instead of ATM. Between the ATU-R and the ATU-C, the IP packets are transported over ATM, which in turn is transported over DSL. An ATM access network provides access to the IP service network. An IP router, having an ATM-interface towards the ATM access network and an IP-interface towards the IP network exists between the ATM and IP networks, and terminates the IPoA protocol.

The protocol stack architecture at the Customer Premises Equipment (CPE)/ATU-R, the ATU-C/ATM Access Concentrator and at the IP router is depicted in Figure 21.6. At the CPE, IP/PPP traffic is carried over AAL5 as defined by the IPoA protocols. At the ATU-C/ATM Access Concentrator, the protocol stacks are similar to those that exist at any intermediate ATM switch. The DSL link serves as the physical layer carrying the ATM traffic between the ATU-R and the ATU-C. The IPoA protocols are terminated at the IP router, which routes the IP traffic from the CPE towards the IP service network and vice-versa.

FIGURE 21.6

Protocol Stack Architecture for IPoA-Based Model

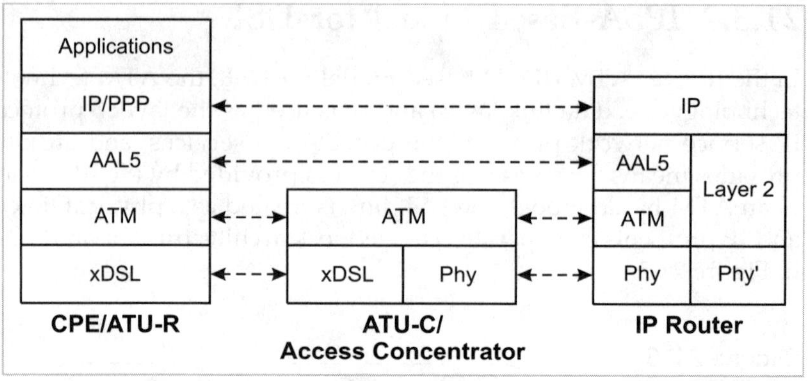

21.4 VOICE OVER DSL (VODSL)

Conventional DSL networks were designed in a manner so as to provide voice-based services via the audible spectrum in the copper cable of the POTS network, while data services were provided over the remaining spectrum of the same copper cable. However, with the introduction of the AAL2 protocol (see Chapter 6 for AAL2 overview and Chapter 13 for description of AAL2 signalling), the capability of the ATM network to efficiently support even voice-based services increased manifolds. This led to an overall enhancement in the DSL technology, and a new variant of DSL network was born, which is called Voice over DSL (VoDSL) network.

Voice over DSL (VoDSL) provides multiple voice communication channels along-with data access over the same single copper cable originally used for POTS service. VoDSL thus makes the POTS-based voice service available in conventional DSL redundant, and hence, the same is not provided in VoDSL. In other words, the entire spectrum of the copper cable is used in VoDSL for the purpose of transporting digitized voice and packet service traffic. The next few sections discuss the VoDSL network architecture, and the role of ATM in VoDSL.

21.4.1 Network Architecture

The VoDSL network architecture (see Figure 21.7) can be considered as an enhancement over the DSL network architecture, depicted in Figure 21.2. As briefly mentioned in the previous section, VoDSL does away with the analog POTS-based voice communication channel available in conventional DSL networks. In VoDSL, the copper cable in instead used for transporting multiple packetized voice channels and data channels. Thus, by using VoDSL, a residential user or a small office can obtain multiple voice communication channels, along with data services, by using the existing single copper cable that was originally installed for POTS service.

As depicted in Figure 21.7, a more sophisticated device, called the Integrated Access Device (IAD) replaces the DSL modem in the customer premises. The IAD provides multiple

FIGURE 21.7

Network Architecture for VoDSL

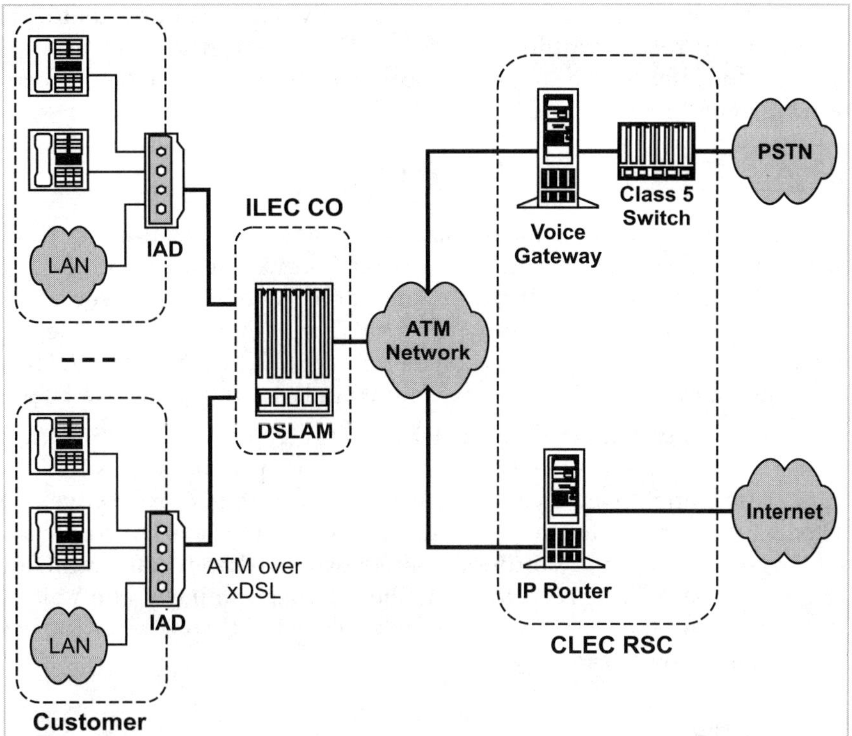

ports for connecting voice-communication devices (e.g. telephones), and also provides one or more ports for data services. Within the IAD, the analog voice is digitized, and carried as packet data over the copper loop using the Voice over ATM (VoATM) protocols. Similarly, packetized data is also transported along with digitized voice over the copper loop using ATM over xDSL. At the central office of the Incumbent Local Exchange Carrier (ILEC), a DSL Access Multiplexer (DSLAM) is placed, which has identical functions to the DSLAM used in conventional DSL networks. Since the copper cable used for VoDSL will mostly belong to the POTS service provider, the ILEC will mostly be the POTS service provider. The VoDSL service provider would therefore require to rent/lease space from the POTS service provider to host the DSLAM at the latter's site.

The DSLAM at the ILEC central office switches the ATM traffic (carrying both voice and data) towards the Competitive Local Exchange Carrier's (CLEC) Regional Switching Center (RSC). The CLEC is the VoDSL service provider, which can be distinct from the ILEC. At the CLEC RSC, a voice gateway terminates the Voice over ATM protocols, and switches the voice traffic towards a Public Switched Telephone Network (PSTN) Class 5 switch. From the PSTN class 5 switch onwards, the voice traffic is switched as in normal PSTN networks. For data traffic, an IP router within the ILEC RSC routes traffic from multiple customer premises towards the internet. The following section elaborates upon the protocol stack architecture for VoDSL in more detail.

21.4.2 Protocol Stack Architecture

ATM has emerged as the most popular protocol used for transporting voice and data traffic within a VoDSL network. While some VoDSL service providers prefer IP for carrying both voice (using Voice over IP, or VOIP) and data traffic in VoDSL, most service providers are choosing ATM instead of IP. Though only practical deployments will be able to prove which amongst ATM or IP is a better transport solution for VoDSL, the fact that ATM was designed to carry multimedia traffic, while IP was designed only to transport data traffic, speaks for itself.

Figure 21.8 depicts the protocol stack architecture for VoDSL, when ATM is used as the transport option for VoDSL. At the customer premises, the IAD is used to carry voice (channels) using AAL2 channel(s), while data traffic is carried over AAL5 virtual circuit(s). The DSLAM terminates the DSL link, and switches the ATM traffic towards the CLEC RSC. At the CLEC RSC, the IP router terminates the AAL5 virtual circuit, and routes the IP traffic carried by the AAL5 circuit towards the Internet. Similarly, the Voice Gateway at the CLEC RSC terminates the AAL2 channels, and subsequently, switches the voice traffic towards the PSTN class 5 switch.

FIGURE 21.8

Protocol Stack Architecture for VoDSL

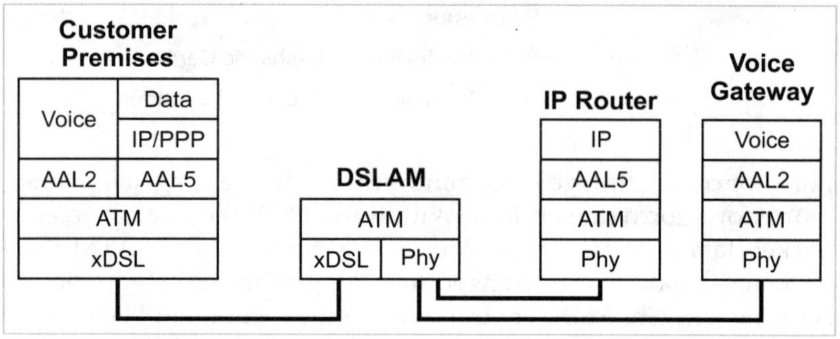

CONCLUSION

For decades, residential users have been using the dial-up modem to access the Internet, with the theoretical upper bound of 56 Kbps on the bandwidth available using dial-up modems. However, recently, Digital Subscriber Line (DSL) has emerged as a popular broadband access technology that will provide high-rate Internet access (up to 2 Mbps) over the existing copper cables installed for the POTS network. DSL technology allows the simultaneous access of the Internet and usage of the telephone/POTS service over the same copper cable by using a frequency spectrum for Internet access that is different from the spectrum used for POTS service. The network architecture for DSL-based broadband access was discussed briefly in this chapter. Multiple flavours of the DSL technology, namely, ADSL, SDSL, VDSL and HDSL were also briefly discussed.

ATM plays an extremely important role as an enabler technology for providing the DSL-based broadband access solution. Two main models proposed for DSL-based broadband access use ATM as the transport protocol. This chapter discussed the two models for DSL-based broadband access, namely, the ATM-based model and the IPoA-based model for DSL. The protocol stack architectures for both models were also discussed in the chapter.

Within the ATM fraternity, the introduction of the AAL2 protocol increased the capability of the ATM networks to efficiently support voice-based services. This lead to an overall enhancement in the DSL technology, and a new variant of DSL network was born, which is called Voice over DSL (VoDSL) network. VoDSL provides multiple voice communication channels along-with data access over the same single copper cable originally used for POTS service. This chapter discussed the VoDSL network architecture and the VoDSL protocol stack architecture. The role of ATM in VoDSL networks was also discussed.

REVIEW QUESTIONS

Q 1. Compare the operations of the dial-up modem and the DSL modem with regard to the means used to provide Internet access. How can DSL technology provide higher access speeds when the theoretical upper limit on usage of dial-up modems is 56 Kbps?

Q 2. The DSL network architecture is an augmentation of the existing POTS network architecture. Explain with the help of a diagram.

Q 3. A DSL subscriber wishes to use the video telephony application over DSL. Which flavour of the DSL technology will you advise him to use and why?

Q 4. Compare and contrast the ATM-based model for DSL with the IPoA-based model for DSL. What are the similarities and differences between the two models?

Q 5. The VoDSL network architecture is an enhancement over the existing DSL network architecture. Discuss if this statement is correct or incorrect.

FURTHER READING

For further reading on the concepts presented in this chapter, one can read [DSL Paradyne] which is an e-book available for free download from the Internet. [DSL Broadxent] is another good white paper on VoDSL. For ATM application in DSL networks, the readers can refer to [DSL V. Bhagavath], [DSL R. Cohen] and [DSL A. Azcorra].

Chapter 22

ATM and 3G Networks

22.1 Introduction

In recent times, there has been gradual migration of voice users from fixed wireline networks to wireless mobile cellular networks. The wireless networks too have evolved from Second Generation (2G) networks like Global System for Mobile communications (GSM) and General Packet Radio Service (GPRS) to newer Third Generation (3G) networks like Universal Mobile Telecommunications System (UMTS). As the networks evolve, there is associated requirement for higher bandwidth and better service capabilities. Towards this end, the ability of ATM to provide speed transfer with service guarantees emerges as a suitable choice for transfer of voice and data in wireless networks. This chapter briefly looks at the role of ATM in 3G UMTS networks.

22.2 Overview of Universal Mobile Telecommunications System (UMTS)

A typical UMTS network can be modeled on its three basic parts or sub-systems (see Figure 22.1), namely User Equipment (UE), Access Network (AN) and the Core Network (CN). The User Equipment (UE) is used by a subscriber/user to access the services provided by the network. To connect to the network, a UE interfaces with the Access Network using the air interface, which is referred to as the Uu interface. The UE is divided into two logical parts: the Mobile Equipment (ME) or the handset and the Universal Subscriber Identity Module (USIM) issued by the operator.

FIGURE 22.1

Basic Structure of UMTS Network

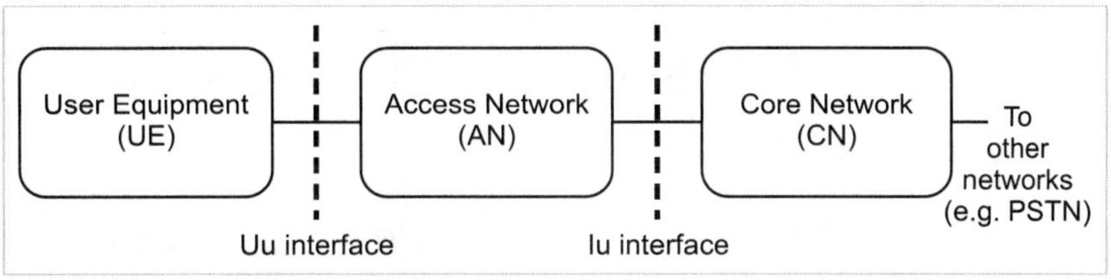

The Access Network (AN) performs functions specific to the UMTS radio access technique. The Access Network in UMTS allows two different types of access network systems to interface with the Core Network (see Figure 22.2). These two systems are the Base Station Sub-system (BSS) and the Radio Network Sub-system (RNS). While BSS is the legacy of the GSM era, the RNS is the newly standardized access network for UMTS networks. The BSS has two different types of entities: the Base Transceiver Station (BTS) that terminates the radio connection with the UE, and a Base Station Controller (BSC) that controls the resources of the BTS. BSC and one or more BTS collectively form the Access Network. The entities in RNS corresponding to that in BSS are Node B (for BTS) and Radio Network Controller (RNC) for BSC. The RNC interfaces with the Core Network over the Iu interface. The RNS comprising the RNC and Node B is also referred to as the Universal Terrestrial Radio Access Network (UTRAN).

The Core Network (CN) performs the core functions of the network, which include mobility management, call control, switching and routing. The Core Network also manages the subscription information of a subscriber and provides services based on this information. Core network handles two classes of traffic, namely, voice and data. The 2G mobile networks, like GSM networks, were designed primarily for voice. The GPRS networks provided capability for data transfer. Based on the fact that the UMTS Core Network is an evolved GSM/GPRS core network, the former is divided into two domains: the Circuit-Switched (CS) domain and the Packet-Switched (PS) domain. The CS domain provides services related to voice transfer, the PS domain to those related to data transfer. The entities of the Core Network and its decomposition into the CS and PS domain are depicted in Figure 22.3.

22.3 UTRAN AND ATM

As stated in the previous chapter, the UTRAN is the access network comprising the RNC and Node B. These entities interface with each other over Iub interface (Node B–RNC) and Iur interface (RNC–RNC). RNC also interface with entities of the core network over Iu_CS interface (RNC–MSC/VLR) and Iu_PS interface (RNC–SGSN). All these Iu based interfaces adhere to the UTRAN protocol architecture. The UTRAN protocol architecture along with the role of ATM in UTRAN is discussed of this section.

The 3GPP specification [3GPP TS 25.401] defines a generic model for UTRAN protocols (see Figure 22.4). The salient point of this model is that it is divided into two horizontal parts,

FIGURE 22.2

Logical Structure of Access Network

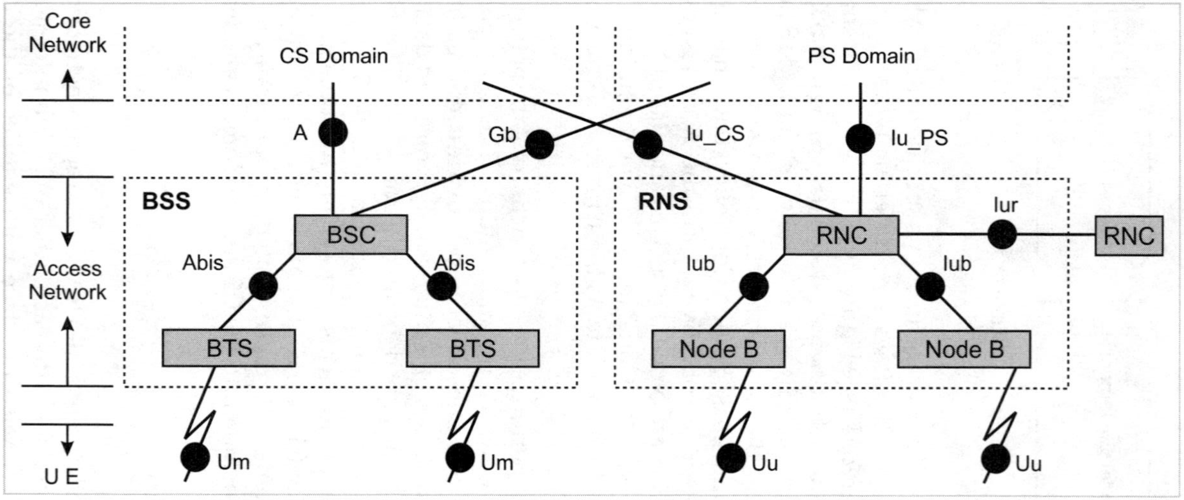

FIGURE 22.3

Logical Structure of Core Network

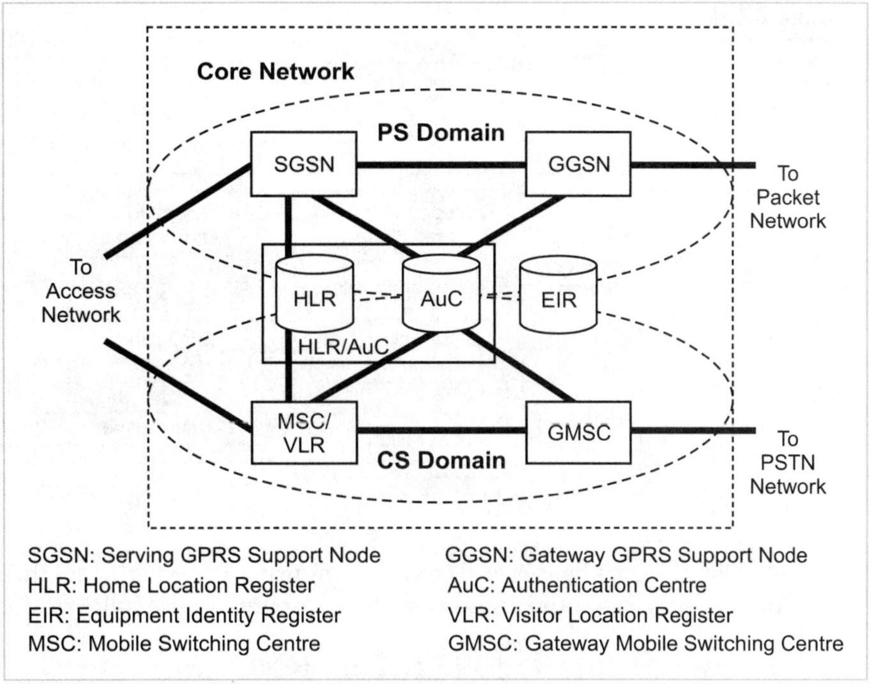

namely the radio network layer and the transport network layer. This enables the altering of one set of protocols without altering the other set.

The radio network layer has a control plane and a user plane. The control plane includes application layer signalling protocols like Radio Access Network Application Part (RANAP) and Node B Application Part (NBAP). In the user plane, the radio network layer includes User Plane (UP) protocols. These protocols facilitate transfer of user data. Examples of these protocols are, Iu User plane Protocol and Iu Framing Protocol. ATM has no role to play in the radio network layer and thus it is not elaborated in any great detail in this chapter.

In the transport network layer, there is again a control plane and a user plane. The control plane, also referred to as Access Link Control Application Part (ALCAP), is used to set bearers for user plane. For certain type of bearers, the ALCAP does not exist. This happens when the bearers are established through configuration (e.g. ATM PVC). The user plane in the transport network layer is used to carry data bearers and signalling bearers. As shown in Figure 22.4, there are three types of bearers in this layer. There are signalling bearers that carry the control plane information of radio network layer. Then there are the signalling bearers that carry control plane information of the transport network layer (i.e. they carry ALCAP signalling messages). Both these types of signalling bearers are pre-configured. The third type of bearers, called the data bearers, are used to carry user plane information of the radio network layer. These bearers are established using ALCAP. In cases where ALCAP

FIGURE 22.4

General Protocol Model for UTRAN

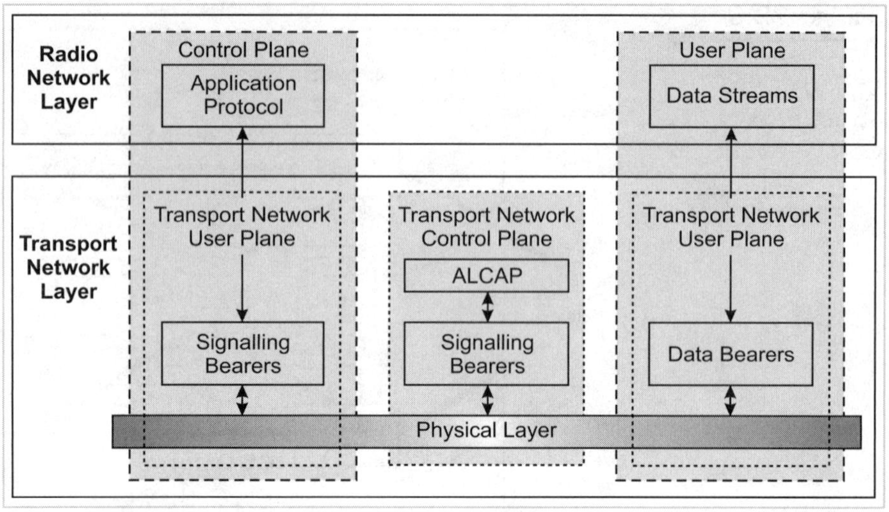

does not exist, the data bearer is also pre-configured (i.e. established through configuration). This model is elaborated through examples presented in the following section.

22.3.1 Iu_CS interface between RNC and MSC/VLR

The Iu_CS interface exists between RNC and MSC/VLR (see Figure 22.5). In the radio network layer, the control plane has the Radio Access Network Application Part (RANAP) protocol and the user plane has the Iu User Plane (UP) protocol for transport of user data.

The key to understanding the role of ATM in UTRAN is to concentrate on the transport network layer. Here, as mentioned in the previous section, three distinct types of information need to be carried. These are as follows:

1. **User plane for actual user traffic:** Since Iu_CS interface is designed for voice, AAL2 protocol is used for this interface. The AAL2 is considered to be a suitable choice for low rate voice between RNC and MSC/VLR. Note that AAL1 has no role in UMTS networks.
2. **User plane for signalling traffic:** Apart from user traffic, RNC and MSC/VLR also have to exchange signalling messages (RANAP messages) to manage the Iu_CS connections. For this, SCCP over MTP3b/SSCF-NNI/SSCOP/AAL5 is used. In this scenario, the SSCF-NNI, SSCOP and AAL5 can be collectively viewed as SAAL (see Chapter 11). The SAAL provides a reliable layer for transfer of higher layer messages.
3. **Control plane for ALCAP signalling:** For both the user planes mentioned above, there may be an associated requirement for signalling. However, all AAL5 applications are

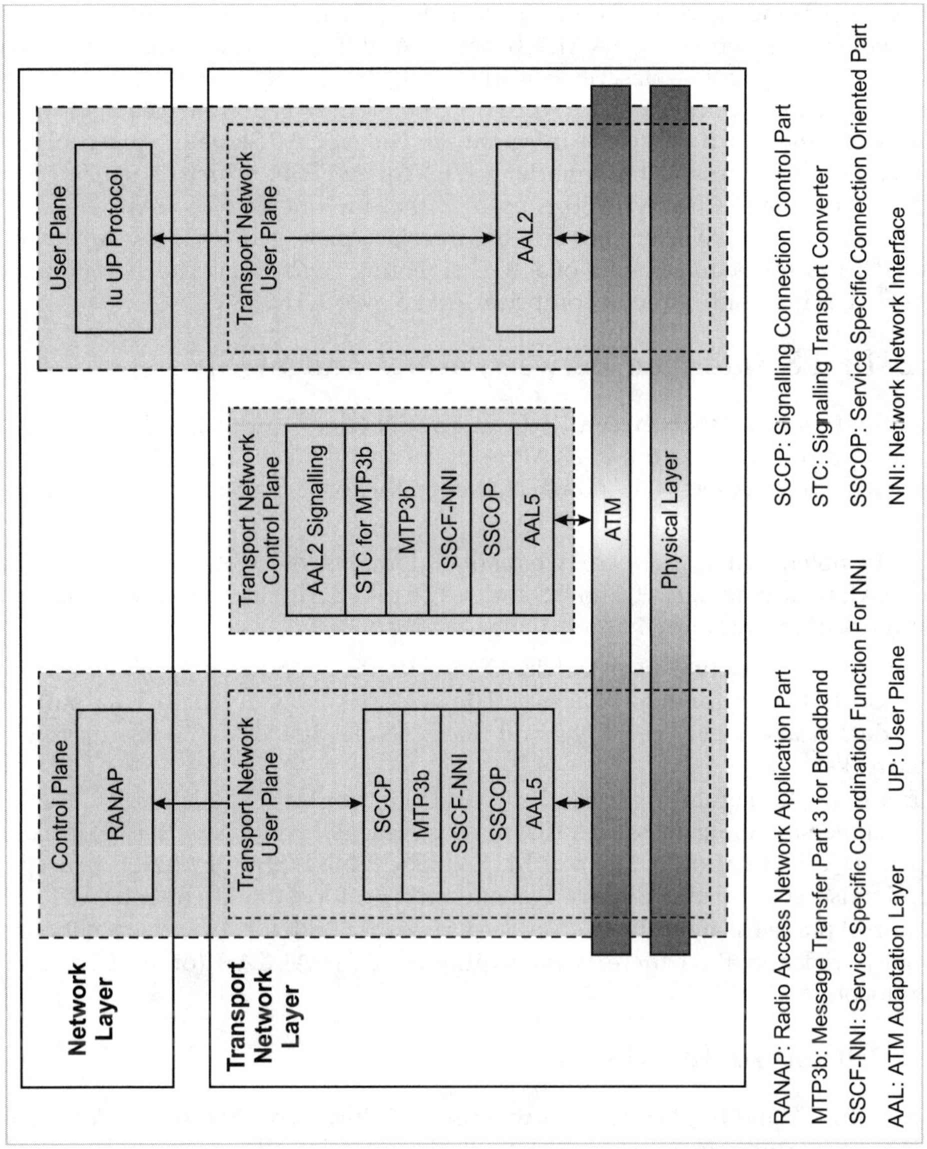

FIGURE 22.5 Protocol Architecture of Iu_CS

PVC-based and there is no requirement for ATM signalling protocols (e.g. Q.2931). However, for setting up AAL2 bearers, AAL2 signalling is required. The AAL2 Signalling protocol as described in [ITU-T Q.2630.1] is also referred to as the Access Link Control Application Part (ALCAP). The ALCAP is used to establish AAL2 bearers that carry the Iu_CS user plane information. The ALCAP Signaling protocol resides over the Signaling Transport Converter (STC) for MTP3b, which is specified in [ITU-T Q.2150.1]. The STC for MTP3b is used so that the interface of AAL2 signaling is kept constant and a converter converts the interface provided by lower layer (in this case MTP3b) to the requirements of AAL2 signaling (Q.2630.1). The signalling bearer for ALCAP is pre-configured. It comprises AAL5 over ATM PVC.

22.3.2 Iu_PS interface between RNC and SGSN

The Iu_PS interface exists between RNC and SGSN (see Figure 22.6). In the radio network layer, the control plane has the RANAP protocol and the user plane has the Iu User Plane (UP) protocol for transport of user data. However, the focus is again on the transport network layer.

It may be noted that as there are two distinct domains in UMTS, CS and PS, there are two types of AAL to cater to it, AAL2 for CS and AAL5 for PS. Again, consider the following three distinct types of information that need to be carried:

1. **User plane for actual user traffic:** Since Iu_PS interface is designed for data AAL5 protocol is used for this interface. The AAL5 is considered to be a suitable choice for data transfer between RNC and SGSN. Note that AAL3/4 has no role in UMTS networks.
2. **User plane for signalling traffic:** Apart from user traffic, RNC and SGSN also have to exchange signalling messages (RANAP messages) to manage the Iu_PS connections. Like Iu_CS scenario, SCCP over MTP3b/SSCF-NNI/SSCOP/AAL5 is used. Another option is to use Stream Control Transmission Protocol (SCTP) over IP.
3. **Control plane for signalling:** The AAL5 used for both the scenarios mentioned are pre-configured and thus there is no requirement for ALCAP (or AAL2 signalling like protocol).

22.3.3 Other Iu Interfaces

Apart from Iu_CS and Iu_PS interfaces that use ATM protocols like AAL2, AAL5, SAAL and ALCAP, there are other Iu interfaces like Iur and Iub. Iur is quite similar to the Iu_CS as it also uses AAL2 for user plane, ALCAP for AAL2 bearer management, and NNI-SAAL for carrying signalling messages between an RNC and another RNC. The Iub interface also has AAL2 for user plane, ALCAP for AAL2 bearer management, and UNI-SAAL for carrying signalling messages between Node B and RNC.

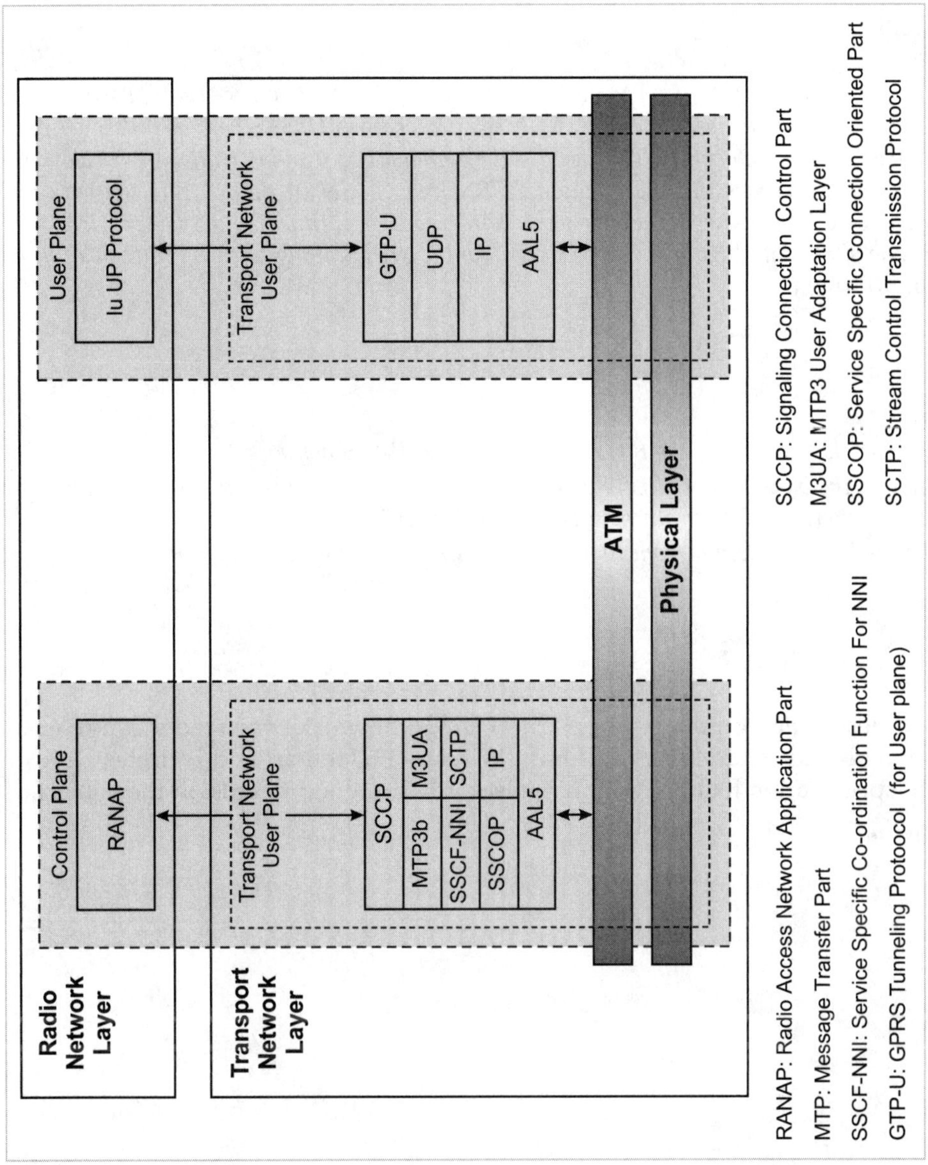

FIGURE 22.6

Protocol Architecture of Iu_PS

Conclusion

This chapter discussed the role of ATM in the UTRAN where AAL2 and AAL5 are used in the Iu-based interfaces. Apart from the direct use of ATM-based protocols in UMTS both as link layer protocols as well as network layer protocols, ATM can also be employed to provide transport network for all IP-based traffic. This includes the use between SGSN and GGSN in the PS core network and in the UTRAN where SCTP is used. In the newly standardized IP Multimedia Sub-system (IMS), ATM can be used as one of the data link layer protocols below IP. MPLS along with ATM too can be employed for carrying IP traffic.

Review Questions

Q 1. *Provide an overview of the three sub-systems of UMTS network.*

Q 2. *What is the basic model of UTRAN interface? How is it applied in Iu_CS interface? What is role of different ATM protocols in Iu_CS interface?*

Q 3. *How is Iu_PS different from Iu_CS? Why is ALCAP not used in Iu_PS?*

Further Reading

For a good description on the use of ATM in UMTS networks, the reader can refer to any of the following books: [Wireless S. Kasera], [Wireless H. Holma] and [Wireless J. Korhonen]. The 3GPP specification [3GPP TS 25.401] also provides good details on the subject.

Glossary*

AAL1 Trunking	A trunking solution that uses AAL1 to support dynamic allocation of resources and call-by-call routing without circuit emulation.
AAL2 Channel	A channel that carries CPS packets of a user identified by the CID on an ATM VCC.
AAL2 Connection	The logical concatenation of one or more AAL2 links between two AAL2 service end-points.
AAL2 Link	The logical user plane communication facility between two adjacent AAL2 switching points or service end-points. An AAL2 link is designated by a single CID value.
AAL2 Node	An AAL2 service endpoint or an AAL2 switch.
AAL2 Path	An ATM VCC between two adjacent AAL2 nodes.
AAL2 Service End-point	A termination point of an AAL2 connection.
AAL2 Signalling	The ITU-T standard that provides means to dynamically establish, manage and release AAL2 connections over ATM VCs.
AAL2 Signalling Protocol	Control plane functions for establishing and releasing AAL2 connections and the maintenance functions associated with the AAL2 signalling.

* *This section has some of the important ATM specific terms used in the book. All the terms are generally with respect to ATM (e.g. control plane refers to ATM control plane). For expansion of a given abbreviation, the abbreviation must first be expanded through the abbreviation section and the same needs to be then referred in this section. This section may not be comprehensive as only important ATM-specific terms are covered here.*

AAL2 Signalling Transport	The underlying transport used for carrying AAL2 signalling messages.
AAL2 Switch	A switch that is capable of switching AAL2 connections.
AAL2 Trunking	A trunking solution that uses AAL2 to provide a broad range of applications involving interconnection of an IWF with narrowband and broadband facilities and interworking with various other telecommunications devices including PBXs and ATM network switches.
ABR Flow Control	A fast resource management mechanism to utilize any leftover bandwidth in the network for connections belonging to ABR service category.
Access Link Control Application Part (ALCAP)	The signalling protocol that is used to set bearer channels for the transport network layer in UMTS network.
Address Resolution	A mechanism to bind a high-level protocol address (like IP) to a low-level hardware address (like Ethernet address).
Address Summarization	The process of using a single reachable address prefix to represent a collection of end systems and/or node addresses that begin with the same prefix.
Alarm Indication Signal (AIS)	A signal sent to a downstream node, when a VPC/VCC failure/defect is detected.
Anycast	ATM Anycast capability allows a user to request a point-to-point connection to a single ATM end system that is part of an ATM group.
Associated Signalling	A scheme in which all the signalling messages for each VP is exchanged on VCI = 5 of that virtual path.
Asynchronous Transfer Mode (ATM)	A transfer mode in which the information is organized into cells. It is asynchronous in the sense that the recurrence of cells containing information from an individual user is not necessarily periodic.
ATM Adaptation Layer (AAL)	The layer that allows multiple applications to have data converted to and from the ATM cell. In other words, it is a protocol used that translates higher layer services into the size and format of an ATM cell.
ATM Adaptation Layer 1 (AAL1)	AAL1 provides transfer of time-sensitive, constant bit rate traffic on an ATM connection.
ATM Adaptation Layer 2 (AAL2)	AAL2 provides transfer of time-sensitive, variable bit rate traffic on an ATM connection.
ATM Adaptation Layer 3/4 (AAL3/4)	AAL2 provides transfer of connectionless/connection-oriented, time-insensitive, variable bit rate data on an ATM connection.

ATM Adaptation Layer 5 (AAL5) AAL5 provides transfer of time-insensitive, variable bit rate data on an ATM connection requiring minimal sequencing or error detection support.

ATM End System Address (AESA) An addressing mechanism based on the Network Service Access Point (NSAP) that identifies the location of one or more ATM interfaces.

ATM Forum It is a consortium of ATM vendors, formed with the objective of accelerating the use of ATM products and services through a rapid convergence of interoperability specifications. It also promotes industry cooperation and awareness.

ATM Group Address An ATM address that identifies a service (rather than an end-system).

ATM Inter-Network Interface (AINI) It is the interface that facilitates interworking of two networks running PNNI internally in disjoint PNNI routing domains.

ATM Layer The layer that performs cell multiplexing/demultiplexing and cell relaying.

ATM Switch A switch that is capable of switching ATM virtual connections.

ATM Trunk Refers to one or more ATM Virtual Connections that carry a number of 64 Kbps narrowband channels and associated signalling between a pair of Inter-Working Functions.

ATMARP A service that is quite similar to the conventional ARP (defined for broadcast-based mediums like Ethernet and token ring) that uses a central server that keeps such a mapping between ATM addresses and IP addresses.

Authority and Format Identifier (AFI) AFI field specifies the authority controlling the IDI and the format of the IDI.

Available Bit Rate (ABR) A service category for which the ATM layer characteristics provided by the network changes dynamically using a flow control mechanism. This mechanism supports several types of feedback to control the source rate in response to changing network conditions.

B-ISDN Inter-Carrier Interface (B-ICI) It is the interface that defines rules for communication between two public ATM networks.

Border nodes Nodes connected via an outside link are called border nodes.

Broadband Integrated Services Digital Network (B-SIDN) A network that supports all types of communication (including voice, video and data). The term broadband is formally defined as 'a service or a system that requires transmission channels capable of supporting rates greater than the primary access rate' (the primary rate depending upon the hierarchy used is equal to 1.544 Mbps or 2.048 Mbps). Informally, broadband technologies refer to an array of high-speed technologies including SONET/SDH, DSL, ATM and Gigabit Ethernet.

Broadcast and Unknown Servers (BUS) The server that handles data sent by an LE Client to the broadcast MAC address (0xFFFFFFFFFFFF), all multicast traffic, and initial unicast frames, which are sent by a LE Client.

Burst Tolerance (BT) BT is a measure of the interval between consecutive bursts during which cells are sent at PCR. It provides the time interval after which an ATM source can again send data at PCR without violating the long-term average of SCR.

Call Reference Value (CRV) Identifies a particular signalling connection between signalling peers.

Cell The lowest unit of information in ATM that is of fixed size (53 bytes) with 5 bytes of header and 48 bytes of payload.

Cell Delay Variation (CDV) It is the part of cell transfer delay that is induced by buffering and cell scheduling leading to inconsistencies in the cell transfer delay.

Cell Delay Variation Tolerance (CDVT) A cell stream arriving at the user network interface experiences some cell delay variation while entering the network. The reasons for this include multiplexing of several incoming cell streams or due to the variable processing delays in switches at the network ingress point. This delay variation or randomness affects the inter-arrival time (i.e. 1/PCR) between consecutive cells of a connection as monitored at the UNI. Due to this, a cell stream injected at PCR may suffer some delay variation, resulting in the rate exceeding PCR. CDVT represents an upper bound on the delay variation (i.e. jitter) of cells at a UNI.

Cell Delineation The identification of cell boundaries in a cell stream.

Cell Error Ratio (CER) It is the ratio of the total number of cells delivered with error to the total number of cells delivered.

Cell Loss Priority (CLP) A single bit field in the cell header that is used to assign a two-level priority to a cell.

Cell Loss Ratio (CLR) It is the fraction of cells that are either not delivered to the destination or delivered after a pre-specified time.

Cell Mis-insertion Ratio (CMR) It is the ratio of cells received at an endpoint that were not originally transmitted by the source to the total number of cells properly transmitted.

Cell Rate Decoupling It refers to the process of inserting dummy cells (or idle cells) to fill the time slots allotted for user cells and to match the rate of outgoing cells to the link rate.

Cell Relaying The process of replacing the VPI/VCI value of an incoming cell to a new VPI/VCI value. Depending upon the type of connection (virtual path connection or virtual channel connection), either only the VPI value or both VPI/VCI values are swapped.

Cell Scrambling A mechanism whereby the undesired patterns in data are changed at the source and the changes reverted back at the destination to improve the security and robustness of the HEC delineation algorithm.

Cell Switching Virtual circuit switching technique in which fixed-size small cells are used instead of variable-sized frames.

Cell Tagging The action taken against a non-conforming cells in which the CLP bit in the ATM header is set to 1. Cell tagging applies to CLP = 0 cells only.

Cell Transfer Delay (CTD) It is the elapsed time between exit of a cell at the measurement point 1 (e.g., at the source UNI) and the corresponding cell entry event at measurement point 2 (e.g., the destination UNI) for a particular connection.

Channel Identifier (CID) The identifier that identifies an AAL2 channel. This field is used for multiplexing a number of AAL2 users on the same ATM virtual channel.

Circuit Emulation Service (CES) A technique in which ATM emulates the behaviour of a TDM circuit by providing a transparent transport pipe for carrying user traffic.

Circuit Switching A technique in which a dedicated circuit (or a channel) is established from source to destination and the information is sent as a bit stream through the fixed bandwidth channel.

Classical IP Over ATM This defines a method to carry IP datagrams over ATM networks using the classical model.

Classical Model This refers to the treatment of the ATM host adapter as a networking interface to the IP protocol stack operating in a LAN-based paradigm.

CLP Significant This category refers to the cell flows in which the CLP bit is not ignored by the network. The CLR objectives apply onto to cells with CLP = 0. Thus, even if cells carrying CLP = 1 are dropped by the network, this is not counted against the performance guarantees provided by the network.

CLP transparent — This category refers to the cell flows in which the CLP bit is ignored by the network. For this category, cells with CLP = 1 and cells with CLP = 0 are given identical treatment.

Common Part Convergence Sub-layer (CPCS) — It is that part of the convergence sub-layer that provides generic functionality common across applications and is fixed for an AAL.

Common Part Sub-layer (CPS) — The sub-layer that provides the basic functionality of AAL2, which includes packaging the variable payload into cells and providing error correction.

Compliance — A connection is compliant if it does not have 'too many' non-conforming cells. The exact definition of 'too many' is network-specific. For a compliant connection, the network provides QoS guarantee for at least the number of cells that are conforming. For a non-compliant connection, the network is not obliged to provide QoS guarantees to any cell on that connection.

Conformance — Each cell is either declared conforming or non-conforming, depending upon the adherence of that cell to the conformance definition. For each service category, a conformance definition is explicitly defined in the ITU-T/ATM Forum specifications.

Connection Admission Control (CAC) — Set of actions taken by the network at the call establishment phase (or during call renegotiation phase) in order to determine whether a request to establish a VCC or VPC can be accepted or rejected.

Constant Bit Rate (CBR) — A service category used by applications that require a constant bandwidth allocated to them throughout the life of the connection. This service category places strict upper bounds on cell delay and cell delay variation.

Continuity Check — A mechanism that is used to identify faults in the link quickly through the use of continuity check OAM cells.

Control Plane — It performs connection control functions, which includes the establishment, monitoring and release of virtual connections.

Control Plane Security — The set of security functions available in the ATM control plane that includes data origin authentication and data integrity.

Convergence — The phenomenon whereby the telecommunication and datacommunication world converge to provide integrated services.

Convergence Sub-layer (CS) — The sub-layer that interfaces directly with the user and accepts data from it. It optionally adds a header and/or a trailer to the user data, and passes it to the SAR sub-layer.

Crankback	A mechanism for partially releasing a connection setup in progress which has encountered a failure. This mechanism allows PNNI to perform alternate routing.
Cut-through Model	A model that involves establishing direct virtual connections with the destination as soon as a significant flow is detected. A significant flow implies that at least a given number of packets are sent to a particular end-system within a specified time interval.
Data Exchange Interface (DXI)	It is a standardized mechanism to interface between frame-based devices (e.g. IP-based routers) and ATM network elements.
Designated Transit List (DTL)	The information element used to carry source routing information.
Digital Signal, Level 0 (DS0)	The 64 Kbps rate that is the basic building block for both the North American and European digital hierarchies.
Digital Signal, Level 1 (DS1)	Also called T1, this is a North American Digital Hierarchy signaling standard for transmission at 1.544 Mbps. This standard supports 24 simultaneous DS-0 signals.
Digital Signal, Level 3 (DS3)	Also called T3, this is a North American Digital Hierarchy signaling standard for transmission at 44.736 Mbps that is used by T3 carrier.
Digital Subscriber Line (DSL)	Refers to the a technology that provides access solution for the home and business environments using the existing copper infrastructure that is installed for the Plain Old Telephone System (POTS) network.
DSL Access Multiplexer (DSLAM)	A device that serves as a packet concentrator, which receives data traffic from multiple customers over DSL links, and sends this aggregate traffic towards the data network (e.g. Internet).
DSL Modem	A device installed within the customer premises that is used to originate and terminate the data traffic over the local loop.
Dynamic Bandwidth Circuit Emulation Services (DBCES)	An improvement over CES scheme in which when an inactive state is detected in a specific time slot, the time slot is dropped from the next ATM structure and the bandwidth it was using is utilized for other services.
E1	This refers to the 2.048 Mbps rate used by European carrier to transmit 30 digital channels for voice or data calls at 64 Kbps, plus a 64 Kbps signaling channel and a 64 Kbps channel for framing and maintenance.
E3	This refers to the 34.368 Mbps rate used by European carrier to transmit 16 E1 plus overhead.

Early Packet Discard (EPD)
A packet discard scheme in which an entire incoming frame is discarded to prevent any possibility of buffer overflow.

End-point Reference Value (ERV)
An identifier used to differentiate between different parties in the same point-to-multipoint call.

Explicit Forward Congestion Indication (EFCI)
A mechanism in which the congestion bit in the ATM header is set to 1 to indicate an impending congestion in the network. Upon receiving an EFCI notification, the destination should somehow communicate to the source to slow down the transmission rate because congestion is prevalent in the downstream direction.

F1 Level OAM Flow
An OAM flow defined at regenerator section level.

F2 Level OAM Flow
An OAM flow defined at digital section level.

F3 Level OAM Flow
An OAM flow defined at transmission path level.

F4 level OAM flow
An OAM flow defined on an end-to-end VPC or on single/group of virtual path links.

F5 Level OAM flow
An OAM flow defined on an end-to-end VCC or on single/group of virtual channel links.

Flooding
The technique used in link-state routing protocols including PNNI that enables a switch/router to have information of other parts of the network.

Frame Discard
A mechanism whereby instead of discarding cells, complete higher layer PDUs are discarded.

Frame Relay
A frame based virtual circuit switching technique that switches variable-sized frames using a single label Data Link Connection Identifier (DLCI).

Frame-based ATM Over SONET/SDH Transport (FAST)
It is a standardized mechanism to carry frame-based ATM traffic over SONET/SDH.

Frame-based ATM Transport Over Ethernet (FATE).
It is a standardized mechanism to carry frame-based ATM traffic over Ethernet.

Frame-based UNI (FUNI)
An enhancement over DXI that pushes SAR function within the ATM network, as part of the ATM switch functionality.

Generic Cell Rate Algorithm (GCRA)
A technique based on leaky bucket algorithm that is used to define conformance of the cells with respect to the negotiated traffic contract.

Generic Flow Control
It refers to the link-level flow control mechanism used to control data offered at a UNI.

Generic Signalling Transport
The technique that that enables an AAL2 signalling entity to communicate with a peer AAL2 signalling entity independently of the underlying signalling transport.

Guaranteed Frame Rate (GFR) A service category, like ABR, that attempts to dynamically utilize the residual bandwidth in the network without any complex flow control mechanism.

Header Error Control (HEC) An error control mechanism for the cell header that provides single-bit error correction and multiple-bit error detection capabilities

Hello Packet A type of PNNI Routing packet that is exchanged between neighboring logical nodes to indicate availability.

Hierarchical Network A structure used in PNNI in which each node in the hierarchy maintains a partial view of the network. This mechanism improves the memory and processing requirements at intermediate switches.

Horizontal Link A logical link within a peer group.

Idle Cells Idle cells apply at the physical layer and are equivalent to unassigned cells at the ATM layer.

InATMARP A mechanism for a node to know the IP address of the destination node, corresponding to an open PVC connection. Using this mechanism, a node can know all the IP addresses reachable through it.

Initial Domain Identifier (IDI) IDI identifies the authority controlling the assignment of the Domain Specific Part (DSP).

Initial Domain Part (IDP) This identifies a particular network-addressing domain that is part of the global network-addressing domain

Integrated Access Device (IAD) A device installed within the customer premises that provides multiple ports for connecting voice-communication devices (e.g. telephones), and also provides one or more ports for data services.

Integrated Layer Management Interface (ILMI) A protocol that provides the status and configuration parameters to an ATM device necessary for it's functioning. These status and configuration parameters primarily relate to Virtual Path Connections, Virtual Channel Connections, ATM Network Prefixes and ATM Addresses associated with an ATM interface.

Integrated Services Digital Network (ISDN) A digital network that supports a host of services including voice, fax, Internet, videoconferencing and LAN interconnection.

Interim Inter Switch Protocol (IISP) A rudimentary inter-switch communication protocol that does not support QoS routing, crankback and most other features of PNNI protocol.

Internet Engineering Task Force (IETF) The IETF is a large and open international community of network designers, operators, vendors and researchers, concerned with the evolution of Internet architecture and the smooth operation of the Internet. It is the main body responsible for the development of new Internet standards.

Label Swapping See 'Virtual circuit switching'

Label Switching Router (LSR) The routers in MPLS network that route packets using labels.

LAN Emulation A technique that provides means for traditional LAN technologies like Ethernet and Token ring to co-exist with ATM LANs in a heterogeneous LAN environment. Using LAN emulation, two non-ATM end-systems can not only communicate with each other, but they can also communicate with end-systems directly attached to an ATM network

Layer Management This layer handles Operation and Maintenance (OAM) flows, specific to the layer concerned. This layer also handles the resources and parameters residing in its protocol entities.

LE Client (LEC) Any network element (e.g. end-systems, bridges and routers) that is attached to an ATM network and participating in LAN emulation, contains a LAN Emulation Client (LEC).

LE Configuration Server (LECS) A server that acts as a central repository of configuration information that assigns individual LE Clients to an Emulated LAN.

LE Server (LES) The server that implements the control coordination function for the Emulated LAN. It provides a facility to register and resolve unicast and multicast MAC addresses to ATM addresses.

Leaf Initiated Join (LIJ) A technique in which leaves can initiate procedures for establishing a new point-to-multipoint call, or can even initiate procedures for adding themselves to an already existing connection, with or without the intervention from the root.

Leaky Bucket Algorithm A scheme in which bursty data generated by a source is stored in the buffer and sent out at a lower rate, thereby transforming bursty traffic with a high data rates to a more stable traffic with lower data rates. It is used for traffic shaping and traffic policing.

Logical Group Node (LGN) An abstract representation of a peer group as a single point for the purpose of representing that peer group at a particular level (but not at the lowest) in the PNNI routing hierarchy.

Logical IP Sub-network (LIS) An LIS consists of ATM attached hosts and routers that is configured in a manner similar to the way any IP sub-net is configured. All members of the LIS are directly connected to the ATM network and have the same IP network/sub-net number and address mask.

Logical Link — Links in PNNI routing hierarchy that represents the connection between two logical nodes.

Logical Link Control (LLC) Encapsulation — A technique in which packets belonging to multiple protocols are multiplexed over a single VC through the use of special purpose header called the LLC/SNAP header.

Logical Node — A logical node refers either to a lowest-level node or to a Logical Group Node (LGN)

Lowest-level Node — A node that is at the lowest level in the PNNI hierarchy.

Management Plane — It is responsible for maintenance and management-related functions. The functions of this plane are categorized into layer management functions and plane management functions.

Maximum Burst Size (MBS) — This is the amount of data that an ATM source can send at its peak cell rate. It gives the number of cells an ATM source can send at the PCR.

Maximum Frame Size (MFS) — This parameter specifies the maximum size of AAL PDU for the Guaranteed Frame Rate (GFR) service category.

Meta-signalling — The signalling procedure used for establishing, managing and releasing signalling channels.

Minimum Cell Rate (MCR) — This parameter is the minimum cell rate that the network must provide to a connection. Its value can even be zero.

MPOA Client (MPC) — This resides in a host attached to the ATM network or in edge devices that forward data from non-ATM networks to ATM networks. The MPOA clients act as entry and exit points in the MPOA network.

MPOA Server (MPS) — A component of the router that provides necessary forwarding information to MPOA Clients.

Multi-Protocol Encapsulation Over AAL5 — The techniques includes LLC encapsulation and VC based multiplexing that allow multiple protocols to be carried over AAL5 channel.

Multi-Protocol Label Switching (MPLS) — A new industry development standardized by the IETF in which the ingress point classifies each packet and attaches a label to it. Subsequent nodes within the MPLS domain only perform a label-table look-up to switch the packet towards the destination.

Multi-Protocol Over ATM (MPOA) — A protocol that uses the NHRP to establish cut-through connections bypassing the intermediate routers in the data path. MPOA also employs the LANE standard for intra-sub-net communication.

Network Node Interface (NNI)	The interface used by a network node to interconnect with another network node.
Network Parameter Control (NPC)	The mechanisms applied at the NNI to monitor the traffic flow and to enforce the traffic contract.
Next Hop Resolution Protocol (NHRP)	An address resolution mechanism that gives the mapping between IP and the ATM address or provides the means to determine the egress point from the ATM network to reach the destination.
Next Hop Server (NHS)	An entity within the logical sub-network that receives NHRP address resolution requests from NHC and responds by sending the mapping between ATM and IP addresses.
NHRP Client (NHC)	An entity that initiates NHRP requests in order to obtain access to various NHRP services. All requests initiated by an NHRP Client are sent to the NHS.
Nodal Information	This refers to the general information stored by a PNNI node (e.g. AESA, leadership priority, etc.).
Node	A single instance of the PNNI routing protocol created in a switching system via configuration.
Non-associated signalling	A scheme in which all the signalling messages of all the virtual paths are exchanged on VPI = 0 and VCI = 5.
Non-Broadcast, Multi-Access (NBMA) networks	A network that consists of a set of nodes interconnected using point-to-point or point-to-multipoint links (e.g. ATM network).
Non-Real-Time Variable Bit Rate (nrt-VBR)	A service category used by applications that have bursty traffic but do not have strict requirements for cell transfer delay and cell delay variation.
OAM Cell	Special purpose cells that are associated with different OAM information flows and are involved in OAM functions.
Operations and Maintenance (OAM)	The set of performance and fault management procedures defined for ATM using OAM cells.
Outside Link	Link connecting nodes of two different peer groups.
Partial Packet Discard (PPD)	A packet discard scheme in which once a cell of a frame is dropped, all the remaining cells (barring the end-of-frame cell) of this frame are also dropped.
Payload Type Identifier (PTI)	A 3-bit field in the cell header identifies the type of the payload in the cell
Peak Cell Rate (PCR)	This is the maximum rate at which a user is allowed to inject data into the network. It defines an upper bound on the traffic that can be submitted by an ATM source.

Peer Group (PG)	A set of logical nodes which are grouped for the purpose of creating a routing hierarchy; the member nodes of the group exchange topology information between themselves, thereby maintaining a single view of the group.
Peer Group Leader (PGL)	A node of a peer group that performs the extra work of collecting, aggregating and building data that is suitable to represent the entire peer group as a single node in a higher-level peer group.
Permanent Virtual Circuit (PVC)	This is a static link that is established through manual setup.
Physical Medium Dependent (PMD) Sub-layer	It is the layer whose functions depend upon the underlying physical media. These functions directly relate to the physical representation of the bits on the physical medium and the interpretation of the bits at the receiver.
Plane Management	This provides coordination between all the planes. It does not contain a layered structure.
Plesiochronous Digital Hierarchy (PDH)	This was developed to carry digitized voice over twisted pair cabling more efficiently. This evolved into the North American, European, and Japanese Digital Hierarchies where only a discrete set of fixed rates is available, namely, nxDS0 (DS0 is a 64 Kbps rate) and then the next levels in the respective multiplex hierarchies.
Point-to-multipoint Signalling	Refers to the signalling for establishment and release of connections between a root and multiple leaves.
Point-to-point Signalling	Refers to the signalling for establishment and release of connection between two entities.
Private Network-Node Interface (PNNI)	A protocol that provides signalling and routing functionality at the NNI interface.
QoS-based Routing	A mechanism of routing connection request through a path that can satisfy the QoS requirements specified in the request.
Quality of Service (QoS)	Providing Quality of Service for an ATM connection refers to providing a bound to the cell loss, the transit delay, and the delay variation induced by the ATM network in transiting cells of that connection.
Random Early Discard (RED)	A packet discard scheme in which whenever the queue size in the switch buffer exceeds certain threshold, each incoming packet is dropped with a certain probability, where the exact probability is a function of average queue size.
Reachability Information	The topology information that binds reachable addresses to nodes within the PNNI routing domain.

Real-Time Variable Bit Rate (rt-VBR) A service category used by applications that have bursty traffic and require strict bounds on cell transfer delay and cell delay variation.

Remote Defect Indication (RDI) A signal sent to an upstream node, when a VPC/VCC failure/defect is detected.

Routing Routing is a hop-by-hop datagram forwarding technique in which the destination address field of the packet header is used to make forwarding decisions

Security Agent (SA) The entities that reside at end systems or at intermediate switches and provide means for secured communication.

Security Association Refers to the distributed contextual information (that includes the algorithm type, security keys, mode of operation, etc.), which controls the nature of the security service to be provided for a given VC.

Security Message Exchange (SME) protocol The mechanism to exchange security information and to negotiate security services with a peer security agent.

Segmentation and Reassembly (SAR) sub-layer The sub-layer segments the data received from CS into smaller data units. The SAR sub-layer may also add a header and/or a trailer to these smaller data units to form a 48-byte SAR-PDU.

Selective Cell Discard It refers to the action of selectively discarding cells with CLP bit set when buffer overflow occurs within the network.

Service Level Agreement (SLA) Set of parameters that define the quantitative benchmarks for the service provided. It indicates the reliability and performance of an ATM VP or VC provided by a service provider.

Service Specific Connection-Oriented Part (SSCOP) The core sub-layer of SAAL that provides connection-oriented transfer of PDUs with retransmission support.

Service Specific Convergence Sub-layer (SSCS) It is that part of the convergence sublayer that provides additional functionality required for specific applications.

Service Specific Coordination Function (SSCF) This sub-layer provides a mapping between the primitives provided by SSCOP to that required by specific-user applications.

Severely Errored Cell Block Ratio (SECBR) It is the ratio of severely errored cell blocks to the total transmitted cell blocks. (A severely errored cell block outcome occurs when more than M error cells, lost cells or mis- inserted cell outcomes are observed in a received cell block of N cells.)

Shared LAN A LAN connectivity model in which multiple hosts share a common communication channel.

Signalling ATM Adaptation Layer (SAAL)	A special AAL used for the purpose of signalling that provides reliable transport for signalling messages.
Signalling Transport Converter (STC)	A function that converts the services provided by a particular signalling transport to the services required by the signalling layer.
Soft PVC	A model in which users have permanent virtual connection but the connection within the network is established using signalling.
Source Routing	The routing technique in which ingress node specifies the complete path to the destination
State Information	The dynamic PNNI information maintained to determine whether a new connection request can be accepted or not
Statistical Time Division Multiplexing (STDM)	A technique in which instead of reserving slots on a permanent basis for an incoming channel, a slot is assigned as and when user data is received.
Structured Data Transfer (SDT)	The mechanism used to support N×64 Kbps circuits (i.e. fractional T1/E1 links) over AAL1.
Sub-net Model	The model in which if two entities connected to two different emulated LANs want to interact, the data is routed through the routers
Sustainable Cell Rate (SCR)	This is the measure of the long-term average of user traffic. It defines an upper bound on the long-term average of conforming cells for an ATM connection.
Switched LAN	A LAN connectivity model in which each host is directly connected to a switch (or an intelligent hub).
Switched Virtual Circuit (SVC)	A connection that is established/released dynamically using signaling procedure.
Synchronous Digital Hierarchy (SDH)	The ITU-T standard for transmitting information over optical fiber.
Synchronous Optical Networks (SONET)	The ANSI standard for transmitting information over optical fiber.
Synchronous Transport Signal Level-1 (STS-1)	SONET standard for transmission over OC-1 optical fiber at 51.84 Mbps.
Synchronous Transport Signal Level-N (STS-N)	SONET standards for transmission over OC-n optical fiber by multiplexing 'N' STS-1 frames, (e.g., STS-3 at 155.52 Mbps, STS-12 at 622.08 Mbps and STS-48 at 2.488 Gbps).
Synchronous Transport Signal Level-N Concatenated (STS-Nc)	SONET standards for transmission over OC-n optical fiber by multiplexing 'N' STS-1 frames, (e.g., STS-3 at 155.52 Mbps, STS-12 at 622.08 Mbps and STS-48 at 2.488 Gbps but treating the information fields as a single concatenated payload).

T1	See 'Digital Signal, Level 1'
T3	See 'Digital Signal, Level 3'
Topology Aggregation	Topology aggregation is the process of reducing topology information by summarizing nodal as well as link information
Traffic Contract	It is defined by two sets of parameters, viz. the Connection Traffic Descriptor and QoS parameters, that collectively defines the traffic characteristics and service requirements.
Traffic Management	Set of procedures used for managing network traffic, providing service guarantees to user connections and ensuring optimal utilization of network resources.
Traffic Parameters	They specify various aspects and characteristics of traffic
Traffic Shaping	It refers to the mechanism that alters the traffic characteristics of a stream of cells on a VCC or a VPC to achieve a desired traffic characteristics, in order to ensure conformance or to achieve better network efficiency whilst meeting the QoS objectives.
Transfer Mode	Transfer mode refers to the techniques used to transmit, switch and multiplex information. In other words, transfer mode is the means of packaging, sending and receiving information on the network.
Transmission Convergence (TC) Sub-layer	It performs functions that are required independent of the underlying physical medium (e.g. cell rate decoupling and cell delineation).
Transmission Frame Adaptation	Mapping of cells into a framed physical layer structure.
Unassigned Cells	The cells generated by ATM layer when it has nothing to send to the lower layer and when the lower layer expects a continuous stream of cells. At the destination, these cells are not passed on to the higher layer.
UNI Signalling	The ATM signalling procedures defined for UNI.
Universal Mobile Telecommunications System (UMTS)	A term coined by ETSI that is synonymous with 3G networks as defined by ETSI.
Universal Terrestrial Radio Access Network (UTRAN)	The access network used in UMTS network that comprises of RNC and Node B.
Universal Test and Operations PHY Interface for ATM (UTOPIA)	It defines an interface between the physical layer (PHY) and the ATM layer.
Unspecified Bit Rate (UBR)	A service category used by applications having no requirement, whatsoever, with regards to cell transfer delay, cell delay variation or cell loss.
Upnode	An upnode is a neighbouring peer of one of the border node's ancestors.

Usage Parameter Control (UPC) The mechanisms applied at the UNI to monitor the traffic flow and to enforce the traffic contract.

User Plane It provides means for the transfer of user information.

User Plane Security The set of security functions available in the ATM user plane that includes identity authentication, confidentiality, data origin authentication and data integrity and access control.

User-Network Interface (UNI) It is the interface between an ATM end-system and a private ATM switch or between a private ATM switch and the public carrier ATM network.

VC Cross Connect A network element which connects virtual channel links, terminates virtual path connections, and is directed by management plane functions.

VC Switch A network element which connects virtual channel links, terminates virtual channel connections, and is directed by control plane functions.

VC-Based Multiplexing A technique in which the protocol of the packets is implicitly identified by the VC connecting two ATM end-systems

Virtual Channel (VC) A concept used to describe unidirectional transport of ATM cells associated by a common unique identifier value called VCI.

Virtual Channel Connection (VCC) A concatenation of virtual channel links that extends between two points where the adaptation layer is accessed.

Virtual Channel Identifier (VCI) A unique 16-bit numerical label in the ATM cell header that identifies a virtual channel, over which the cell is to travel.

Virtual Channel Link A means of unidirectional transport of ATM cells between a point where a virtual channel identifier value is assigned and the point where that value is translated or removed.

Virtual Circuit Switching A technique in which virtual circuits are established and the packets are forwarded using the virtual circuit numbers.

Virtual Path (VP) A concept used to describe unidirectional transport of ATM cells belonging to virtual channels that are associated by a common identifier value called VPI.

Virtual Path Connection (VPC) A concatenation of virtual path links that extends between the point where the virtual channel identifier values are assigned and the point where those values are translated or removed.

Virtual Path Identifier (VPI) A unique 8-bit or 12-bit numerical label in the ATM cell header that identifies a virtual path, over which the cell is to travel.

Virtual Path Link — The group of virtual channel links, identified by a common value of the virtual path identifier, between the point where the VPI value is assigned and the point where the VPI value is translated or removed.

Virtual Scheduling — It is a method used to determine the conformance of an arriving cell. The algorithm updates a Theoretical Arrival Time (TAT), which is the 'nominal' arrival time of the cell assuming that the active source sends equally spaced cells. If the actual arrival time of a cell is not 'too' early relative to the TAT, then the cell is conforming. Otherwise the cell is non-conforming.

Voice and Telephony Over ATM (VTOA) — The ATM Forum Voice and Telephony over ATM service interoperability specifications that provide means for carrying voice over ATM networks.

Voice Over DSL (VoDSL) — A technique that provides multiple voice communication channels along-with data access over the same single copper cable originally used for POTS service.

VP Cross Connect — A network element that connects virtual path links, translates VPI values and is directed by management plane functions.

VP Switch — A network element that connects virtual path links, translates VPI values and is directed by control plane functions.

xDSL — Jointly refer to varios flavours of DSL like ADSL, SDSL, VDSL and HDSL.

References

[3GPP TS 25.401] 3GPP Technical Specification, "UTRAN Overall Description".

[AAL2 C. Liu] Chunlei Liu, *et. al.*, "Packing Density of Voice Trunking using AAL2", GlobeCom'99, December 5-9, 1999, Rio de Janeiro, Brazil (url: http://www.cis.ohio-state.edu/~jain/papers/density.htm).

[AAL2 G. Eneroth] G. Eneroth, *et. al.*, "Applying ATM/AAL2 as a Switching Technology in Third-Generation Mobile Access Networks", *IEEE Communications*, Vol. 37, No. 6, June 1999, pp. 112-122.

[AAL2 GDC] "Adapting Voice for ATM Networks: A Comparison of AAL1 versus AAL2", General DataComm white paper, (url: http://www.gdc.com/inotes/pdf/aal1vs2.pdf).

[AAL2 J. Baldwin] John H. Baldwin, *et. al.*, "AAL2- A new ATM Adaptation Layer for Small Packet Encapsulation and Multiplexing", *Bell Labs Technical Journal*, Spring 1997.

[ATM A. Alles] Anthony Alles, "ATM Internetworking", Cisco white paper (url: http://cio.cisco.com/warp/public/614/12.html).

[ATM D. McDysan] David E. McDysan and Darren L. Spohn, "Atm Theory and Application", McGraw-Hill.

[ATM IBM Redbook] IBM Redbook, "Asynchronous Transfer Mode (ATM) Technical Overview" (url: http://www.redbooks.ibm.com/redbooks.nsf/0/f0045a5a9adcc3758525659d002a576d?OpenDocument)

[ATM M. Prycker] Martin De Prycker, "Asynchronous Transfer Mode: Solution for Broadband ISDN", Prentice-Hall.

[ATM R. Handel] Rainer Handel, *et. al.*, "Introduction to ATM Networking", McGraw-Hill.

[ATM U. Black] Uyless D. Black, "ATM Volume I: Foundation for Broadband Networks", Prentice Hall.

[ATM W. Goralski] Walter J. Goralski, "Introduction to ATM Networking", McGraw-Hill.

[ATMF AAL1 Trunk 1.1] ATM Forum specification, "ATM Trunking Using AAL1 for Narrowband Services Version 1.1", af-vtoa-0089.001, February, 2000.

[ATMF AAL2 LES] ATM Forum specification, "Loop Emulation Service Using AAL2 Rev 1", af-vmoa-0145.001, February, 2003.

[ATMF AAL2 Trunk] ATM Forum specification, "ATM Trunking Using AAL2 for Narrowband Services", af-vtoa-0113.000, February, 1999.

[ATMF Addr ANS 2.0] ATM Forum specification, "ATM Name System V2.0", af-dans-0152.000, July, 2000.

[ATMF Addr BA 1.0] ATM Forum specification, "ATM Forum Addressing: Reference Guide", af-ra-0164.000, April, 2001.

[ATMF Addr RG 1.0] ATM Forum specification, "ATM Forum Addressing: Reference Guide", af-ra-0106.000, February, 1999.

[ATMF Addr UG 1.0] ATM Forum specification, "ATM Forum Addressing: User Guide Version 1.0", af-ra-0105.000, January, 1999.

[ATMF AINI 1.1] ATM Forum specification, "ATM Inter-Network Interface Specification Version 1.1", af-cs-0125.002, September, 2002.

[ATMF BICI 2.0] ATM Forum specification, "BISDN Inter-Carrier Interface (B-ICI) Specification Version 2.0", af-bici-0013.003, December, 1995.

[ATMF CES] ATM Forum specification, "Circuit Emulation Service Interoperability Specification Version 2.0", af-vtoa-0078.000, January, 1997.

[ATMF CNM] ATM Forum specification, "Customer Network Management (CNM) for ATM Public Network Service", af-nm-0019.000, October, 1994.

[ATMF DBCES] ATM Forum specification, "Specifications of (DBCES) Dynamic Bandwidth Utilization—in 64 Kbps Time Slot Trunking over ATM - Using CES", af-vtoa-0085.000, July, 1997.

[ATMF DS1] ATM Forum specification, "DS1 Physical Layer Specification", af-phy-0016.000, September, 1994.

[ATMF DS3] ATM Forum specification, "DS3 Physical Layer Interface Specification", af-phy-0054.000, March, 1996.

[ATMF DXI] ATM Forum specification, "Data Exchange Interface (DXI) Specification", af-dxi-0014.000, August, 1993.

[ATMF E1] ATM Forum specification, "E-1 Physical Layer Interface Specification", af-phy-0054.000, September, 1996.

[ATMF E3] ATM Forum specification, "E3 UNI", af-phy-0034.000, August, 1995.

[ATMF FAST] ATM Forum specification, "Frame-based ATM over SONET/SDH Transport (FAST)", af-fbatm-0151.000, July, 2000.

[ATMF FATE] ATM Forum specification, "Frame-based ATM Transport over Ethernet (FATE)", af-fbatm-0139.001, July, 2002.

[ATMF FRA E1/T1] ATM Forum specification, "ATM on Fractional E1/T1", af-phy-0130.000, October, 1999.

[ATMF FUNI 2.0] ATM Forum specification, "Frame-based User-To-Network Interface (FUNI) Specification", af-saa-0088.000, July, 1997.

[ATMF FUNI MSE] ATM Forum specification, "Multi-service Extensions to FUNI v2.0 Specification", af-saa-0109.000, February, 1999.

[ATMF ILMI 4.0] ATM Forum specification, "Customer Network Management (CNM) for ATM Public Network Service", af-nm-0019.000, October, 1994.

[ATMF LNNI 2] ATM Forum specification, "LAN Emulation over ATM version 2-LNNI Specification", af-lane-0112.000, February, 1999.

[ATMF LSCES] ATM Forum specification, "Low Speed Circuit Emulation Service (LSCES)", af-vtoa-0119.000, May, 1999.

[ATMF LUNI 2] ATM Forum specification, "LAN Emulation over ATM version 2-LUNI Specification", af-lane-0084.000, July, 1997.

[ATMF MPLS 2.0] ATM Forum specification, "ATM-MPLS Network Interworking Version 2.0", af-aic-0178.001, August, 2003.

[ATMF MPOA 1.1] ATM Forum specification, "Multi-Protocol Over ATM Specification, Version 1.1", af-mpoa-0114.000, May, 1999.

[ATMF PHY 2.4G] ATM Forum specification, "2.4 Gbps Physical Layer Specification", af-uni-0133.000, October, 1999.

[ATMF PNNI 1.0 SEC] ATM Forum specification, "PNNI Version 1.0 Security Signaling Addendum," af-cs-0116.000, May 1999.

[ATMF PNNI 1.1] ATM Forum specification, "P-NNI V1.1", af-pnni-0055.000, April, 2002.

[ATMF SEC C-PLANE] ATM Forum specification, "Control Plane Security", af-sec-0096.000, February, 1998.

[ATMF SEC FW 1.0] ATM Forum specification, "ATM Security Framework Version 1.0", af-sec-0096.000, February, 1998.

[ATMF SEC SPEC 1.1] ATM Forum specification, "ATM Security Specification Version 1.1", af-sec-0100.002, March, 2001.

[ATMF TM 4.1] ATM Forum specification, "Traffic Management Specification 4.1", af-tm-0121.000, March, 1999.

[ATMF UNI 3.1] ATM Forum specification, "ATM User-Network Interface Specification V3.1", af-uni-0010.002, 1994.

[ATMF UNI 4.0 SEC] ATM Forum specification, "UNI Signalling 4.0 Security Addendum," af-cs-0117.000, May 1999.

[ATMF UNI 4.0] ATM Forum specification, "UNI Signalling 4.0", af-sig.0061.000, July, 1996.

[ATMF UNI 4.1] ATM Forum specification, "ATM User Network Interface (UNI) Signalling Specification Version 4.1", af-sig.0061.002, April, 2002.

[ATMF UTOPIA L1] ATM Forum specification, "Utopia", af-phy-0017.000, March, 1994.

[ATMF UTOPIA L2] ATM Forum specification, "Utopia Level 2", af-phy-0039.000, June, 1995.

[ATMF UTOPIA L3] ATM Forum specification, "Utopia Level 3", af-phy-0136.000, November, 1999.

[ATMF UTOPIA L4] ATM Forum specification, "Utopia Level 4", af-phy-0144.001, March, 2000.

[ATMF VTOA Desktop] ATM Forum specification, "Voice and Telephony over ATM to the Desktop", af-vtoa-0083.001, February, 1999.

[BCR R. Cochran] Rosemary Cochran, "ATM Sales Finally Match the Hype", Business Communications Review, pp 40-44, January, 1999.

[Bri R.Perlman] R. Perlman "Interconnections: Bridges and Routers", Addison Wesley.

[DSL A. Azcorra] A. Azcorra, *et. al.*, "IP/ATM Integrated Services over Broadband Access Copper Technologies" *IEEE Communications*, pp. 90-97, Vol. 37, No. 5, May 1999

[DSL Broadxent] Broadxent, "Mastering Voice over DSL: Network Architecture", Broadxent white paper (url: www.broadxent.com/pdfs/wp_network_architecture.pdf).

[DSL Paradyne] Paradyne, "The DSL Sourcebook - Third Edition", Paradyne e-book, (url: http://www.paradyne.com/solutions/whitepapers/dsl_sourcebook/).

[DSL R. Cohen] R. Cohen, "Service provisioning in an ATM-over-ADSL access network", *IEEE Communications Magazine*, Vol. 37 No. 10, pp. 82-87, 1999.

[DSL V. Bhagavath] Vijay K. Bhagavath, "Emerging High-Speed xDSL Access Services: Architectures, Issues, Insights and Implications," *IEEE Communications Magazine*, pp. 106-114, November, 1999.

[FIPS 197] Federal Information Processing Standard (FIPS) 197, "Advanced Encryption Standard".

[Gen A. Tanenbaum] Andrew Tanenbaum. "Computer Networks", Prentice-Hall of India.

[Gen D. Comer] Douglas Comer, "Computer Networks and Internets", Pearson Education India.

[Gen J. Kurose] James F Kurose, "Computer Networking: A Top-down Approach Featuring the Internet", Pearson Education India.

[Gen S. Kasera] Sumit Kasera, Nishit Narang and Sumita Narang, "Communication Networks: Principles and Practice", Tata McGraw-Hill.

[Gen S. Keshav] Srinivasan Keshav, "An Engineering Approach to Computer Networks", Pearson Education India.

[Gen W. Stallings] William Stallings "Data and Computer Communications", Prentice-Hall of India.

[ITU-T E.164] ITU-T specification E.164, "The International Public Telecommunication Numbering Plan".

[ITU-T G.703] ITU-T specification G.703, "Physical/Electrical Characteristics of Hierarchical Digital Interfaces".

[ITU-T G.704] ITU-T specification G.704, "Synchronous Frame Structures Used at 1544, 6312, 2048, 8448 and 44 736 kbit/s hierarchical levels".

[ITU-T G.804] ITU-T specification G.804, "ATM Cell Mapping into Plesiochronous Digital Hierarchy (PDH)".

[ITU-T I.113] ITU-T specification I.113, "Vocabulary of Terms for Broadband Aspects of ISDN".

[ITU-T I.150] ITU-T specification I.150, "B-ISDN Asynchronous Transfer Mode Functional Characteristics".

[ITU-T I.211] ITU-T specification I.121, "B-ISDN Service Aspects".

[ITU-T I.311] ITU-T specification I.311, "B-IS DN General Network Aspects".

[ITU-T I.321] ITU-T specification I.321, "B-ISDN Protocal Reference Model and its Application".

[ITU-T I.356] ITU-T specification I.356, "B-ISDN Layer Cell Transfer Performance".

[ITU-T I.361] ITU-T specification I.361, "B-ISDN ATM Layer specification".

[ITU-T I.363.1] ITU-T specification I.363.1, "B-ISDN ATM Adaptation Layer Specification: Type 1 AAL".

[ITU-T I.363.2] ITU-T specification I.363.2, "B-ISDN ATM Adaptation Layer Specification: Type 2 AAL".

[ITU-T I.363.3] ITU-T specification I.363.3, "B-ISDN ATM Adaptation Layer Specification: Type 3/4 AAL".

[ITU-T I.363.5] ITU-T specification I.363.5, "B-ISDN ATM Adaptation Layer Specification: Type 5 AAL".

[ITU-T I.363] ITU-T specification I.363, "B-ISDN ATM Adaptation Layer Specification".

[ITU-T I.366.1] ITU-T specification I.366.1, "Segmentation and Reassembly Service-specific Convergence Sub-layer for the AAL Type 2".

[ITU-T I.366.2] ITU-T specification I.366.2, "AAL Type 2 Service-specific Convergence Sub-layer for Trunking".

[ITU-T I.371] ITU-T specification I.371, "Traffic Control and Congestion Control in B-ISDN".

[ITU-T I.411] ITU-T specification I.411, "ISDN User-Network Interfaces - Reference Configurations".

[ITU-T I.413] ITU-T specification I.413, "B-ISDN User-Network Interface".

[ITU-T I.430] ITU-T specification I.430, "Basic User-Network Interface - Layer 1 specification".

[ITU-T I.431] ITU-T specification I.431, "Primary Rate User-Network Interface - Layer 1 specification".

[ITU-T I.432] ITU-T specification I.432, "B-ISDN User-Network Interface—Physical Layer specification".

[ITU-T I.610] ITU-T specification I.610, "B-ISDN Operation and Maintenance Principles and Functions".

[ITU-T I.630] ITU-T specification I.610, "ATM Protection Switching".

[ITU-T Q.2110] ITU-T specification Q.2110, "B-ISDN ATM Adaptation Layer-Service-specific Connection-oriented Protocol (SSCOP)".

[ITU-T Q.2119] ITU-T specification Q.2119, "B-ISDN ATM Adaptation Layer - Convergence Function for SSCOP above the Frame Relay Core Service".

[ITU-T Q.2130] ITU-T specification Q.2130, "B-ISDN Signalling ATM Adaptation Layer - Service-specific Co-Ordination Function for Support of Signalling at the User-network-Interface (SSCF at UNI)".

[ITU-T Q.2140] ITU-T specification Q.2130, "B-ISDN Signalling ATM Adaptation Layer - Service-specific Co-ordination Function for Support of Signalling at the User-network-Interface (SSCF at UNI)".

[ITU-T Q.2150.1] ITU-T specification Q.2150.1, "Signalling Transport Converter on MTP3 and MTP3b."

[ITU-T Q.2150.2] ITU-T specification Q.2150.2, "Signalling Transport Converter on SSCOP".

[ITU-T Q.2150.3] ITU-T specification Q.2150.3, "Signalling Transport Converter on SCTP".

[ITU-T Q.2630.1] ITU-T specification Q.2630.1, "AAL Type 2 Signalling Protocol–Capability Set 1".

[ITU-T Q.2630.2] ITU-T specification Q.2630.2, "AAL Type 2 Signalling Protocol–Capability Set 2".

[ITU-T Q.2630.3] ITU-T specification Q.2630.3, "AAL Type 2 Signalling Protocol–Capability Set 3".

[ITU-T Q.2931] ITU-T specification Q.2931, "Digital Subscriber Signalling System No. 2—User-Network Interface (UNI) Layer 3 Specification for Basic Call/Connection Control".

[ITU-T Q.2971] ITU-T specification Q.2971, "Digital Subscriber Signalling System No. 2—User-network Interface Layer 3 Specification for Point-to-Multipoint Call/Connection Control".

[ITU-T Q.921] ITU-T specification Q.921, "ISDN User-network Interface–Data Link Layer Specification".

[ITU-T Q.922] ITU-T specification, "ISDN Data Link Layer Specification for Frame Mode Bearer Services".

[ITU-T Q.931] ITU-T specification Q.931, "ISDN User-network Interface Layer 3 Specification for Basic Call Control".

[ITU-T Q.933] ITU-T specification Q.933, "Signalling Specifications for Frame Mode Switched and Permanent Virtual Connection Control and Status Monitoring".

[ITU-T X.21 bis] ITU-T specification X.21, "Use on Public Data Networks of Data Terminal Equipment (DTE) which is Designed for Interfacing to Synchronous V-Series Modems".

[ITU-T X.21] ITU-T specification X.21, "Interface between Data Terminal Equipment and Data Circuit-terminating Equipment for Synchronous Operation on Public Data Networks".

[ITU-T X.213] ITU-T specification X.213, "Information Technology-Open Systems Interconnection-Network service Definition".

[ITU-T X.25] ITU-T specification X.25, "Interface between Data Terminal Equipment (DTE) and Data Circuit-terminating Equipment (DCE) for Terminals Operating in the Packet mode and Connected to Public data networks by dedicated circuit".

[ITU-T X.700] ITU-T specification X.700, "Management Framework for Open Systems Interconnection (OSI) for CCITT Applications".

[LAN B. Vickers] B. J. Vickers and T. Suda, "Connectionless Service for Public ATM Networks," IEEE *Communications Magazine*, Vol. 32, No. 8, pp. 34-42, August, 1994.

[LAN E. Biagoini] E. Biagoini, *et. al.*, "Designing a Practical ATM LAN," *IEEE Network*, Vol. 7, No. 2, pp. 32-39, March, 1993.

[LAN G. Parulkar] G. Parulkar, "Local ATM Networks," *IEEE Network*, Vol. 7, No. 2, pp. 8-9, March, 1993.

[LAN H. Troung] H.L. Truong *et. al.*, "LAN Emulation on an ATM Network," *IEEE Communications Magazine*, Vol. 33, No. 5, pp. 70-85, May 1995.

[LAN I. A. Akylidiz] I. A. Akylidiz *et. al.*, "ATM Local Area Networks: A Survey of Requirements, Architectures, and Standards," *IEEE Communications Magazine*, Vol. 35, No. 7, pp. 72-80, July, 1997.

[LAN N. Finn] N. Finn, "ATM LAN Emulation," *IEEE Communications Magazine*, Vol. 34, No. 6, pp. 96-100, June, 1996.

[NM M. Subramanian] Mani Subramanian, "Network Management: Principles and Practice", Pearson Education Asia.

[RFC 1034] P.V. Mockapetris, "Domain Names—Concepts and Facilities", November, 1987.

[RFC 1155] M.T. Rose, K. McCloghrie, "Structure and Identification of Management Information for TCP/IP-based Internets", May, 1990.

[RFC 1157] J. D. Case *et. al.*, "Simple Network Management Protocol (SNMP)", May, 1990.

[RFC 1191] J.C. Mogul, S.E. Deering, "Path MTU discovery", November, 1990.

[RFC 1212] M.T. Rose, K. McCloghrie, "Concise MIB definitions", May, 1991.

[RFC 1213] M.T. Rose, K. McCloghrie, "Management Information Base for Network Management of TCP/IP-based Internets: MIB-II", May, 1991.

[RFC 1321] R. Rivest, "The MD5 Message-Digest Algorithm", April 1992.

[RFC 1406] F. Baker *et. al.*, "Definitions of Managed Objects for the DS1 and E1 Interface Types", January, 1993.

[RFC 1407] T. Cox, K. Tesink, "Definitions of Managed Objects for the DS3/E3 Interface Type", January, 1993.

[RFC 1441] J. Case, *et. al.*, "Introduction to Version 2 of the Internet-standard Network Management Framework", April, 1993.

[RFC 1483] J. Heinanen, "Multi-Protocol Encapsulation over ATM Adaptation Layer 5," July, 1993.

[RFC 1577] M. Laubach, "Classical IP and ARP over ATM," January, 1994.

[RFC 1626] R. Atkinson, "Default IP MTU for User over ATM AAL5," May, 1994.

[RFC 1755] M. Perez, *et. al.*, "ATM Signalling Support for IP over ATM," May, 1996.

[RFC 1771] Y. Rekhter, T. Li, "Border Gateway Protocol 4", March, 1995.

[RFC 1902] J. Case, *et. al.*, "Structure of Management Information for Version 2 of the Simple Network Management Protocol (SNMPv2)", January, 1996.

[RFC 1903] J. Case, *et. al.*, "Textual Conventions for Version 2 of the Simple Network Management Protocol (SNMPv2)", January, 1996.

[RFC 1904] J. Case, *et. al.*, "Conformance Statements for Version 2 of the Simple Network Management Protocol (SNMPv2)", January, 1996.

[RFC 1907] J. Case, *et. al.*, "Management Information Base for Version 2 of the Simple Network Management Protocol (SNMPv2)", January, 1996.

[RFC 1937] Y. Rekhter and D. Kandlur, "Local/Remote Forwarding Decision in Switched Data Link Sub-networks," November 1996.

[RFC 2022] G. Armitage, "Support for Multicast over UNI 3.0/3.1 based ATM Networks," November, 1996.

[RFC 2225] M. Laubach, J. Halpern, "Classical IP and ARP over ATM," April, 1998.

[RFC 2328] J. Moy, "OSPF version 2", April, 1998.

[RFC 2331] M. Maher, "ATM Signalling Support for IP over ATM - UNI Signalling 4.0", April, 1998.

[RFC 2332] J. Luciani, *et. al.*, "NBMA Next Hop Resolution Protocol (NHRP)," April, 1998

[RFC 2390] T. Bradley, *et. al.*, "Inverse Address Resolution Protocol," September, 1998.

[RFC 2453] G. Malkin, "RIP Version 2", November, 1998.

[RFC 2460] S. Deering, R. Hinden, "Internet Protocol, Version 6 (IPv6)", December, 1998.

[RFC 2515] K. Tesink, "Definitions of Managed Objects for ATM Management", February, 1999.

[RFC 2558] K. Tesink, "Definitions of Managed Objects for the SONET/SDH Interface Type", March, 1999.

[RFC 2684] D. Grossman and J. Heinanen, "Multi-Protocol encapsulation over ATM Adaptation Layer 5," September, 1999.

[RFC 3031] E. Rosen, A. Viswanathan, R. Callon, "Multi-Protocol Label Switching Architecture", January, 2001.

[RFC 3032] E. Rosen, *et. al.*, "MPLS Label Stack Encoding", January, 2001.

[RFC 3035] B. Davie, "MPLS using LDP and ATM VC Switching", January, 2001.

[RFC 3038] K. Nagami, "VCID Notification over ATM link for LDP", January, 2001.

[RFC 3174] D. Eastlake Third, P. Jones, "US Secure Hash Algorithm 1", September, 2001.

[RFC 826] D.C. Plummer, "Ethernet Address Resolution Protocol: Or converting network protocol addresses to 48.bit Ethernet address for Transmission on Ethernet hardware," November, 1982.

[Rou M. Sportack] Mark Sportack, "IP Routing Fundamentals", Cisco Press.

[Secu Kahate] Kahate, "Cryptography and Network Security", Tata McGraw-Hill.

[Secu R. Bragg] Roberta Bragg, *et al.*, "Network Security (The Complete Reference)", Tata McGraw-Hill.

[Secu W. Stallings 2] William Stallings, "Network Security Essentials", Pearson Education India.

[Secu W. Stallings] William Stallings, "Cryptography and Network Security: Principles and Practice", Pearson Education Asia.

[Sig R. Onvural] Raif. O. Onvural and Rao Cherakuri, "Signalling in ATM Networks", Artech House.

[Swi A. Huang] A. Huang and S. Knauer, "Starlite, a Wideband Digital Switch", *GLOBECOM '84*, Atlanta, 1984.

[Swi Fouad A. Tobagi] Fouad A. Tobagi, "Fast Packet Switch Architectures for B-ISDN," Proceedings of the IEEE, Vol. 78, No. 1, pp. 133-167, January, 1990.

[Swi H. Ahmadi] H. Ahmadi and W. E. Denzel, "A Survey of Modern High-performance Switching Techniques," *IEEE Journal on Selected Areas in Communication*, Vol 7, No. 7, pp. 1091-1102, September, 1989.

[Swi I.S. Gopal] I. S. Gopal, *et. al.*, "PARIS: An Approach to Integrated Private Networks", *ICC '87*, 1987, pp. 764-773.

[Swi J.S. Turner] J.S. Turner "Design of a Broadcast Packet Network," INFOCOM '86, Miami, 1986.

[Swi J.Y. Hui] Joseph Y. Hui and Edward Arthurs, "A Packet Switch for Integrated Transport", *IEEE Journal on Selected Areas in Communication*, Vol. 5, No. 8, pp. 1264-1273, October, 1987.

[Swi M. Devault] M. Devault, *et. al.*, "The Prelude ATD Experiment: Assessment and Future Prospects," *IEEE Journal on Selected Areas in Communication*, Vol. 6, No. 9, pp 1528-1537, December, 1988.

[Swi M.G. Hluchyj] M.G. Hluchyj and Mark Karol, "Queueing in High-performance Packet Switching," *IEEE Journal on Selected Areas in Communication*, Vol. 6, No. 9, pp. 1587-1597, December, 1988.

[Swi Nick McKeown] Nick McKeown and Thomas E. Anderson, "A Quantitative Comparison of Scheduling Algorithms for Input-queued Switch", *Computer Networks*, Vol. 30, No. 24, December, 1998.

[Swi Shiv Kalyanraman] Shiv Kalyanraman, Rohit Goyal, Sonia Fahmy, and Ram Viswanathan, "ERICA Switch Algorithm: A Complete Description," *AF-TM 96-1172*, August, 1996.

[Swi Y.S. Yeh] Y.S. Yeh, *et al.*, "The Knockout Switch: A Simple, Modular Architecture for High-performance Packet Switching," *IEEE Journal on Selected Areas in Communication*, Vols. 5, No. 8, pp. 1274-1283, October 1987.

[TCP/IP D. Comer] Douglas Comer, "Internetworking with TCP/IP, Volume I", Prentice Hall.

[TCP/IP R. Stevens] Richard Stevens and Gary R Wright, "TCP/IP Illustrated", Vols. 1 and 2, Pearson Education India.

[TM Allyn Romanow] Allyn Romanow and Sally Floyd, "Dynamics of TCP Traffic over ATM networks," *IEEE Journal on Selected Areas of Communication*, Vol. 13, No. 4, pp. 633-641, May, 1995.

[TM Bob Simcoe] Bob Simcoe, "Generic Flow Control (GFC) Design Considerations," Digital Equipment Corporation, March, 1996.

[TM David Hughes] David Hughes, Kambiz Hooshmand, "ABR stretches ATM Network Resources", *Data Communication*, April, 1995.

[TM Eulàlia Mèlich] Eulàlia Mèlich, Antonio Barba, "Congestion Control Algorithm in ATM Networks," Yuforic 1997.

[TM H.G. Peros] H.G. Peros and K.M. Elsayed, "Call admission control Schemes: A Review," *IEEE Communication Magazine*, Vol. 34, No. 11, pp. 82-91, November, 1996.

[TM J. Roberts] J. Roberts, U. Mocci and J. Virtamo, "Broadband Network Teletraffic," Berlin, Heidelberg: Springer, 1996.

[TM K. Shiomoto] K. Shiomoto *et. al.*, "Overview of Measurement-based Connection Admission Control Methods in ATM Networks," *IEEE Communication Surveys*, (url:http://www.comsoc.org/pubs/surveys, First Quarter 1999.

[TM K.K. Ramakrishnan] K.K. Ramakrishnan, Peter Newman, ATM Flow Control: Inside the Great Debate, *Data Communication*, June 1995.

[TM L. Kalampoukas] L. Kalampoukas, A. Varma, K.K. Ramakrishnan, "An Efficient Rate Allocation Algorithm for ATM Networks Providing Max-min Fairness," Technical report UCSC-CRL-95-29, Computer Engineering Dept., University of California, Santa Cruz, June, 1995.

[TM M. Labrador] Labrador M. and Banerjee S., 'Packet Dropping Policies for ATM and IP Networks, *IEEE Communication Surveys*, (url: http://www.comsoc.org/pubs/surveys), Third Quarter, 1999.

[TM Raj Jain 2] Raj Jain, S. Kalyanaraman, R. Vishwanathan, The OSU Scheme for Congestion Avoidance in ATM Networks Using Explicit Rate Indication, *Proceedings, WATM'95 First Workshop on ATM Traffic Management*, December, 1995.

[TM Raj Jain] Raj Jain, Congestion Control and Traffic Management in ATM Networks: Recent Advances and A Survey, *Computers Networks and ISDN Systems*, Vol. 28, pp. 1723-1738, 1998.

[TM Rohit Goyal] Rohit Goyal, Raj Jain, Sonia Fahmy, Bobby Vandalore, Mukul Goyal, "Buffer Management for TCP over the ATM GFR Service", ATM Forum contribution, ATM 98-0405, July 1998.

[TM S. Floyd] S. Floyd and V. Jacobson, "Random Early Detection Gateways for Congestion Avoidance", *IEEE/ACM Transactions on Networking*, Vol. 1, No. 4, p. 397-413, August 1993.

[TM V. Jacobson] V. Jacobson, Congestion Avoidance and Control, *Proceedings of SIGGCOMM*, 1988.

[TM Yoon Chang] Yoon Chang, David Su, Shukri Wakid, Xiomei Qian, "The Generic Flow Control (GFC) Protocol: A Performance Assessment".

[Trends K.G. Coffman] K.G. Coffman and Andrew Odiyzko, "The Size and Growth Rate of Internet", (url: http://www.firstmonday.dk/issues/issue3_10/coffman).

[Trends M. Betker]Betker, M.R, *et. al.*, "The History of the Microprocessors", *Bell Labs Technical Journal*, Autumn 1997.

[VTOA ATM Forum] ATM Forum, "Speaking Clearly with ATM—A Practical Guide to Carrying Voice over ATM", ATM Forum White Paper, September 99, (url: http://www.atmforum.com/aboutatm/2.html).

[VTOA D. Minoli] Daniel Minoli, Emma Minoli, "Delivering Voice over Frame Relay and ATM", John Wiley and Sons.

[VTOA G. Biran] Gil Biran, "Voice over Frame Relay, IP and ATM", a white paper (url: http://www.protocols.com/papers/voe.htm).

[VTOA R. Jagannathan] Ramesh Jagannathan, "Voice Over ATM", a white paper (url: http://www.cse.ohio-state.edu/~jain/cis788-99/ftp/vtoa.pdf).

[VTOA S. Kasera] Sumit Kasera, "Voice Over ATM with Special Focus on AAL2 Signaling" a presentation in ATM Forum Seminar on 'Broadband Networking', Delhi, August 2001 (url: http://skasera.tripod.com/ppt/aal2sig.pdf).

[Wireless B. Walke] B. Walke, "Mobile Radio Networks", John Wiley and Sons.

[Wireless C. Bettstetter] C. Bettstetter, H.J. Vogel, J. Eberspacher, "General Packet Radio Service (GPRS): Architecture, Protocols, and Air Interface", *IEEE Communications Survey*, Third Quarter 1999, Vol. 2, No. 3.

[Wireless H. Holma] Harri Holma and Antti Toskala, "WCDMA for UMTS: Radio Access for Third Generation Mobile Communications", John Wiley and Sons.

[Wireless J. Korhonen] Juha Korhonen, "Introduction to 3G Mobile Communications", Artech House.

[Wireless M. Mouly] M. Mouly and M.B. Pautet, "The GSM System for Mobile Communications", Published by the Authors.

[Wireless S. Kasera] Sumit Kasera and Nishit Narang, "3G Networks: Architecture, Protocol and Procedures", Tata McGraw-Hill.

Index

K

L

M

N

T

U